Structural Molecular Biology
Methods and Applications

NATO ADVANCED STUDY INSTITUTES SERIES

A series of edited volumes comprising multifaceted studies of contemporary scientific issues by some of the best scientific minds in the world, assembled in cooperation with NATO Scientific Affairs Division.

Series A: Life Sciences

Recent Volumes in this Series

Volume 37—Active Defense Mechanisms in Plants
edited by R. K. S. Wood

Volume 38—Cell Growth
edited by Claudio Nicolini

Volume 39—Drug Development, Regulatory Assessment, and Postmarketing Surveillance
edited by William M. Wardell and Giampaolo Velo

Volume 40—Chromosome Damage and Repair
edited by Erling Seeberg and Kjell Kleppe

Volume 41—Protein Biosynthesis in Eukaryotes
edited by R. Pérez-Bercoff

Volume 42—Nitroimidazoles: Chemistry, Pharmacology, and Clinical Application
edited by A. Breccia, B. Cavalleri, and G. E. Adams

Volume 43—Advanced Topics on Radiosensitizers of Hypoxic Cells
edited by A. Breccia, C. Rimondi, and G. E. Adams

Volume 44—Cell Regulation by Intracellular Signals
edited by Stéphane Swillens and Jacques E. Dumont

Volume 45—Structural Molecular Biology: Methods and Applications
edited by David B. Davies, Wolfram Saenger, and Steven S. Danyluk

Volume 46—Post-Harvest Physiology and Crop Preservation
edited by Morris Lieberman

This series is published by an international board of publishers in conjunction with NATO Scientific Affairs Division

A Life Sciences B Physics	Plenum Publishing Corporation London and New York
C Mathematical and Physical Sciences	D. Reidel Publishing Company Dordrecht, The Netherlands and Hingham, Massachusetts, USA
D Behavioral and Social Sciences E Applied Sciences	Martinus Nijhoff Publishers The Hague, The Netherlands

Structural Molecular Biology

Methods and Applications

Edited by

David B. Davies

University of Birkbeck
London, England

Wolfram Saenger

Free University of Berlin
West Berlin, Federal Republic of Germany

and

Steven S. Danyluk

Domtar Inc.
Senneville, Quebec, Canada

PLENUM PRESS • NEW YORK AND LONDON
Published in cooperation with NATO Scientific Affairs Division

Library of Congress Cataloging in Publication Data

NATO Advanced Study Institute on Current Methods in Structural Molecular Biology (1981: Maratea, Italy)
Structural molecular biology.
(NATO advanced study institute series. Series A, Life sciences; v. 45)
"Proceedings of a NATO Advanced Study Institute (FEBS advanced course no. 78) on Current Methods in Structural Molecular Biology, held May 3–16, 1981, in Maratea, Italy."
"Published in cooperation with NATO Scientific Affairs Division."
Bibliography: p.
Includes index.
1. Molecular biology–Methodology–Congresses. I. Davies, David B. II. Saenger, Wolfram. III. Danyluk, Steven S. IV. Federation of European Biochemical Societies V. North Atlantic Treaty Organization. Division of Scientific Affairs. VI. Title. VII. Series. [DNLM: 1. Molecular biology–Congresses. 2. Structure–activity relationship–Congresses. QH 506 S926 1981]
QH506.N38 1981 574.8'8 81-23540
ISBN 978-1-4684-4222-9 ISBN 978-1-4684-4220-5 (eBook) AACR2
DOI 10.1007/978-1-4684-4220-5

Proceedings of a NATO Advanced Study Institute and FEBS Advanced Course No. 78 on Current Methods in Structural Molecular Biology, held May 3–16, 1981, in Maratea, Italy

Softcover reprint of the hardcover 1st edition 1982

A Division of Plenum Publishing Corporation
233 Spring Street, New York, N.Y. 10013

ACKNOWLEDGEMENTS

The Advanced Study Institute was sponsored by NATO and co-sponsored by the Federation of European Biochemical Societies (FEBS Advanced Course No. 78). The course directors also acknowledge the financial assistance of the following companies:

Enraf-Nonius, Delft
Information Retrieval Ltd., U.K.
JEOL (U.K.) Ltd.
Nicolet Technology Corp., U.S.A.
STOE and CIE, GMBH
Varian Associates, U.S.A.

The following publishers kindly provided books for the institute library and then donated them to Birkbeck College library:

Academic Press
Cambridge University Press
Cornell University Press
Croom Helm Ltd.
Elsevier/North Holland
W.H. Freeman and Co.
Global Book Resources Ltd.
Gordon and Breach Ltd.
Heyden and Son
Information Retrieval Ltd.
IPC Science and Technology Press Ltd.
Kreiger Pub. Co.
McGraw-Hill Book Co.
M.T.P. Press Ltd.
Oxford University Press
Plenum Pub. Co.
Springer-Verlag
Van Nostrand Rheinhold
Walter de Gruyter
J. Wiley and Son

PREFACE

Structural biology is undergoing a revolution in both the sophistication of new biophysical methods and the complexity of problems in biomolecular structure and organization opened up for study. These changes are directly attributable to major advances in computer technology, computational methods, development of high-intensity synchrotron radiation sources, new magnetic resonance methods, laser optical techniques, etc. Structure-function problems previously considered intractable may now be solved. As this area of specialisation continues to expand, there is a need to review the various physical methods currently being used and developed in structural molecular biology. At the same time that individual techniques and their applications become more specialized, the need for effective communication between investigators gains in imperative. It is vital to forge links among sub-disciplines and to emphasise the complementary nature of results observed by different biophysical methods. This publication contains the review lectures given at a meeting on "Current Methods in Structural Molecular Biology" sponsored by NATO as an Advanced Study Institute and by FEBS as Advanced Course No. 78. The aim of the meeting was to bring together, in a teaching environment, students and specialists in diverse biophysical methodologies with the specific purpose of exploring, questioning and critically assessing the present and future state of biological structure research.

The scientific content of the interdisciplinary Study Institute centred around three interrelated aspects; biophysical methods and instrumentation, their application to biological structure problems, and derivation of structural information and insights. Within this framework the course covered:

- methods for theoretical computation of structural and conformational properties of biological systems;
- methods for structure elucidation in the solid state (X-ray and neutron diffraction, especially as applied to single crystal and fibre diffraction studies; electron microscopy, as applied to crystalline arrays of molecules and individual particles);

- methods used for solution studies (NMR spectroscopy, optical and chiroptical spectroscopy, neutron and X-ray and light scattering methods, fast kinetic methods);
- and methods applicable to both solution and solid state studies (Resonance Raman Spectroscopy, NMR spectroscopy, calorimetry).

It will be seen from the content of the chapters in the book that it was possible not only to stress the unique contributions of each biophysical method, but also to show how information generated by one method can complement that of another to enhance our overall understanding of biological structure and function. From the large field of structural molecular biology we focussed attention on a few topics of current interest in order to highlight the complementary nature of results obtained by different physical methods. The topics represented investigations of different aspects of biomolecular organisation from the structure and dynamics of water in biological systems, to the dynamics of peptides and proteins, protein-nucleic acid interactions and the structure and assembly of viruses.

The scientific meeting was held at Maratea, Italy from May 3-16, 1981 and was attended by one hundred participants. In addition to the main lectures and associated tutorial sessions, the active participation of students in the course was reflected by the number of research seminars (33) and research posters presented (34). The outcome of bringing together experts from different disciplines was the positive interchange of ideas in structural molecular biology, the initiation of collaborative research projects by participants, and the emphasis for younger participants on the need for an interdisciplinary perspective at the start of their research careers. Among the other lessons to be learned from the course are that biophysical techniques currently in use for biological structure investigations continue to be developed and applied to increasingly more complex systems and that new multidisciplinary methodologies are evolving which require information to be determined from a number of different biophysical approaches in order to understand the structure and organisation of complex systems.

In conclusion, the organisers would like to thank all participants for their contributions to such a stimulating course and, in particular, would like to thank the lecturers for producing excellent review chapters which not only will serve as essential sources of reference material, but also develop into important interdisciplinary teaching texts for investigators in the general areas of structural biology, molecular biology, biochemistry, and biophysics.

D.B. DAVIES
W. SAENGER
S.S. DANYLUK

SEPTEMBER 1981

CONTENTS

A KEYNOTE ADDRESS

B METHODS

PRINCIPLES OF BIOLOGICAL ORGANIZATION AND THEIR APPLICATIONS TO MOLECULAR STRUCTURES

Manfred Eigen

Max-Planck-Institute für
biophysikalische chemie
34 Göttingen, F.R.G.

Abstract of the introductory talk

What is particular about biological structure? Consider a protein molecule. It appears with a given stable conformation, without which it would never function reproducibly. Erwin Schrödinger[1] called such structures aperiodic crystals, and this is, indeed, a most illustrative characterization of their three-dimensional architecture, which we nowadays can reconstruct precisely from X-ray diffraction patterns[2]. Particular about these "aperiodic crystals" is that they fulfil their assigned functions with optimal efficiency. Stability of structure requires each atom to assume - at least locally - a position of minimum potential energy. This certainly holds true for the few subgroups which form the active centre of the molecule and which are directly involved in its catalytic function. However, there is no inherent linkage between structure and function. The internal interactions which stabilize a structure have no causal relation to the final function. The structure is designed in order to fulfil a particular task. Obviously that is why a protein molecule requires hundreds of residues while only very few of them are actually involved in function.

If we ask how the protein molecule is made in the first place, the answer is: it is translated from a gene. The gene itself is a stable structure, too, i.e. the Watson-Crick double helix. The stability, on the other hand, is essentially independent of individual structural features, such as its specific sequence of nucleotides. There are certainly some structural details associated with sequence. For instance: GC-rich regions melt less easily than AU-rich ones, palindromic sequences allow for recognition at certain sites, and specific interactions with proteins cause formation of superstructures allowing the packing of millimeter long threads into the micrometer dimension of the nucleus. In general, however, the structure of DNA as such does not provide any clue to the function of its own translation products. The particular primary structure, namely the sequence of nucleotides, has been selected solely on the basis of the phenotypic efficiency of its translation products. Each stable structure resulted because it is one of the few that are consistent with optimal function. This "external" design procedure is a protracted dissipative process, involving reproduction, mutation and selective fixation of those structures that offer functional advantage. Hence the essence of structure in biology is its function. The word crystal in this respect is misleading as it implies a spatial rather than a dynamic principle. Structural symmetries in biology are not due to apriori principles; rather, they are a posteriori symmetries, distinguished by a selective advantage.

The process of molecular optimization today can be studied in test-tube experiments. Isolated viral replicases proved to be especially useful for producing artificial templates and adapting them to certain optimal function.[3,4] The laws behind these self-organization processes have been identified[5,6], and their consequences tested with microorganisms[7,8]. Molecular paleontology provides hints for a synthetic reconstruction of ancestral structures.[9,10] If we study biological organization - and it is the aim of this meeting to review the various methods which can be used for this purpose - we should be aware of the basic functional constraints, which are common to all "living" structure.

REFERENCES

1. E. Schrödinger: "What is Life?" Cambridge University Press 1944.
2. R.E. Dickerson and I. Geis, "The Structure and Action of Proteins", Harper and Row, Publishers, New York-Evanston-London 1969.
3. Ch. K. Biebricher, M. Eigen and R. Luce, J. Mol. Biol. 148: 369 (1981).
4. Ch. K. Biebricher, M. Eigen and R. Luce, J. Mol. Biol. 148: 391 (1981).
5. M. Eigen and P. Schuster, "The Hypercycle", Springer-Verlag, Berlin, Heidelberg, New York 1979.
6. M. Eigen, Angewandte Chemie Int. Edition 20: 233 (1981).
7. M. Eigen, W. Gardiner, P. Schuster and R. Winkler-Oswatitsch, Scientific American, April 1981, pg. 88.
8. E. Domingo, R.A. Flavell and Ch. Weissmann, Gene 1: 3 (1981).
9. M. Eigen and R. Winkler-Oswatitsch, Naturwissenschaften 68: 217 (1981).
10. M. Eigen and R. Winkler-Oswatitsch, Naturwissenschaften 68: 282 (1981).

X-RAY FIBER DIFFRACTION

D.A. Marvin and C. Nave

European Molecular Biology Laboratory
Heidelberg, Federal Republic of Germany

1. INTRODUCTION

X-ray fibre diffraction is used to study the molecular structure of long assemblies of identical subunits. Such an assembly will normally have minimum energy when all subunits have the same environment. This means that subunits in elongated assemblies follow a helix of pitch $\underline{P}$ in which subunits are related to one another by integer multiples of a unit rise $\underline{p}$ parallel to the helix axis and a unit twist $\underline{p}/\underline{P}$ around this axis. Both the experimental techniques and the methods of data analysis lie somewhere between single-crystal diffraction and solution scattering, and fibre diffraction draws on both these techniques.

X-ray fibre diffraction can be applied to structures such as DNA or poly-α-amino acids, where each subunit has only a few tens of atoms; or to more complicated structures, where the subunit is a whole protein molecule or even a nucleoprotein complex. In the former case it may be possible to determine positions of the atoms in the asymmetric unit by combining X-ray diffraction data with known chemical information about the nature of the unit. In the latter case it may be possible to determine the helix symmetry and the general shape of the subunit from the fibre pattern; but more extensive chemical or crystallographic data are necessary to determine the detailed structure of the subunit.

Helical symmetry is most easily represented in cylindrical-polar coordinates. Representation of the Fourier transform in cylindrical-polar coordinates introduces Bessel functions which simplify the relationship between the structure and its diffraction pattern. Thus small changes in the diffraction pattern can often be related directly to corresponding changes in the structure. This feature of helix diffraction enables the study of molecular motion in structures not yet solved at atomic resolution.

X-ray fibre diffraction has been reviewed by many authors[1–12]. Here we give an outline of the basic principles of X-ray fibre diffraction, with special emphasis on those topics which have not appeared elsewhere in reviews.

2. EXPERIMENTAL TECHNIQUES

2.1 Specimen preparation

One of the most important but difficult tasks in any X-ray fibre diffraction study is to induce elongated molecules to align parallel to one another. Although methods of extracting intensity data from poorly aligned specimens have been developed (Section 2.4), they are no substitute for well-aligned specimens. In some biological fibres the molecules are naturally aligned, for instance in hair, collagen, and muscle[10]. In many other cases the starting material is a solution of randomly oriented purified molecules, and methods must be found to align the molecules.

Elongated molecules can form a liquid crystal phase, and it is important to understand the properties of liquid crystals[13,14] when preparing aligned specimens. The liquid crystal phase of matter (also called the mesophase) is intermediate between the solid phase and the liquid (or isotropic) phase. Liquid crystals flow like liquids, but they have anisotropic properties like solid crystals.

As the concentration of a solution of elongated molecules decreases it passes from an isotropic phase, to an equilibrium mixture between an isotropic phase and a liquid crystal phase, and thence to a pure liquid crystal phase. The concentration at which the isotropic phase is in equilibrium with the liquid crystal phase depends on the competition between the "orientational entropy" (which is a minimum in an isotropic phase) and the "translational entropy" (which is a minimum in a parallel array), and the equilibrium concentration is therefore lower for rods of larger axial ratio[15].

Liquid crystals or mesophases have been subdivided into several types. In the smetic mesophase, the molecules are arranged in layers, with their long axes perpendicular to the planes of the layers and their ends in register in the planes. In the nematic mesophase, the molecules lie parallel but there is no coherence between their ends. For the twisted nematic or cholesteric mesophase it is convenient to define a system of unit planes within which molecules lie parallel. These planes lie a unit distance apart and consecutive planes are twisted with respect to one another about an axis perpendicular to the planes.

The cholesteric phase is found for chiral molecules, so this phase is important in studying biological molecules, which are usually chiral. Examples of molecules forming cholesteric mesophases are poly-α-amino acids, DNA, tobacco mosaic virus (TMV), and filamentous bacterial viruses (FV)[15,16,17,18]. The twist between unit planes may be very small, but the cumulative macroscopic rotation can be very large. This cholesteric superstructure may appear as retardation lines spaced regularly every half-pitch, which are seen when the liquid crystal is viewed under crossed polars perpendicular to the axis of twist. The pitch can also be determined by measuring the rotation of polarized light directed along the axis of rotation[18]. The cholesteric twist depends on temperature and on the relationship between the dielectric permittivity of the solvent and that of the molecule along the major and minor axes[15]. Therefore an appropriate choice of temperature and/or solvent may remove the cholesteric twist, allowing samples for X-ray fibre diffraction to align. Cholesteric behaviour can also give rise to spherulites, which are birefringent structures embedded in the isotropic phase. In spherulites, the axis of twist between planes is directed radially in a sphere, so the molecules are everywhere tangential to spherical shells[19]. Spherulites under crossed polars show a characteristic maltese cross.

Even in the nematic phase, although the molecules lie parallel to one another locally, there is no other local order, and there is long-range disorder. Therefore whether or not the molecules form liquid crystals, they must be induced to lie parallel over the whole specimen for fibre diffraction. Many mechanical techniques have been used for this purpose. From concentrated gels of DNA, long fibres may be drawn by dipping the tip of a glass rod into the gel and withdrawing it quickly[20]. The thickness of the fibre depends on the speed of drawing. Slow drawing can give a fibre suitable for X-ray experiments (about 0.1 mm in diameter). Faster drawing gives smaller fibres, which may show better alignment but which may need to be mounted in bundles to give specimens large enough for X-ray experiments. Even faster drawing may give very thin (1-2 μm diameter) threads which must be spun into bundles to give sufficiently large specimens. Solutions of molecules that cannot be pulled into long fibres may nevertheless be aligned by hanging a drop of a concentrated solution between two glass rods with their tips one or two millimeters apart. This technique has been used for polynucleotides[21] and for FV[22]. Specimens of polysaccharides have been prepared by casting flat sheets and then stacking the sheets to build a sufficiently thick specimen. Any of these kinds of specimen may be further improved by annealling under high humidity[21]. The specimen may be mounted on the end of a glass rod; in a sealed capillary in equilibrium with a drop of solvent to maintain the humidity; or between the jaws of a stretcher to maintain tension[20,21,22]. Too much tension can cause structural changes in the material. Some molecules which are rigid rods, like TMV, can be induced to align by flowing concentrated solutions through small (0.7 mm diameter) quartz capillaries[17].

Some biological materials are diamagnetically anisotropic[23,24,25,26] and therefore can be aligned in a magnetic field for X-ray diffraction[27,28,29,30]. The effect on a single molecule is small, so it is necessary to use a liquid crystalline array above a critical concentration[25], where many molecules act in concert, and to use fields of the strength available with a superconducting magnet. The effect of a magnetic field on a liquid crystal can be monitored by the appearance of uniform birefringent areas, or by optical techniques that detect the transition from the twisted cholesteric phase to the parallel nematic phase[15]. Specimens aligned in a magnetic field may become misaligned when they are removed from the field. Alignment can be fixed in the specimen by slow drying while it is still in the field[30]. In principle, wet liquid crystalline specimens could also be studied by fibre diffraction while they were still in the magnetic field. For molecules aligning parallel to the field, which therefore must be viewed perpendicular to the field, this would require a special split-pair superconducting magnet geometry and special cameras. Such an experimental arrangement would permit the study of changes in molecular structure by X-ray diffraction in a wet state, even when changes are inhibited in dry fibres.

Electric fields also have an effect on the orientation of elongated macromolecules. Probably the best explanation for the observed orientation of macromolecules by electric fields is anisotropic ion flow. A torque is exerted on the ionic macromolecule by an asymmetric counter-ion atmosphere created by counter-ion flow in the field[31]. Flow of counter-ions along the macromolecule leads to a deficiency of counter-ions behind the rod at one end, and an excess of counter-ions in front of the rod at the other end.

Appropriately tuned AC fields may be useful for orienting some systems[32]. In other systems a combination of electric fields and magnetic fields may be useful. For instance DNA molecules tend to lie with their axes in a plane perpendicular to a magnetic field but with no orientation in the plane[24,25]. An electric field applied perpendicular to the magnetic field may then induce the DNA molecules to align parallel to one another.

Heavy atom derivatives can be helpful for phasing, determining the order of Bessel functions, and for separating overlapping Bessel function terms. The usual way to obtain a heavy atom derivative of a protein single crystal is to soak the crystal in a solution of the heavy atom, and then to determine the heavy atom site in the unit cell by crystallographic methods[33]. Attaching a heavy-atom compound to a chemically-defined site in the protein before growing the crystal may yield a different crystal form, which is not useful for multiple isomorphous replacement (MIR) studies. In the case of X-ray fibre diffraction, there are different problems. It is difficult to diffuse heavy atom compounds into fibres: either the compound does not enter the fibre at all, or soaking the fibre causes loss of orientation. On the other hand, aligned specimens of chemically modified virus have been useful in studies of TMV[34] and FV[30]. Chemical determination of the site in the protein sequence to which the heavy atom is bound constrains models by constraining the position of this site in space.

2.2 X-ray cameras

The weak scattering given by many fibrous specimens means that long exposures are necessary to record the diffraction pattern. This has led to the design of several types of X-ray camera to maximize the X-ray flux on the film. The optimum camera arrangement depends on factors such as the size of the focal spot on the anode, the size of the specimen, the sharpness of the diffraction peaks, and the resolution required. Use of an unsuitable camera can result in long exposure times, loss of resolution, and high background. A study of the optics of the experimental arrangement (by means of ray diagrams) is therefore worthwhile. Many specimens are only ordered over a relatively small region and this should influence the design of the camera.

The simplest camera consists of a series of collimation pinholes or slits followed by guard apertures to cut off radiation scattered by the collimator[20]. This camera is suitable for specimens which give diffuse diffraction peaks where the divergence of the X-ray beam from the collimator is not important. For specimens which give sharper diffraction, a small focal spot on the anode should be used in conjunction with a focussing camera. In the Franks focussing camera[35], X-rays are reflected at a glancing angle off two mutually perpendicular curved mirrors which produce an image of the X-ray source at the film. Using 6 cm long glass mirrors at a glancing angle of 10′, the X-ray beam is about 160 μm wide near the mirrors and is therefore suitable for small specimens. A first order resolution of about 500 Å is attainable with a resolution between orders greater than 2000 Å. An additional advantage is that, when the camera is properly adjusted, short wavelength radiation (for which a nickel filter is transparent) is not reflected by the mirrors. A camera employing gold-coated toroidal mirrors[36] has a much larger aperture than the Franks camera. This can be an advantage for larger specimens if it is not necessary to record spacings greater than 70 Å.

Bragg reflection off a curved crystal monochromator[37] provides an even greater aperture and spacings of more than 2000 Å can be observed. However, this type of camera is only suitable for large specimens. Any combination of the above optics can be used to define the X-ray beam horizontally and vertically. Two additional points should be noted concerning focussing cameras. In commercial versions the X-ray beam is often not exactly normal to the flat film after reflection by the mirrors or monochromators. This is not important for low angle diffraction but should be considered in high angle studies. In addition, the X-ray beam only remains in focus on a flat film at low diffraction angles. Use of a conical film[36] can be advantageous if a sharp focus is required over a fairly wide range of Bragg angles.

Cameras should be evacuated or else filled with helium to reduce scatter from air. The relative humidity of the helium and thus the water content of the fibre can be controlled by bubbling the helium through a saturated solution of the appropriate salt[20,22,38] before it enters the camera.

2.3 Recording intensity

A careful choice of the means of recording a fibre diffraction pattern can save much trouble during data reduction. It is common to expose a pack of several films to record a large range of intensities. Strong intensities which are overexposed on the film nearest the specimen will be recorded on other films in the pack. The intensities on films in the pack must then be scaled together. A slow film with a lower fog level will generally record a greater range of intensities than a fast film, and fewer scaling parameters will be required. Many fibrous specimens are relatively insensitive to beam damage so longer exposure times can be tolerated. In our work we use Agfa D10 for monitoring specimens and Agfa D7, which has about 1/6 the speed of D10, for data collection. The characteristics of X-ray films have been described[39]. Freshly manufactured films should be obtained and stored cold to reduce background due to chemical fog.

When accurate techniques of data reduction are used, films are sufficiently accurate for most data collection purposes. High-resolution two-dimensional detectors[40,41] promise substantially more accurate measurements, notably of weak (near background) intensity, small differences between intensities, and large ranges of intensity (without scaling problems). These kinds of measurement are especially important in separating partially overlapping layer lines or using data from isomorphous derivatives. Detectors are also necessary when measuring time-resolved changes in structure.

2.4 Data reduction

Measuring intensities on a single crystal diffraction pattern can be automatic, but measuring intensities on a fibre diffraction pattern is not, and computer programs developed for this purpose require user interaction. The fibre diffraction pattern shows a series of lines of intensity, the layer lines, running perpendicular to the fibre axis. With the fibre axis vertical, the zero layer line, or equator, runs horizontally through the centre of the diffraction pattern, and the meridian is an imaginary line running vertically through the diffraction pattern (Figures 1 and 2). Intensities on closely-spaced layer lines may overlap and must then be resolved. In imperfectly aligned fibres the intensity will be spread out along arcs, with a half-width proportional to the distance from the centre of the pattern, which may also cause

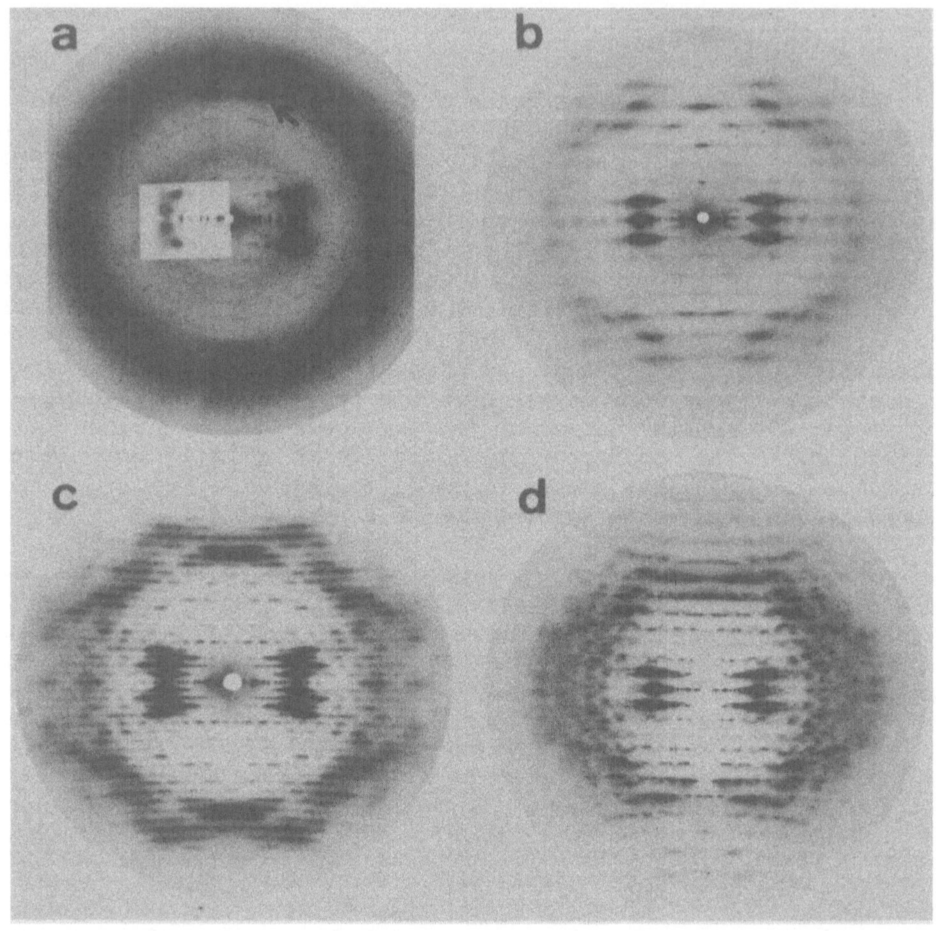

Fig. 1. X-ray fibre diffraction patterns of FV. Fibre axis vertical. (a) An early fibre of the Pf1 species[43], 5.40 units/turn. There is splitting of the strong near-meridional intensity at about 5 Å, arrowed, showing evidence of perturbation (perhaps a coiled-coil). (b) A fibre of the fd species[116]. (c) fibre of Pf1, 5.40 units/turn[30]. (d) Magnetically oriented fibre prepared at 4°C, 5.46 units/turn[30].

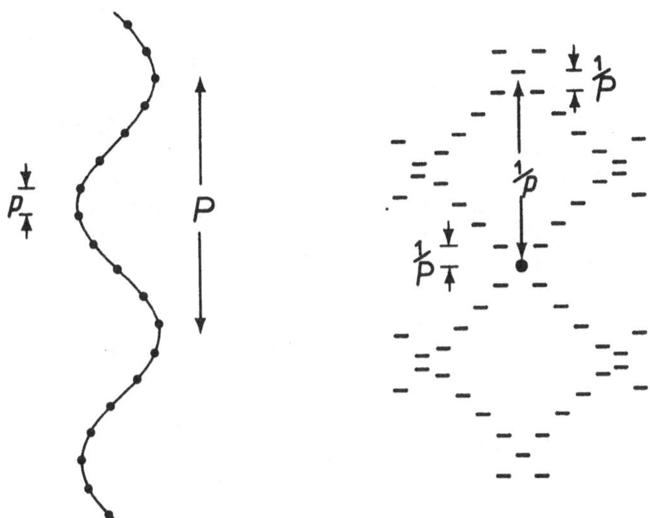

Fig. 2. A simple helix (left) and its diffraction pattern (right). The helix has 9.33 units/turn. There are two major sets of layer lines on the diffraction pattern, with a spacing of $1/\underline{P}$ between the layer lines of each set. One set ($\underline{m}$ = 0 in Equation 2) is centred at the origin and has $\underline{Z}$ spacings $1/\underline{P}$, $2/\underline{P}$, $3/\underline{P}$, etc. The other set ($\underline{m}$ = 1 in Equation 2) is centred at $1/\underline{p}$ and has spacings $1/\underline{p} \pm 1/\underline{P}$, $1/\underline{p} \pm 2/\underline{P}$, $1/\underline{p} \pm 3/\underline{P}$, etc. An equivalent set ($\underline{m}$ = -1) is centered at $-1/\underline{p}$. Further sets of layer lines are centred at $2/\underline{p}$, $3/\underline{p}$, etc., but these are not shown on this Figure.

overlapping layer lines. Because of the complicated corrections that must be applied to the data, intensities measured on different diffraction patterns of the same material may differ considerably. This raises problems when measuring small intensity differences, for instance those induced by heavy atom substitution.

The first step in analysing the data is to digitize intensities on the film, typically on a 50 μm by 50 μm raster, using an Optronics rotating drum scanner or similar device. With the program GUCKMAL[30] developed in our laboratory, the digitized plot can be displayed, all or in part, at any magnification, on the screen of a Tektronix computer terminal using threshold plots, contours, or shades of grey. Positions of regions of intensity can be identified either by setting the cursor at the centre of the region, or by drawing around the region and calculating the centre of gravity. The tilt of the fibre with respect to the X-ray beam is calculated by GUCKMAL from the coordinates of spots in the four equivalent quadrants of the diffraction pattern[42]. Positions of intensity are converted by GUCKMAL from x,y coordinates on the film image to the corresponding $\underline{R}$,$\underline{Z}$ coordinates in reciprocal space[10,43], using for calibration the diffraction ring at 3.029 Å given by a fine dust of calcite on the fibre.

The $\underline{n}$ $\underline{m}$ indices for each layer line (Figure 2) can be assigned graphically[44] and the helix parameters refined by least squares[45].

Measurement of the continuous intensity distribution along the layer lines [30,46,47,48] can most easily be done if the square raster of intensity is first converted to a polar raster. The observed intensity along an arc of constant radius is then fitted to a background term plus a set of Gaussian functions centred on each theoretical layer line position (which depends not only on the layer line spacing but also on the tilt of the fibre). Displaying the observed intensity data and the calculated distributions along an arc makes it convenient to optimize the tilt angle and the standard deviation of disorientation by trial-and-error. The background is calculated for each radial shell independently of the others, which means that the calculated background distribution may have high-frequency fluctuations, especially at radii that include strong intensity. Since the true background distribution must vary slowly with radius, a smoothly varying background distribution can be defined from the initial calculated background distribution and included in the final calculation. The height of the Gaussian distribution multiplied by reciprocal space radius $\underline{R}$ gives the integrated intensity at that point on the layer line.

Where there is considerable overlap between adjacent layer lines, this procedure may not be appropriate. Even here it is possible to resolve layer lines if calculated intensities are available, for instance from a trial model. Then the calculated layer line intensities can be distributed along an arc according to the known disorientation function[48], and this calculated distribution can be used to divide the intensity of the observed distribution in the ratio of the calculated (Figure 3). This procedure should converge if it is applied at each cycle during the refinement of a model. A computer program which calculates the disorientation function has been written[49].

Somewhat different considerations apply to analysis of crystalline reflexions instead of continuous transforms[10,50,51]. Once the $\underline{h}\ \underline{k}\ \underline{l}$ indices have been assigned to crystalline reflexions, their positions are refined by least squares as for single crystals. Measured intensity is corrected for the Lorentz factor, which can vary across the reflexion, especially near the meridian. It may be necessary to correct the intensity at reciprocal lattice points which superpose on the film because they have identical reciprocal space radii $\underline{R}$ (Section 3.2).

3. DIFFRACTION BY A HELIX

3.1 Helical transform

Analysis of X-ray fibre patterns is greatly simplified by expressing the structure and its Fourier transform in cylindrical-polar coordinates. The symbols used are defined in Table 1. The Fourier transform[52] is then

$$F_{\underline{Z}}(\underline{R},\Psi) = \sum_j \sum_{\underline{n}} f_j J_{\underline{n}}(2\pi \underline{R} r_j) \exp 2\pi i(\underline{Z} z_j - \underline{n}\phi_j) \exp 2\pi i \underline{n}(\Psi + 1/4) \quad (1)$$

where the sum over j refers to the atoms in the subunit, and the reciprocal space coordinate $\underline{Z}$ is related to the integral indices $\underline{n}$ and $\underline{m}$ by the selection rule

$$\underline{Z} = \underline{n}/\underline{P} + \underline{m}/\underline{p} = \underline{l}/\underline{c} \quad (2)$$

We use the conventions of Ramachandran[3] which avoid integral layer line indices. If there is a two-fold rotation axis perpendicular to the helix axis, as for DNA, the first exponential term in Equation 1 is replaced by cos $2\pi i(\underline{Z} z_j - \underline{n}\phi_j)$. If there is an N-fold rotation axis along the helix axis, then $\underline{n}$ must be an integral multiple of N. The continuous transform along one layer line $\underline{Z}$ for one Bessel function order $\underline{n}$ is[2]

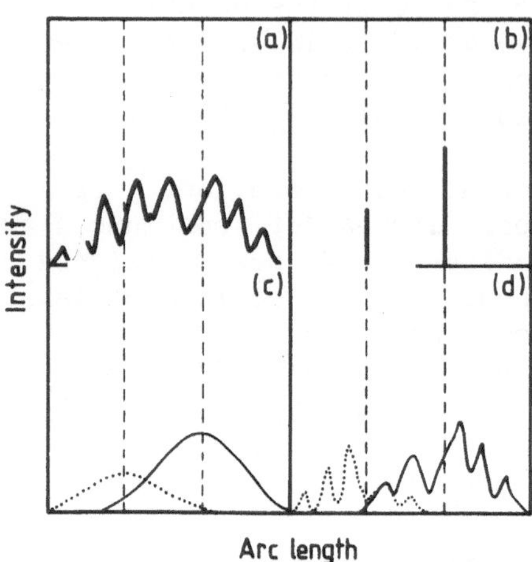

Fig. 3. Resolution of overlapping layer lines by comparison with the calculated transform (schematic). Tracings along arcs of constant distance from the origin are shown, cutting across two layer lines, whose positions are indicated by vertical dashed lines. (a) Observed noisy intensity distribution. (b) Calculated intensity on the two (infinitesimally narrow) layer lines. (c) Calculated intensity of the two layer lines, shown dotted and solid, distributed according to the measured disorientation function. (d) Observed intensity from (a), divided point-by-point in the ratio of the two intensities at corresponding points in (c). Integration under the dotted and dashed curves then gives the observed intensity divided between the two layer lines in the ratio of the calculated.

Table 1. Symbols Used in Diffraction Formulae

Symbol	Meaning
$\underline{P}$	Pitch of helix.
$\underline{p}$	Axial rise between subunits.
$\underline{p}/\underline{P}$	Rotation between subunits (in fractions of a turn).
c	Axial repeat distance along helix (where $\underline{p}/\underline{P}$ is rational).
r,ϕ,z	Cylindrical polar coordinates in real space.
$r_j\phi_j z_j$	Coordinates of jth atom in asymmetric unit.
$\underline{R},\psi,\underline{Z}$	Cylindrical polar coordinates in reciprocal space. ϕ and ψ are in fractions of a turn.
f_j	Scattering factor of jth atom.
$\underline{h},\underline{k},\underline{l}$	Reciprocal space indices of lattice point. When there is an exact repeat c in z then $\underline{l}$ is also used as a layer line index.
$\underline{m}$	Integer representing periodicity along the basic helix.
$\underline{n}$	Integer representing rotational periodicity.
$J_{\underline{n}}(X)$	Bessel function of order $\underline{n}$ and argument X.
$F_{\underline{Z}}(\underline{R},\psi)$	Fourier transform in polar coordinates for a layer plane at $\underline{Z}$ in reciprocal space.
$F_{\underline{hkl}}$	Structure factor for a crystalline reflection.
$G_{\underline{nZ}}$	Amplitude along a layer line in a cylindrically averaged specimen. $F_{\underline{Z}}$, $F_{\underline{hkl}}$ and $G_{\underline{nZ}}$ are on a scale corresponding to the number of electrons in a subunit.
$\rho(r,\phi,z)$	Electron density.
α	Phase angle.
α_c	Phase angle calculated from a model.
F_o	Observed amplitude.
F_c	Amplitude calculated from a model.
$\langle I_u \rangle$	Average discrepancy between observed and calculated intensities.
$P(\alpha)$	Overall phase probability distribution.
$P_{MIR}(\alpha)$	Phase probability distribution from multiple isomorphous replacement.
$P_{SIM}(\alpha)$	Phase probability distribution derived from model.

$$G_{\underline{n}\,\underline{Z}}(\underline{R}) = \sum_j f_j J_{\underline{n}}(2\pi \underline{R} r_j) \exp 2\pi i(\underline{Z} z_j - \underline{n}\phi_j) \qquad (3).$$

Equation 1 describes a diffraction pattern with intensity on lines of constant $\underline{Z}$, the layer lines (Figures 1 and 2). As a rule of thumb, the value of the Bessel function factor $J_n(2\pi \underline{R}r)$ is only appreciable on a given layer line for $2\pi\underline{R}r > \underline{n}$. This gives rise to the classic central cross on the diffraction pattern of B-DNA. Near the centre of the pattern, only layer lines for $\underline{m} = 0$ are appreciable, so $\underline{Z} = \underline{n}/\underline{P}$. For successively larger $\underline{n}$, layer lines at successive orders of $1/\underline{P}$ are defined, with the region of strong intensity progressively farther from the meridian. If electron density does not follow the basic helix, the central cross of intensity may not be observed.

Equation 2 shows that the same set of layer lines, with different Bessel function order, will be observed for $\underline{P}/\underline{p}$ units per turn as for $\underline{P}/\underline{p} \pm \underline{s}$ units per turn, where $\underline{s}$ is any integer. This discrete ambiguity may make it difficult to determine $\underline{P}/\underline{p}$ directly from the diffraction pattern, but it is often possible to determine the general shape of the subunit even when there is some uncertainty about $\underline{P}/\underline{p}$. For both double-stranded RNA[21] and for the Pf1 species of FV[43] it was difficult to distinguish between two adjacent options for the number of units per turn on the basis of the fit between the Fourier transform of similar models for the subunit and the observed data. For reconstituted helices of TMV coat protein, both 16 1/3 and 17 1/3 units per turn have been observed, depending on subtle changes in conditions, and the diffraction patterns of the two forms can scarcely be distinguished[53].

The unit rise $\underline{p}$ can in principle be determined from the layer line at $\underline{Z} = 1/\underline{p}$, but often it is difficult to define this layer line. The zero order Bessel function J_o appears on this layer line, and since J_o is the only Bessel function giving intensity on the meridian, inspection of the meridian should immediately give $\underline{p}$. The J_o layer line may however be weak because the molecular transform of the subunit is low, or disoriented layer lines may falsely appear to have intensity on the meridian. In this case ancillary information must be used to determine $\underline{p}$. The mass of the subunit divided by the mass per length of the helical assembly would give $\underline{p}$ directly. The mass of the subunit may be determined accurately by chemical means, and scanning transmission electron microscopy of unstained samples[54] can give the mass per length.

A single isomorphous heavy atom derivative can also be used to determine $\underline{p}$[30,55]. The radius r of a heavy atom is defined by the difference between the native and derivative equatorial (J_o) amplitude distribution as a function of reciprocal space radius $\underline{R}$. This difference gives $J_o(2\pi\underline{R}r)$ and therefore gives the radius r of the heavy atom. The difference distribution on the layer lines for

$\underline{m} \neq 0$ defines the orders of Bessel functions on these layer lines, and hence the helix symmetry. Where derivative amplitude is strong and native amplitude is weak, the Bessel function order can be determined independently of the phase. At the nodes of the heavy atom Bessel function term, the native and deriviative amplitude must be equal. This gives a further constraint on the Bessel function order.

The intensity distribution along each layer line depends on the Bessel function $J_n(2\pi \underline{R}r)$, so the radius r of atoms contributing to diffraction on the layer line defines the reciprocal space radius $\underline{R}$ at which strong intensity is observed. Where observed intensity is very strong, scattering from most atoms in the subunit must add in phase. That is, the argument $(\underline{Z}z_j - \underline{n}\phi_j)$ in the first exponential factor of Equation 1 must be about the same for most atoms in the subunit, or $\phi_j/z_j = \underline{Z}/\underline{n}$ for most atoms. Thus when certain regions of the diffraction pattern are very strong, it is possible to determine the radius and orientation of domains of electron density within the subunit by qualitative considerations.

Such considerations, in effect,give low resolution electron density distributions by inspection of the intensity distribution. In a simple case one may start to build atomic models and calculate transforms at the very beginning of a structure analysis, since this is a quick way to define domains of electron density. Therefore, the phrase "model building" is applied to the technique of structure determination by consideration of strong regions of intensity.

3.2 Crystal lattices

Helices with an integral number of units in a small number of turns can form three-dimensional crystalline arrays similar to crystal lattices, except that these helices pass continuously from one unit cell to the next along the helix axis in the z direction. The helices form local crystalline regions, or crystallites, randomly oriented around the z axis with respect to one another. The resulting diffraction patterns are similar to single-crystal rotation patterns, but they may also show the disorientation arcs typical of fibre patterns. The continuous Fourier transform $F_Z(\underline{R},\Psi)$ of a single helix is sampled by the crystalline reflexions at lattice points $\underline{h}\ \underline{k}\ \underline{l}$. On a given layer line $\underline{l}$, the crystalline reflexions $\underline{h}\ \underline{k}$ are observed only as a function of $\underline{R}$, and two different $\underline{h}\ \underline{k}$ with identical $\underline{R}$ will superpose. When the superposed reflexions sample symmetry-related parts of the molecular transform, this multiplicity effect can be corrected by simply dividing the observed intensity by the number of superposed reflexions. When the relationship between the intensities in the different superposed reflexions is not known, the intensity can be divided in the ratio calculated for a model.

Two different $\underline{h}\ \underline{k}$ with similar but not identical $\underline{R}$ will appear near each other on the layer line even if they are quite far apart in Ψ on the layer plane. If more than one $G_{\underline{n}\ \underline{Z}}$ has significant intensity on the layer plane at a given $\underline{R}$, the $G_{\underline{n}\ \underline{Z}}$ will interfere around Ψ according to Equation 1. Thus neighbouring $\underline{h}\ \underline{k}$ may differ in intensity by far more than would be possible if they were sampling a single $G_{\underline{n}\ \underline{Z}}$. This effect gives a more stringent test of models than that provided by the cylindrically averaged transform[56], and it can also be used to distinguish between discretely different choices for $\underline{P}/\underline{p}$ with similar subunit conformations[57].

Crystalline lattices with several helices passing through the unit cell may show interference between the helices. If the helices have fractional unit cell coordinates x_M, y_M, z_M, the amplitude at $\underline{h}\ \underline{k}\ \underline{l}$ is[52]

$$F_{\underline{h}\ \underline{k}\ \underline{l}} = \frac{1}{M}\sum_M F_{\underline{Z}}(\underline{R},\Psi)\ \exp 2\pi i(\underline{h}x_M + \underline{k}y_M + \underline{l}z_M - \underline{n}\phi_M) \tag{4}$$

If crystalline fibre patterns show interference effects, the relative positions of the molecules must be determined and the observed intensities corrected according to Equation 4 to give the transform of a single molecule. The relative x,y positions are most easily determined from the crystalline reflexions on the equator, and the relative z positions can then be determined by analysis of the systematically absent or weak intensities on the higher layer lines.

The terms $\underline{l}z_M - \underline{n}\phi_M$ in the argument of the exponential in Equation 4 relate $\underline{l}$ and $\underline{n}$ if z_M and ϕ_M are known. It has been suggested that this could be used to distinguish between discrete alternatives for the number of units per turn $\underline{P}/\underline{p}$ for β-reovirus double-stranded RNA[58] and for the Pf1 species of FV[43,59]. In both these cases the helices crystallize in a hexagonal unit cell with three helices passing through the unit cell, and it was thought that consideration of systematic absences of crystalline reflexions would define the number of units per turn $\underline{P}/\underline{p}$. But we find (unpublished) that in the case of reovirus RNA, the observed systematic absences can be explained either by helices with 9 or 12 subunits per turn lying on the 3_1 screw axes of the hexagonal space group, or by helices with 10 or 11 subunits per turn related to one another by 3_1 screw axes between them. Similar considerations apply to Pf1. These examples show that a thorough search of possible packing schemes is necessary when using systematic absences to determine $\underline{P}/\underline{p}$.

Crystalline reflexions may be observed in only certain regions of a diffraction pattern, with continuous transform observed elsewhere on the pattern. This indicates that the helices are arranged so that they diffract as a crystal when viewed along certain directions but not when viewed along other directions. A common example is a diffraction pattern with crystalline reflexions on the equator but continuous transform elsewhere. This is typical of helices acting like close-packed cylinders, arranged on a defined lattice (usually hexagonal) in the xy plane, but with random relationships in ϕ and z. If the molecules are randomly translated by $\underline{c}/2$ parallel to z, there will be crystalline reflexions on even layer lines and continuous transform on odd layer lines[60]. Another common example is screw disorder[2,61,62], which is found for helices like TMV or DNA with a pronounced helical groove. The helix can screw along the direction of the groove and, if $\Delta\phi$, Δz defines the screw, $Zz - n\phi$ in Equation 1 remains constant for $Z = n\,\Delta\phi/\Delta z$. Therefore crystalline reflexions are observed for regions of the diffraction pattern where $Z = n\,\Delta\phi/\Delta z$ but not elsewhere. Yet another common kind of disorder is up-down disorder, where helices with a sense (unlike DNA which looks the same either way up) may be oriented in either direction in the fibre, and crystalline reflexions are seen only where the transforms of the "up" and "down" chains add in phase[63].

3.3 Perturbed helices

In an ideal helix, each subunit is exactly related to the next by a unit twist around an axis and a unit rise parallel to this axis, but in many real helices these relationships are only approximately true, and corresponding modifications of the simple diffraction pattern are observed. In the coiled-coil[64], the axis of a minor helix may itself follow a longer pitch major helix. In this case the diffraction pattern of a coiled-coil looks roughly like the diffraction pattern of the minor helix, but modified by additional closely-spaced near-meridional satellite layer lines coming from the major helix (Figure 1). In the limit, as the pitch of the major helix gets longer, and its radius smaller, these satellite layer lines get closer to the primary layer lines, and the diffraction pattern becomes identical to that of a simple helix[3].

Perturbations reflect improved packing between neighbouring helices in the perturbed structure. An example is the α-helix coiled-coil, where two α-helical strands wrap around one another, thereby improving the close packing of sidechains[64]. In consequence the local helix geometry is distorted from the ideal regular unit rise plus unit twist. Interactions with neighbouring helices in a crystalline unit cell may also distort the helix[7,63]. Macroscopic considerations distort the local environment.

A different kind of perturbation arises when a simple helix cannot reach its minimum energy state by a simple unit rise and unit twist, and the local conformation of the subunit differs from one unit to the next along the helix, often in a periodic way. This kind of perturbation is independent of any interaction between different helices in the fibre. For example, contacts between neighbouring subunits might be optimized at one radius by one set of helix parameters, but at another radius by another set, and the resultant compromise structure is perturbed away from an ideal helix[65]. For subunits in close contact along more than one helical line, optimal neighbour contacts along one of the helices may be incompatible with optimal neighbour contacts along another helix, and perturbations result[43,66,67]. Local considerations give rise to an emergent macroscopic change.

Perturbations complicate the analysis of a fibre pattern, but they may have functional significance, and they should be carefully analysed when they arise.

3.4 Electron density

The electron density distribution $\rho(r,\phi,z)$ within the subunit of a simple helix can be calculated by inverse Fourier transformation of the diffracted amplitudes $G_{n\,Z}(R)$ if the phases of the diffracted amplitudes are known. The relevant expressions in cylindrical polar coordinates are[2]

$$\rho(r,\phi,z) = \frac{1}{p}\sum_m\sum_n g_{n\,Z}(r)\exp - 2\pi i(Zz - n\phi) \tag{5}$$

where Z is related to m and n by Equation 2, and

$$g_{n\,Z}(r) = \int_0^\infty G_{n\,Z}(R)J_n(2\pi Rr)2\pi R\,dR \tag{6}$$

For each layer line Z and Bessel function order n, $g_{n\,Z}(r)$ is calculated from Equation 6. $G_{n\,Z}(R)$ is complex (Equation 3) but we only observe $G^2_{n\,Z}$ in the diffraction experiment, so it is necessary to determine the phases of $G_{n\,Z}$ in order to calculate $\rho(r,\phi,z)$. This can be done by MIR, as with single-crystals. But in fibre diffraction, unlike single-crystal diffraction, it may be possible to measure the continuous distribution of diffracted intensity along a layer line. Both the amplitude and the phase of $G_{n\,Z}$ must be continuous along a layer line, and this introduces important constraints on the phases (Section 5.2).

4. MODEL BUILDING

4.1 General principles

An X-ray diffraction pattern of an unknown structure does not lead directly to the structure, because only amplitudes and not phases can be measured from the diffraction pattern. One of the most venerable methods of solving this phase problem is by trial-and-error. From consideration of the diffraction pattern, symmetry, and chemical information, a model is postulated. This model is refined or replaced until a model is found that predicts the observed diffraction pattern so well that the possibility of a homometric structure (a different model that also predicts the diffraction pattern) is virtually eliminated. Many inorganic and small organic structures have been solved by this method. Some biological structures (notably the α-helix and DNA) have also been solved by this method, but in general it has not been used for large structures since the development of MIR, which gives phases directly if heavy atom derivatives can be found.

X-ray fibre diffraction data are rarely of sufficiently high quality to enable structure determination at atomic resolution, even when heavy atom derivatives can be found. The effective resolution of the data can be extended by using model building. Atoms are covalently linked in certain chemically-defined ways, and even unlinked atoms may not approach more closely than the sum of their van der Waals radii. These constraints are so strong that even low resolution fibre diffraction data are sufficient to solve structures such as DNA and polysaccharides where the subunit consists of relatively few atoms which are covalently linked to the atoms in the next subunit along the helix.

Even when a subunit contains many hundreds of atoms, as for the protein subunit of helical viruses, trial-and-error is still feasible provided enough information about the protein subunit can be obtained. Then the problem can be reduced to one of solving the relative position and orientation of only one, or a few, known units. If the X-ray data extend to dimensions smaller than the subunit, the problem is solvable in principle. In the ideal case, the subunit can be crystallized and its structure solved separately by single crystal methods, as for TMV[68]. But considerable structural information can be obtained even on large subunits that cannot be crystallized. The amino acid sequence of a protein may define secondary structure[69], and domains of secondary structure combined with energy calculations[70] suggest specific domains of tertiary structure within proteins[71,72,73,74]. These stereochemical constraints, when taken together with fibre diffraction data, may enable structure determination in cases where determination of structure from the protein sequence alone is not yet possible.

4.2 Symmetry constraints

The most general constraint in a helix is that on packing of subunits, or the symmetry constraint. When each subunit is covalently linked to the next along a one-start helix (as for DNA), the subunits will be closer together in this unique direction than in any other helical direction. When the subunits are linked by non-covalent bonds (as for helical viruses), there may be equally close contacts along two, or sometimes more, helical directions. In this latter case it may be helpful to visualize the packing of subunits by representing each subunit as a smooth space curve. This curve may be thought of as approximating the major axis of the ellipsoid of revolution representing the subunit; or, more precisely, as a smooth curve roughly defined by the centres of gravity of sections cut through the subunit by cylindrical shells of successively smaller radii. Each of these space curves is associated with one lattice point on the surface lattice. For TMV[75], these curves run more or less radially, whereas for FV[76] they run roughly parallel to the virus axis but slew around this axis and also slope radially (Figure 4; Figure 5).

Rod-like subunits tend to interleave between subunits on the turn below, giving an odd number of subunits in about two turns of the helix. If the α-helical subunits of FV adopt a minimum energy arrangement controlled by symmetry rules, each space curve following the axis of the α-helix will be equidistant from its nearest neighbours along its length. Tangents to the two curves drawn at the ends of the shortest line segments joining them will make a constant angle. Curves with these properties are called Bertrand curves[77]. Myosin thick filaments[37] are another structure in which the subunits are rod-like, so these considerations would also apply there.

Slightly different local interactions between nearest neighbours in FV yield different helix symmetries (Figure 4). These changes in symmetry are a sensitive test for small changes in interactions between subunits[66].

Model-building is the construction of three-dimensional representations of a structure to fit all the known chemical facts, to obey any symmetry constraints, and to be used to calculate Fourier transforms for comparison with the observed X-ray fibre diffraction data. The helix symmetry can constrain the shape of subunits if they pack together to fill space. The symmetry constraint is an example of the "powerful but dangerous weapon, the principle of simplicity"[78].

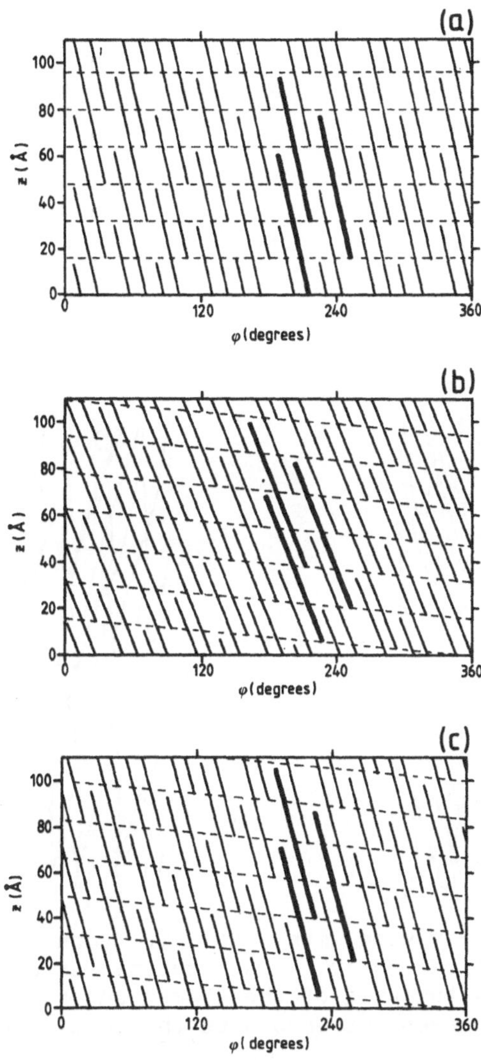

Fig. 4. Surface lattices of FV. The surface lattice represents equivalent points in the helix subunits, projected along radii onto a cylindrical surface of radius 26 Å, coaxial with the helix. This surface is then cut vertically along $\phi = 0$, opened out flat, and viewed from the outside of the helix. The lines represent the direction followed by rods of high electron density defined by inspection of the diffraction pattern (Section 4.3). The three bold lines represent the nearest neighbours that are in close contact. The dashed line is the basic helix (degenerate in (a)). (a) The fd species[116], with a five-fold rotation axis and a two-fold screw axis of pitch 32 Å along the helix axis. (b) The Pf1 species[30] at room temperature, with pitch 15.7 Å and 5.40 units/turn. (c) The Pf1 species at 4°C, with pitch 16.6 Å and 5.46 units/turn.

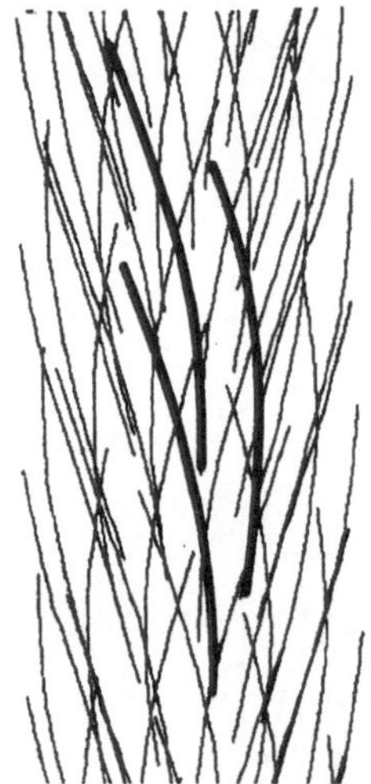
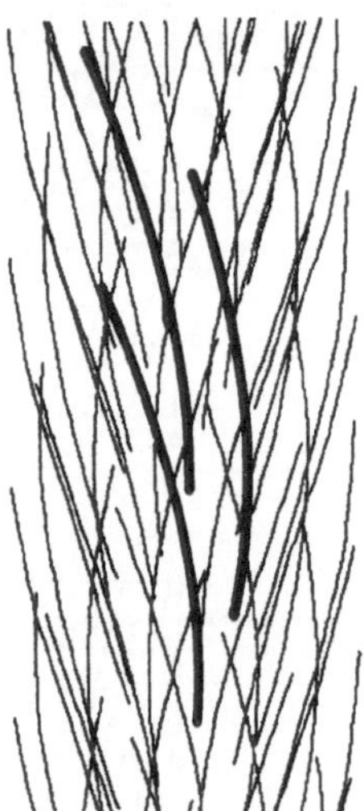

Fig. 5. The Pf1 virion with 5.46 units/turn. Stereo pair of a segment of the helical array. Each subunit is represented by a curve running along the axis of the 46-residue continuous α-helix model. The bold curves correspond to the bold lines in Figure 4(c).

4.3 Low-resolution models

Even if there is little chemical information about the subunit, it may be possible to build a model at low resolution. For instance something may be known about the shape of the subunit, either from X-ray scattering and hydrodynamic studies on subunits in solution, as for bacterial flagellin[79], or from electron microscopy, as for myosin molecules[80]. Then these subunit shapes can be arranged in the surface lattice determined for the corresponding helix with regard to the symmetry constraints discussed in Section 4.2. In the examples given, the corresponding assemblies are flagella or myosin thick filaments. The Fourier transform of this low resolution arrangement can be calculated using Equation 1, by filling the envelope of the subunit with an array of points on an arbitrarily fine raster, and the orientation of the subunit can be refined to fit the observed X-ray data at low resolution. The shape of the subunit might instead be represented analytically, as in solution scattering[81], and the orientation refined. For an open helix, electron microscopy can give information about the path followed by the bulk of the material in the helix[82].

In some cases the shape of the subunit can be determined directly from the distribution of strong regions of intensity on the diffraction pattern. If the subunits were spherical, the intensity on each layer line would be proportional to the square of the amplitude of the Bessel function contributing to that layer line, decreasing slowly with distance from the centre of the diffraction pattern (Section 3.1). Unusually strong diffraction on layer line $\underline{Z}$ for Bessel function order $\underline{n}$ indicates that $\exp 2\pi i(\underline{Z}z_j - \underline{n}\phi_j)$ has constant phase for most atoms in the structure. The classic example is DNA, where strong intensity at $\underline{Z} = 1/3.4$ Å^{-1}, $\underline{n} = 0$ indicates planes of electron density 3.4 Å apart, oriented perpendicular to the helix axis. If there is no perpendicular two-fold axis the origin is arbitrary, and we can set $\underline{Z}z_j - \underline{n}\phi_j = 0$, or $z_j/\phi_j = \underline{n}/\underline{Z}$. Strong intensity on layer line $\underline{Z}$ assigned to Bessel function J_n then means that there is electron density running along the direction for which $\underline{z}/\phi = \underline{n}/\underline{Z}$. For the Pf1 species of FV, this argument was used to interpret the strong intensity in the near-equatorial region at about 10 Å spacing[43] (Figure 1c). There is strong intensity at $(\underline{n},\underline{Z})$ of (-11, 0.013 Å^{-1}) and (-6, 0.039 Å^{-1}). The first of these regions of strong intensity indicates a rod of electron density at about $\underline{r}$ = 20 Å (the maximum of $J_{11}(2\pi \underline{Rr})$ corresponding to the observed intensity maximum at $\underline{R}$ = 0.105 Å^{-1}) with pitch z/ϕ = -846 Å/turn. The second region of intensity indictes a rod of electron density at about $\underline{r}$ = 13 Å with pitch -154 Å/turn. This distribution of electron density can most simply be represented by a single curved α-helix rod[43,83], but two-rod representations have also been considered[84]. The radial projection of a rod model with an intermediate pitch of -440 Å is shown in Figure 4(b). The helix parameters of this virus change with temperature (Figure 1d) and there is a corresponding change in the orientation of the α helical rods (Figure 4c). A stereo representation of the curved rods is shown in Figure 5. Similar arguments were used to give a low resolution model for the structure of bacterial pili[85,86].

4.4 Atomic models

The simplest way to include stereochemical constraints in model-building is to construct a wire model, in which covalent bonds are represented by lengths of wire to scale, and atoms are represented by the junctions of the wires. Wire models are increasingly being superseded by interactive computer graphics[87] where lines drawn on a computer screen represent a projection of a model. With such an arrangement it is possible to view the model in stereo and colour, and to rotate, change scale, move part of the structure, measure non-bonded distances, optimize stereochemistry, or display van der Waals surfaces.

For simple structures, the stringent stereochemical constraints of a precise atomic model should be included at the start of the structure analysis. For instance, both the α-helix and DNA structures were solved primarily by building models. X-ray fibre patterns gave indications of the types of model to be considered, but in both these cases, the stereochemical sense made by one model was the critical step in the structure determination. In fact the Fourier transforms calculated for the α-helix and for the original Watson-Crick DNA model fit the X-ray data so poorly that they would have been rejected had stringent crystallographic standards been applied. In the case of the α-helix it was later shown that the fibre pattern was modified from that expected for a single helix because α-helices tend to coil around one another in pairs. In the case of DNA, it was later shown that another model of the Watson-Crick type gave a better fit to the diffraction data. These examples show that simple and elegant models that explain the main features of the fibre diffraction pattern should not be discarded just because they do not explain all the details of the pattern.

As an example of a more complex model-building analysis, consider the analysis of the structure of the Pf1 species of FV[30,43,76,83,88,89] (Figures 4 and 5). The orientation of the α-helical protein subunit was defined by consideration of the strong regions of intensity on the diffraction pattern (Section 4.3). The subunit position was refined by trial-and-error and by symmetry considerations. Distances between α-carbons on adjacent subunits could be plotted as a function of the rotation of the α-helix around its own axis, to test symmetrical packing of nearest-neighbour α-helices and to define optimum knobs-into-holes packing[64]. Sidechains were added to the backbone in standard orientations, and stereochemistry was refined[90,91]. An intermediate state in the structure analysis is illustrated in Figure 6. Here the orientation of the subunit has been confirmed from an electron density map (Figure 6c) based on MIR phases. The position of the α-helix backbone (Figure 6a) is in agreement with this map. Addition of sidechains to the backbone (Figure 6b) gives a molecular envelope. In some positions within this envelope there are overlapping sidechain atoms and elsewhere holes are present. As a result, a map (Figure 6d), based on amplitudes and phases calculated from the model, is not in precise agreement with the MIR map. Refinement of the sidechain conformations, or an alternative chain tracing through the MIR map, might resolve these discrepancies.

The correct treatment of solvent is important when the molecular transform is calculated, because at low resolution the molecule scatters relative to surrounding solvent, whereas at high resolution the molecule scatters relative to vacuum. Although some water molecules may be specifically bound to the structure, most water will be randomly arranged, and the solvent effect can be approximated by excluding uniform density solvent (0.334 e/Å^3 for water) from the molecular envelope. Then by Babinet's principle the Fourier transform of the molecule immersed in solvent is equal to the transform of the molecule in vacuum minus the transform of the molecular envelope filled with uniform density solvent. This approach can be applied to the individual atoms by appropriately modifying the atomic scattering factors to give "water-weighted" scattering factors [56,92,93]. The displaced volume is then represented by an array of spheres centred at each of the atomic centres. The method is easy to apply, because the Fourier transform can be calculated directly from Equation 1 once the normal scattering factors f_j have been replaced by the water-weighted scattering factors. It has the flaw that an array of spheres does not define a uniform density distribution within the molecular boundaries.

A more accurate way to allow for the displaced solvent is to calculate directly the transform of a uniform density solvent within the molecular boundary. The displaced solvent can conveniently be represented on a grid of points within the molecular boundaries defined by the van der Waals outline[94], so we call this the grid method[89] (Figure 6(b)). A cubic grid about 1.75 Å on a side was satisfactory to calculate the water transform for Pf1 virus. The transform of the water defined by this grid of points is calculated and added to the transform of the model calculated with normal (not water-weighted) scattering factors, supplemented to include hydrogens at the centres of the atoms with which they are associated. A high-density shell of water has been inferred for proteins from the fact that the volume of the

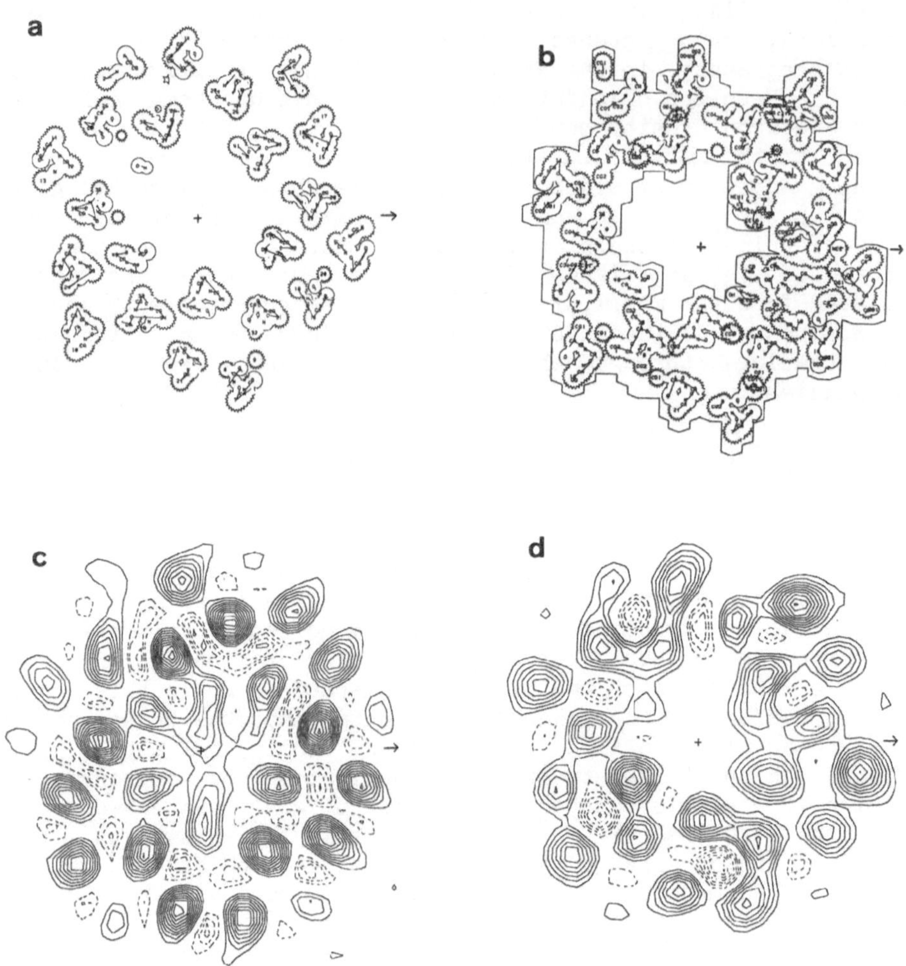

Fig. 6. Model-building of the Pf1 species of FV. Sections perpendicular to the helix axis are shown. The model is equivalent to that represented in Figure 4(c) and Figure 5. The cross marks the helix axis and the horizontal arrow is at $\phi = 0$, $r = 33$ Å. (a) Van der Waals outline of a polyalanine representation of the protein subunit, to show the symmetrical arrangement of the α-helix backbone. (b) Van der Waals outline of the full Pf1 model, including sidechains. The straight lines indicate the volume defined to include water, for water-weighting calculations. (c) Electron density distribution calculated from observed amplitudes and refined MIR phases. (d) Electron density distribution calculated from amplitudes and phases of the model shown in (b), at the same resolution as (c).

molecular envelope is larger than the volume calculated from partial specific volume and molecular weight[95]. We find that the transform of a Pf1 model including a solvent shell of density 0.35 e/Å^3 and thickness 2.8 Å surrounding the molecular envelope does not differ appreciably from the transform without this shell (unpublished).

Once a model has been built to explain the main features of the diffraction data and to satisfy stereochemical criteria, the model must be refined to improve the fit of the Fourier transform to the diffraction data while retaining acceptable stereochemistry. This can be done by trial-and-error. Several models are constructed and their transforms calculated, to get an indication of how changes in the model are related to changes in the transform. Almost every modification of the model which improves the fit of calculated to observed transform in one region of the transform will worsen the fit in another region, and considerable ingenuity is required to optimize both the stereochemistry and the transform. A description of the application of this technique to refining DNA is given in ref. 56. When refinement is sufficiently close to the final stages, a more automatic version of the same model-building technique, called linked atom least-squares[12,96] can be used. Covalent bonds are fixed and other parameters are refined by least squares against both weighted stereochemical functions and the X-ray data. This technique must not be applied blindly or it is possible to get into local minima which could suggest false solutions. For instance both left-handed and right-handed models for DNA have been refined by this technique, but it is only possible to choose between them on the basis of a goodness-of-fit parameter determined at the end of each refinement pathway[12]. Even here some doubt may remain about the choice between the two models[97].

It may be helpful to consider all or part of the subunit as a rigid body when building models. If a group of atoms can be treated together, they become in effect a new larger "atom", with a non-spherical "scattering factor". Data which cannot resolve atoms can resolve these new groupings. This approach has been used in the refinement procedure called constrained-restrained least-squares refinement[98]. Rigid groups, for instance planar DNA bases or aromatic protein sidechains, are constrained to have always the same shape, and distances between groups are restrained to maintain proper stereochemistry. The structure is then refined by least-squares against both stereochemistry and diffraction data. Reducing the number of independent variables in this way increases the chance of convergence to the correct solution using low resolution fibre data.

To examine the van der Waals contacts during refinement, the van der Waals outlines of the model may be examined. This can be done using a computer program[94] which draws the van der Waals outline on a plane section through a molecule. Helical symmetry can result in the sampling of several subunits by one section perpendicular to z, giving all the information about a subunit but using only a limited number of sections. Overlaps between non-bonded portions of one subunit and between different subunits can be seen by this method (Figure 6(b)).

Another stereochemical constraint is the total volume occupied by the model. Analysis of a number of protein structures shows that each amino acid occupies a volume that is the same from one protein to another to within a few percent[99]. From the amino acid sequence of Pf1, for instance, a volume of 5935 $Å^3$ was calculated for the protein subunit[30]. For any trial model the van der Waals outline of the subunit can be plotted, and the included volume corresponding to one subunit can be calculated. If this volume is substantially less than 5935 $Å^3$, the model is stereochemically impossible; if it is substantially greater than 5935 $Å^3$, it is stereochemically unlikely, since large holes within the protein shell are unlikely. A corresponding volume for the viral DNA was calculated to give a total volume for the subunit and associated DNA[30]; this constrains models further, because this volume cannot be greater than the volume available in the unit cell of dry fibres.

Towards the end of a structure determination, it should be kept in mind that it may be inaccurate to represent the detailed structure of a helical molecule by a precise model with perfect stereochemistry. Slight differences in conformation between subunits in the helix are likely in a fibre, because the environment of each subunit as it interacts with subunits in neighbouring helices will differ[7,63] except in the rare cases when the helix symmetry is also a crystallographic symmetry. The longer sidechains on protein subunits may not be in defined positions, and it is incorrect to model them as if they were. A good strategy is to find the best model with precise stereochemistry to fit the data, and then to consider physically reasonable ways in which this model might be modified to improve the fit to the data.

When it is difficult to define sidechain conformations from the diffraction data, it may be necessary to build models using data on observed sidechain conformations in globular proteins[100]. Two different kinds of problem should be distinguished. First, there may be sidechains which have fixed conformations in the structure, but these conformations cannot be determined from the diffraction data alone. The hydrophobic sidechains involved in the interactions between subunits in FV appear to have fixed positions as measured by NMR[101], but diffraction data do not enable the determination of these positions. Secondly, sidechains which have no fixed conformations must be modelled in order to calculate the best Fourier transforms. Here we suggest a possible method to treat both these problems. The data on sidechain conformations in globular proteins summarized in Table 6 of ref. 100 show probabilities of various sidechain conformations. Models can be constructed with sidechain conformations weighted according to their probabilities. These models can then be used to study packing or to calculate transforms.

To study packing, a "probability transform" can be calculated, using multiple coordinates and weighted "scattering factors" to account for partial occupancy, as in calculating transforms of DNA with different bases[56]. Fixed atoms have "scattering factors" that are proportional to their atomic volumes[95] (their probability of occupancy is 1). Disordered sidechains are represented by different coordinate sets for the different conformations, each with "scattering factors" proportional to the atomic volume times the probability of that conformation. The transform calculated from these coordinates is then inverted to give a "probability density distribution" analogous to an electron density distribution. In the backbone region, this density will be uniform. Extended regions with probability density substantially greater than this uniform density would indicate a high probability of overlapping sidechains, and extended regions with probability density substantially less than this uniform density would indicate a high probability of holes. Either of these states would be taken as defects in the stereochemistry of the model, and could be used to discard unlikely models without the time-consuming step of trying all possible conformations of sidechains.

To calculate transforms of models with disordered sidechains, multiple coordinates can be used with scattering factors weighted by probabilities, analogous to the packing calculation. These coordinates and scattering factors can then be used to calculate the transform of the model with the most probable distribution of disordered sidechains.

5. ELECTRON DENSITY CALCULATION

5.1 General principles

To calculate the distribution of electron density in the subunit of the helix from Equation 5, the phases of the diffracted intensity must be determined. If a model of the structure has been built, phases may be calculated from the model and used with the observed amplitudes to calculate an electron density distribution. The transform of the resulting distribution can be refined by passing it through simple mathematical constraints, calculating the transform of the constrained distribution, and using the phases of this new distribution with the observed amplitudes to calculate a new electron density distribution. Examples of simple constraints are to constrain the electron density to equal uniform solvent density outside the molecular envelope, or to constrain the absolute value of the electron density to be within physically feasible bounds. It is important to realize two things when using such constraints. First, the starting phases must be sufficiently accurate to refine to the correct result. If the model predicts completely wrong phases for part of the transform, these phases are unlikely to be improved by the constraints. Secondly, solvent flattening by simply multiplying the electron density by a cylindrical box function[84] surrounding the helix, with a value of unity inside the cylinder and zero outside, will only constrain phases along a layer line. It will not constrain the relationships between phases on different layer lines. A more accurate definition of the molecular envelope may help to relate phases on different layer lines.

A preliminary atomic model can be used to put observed amplitudes on an absolute scale. The $F_0(00)$ of a water-weighted model gives the total number of electrons in the subunit, relative to water. The observed amplitudes F_o can be put on the same scale[30] as the calculated amplitudes F_c by multiplying the observed amplitudes by $(\sum F_c^2/\sum F_o^2)^{1/2}$. The height of the electron density calculated from these scaled observed amplitudes is relative to water. Negative electron density may be physically meaningful here. It may indicate concentrations of hydrophobic sidechains whose density goes below that of water.

The radial electron density distribution can be calculated from Equation 5 with $\underline{n} = 0$ and $\underline{Z} = 0$. In this case the phases are real, and it is only necessary to determine the signs. This can be done using the minimum wavelength principle[102]. If two amplitude peaks are closer together than $2/\underline{d}$, where $\underline{d}$ is the maximum diameter of the helix, then they must have opposite signs. The central maximum at $\underline{R} = 0$ will be positive unless the average density of the molecule is less than that of solvent, and peaks at successively larger $\underline{R}$ can then be assigned signs -, +, -, etc., with sign changes at each zero in the amplitude distribution[50]. In order to determine the positions of the zeros it is necessary to know the continuous transform distribution, either by measuring it directly or by reconstructing it from crystalline reflexions, preferably at successive swelling states[50].

The radial density distribution can also be calculated using a constrained regularization method[30,86,103]. This method minimizes artifacts caused by limited and noisy data by constraining the radial density to be the smoothest non-negative density consistent with the data. No extrapolation of the data to $\underline{R} = 0$ is required, and statistical weights can be assigned to each data point. A series of possible solutions of different smoothness is produced, and a probability is assigned to each solution. By examining all solutions with high probability, an estimate of the error in the radial density distribution can be made.

The electron density distribution in the entire unit cell of a crystalline fibre can be calculated using standard crystallographic techniques. This distribution makes no assumption that the structure is a helix, and can therefore detect either deviations from precise helical symmetry or the positions of water or ion molecules in the unit cell[104].

5.2 Multiple isomorphous replacement

Whenever possible, heavy atom derivatives of the structure should be prepared and used to determine phases. The positions of the heavy atoms in the derivatives may be determined directly by triangulation[34], or indirectly, using a preliminary model to give phases for the stronger intensities. These phases are then used to calculate an electron density distribution of the difference between the native and the derivative structures[30]. Once the position of the heavy atom has been determined, the phase probability distributions are calculated at each point along the layer lines in the same way as for individual reflexions in protein crystals[33]. If more than one Bessel function term contributes to a layer line, further heavy atom derivatives can be used to separate the Bessel function terms (two for each term). Any layer-line splitting should be exploited to separate the Bessel function terms[46,47].

Analysis of heavy atom derivatives in practice does not define a unique phase at each point along the layer line, but rather a phase probability distribution. But there is an additional important constraint in fibre diffraction which is not available in single crystal studies, namely the continuity of the Fourier transform along the layer line. Points on the layer line where the phase is well-defined can be used to resolve ambiguities at neighbouring points on the layer line (Figure 7), either graphically, by plotting the phase probabilities point by point along the layer line and then drawing a smooth curve through them, or semi-automatically, by solvent flattening outside the molecular boundary.

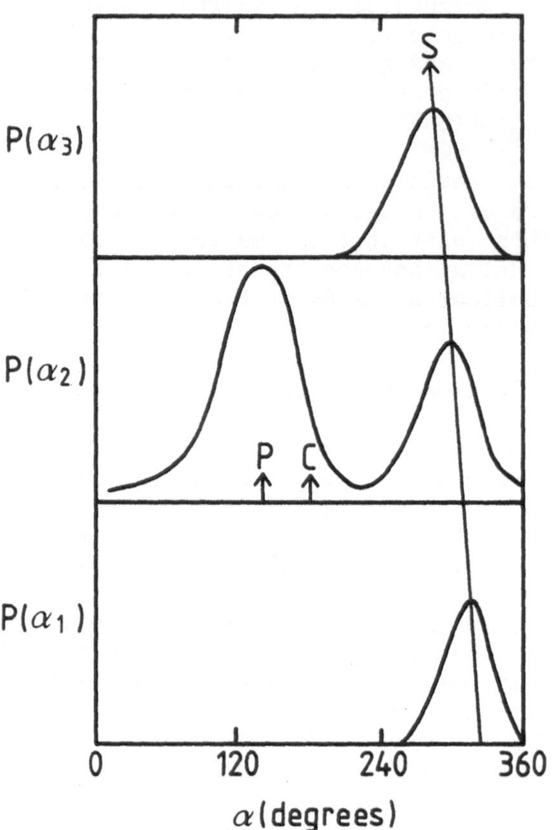

Fig. 7. Application of the phase continuity constraint (schematic). The probability P(α) of the phase α as determined from MIR is plotted for successive reciprocal space radii $\underline{R}_1$, $\underline{R}_2$, $\underline{R}_3$ along a layer line. P(α_2) determined at $\underline{R}_2$ has a bimodal distribution, whereas P(α_1) at $\underline{R}_1$ and P(α_3) at $\underline{R}_3$ are unimodal distributions. The most probable value (P) of α_2 is 140° and the centroid (C) of α_2 is 180°. However, the most probable value of α_1 is 320° and the most probable value of α_3 is 280°, so the smoothed best value (S) for α_2 is 300°.

To refine phases for the Pf1 species of FV, we have developed an iterative method of applying the continuity constraint. We start in the same way as for a single crystal, using the centroid of the phase probability distribution[33], weighted by a figure of merit, for each point along the layer line. We then progress from an electron density distribution calculated from centroid phases to one having continuous phases, using a method similar to that developed[105] to combine phases from isomorphous replacement with those defined by non-crystallographic symmetry. An initial electron density distribution is calculated and constrained by the box function to give calculated amplitudes F_c having phases α_c. The agreement between F_c and the observed amplitudes F_o gives a phase probability distribution[106]

$$P_{SIM}(\alpha) \propto \exp(2|F_o|\cdot|F_c|\cos(\alpha - \alpha_c)/\langle I_u\rangle) \qquad (7)$$

where $\langle I_u\rangle$ is the average difference between the observed and calculated intensities. This probability distribution is combined with the probability distribution determined by MIR[33] to give a combined probability distribution

$$P(\alpha) = P_{MIR}(\alpha)[P_{SIM}(\alpha)]^v \qquad (8)$$

Varying the exponent v varies the weights of the different phase probabilities as the refinement of the phases progresses. For Pf1, v = 0.1 was used in the initial cycles of refinement, and this was gradually increased to v = 1 in the later cycles.

5.3 Refinement

Electron density distributions are used in various ways to refine structures. Given a model with amplitudes $|F_c|$ and phases α_c, an electron density distribution calculated from amplitudes $|F_o|-|F_c|$ and phases α_c will give the electron density difference between the model and the true structure, provided that the model is sufficiently close to the truth. To refine the structure, atoms in the model are moved in the direction of increasing difference density while retaining stereochemical constraints. The resulting new model is used to calculate a new difference distribution. Difference syntheses have been used to determined the positions of ions in DNA fibres[104] and to distinguish between different possible systems of base pairing in DNA[107].

When an electron density distribution calculated from MIR phases is available, it may be useful to take the difference between the electron density distribution calculated from the observed data with MIR phases, and the electron density distribution calculated from the model. This approach avoids the difficulties that may be introduced by using model phases with observed data when the model is still quite far from correct.

Given an electron density distribution calculated by the MIR technique, but no model, the first step is to build an atomic model into the electron density distribution, as in the single crystal case. The electron density distribution may be less well defined than for single crystals, and supplemental information may be needed. For TMV, for instance, the interpretation of the electron density distribution in the helical virus[75] was greatly aided by the high resolution single-crystal analysis of the viral protein structure[68]. Models can then be refined using the difference synthesis technique and/or using Equation 8 with P_{SIM} calculated with F_c and α_c from the model.

6. MOLECULAR MOTION

X-ray fibre diffraction can give information about molecular motion in two different ways. The first is indirect. Some structures are found to have different conformational states which can be converted from one to another, sometimes even within a fibre. By solving the structures of the two different structural forms, studying the transition in solution, and modelling the transition by energy minimization techniques, it is possible to describe the dynamics of the transition. Some examples are the change of DNA symmetry from the A to B forms[108] or B to C forms[62]; the two states of a DNA-RNA polymer[109]; and the transmission of solitons of altered structure along a DNA molecule[110]. Assemblies of subunits that are in contact along two helices may be incommensurate structures[111] because the symmetry operators along the two helical directions are mutually incompatible. An example is bacterial flagella, which form a supercoiled structure delicately poised between different structural states[67,112,113]. Another example is the Pf1 species of FV, which is found in two different structural forms and can be converted from one to the other by a change of temperature[66,114]. The transition from one form to another has been mapped by calculating the minimum energy for a series of models along the transition pathway[114].

The second kind of information on molecular motion can be obtained directly from time-resolved X-ray diffraction experiments on a transition as it takes place. An example is the change in orientation of myosin head-groups in muscle[115]. Muscle can be induced to twitch repeatedly between two well-defined structural states which show different X-ray fibre diffraction patterns. Using high flux X-rays produced by a synchrotron source, combined with a linear detector coupled to a memory and synchronized with a stimulator and a fast shutter, a diffraction pattern of frog muscle was recorded in 10 msec time frames over a few hundred stimulations spaced at 30 sec intervals. The results describe the change in the X-ray fibre pattern with time over the several hundred milliseconds required for one twitch. These data can be used to test models for the motion of muscle during contraction.

References

1. A.R. Stokes, The theory of X-ray fibre diagrams, Prog. Biophys. 5: 140 (1955).
2. A. Klug, F.H.C. Crick, and H.W. Wyckoff, Diffraction by helical structures, Acta Cryst. 11:199 (1958).
3. G.N. Ramachandran, Analysis of the X-ray diffraction pattern of helical structures, Proc. Indian Acad. Sci. 52: 240 (1960).
4. R.E. Dickerson, X-ray analysis and protein structure, in: "The Proteins," H. Neurath, ed., vol. II, p.603, Academic Press, New York (1964).
5. K.C. Holmes and D.M. Blow, "The Use of X-ray Diffraction in the Study of Protein and Nucleic Acid Structure," Interscience, New York (1966).
6. B.K. Vainshtein, "Diffraction of X-rays by Chain Molecules," Elsevier, Amsterdam (1966).
7. A. Elliott, X-ray diffraction by synthetic polypeptides, in: "Poly-α-Amino Acids," G.D. Fasman, ed., p.1, Marcel Dekker, New York (1967).
8. L.E. Alexander, "X-ray Diffraction Methods in Polymer Science," Wiley, New York (1969).
9. M. Kakudo and N. Kasai, "X-ray Diffraction by Polymers," Elsevier, Amsterdam (1972).
10. R.D.B. Fraser and T.P. MacRae, "Conformation in Fibrous Proteins," Academic Press, New York (1973).
11. Y. Mitsui and Y. Takeda, X-ray diffraction studies of helical biopolymers and biological structures, Adv. Biophys. 12: 1 (1979).
12. A.D. French and K.H. Gardner, eds., "Fiber Diffraction Methods," American Chemical Society, Washington (1980).
13. P.G. de Gennes, "The Physics of Liquid Crystals," Clarendon Press, Oxford (1974).
14. A. Blumstein, ed., "Liquid Crystalline Order in Polymers," Academic Press, New York (1978).
15. E.T. Samulski, Liquid crystalline order in polypeptides, in: "Liquid Crystalline Order in Polymers," A. Blumstein, ed., p.167, Academic Press, New York (1978).
16. C. Robinson, Liquid-crystalline structures in polypeptide solutions, Tetrahedron 13: 219 (1961).
17. J. Gregory and K.C. Holmes, Methods of preparing orientated tobacco mosaic virus sols for X-ray diffraction, J. Mol. Biol. 13: 796 (1965).

18. J. Lapointe and D.A. Marvin, Filamentous bacterial viruses VIII. Liquid crystals of fd, Mol. Cryst. and Liquid Cryst. 19: 269 (1973).
19. M.H.L. Pryce and F.C. Frank, The spherulitic texture, Faraday Soc. Discuss. No. 25: 41 (1958).
20. R. Langridge, H.R. Wilson, C.W. Hooper, M.H.F. Wilkins, and L.D. Hamilton, The molecular configuration of deoxyribonucleic acid I. X-ray diffraction study of a crystalline form of the lithium salt, J. Mol. Biol. 2: 19 (1960).
21. W. Fuller, F. Hutchinson, M. Spencer, and M.H.F. Wilkins, Molecular and crystal structures of double-helical RNA I. An X-ray diffraction study of fragmented yeast RNA and a preliminary double-helical RNA model, J. Mol. Biol. 27: 507 (1967).
22. D.A. Marvin, X-ray diffraction and electron microscope studies on the structure of the small filamentous bacteriophage fd, J. Mol. Biol. 15: 8 (1966).
23. F.T. Hong, D. Mauzerall, and A. Mauro, Magnetic anisotropy and the orientation of retinal rods in a homogeneous magnetic field, Proc. Natl. Acad. Sci. USA 68: 1283 (1971).
24. G. Maret and K. Dransfeld, Macromolecules and membranes in high magnetic fields, Physica, 86-88B: 1077 (1977).
25. G. Maret, J. Torbet, E. Senechal, A. Domard, M. Rinaudo, and H. Milas, Polyelectrolytes in high magnetic fields, in: "Nonlinear Behaviour of Molecules, Atoms and Ions in Electric, Magnetic or Electromagnetic Fields," L. Neel, ed., p. 477, Elsevier, Amsterdam (1979).
26. D.L. Worcester, Structural origins of diamagnetic anisotropy in proteins, Proc. Natl. Acad. Sci. USA 75: 5475 (1978).
27. E.T. Samulski and A.V. Tobolsky, Distorted α-helix for poly (-benzyl L-glutamate) in the nematic solid state, Biopolymers, 10: 1013 (1971).
28. M. Chabre, X-ray diffraction studies of retinal rods I. Structure of the disc membrane, effect of illumination, Biochim. biophys. Acta, 382: 322 (1975).
29. J. Torbet and G. Maret, Fibres of highly oriented Pf1 bacteriophage produced in a strong magnetic field, J. Mol. Biol. 134: 843 (1979).
30. C. Nave, R.S. Brown, A.G. Fowler, J.E. Ladner, D.A. Marvin, S.W. Provencher, A. Tsugita, J. Armstrong, and R.N. Perham, Pf1 filamentous bacterial virus: X-ray fibre diffraction analysis of two heavy-atom derivatives, J. Mol. Biol. in press.
31. M. Hogan, N. Dattagupta, and D.M. Crothers, Transient electric dichroism of rod-like DNA molecules, Proc. Natl. Acad. Sci. USA, 75: 195 (1978).

32. M. Sakamoto, T. Fujikado, R. Hayakawa, and Y. Wada, Low frequency dielectric relaxation and light scattering under AC electric field of DNA solutions, Biophys. Chem. 11: 309 (1980).
33. T.L. Blundell and L.N. Johnson, "Protein Crystallography," Academic Press, New York (1976).
34. K.C. Holmes, E. Mandelkow, and J. Barrington Leigh, The determination of the heavy atom positions in tobacco mosaic virus from double heavy atom derivatives, Naturwiss. 59: 247 (1972).
35. A. Franks, X-ray optics, Sci. Prog., Oxf. 64: 371 (1977).
36. A. Elliott, The use of toroidal reflecting surfaces in X-ray diffraction cameras, J. Sci. Instrum. 42: 312 (1965).
37. H.E. Huxley and W. Brown, The low-angle X-ray diagram of vertebrate striated muscle and its behaviour during contraction and rigor, J. Mol. Biol. 30: 383 (1967).
38. H.M. Spencer, Laboratory methods for maintaining constant humidity, International Critical Tables, 1: 67 (1926).
39. C.G. Vonk and A.P. Pijpers, The use of film methods in small-angle X-ray scattering, J. Appl. Cryst. 14: 8 (1981).
40. U.W. Arndt and D.J. Gilmore, X-ray television area detectors for macromolecular structural studies with synchrotron radiation sources, J. Appl. Cryst. 12: 1 (1979).
41. R. Hamlin, C. Cork, A. Howard, C. Nielsen, W. Vernon, D. Matthews, and Ng. H. Xuong, Characteristics of a flat multiwire area detector for protein crystallography, J. Appl. Cryst. 14: 85 (1981).
42. R.E. Franklin and R.G. Gosling, The structure of sodium thymonucleate fibres. II. The cylindrically symmetrical Patterson function, Acta Cryst. 6: 678 (1953).
43. D.A. Marvin, R.L. Wiseman, and E.J. Wachtel, Filamentous bacterial viruses XI. Molecular architecture of the class II (Pf1,Xf) virion, J. Mol. Biol. 82: 121 (1974).
44. Y. Mitsui, The correlation between helical parameters and layer line distribution in the diffraction pattern of helical polymers, Acta Cryst. 20: 694 (1966).
45. Y. Mitsui, General method of obtaining best helical parameters from the diffraction pattern, Acta Cryst. A26: 658 (1970).
46. L. Makowski, Processing of X-ray diffraction data from partially oriented specimens, J. Appl. Cryst. 11: 273 (1978).
47. L. Makowski, Resolution of X-ray intensities by angular deconvolution, in: "Fibre Diffraction Methods," A.D. French and K.H. Gardner, eds., p. 139, American Chemical Society, Washington (1980).

48. R.D.B. Fraser, T.P. MacRae, A. Miller, and R.J. Rowlands, Digital processing of fibre diffraction patterns, J. Appl. Cryst. 9: 81 (1976).
49. S.W. Provencher and J. Glockner, Users guide for CIN- A Fortran routine for the rapid approximation of disorientation integrals in fiber diffraction, Technical Report EMBL-DA01, EMBL, Heidelberg (1980).
50. E.J. Wachtel, R.L. Wiseman, W.J. Pigram, D.A. Marvin, and L. Manuelidis, Filamentous bacterial viruses XIII. Molecular structure of the virion in projection, J. Mol. Biol. 88: 601 (1974).
51. D. Meader, E.D.T. Atkins, M. Elder, P.A. Machin, and M. Pickering, AXIS: A semi-automated X-ray intensity and d-spacing analyser for fiber diffraction patterns, in: "Fiber Diffraction Methods," A.D. French and K.H. Gardner, eds., p. 113, American Chemical Society, Washington (1980).
52. W. Cochran, F.H.C. Crick, and V. Vand, The structure of synthetic polypeptides. I. The transform of atoms on a helix, Acta Cryst. 5: 581 (1952).
53. E. Mandelkow, K.C. Holmes, and U. Gallwitz, A new helical aggregate of tobacco mosaic virus protein, J. Mol. Biol. 102: 265 (1976).
54. R. Freeman and K.R. Leonard, Comparative mass measurement of biological macromolecules by scanning transmission electron microscopy, J. Microscopy, in press.
55. R.E. Franklin and K.C. Holmes, Tobacco mosaic virus: Application of the method of isomorphous replacement to the determination of the helical parameters and radial density distribution, Acta Cryst. 11: 213 (1958).
56. R. Langridge, D.A. Marvin, W.E. Seeds, H.R. Wilson, C.W. Hooper, M.H.F. Wilkins, and L.D. Hamilton, The molecular configuration of deoxyribonucleic acid II. Molecular models and their Fourier transforms, J. Mol. Biol. 2: 38 (1960).
57. S. Arnott, M.H.F. Wilkins, W. Fuller, and R. Langridge, Molecular and crystal structures of double-helical RNA III. An 11-fold molecular model and comparison of the agreement between the observed and calculated three-dimensional diffraction data for 10- and 11-fold models, J. Mol. Biol. 27: 535 (1967).
58. S. Arnott, M.H.F. Wilkins, W. Fuller, and R. Langridge, Molecular and crystal structures of double-helical RNA II. Determination and comparison of diffracted intensities for the α and β crystalline forms of reovirus RNA and their interpretation in terms of groups of three RNA molecules, J. Mol. Biol. 27: 525 (1967).
59. L. Makowski and D.L.D. Caspar, Filamentous bacteriophage Pf1 has 27 subunits in its axial repeat, in: "The Single-Stranded DNA Phages," D.T. Denhardt, D. Dressler, and D.S. Ray, eds., p. 627, Cold Spring Harbor Laboratory, Cold Spring Harbor (1978).

60. M. Feughelman, R. Langridge, W.E. Seeds, A.R. Stokes, H.R. Wilson, C.W. Hooper, M.H.F. Wilkins, R.K. Barclay, and L.D. Hamilton, Molecular structure of deoxyribose nucleic acid and nucleoprotein, Nature, 175: 834 (1955).
61. R.E. Franklin and A. Klug, The nature of the helical groove on the tobacco mosaic virus particle, Biochim. biophys. Acta, 19: 403 (1956).
62. D.A. Marvin, M. Spencer, M.H.F. Wilkins, and L.D. Hamilton, The molecular configuration of deoxyribonucleic acid III. X-ray diffraction study of the C form of the lithium salt, J. Mol. Biol. 3: 547 (1961).
63. A. Elliott and B.R. Malcolm, Chain arrangement and sense of the α-helix in poly-L-alanine fibres, Proc. Roy. Soc. Lond. A249: 30 (1959).
64. F.H.C. Crick, The packing of α-helices: Simple coiled-coils, Acta Cryst. 6: 689 (1953).
65. D.L.D. Caspar and K.C. Holmes, Structure of Dahlemense strain of tobacco mosaic virus: A periodically deformed helix, J. Mol. Biol. 46: 99 (1969).
66. C. Nave, A.G. Fowler, S. Malsey, D.A. Marvin, H. Siegrist, and E.J. Wachtel, Macromolecular structural transitions in Pf1 filamentous bacterial virus, Nature, 281: 232 (1979).
67. C.R. Calladine, Design requirements for the construction of bacterial flagella, J. theor. Biol. 57: 469 (1976).
68. A.C. Bloomer, J.N. Champness, G.Bricogne, R. Staden, and A. Klug, Protein disk of tobacco mosaic virus at 2.8 Å resolution showing the interactions within and between subunits, Nature, 276: 362 (1978).
69. P.Y. Chou and G.D. Fasman, Empirical predictions of protein conformation, Ann. Rev. Biochem. 47: 251 (1978).
70. G. Nemethy and H.A. Scheraga, Protein folding, Quart. Rev. Biophys. 10: 239 (1977).
71. G.E. Schulz and R.H. Schirmer, "Principles of Protein Structure," Springer-Verlag, New York (1979).
72. M.J.E. Sternberg and J.M. Thornton, Prediction of protein structure from amino acid sequence, Nature, 271: 15 (1978).
73. F.E. Cohen, M.J.E. Sternberg, and W.R. Taylor, Analysis and prediction of protein β-sheet structures by a combinatorial approach, Nature, 285: 378 (1980).
74. J. Janin and C. Chothia, Packing of α-helices onto β-pleated sheets and the anatomy of α/β proteins, J. Mol. Biol. 143: 95 (1980).
75. K.C. Holmes, Protein-RNA interactions during TMV assembly, J. Supramol. Struct. 12: 305 (1979).
76. D.A. Marvin, Structure of the filamentous phage virion, in: "The Single-Stranded DNA Phages," D.T. Denhardt, D. Dressler, and D.S. Ray, eds., p. 583, Cold Spring Harbor Laboratory, Cold Spring Habor (1978).

77. D.J. Struik, "Differential Geometry," Addison-Wesley, Reading (1961).
78. F.H.C. Crick and J.D. Watson, Virus structure: general principles, in: "The Nature of Viruses" (Ciba Foundation Symposium), G.E.W. Wolstenholme and E.C.P. Millar, eds., p.5, Churchill, London (1957).
79. W. Bode, J. Engel, and D. Winklmair, A model of bacterial flagella based on small-angle X-ray scattering and hydrodynamic data which indicate an elongated shape of the flagellin protomer, Eur. J. Biochem. 26: 313 (1972).
80. S. Lowey, Myosin: Molecule and filament, in: "Subunits in Biological Systems" Part A, S.N. Timasheff and G.D. Fasman, eds., p.201, Marcel Dekker, New York (1971).
81. H.B. Stuhrmann, Interpretation of small-angle scattering functions of dilute solutions and gases. A representation of the structures related to a one-particle-scattering function, Acta Cryst. A26: 297 (1970).
82. C.W. Gray, G.G. Kneale, K.R. Leonard, H. Siegrist, and D.A. Marvin, A nucleoprotein complex in bacteria infected with Pf1 filamentous virus 1. Identification and electron microscopic analysis, Virology, in press.
83. D.A. Marvin and E.J. Wachtel, Structure and assembly of filamentous bacterial viruses, Nature, 253: 19 (1975).
84. L. Makowski, D.L.D. Caspar, and D.A. Marvin, Filamentous bacteriophage Pf1 structure determined at 7 Å resolution by refinement of models for the α-helical subunit, J. Mol. Biol. 140: 149 (1980).
85. W. Folkhard, K.R. Leonard, S. Malsey, D.A. Marvin, J. Dubochet, A. Engel, M. Achtman, and R. Helmuth, X-ray diffraction and electron microscope studies on the structure of bacterial F-pili, J. Mol. Biol. 130: 145 (1979).
86. W. Folkhard, D.A. Marvin, T.H. Watts, and W. Paranchych, Structure of polar pili from Pseudomonas aeruginosa strains K and O, J. Mol. Biol. in press.
87. R. Langridge, T.E. Ferrin, I.D. Kuntz, and M.L. Connolly, Real-time color graphics in studies of molecular interactions, Science, 211: 661 (1981).
88. Y. Nakashima, R.L. Wiseman, W. Konigsberg, and D.A. Marvin, Primary structure and sidechain interactions of Pf1 filamentous bacterial virus coat protein, Nature, 253: 68 (1975).
89. D.A. Marvin and E.J. Wachtel, Structure and assembly of filamentous bacterial viruses, Phil. Trans. R. Soc. Lond. B276: 81 (1976).
90. R. Diamond, A mathematical model-building procedure for proteins, Acta Cryst. 21: 253 (1966).

91. J. Hermans Jr. and J.E. McQueen Jr., Computer manipulation of macromolecules with the method of local change, Acta Cryst. A30: 730 (1974).
92. R.D.B. Fraser, T.P. MacRae, and E. Suzuki, An improved method for calculating the contribution of solvent to the X-ray diffraction pattern of biological molecules, J. Appl. Cryst. 11: 693 (1978).
93. Phillips, S.E.V., Structure and refinement of oxymyoglobin at 1.6 Å resolution, J. Mol. Biol. 142: 531 (1980).
94. B. Lee and F.M. Richards, The interpretation of protein structures: Estimation of static accessibility, J. Mol. Biol. 55: 379 (1971).
95. F.M. Richards, The interpretation of protein structures: Total volume, group volume distributions and packing density, J. Mol. Biol. 82: 1 (1974).
96. S. Arnott and A.J. Wonacott, The refinement of the crystal and molecular structures of polymers using X-ray data and stereochemical constraints, Polymer, 7: 157 (1966).
97. G. Gupta, M. Bansal, and V. Sasisekharan, Conformational flexibility of DNA: Polymorphism and handedness, Proc. Natl. Acad. Sci. USA, 77: 6486 (1980).
98. J.L. Sussman, S.R. Holbrook, G.M. Church, and S.-H. Kim, A structure-factor least-squares refinement procedure for macromolecular structures using constrained and restrained parameters, Acta Cryst. A33: 800 (1977).
99. C. Chothia, Structural invariants in protein folding, Nature, 254: 304 (1975).
100. J. Janin, S. Wodak, M. Levitt, and B. Maigret, Conformation of amino acid side-chains in proteins, J. Mol. Biol. 125: 357 (1978).
101. S.J. Opella, T.A. Cross, J.A. DiVerdi, and C.F. Sturm, Nuclear magnetic resonance of the filamentous bacteriophage fd, Biophys. J. 10: 531 (1980).
102. L. Makowski, The use of continuous diffraction data as a phase constraint. I. One-dimensional theory, J. Appl. Cryst. in press.
103. S.W. Provencher, Inverse problems in polymer characterization: Direct analysis of polydispersity with photon correlation spectroscopy, Makromol. Chem. 180: 201 (1979).
104. D.A. Marvin, M.H.F. Wilkins, and L.D. Hamilton, Application of Fourier synthesis technique to low-resolution fibre diffraction data: Preliminary study of deoxyribonucleic acid, Acta Cryst. 20: 663 (1966).
105. G. Bricogne, Methods and programs for direct-space exploitation of geometric redundancies, Acta Cryst. A32: 832 (1976).

106. G.A. Sim, The distribution of phase angles for structures containing heavy atoms. II. A modification of the normal heavy-atom method for non-centrosymmetrical structures, Acta Cryst. 12: 813 (1959).
107. S. Arnott, M.H.F. Wilkins, L.D. Hamilton, and R. Langridge, Fourier synthesis studies of lithium DNA Part III: Hoogsteen models, J. Mol. Biol. 11: 391 (1965).
108. R.E. Franklin and R.G. Gosling, The structure of sodium thymonucleate fibres. I. The influence of water content, Acta Cryst. 6: 673 (1953).
109. S.B. Zimmerman and B.H. Pheiffer, A RNA-DNA hybrid that can adopt two conformations: An X-ray diffraction study of poly(rA).poly(dT) in concentrated solution or in fibers, Proc. Natl. Acad. Sci. USA, 78: 78 (1981).
110. S.W. Englander, N.R. Kallenbach, A.J. Heeger, J.A. Krumhansl, and S. Litwin, Nature of the open state in long polynucleotide double helices: Possibility of soliton excitations, Proc. Natl. Acad. Sci. USA, 77: 7222 (1980).
111. J.D. Axe, Incommensurate structures, Phil. Trans. Roy. Soc. Lond. B290: 593 (1980).
112. A. Klug, The design of self-assembling systems of equal units, Symp. Int. Soc. Cell Biol. 6: 1 (1968).
113. R.M. Macnab and M.K. Ornston, Normal-to-curly flagellar transitions and their role in bacterial tumbling. Stabilization of an alternative quaternary structure by mechanical force, J. Mol. Biol. 112: 1 (1977).
114. E.J. Wachtel, F.J. Marvin, and D.A. Marvin, Structural transition in a filamentous protein, J. Mol. Biol. 107: 379 (1976).
115. H.E. Huxley, A.R. Faruqi, J. Bordas, M.H.J. Koch, and J.R. Milch, The use of synchrotron radiation in time-resolved X-ray diffraction studies of myosin layer-line reflections during muscle contraction, Nature, 284: 140 (1980).
116. D.W. Banner, C. Nave, and D.A. Marvin, Structure of the protein and DNA in fd filamentous bacterial virus, Nature, 289: 814 (1981).

RECENT DEVELOPMENTS IN THE CRYSTALLOGRAPHY OF GLOBULAR PROTEINS

T.L. Blundell

Department of Crystallography, Birkbeck College,
University of London,
Malet Street, London WC1E 7HX, United Kingdom

INTRODUCTION

Although it is over twenty years since the first successful crystal structure analysis of a protein at medium resolution, many developments are being stimulated by the introduction of new technologies such as dedicated synchrotron X-radiation sources, interactive computer graphics and vector and array processors. However, much progress has also come from the careful improvement of well-established techniques such as those of crystallisation, anomalous scattering and least-squares refinement.

In this chapter I discuss those areas where protein crystallographers have been most innovative, especially with respect to study of proteins available in microquantities and with high molecular weights. My starting point is the state of the art in 1976, described in "Protein Crystallography"(1). I begin with a description of microcrystallisation techniques of current interest, especially those involving membrane proteins. I then discuss new developments made possible by synchrotron sources; these include not only fast data collection with improved resolution but also the use of anomalous dispersion techniques to solve the phase problem now that the study of extended X-ray absorption fine structure (EXAFS) has shown that anomalous dispersion components can be very large. However, the recent solution of the phase problem using the anomalous scattering of sulphurs in the small protein, crambin(2) shows that anomalous dispersion methods can be very useful if the data are carefully measured even when the anomalous components are small. In the final part of the discussion I consider recent developments in model building with interactive computer graphics and improvement of the model with "constrained" or "restrained" least-squares refinement in reciprocal space. Significant developments have occurred in other areas of the subject, notably in low temperature[3] X-ray crystallography, and in phase refinement using symmetry averaging[4,5] and other direct methods[6]. Our understanding of protein structure has also been revolutionised by analysis in terms of hierarchies of structure - secondary structure, folding units, supersecondary structures and domains - as well as moves towards understanding the dynamic nature of the protein through analysis of the amplitudes of vibrations and librations of the polypeptide. Unfortunately, the space available to the author here prevents discussion of these important topics.

CRYSTALLISATION

Of the many microcrystallisation techniques, vapour diffusion using the hanging drop method has proved to be generally useful (see Chapter 3 of ref. (1)). A convenient experimental set-up involves cell culture trays comprising approximately 40 wells each with a raised lip allowing easy manipulation of the microscope cover slip from which the drop hangs. A transparent dust cover gives safe storage and straightforward observation of the crystals without disturbing their growth. The method can be used with either volatile precipitants, such as ethanol, which are placed in the reservoir, or involatile agents such as salts or polyethylene glycols which are placed in the drop at a lower concentration than in the reservoir. Polyethylene glycols (PEG) have found wide applications[7]; the optimal PEG sizes appear to be 1000, 4000 and 6000 and the concentration of the PEG required to precipitate the protein tends to decrease in that order. Generally, smaller sizes - for example 400 - are either not useful or give crystals of equivalent quality. However, although larger PEG sizes such as 20,000 are less effective a few proteins may be crystallised only with this size and it should therefore be included in any screen. McPherson[7] contends that PEG exhibits greater generality than any previously-used reagents. Furthermore, although the crystals appear to be of good quality and often identical to those prepared in the presence of salts, the concentrations of PEG required to give good crystals fall in a narrower range (2 to 20%) and are anyway less critical than for salts such as ammonium sulphate.

After many years of unsuccessful attempts to crystallise membrane proteins, several laboratories have now developed methods leading to crystals which may be suitable for medium resolution X-ray analysis[8]. The first paper[9] described crystals of the cytochrome $\underline{c}$ oxidase - cytochrome $\underline{c}$ complex which were produced from a cholate solubilised complex of lipid-depleted protein from which the cholate detergent was removed by dialysis. The resulting crystals were detergent- and lipid-free, but rather small. Larger crystals diffracting to 8Å resolution have been obtained by

Michel and Oesterhelt(10) from bacteriorhodopsin by adding 25M salt to the octylglucoside-solubilised protein. A similar method was used by Garavito and Rosenbusch(11) to crystallise porin, the pore-forming protein from E. coli outer membranes. They obtained crystals from a 0.5%-1.0% solution of β-octylglucoside by equilibration with 20% polyethylene glycol (PEG). At pH7 tetragonal prisms, and at pH9.8 hexagonal plates crystallise from small droplets containing the protein and the octylglucoside. The search for more effective detergents is now on; problems arise as a result of a tendency of micelles to aggregate, the effect of the detergent on the crystals leading to disordering, and the lack of surface tension giving rise to movement of crystals (J. A. Jenkins, M. Garavito, unpublished results). But the X-ray analysis of a membrane protein at high resolution is now a real possibility.

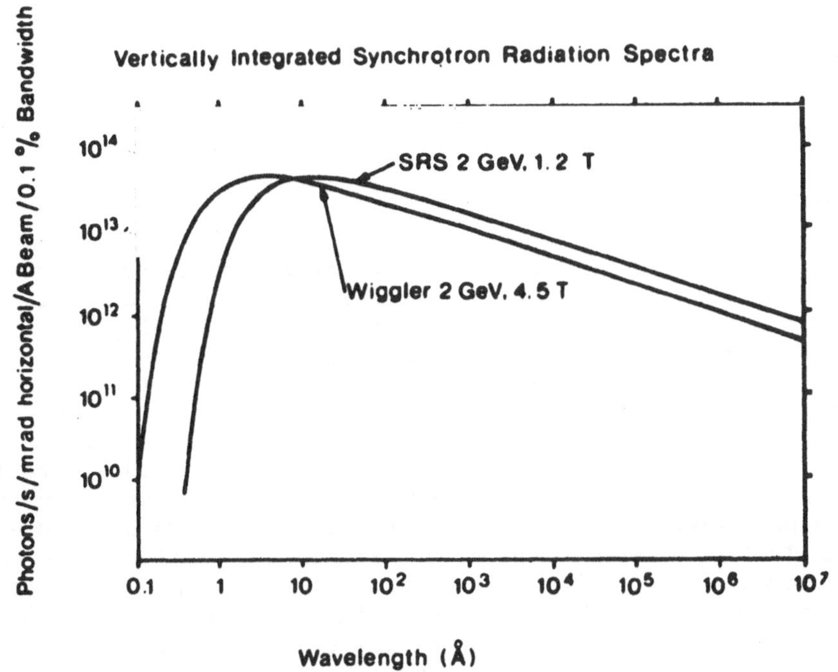

Fig. 1. Spectral curves from normal bending magnet (1.2T) and from 4.5T wiggler magnet, for a 2GeV 1Å beam in synchrotron radiation source, Daresbury Laboratory, United Kingdom. The wavelength in Å may be converted to other units by use of the relations: 1000Å = 100nm = 0.1μ ≡ 12.399eV. (Reproduced with permission from Drs K. R. Lea and I. H. Munro, Daresbury Laboratory, Science and Engineering Research Council, United Kingdom.)

Table 1. Machine specification and parameters relevant to the spectra of various synchrotron radiation sources

Machine (location)	E(GeV)	I(mA)	R_{meters}	$\varepsilon_c$keV	λ_c	Remarks
SRS (Daresbury, UK)	2.0	(500- (1000	5.55 1.33	3.2 13.3	3.87 0.93	dedicated (from 5T (wiggler
DCI (Orsay)	1.8	500	3.8	3.4	3.65	very similar to Daresbury SRS
DORIS (Hamburg)	4.0	100	12.1	11.7	1.06	optimal parameters for hard X-rays

(Reproduced with permission from Dr J. R. Helliwell, Daresbury Laboratory)

SYNCHROTRON RADIATION

During the last decade synchrotron radiation sources have been or are being established at a number of centres - Hamburg, Paris, Daresbury, Stanford, Tokyo, Brookhaven - providing the protein crystallographer with an intense and tunable beam of X-rays[(13)]. For example, the Daresbury synchrotron in the United Kingdom comprises sixteen bending magnets of field 1.2T. The vertically integrated synchrotron radiation spectrum emitted in the forward direction centred on the target line is given in Figure 1. The critical wavelength, λc, is 3.87Å; other machine specifications and some parameters relevant to the spectra at various sources are given in Table 1. For the Daresbury source the critical wavelength may be decreased to 0.93Å by use of a wiggler magnet of $\sim$5T which effectively increases the curvature of the electron path and thus changes the spectral distribution of the synchrotron radiation. On such a storage ring the wiggler is most important for anomalous dispersion experiments where diffraction experiments need to be carried out in the region of the L-absorption edges of metal atoms such as Pt, Au and Hg.

X-ray photography (Figure 2) with protein crystals at the LURE facility near Paris has shown a gain in intensity by a factor of about fifty in the intensity pattern compared to equivalent X-ray photographs on a rotating anode generator (GX6 Ni-filtered

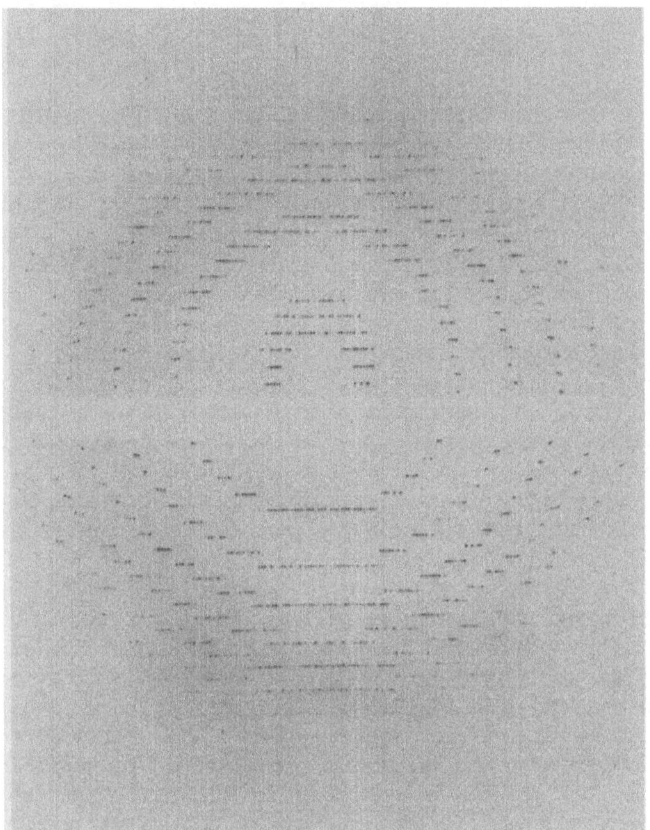

Fig. 2. X-ray photograph taken on an Arndt-Wonacott oscillation camera (2° oscillation) from crystals of hexagonal pepsin using the LURE synchrotron 1.8GeV, 170mA, exposure time = 25 minutes. (Reproduced with permission from Dr G. L. Taylor, Dr G. Khan and Dr R. Fourme.)

radiation, 1.6kW)[13]. Moreover, certain crystals such as those of phosphorylase appear to give exceptionally high resolution diffraction patterns[13,14]. This rather unexpected consequence may result partly from the very clean beam which is used after careful monochromatisation and collimation, but more importantly it may arise from the fact that radiation decay is probably mediated by solvated electrons and is, therefore, time- as well as dose-dependent. Synchrotron radiation may allow a recording of the high angle diffraction pattern before the highly ordered nature is disturbed by the radiation damage.

Apart from the advantages of rapid, high resolution X-ray data collection, synchrotron radiation is tunable and offers the possibility of exploiting anomalous dispersion effects in the solution of the phase problem.

ANOMALOUS DISPERSION AND THE SOLUTION OF THE PHASE PROBLEM

Since the preparation of the first heavy atom derivatives of haemoglobin in 1954 by Perutz and his colleagues[15], the method of multiple isomorphous replacement (MIR) has proved to be a versatile method in the solution of the phase problem. However, heavy atom derivatives which are truly isomorphous have often proved difficult to obtain, and a number of successful analyses have involved single isomorphous replacment with anomalous scattering (SIRAS) to break the phase ambiguity which arises from the use of a single isomorphous derivative. Although the anomalous scattering effects, which give rise to the Friedel differences, are an order of magnitude less than isomorphous differences given by elements such as Pt, Hg or Pb (see Figure 3) with CuKα radiation, they can be measured quite accurately as they are free from factors such as lack of isomorphism, and absorption differences can be minimised.

With small proteins the possibility arises of gaining meaningful phase information from the anomalous differences alone. Although this gives rise to ambiguity in the phase determination, for vitamin B_{12} (comprising one Co atom and about 100 carbon, nitrogen and oxygen atoms) Dorothy Hodgkin and her colleagues[16] had demonstrated in 1963 that this ambiguity can be usefully resolved by choosing the phase closest to that of the heavy atom contribution on the assumption that this would dominate the structure factor in a sizeable proportion of the reflections. Although this method was never used successfully for a first determination of the structure of a metalloprotein, Hendrickson[2] has shown that the much smaller anomalous scattering of the sulphurs can be exploited in the same way.

The small protein, crambin (46 amino acid residues) has three disulphide bridges and crystals (space group $P2_1$ and cell dimensions $\underline{a}$ = 40.96, $\underline{b}$ = 18.65, $\underline{c}$ = 22.52Å and β = 90.77°) diffract strongly to high resolution[2]. Hendrickson showed first that the sulphur positions could be determined at 3.0Å resolution from an anomalous differences Patterson (see ref. 1, page 165 for review of the method). At this resolution the two sulphurs of each cystine bridge diffract in phase and so the Patterson is less complicated and the peaks larger than at higher resolutions (∿1.5Å). Having found the positions of the cystine bridges at a resolution of 3Å, a Patterson function at higher resolution (1.5Å) allowed positioning of the sulphurs. The sulphur positions were then refined against the largest anomalous differences. The phase information obtained from the heavy atom method using the sulphur atoms was then combined with that from anomalous scattering, giving rise to phases and an electron

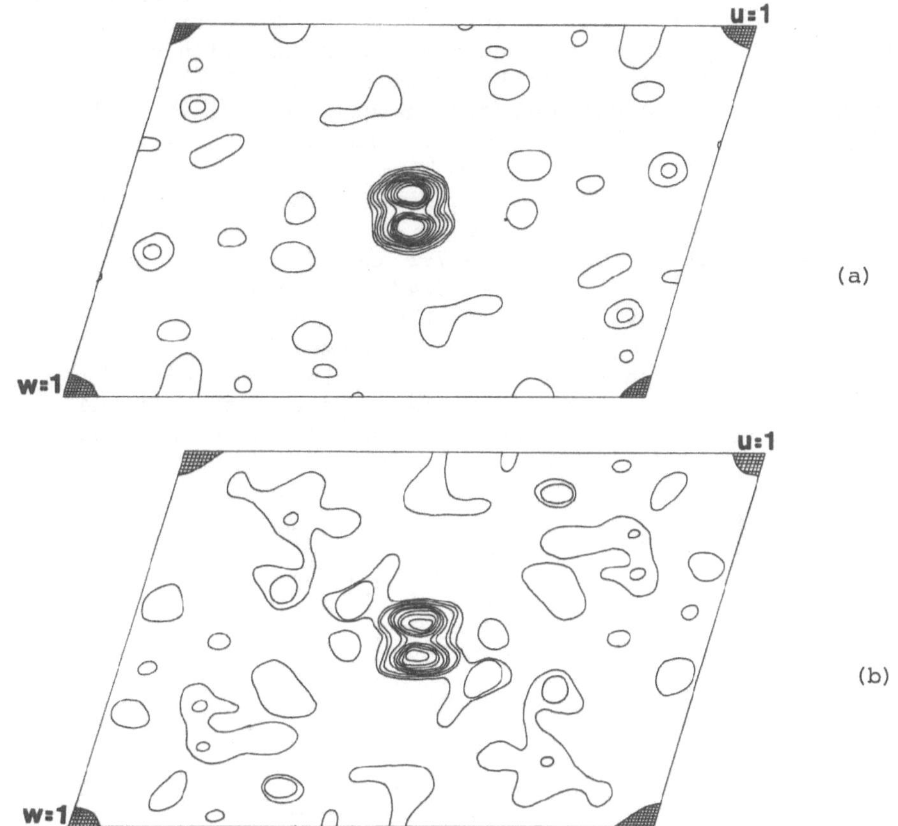

Fig. 3. Harker sections of Pattersons of a mercury derivative of pancreatic polypeptide (36 amino acids, space group C2) at 2.1Å resolution with coefficients (a) isomorphous differences ($|F_{PH} - F_P|^2$) and (b) anomalous differences, $k^2/4\,|F_{PH}(+) - F_{PH}(-)|^2$). (Reproduced with permission from Dr J. E. Pitts, Dr I. J. Tickle and Dr S. P. Wood.)

density map of sufficient quality to be useful as a starting point in refinement and difference Fourier calculation; this revealed the molecular structure. Precisely measured data and the use of local scaling procedures ensured meaningful anomalous differences. Phase calculation using the methods earlier developed by Hendrickson and Lattman(17), allowing combination of anomalous data with that of calculated structure factors from the partial structure (Sim weighted) ensured a proper choice of phase and figure of merit for inclusion in the electron density calculations. This remarkable crystallographic achievement is the first successful structure analysis of a protein using neither isomorphous nor molecular replacement.

With the advent of synchrotron radiation the wavelength can be tuned to optimise the anomalous absorption and dispersion components; and multiple wavelength measurements allow the solution of the phase problem without recourse to the heavy atom method(18-21). Recent EXAFS measurements have demonstrated the existence of white lines for metal ions with unfilled 'd' or 'f' shells. Optimisation of the wavelength and band width then makes the method of anomalous dispersion more powerful than was originally thought. For instance, for the L_{III} edge of Cs where f_o is 38 at $\sin\theta/\lambda = 0.3 Å^{-1}$, Phillips et al(19-21) have shown that $f' = -27.1$ and $f'' = 5.9$ at $\lambda = 2.473 Å$ and $f' = -16.1$ and $f'' = 11.1$ at $\lambda = 2.470$ as shown in Figures 4 and 5. Several workers have considered the optimal choice of wavelengths; the minimal requirement is measurement at two wavelengths including the Friedel pair where the absorption component is large. In general the wavelengths should allow maximal changes of the absorption and dispersion components so that R_1 and R_2 are approxi-

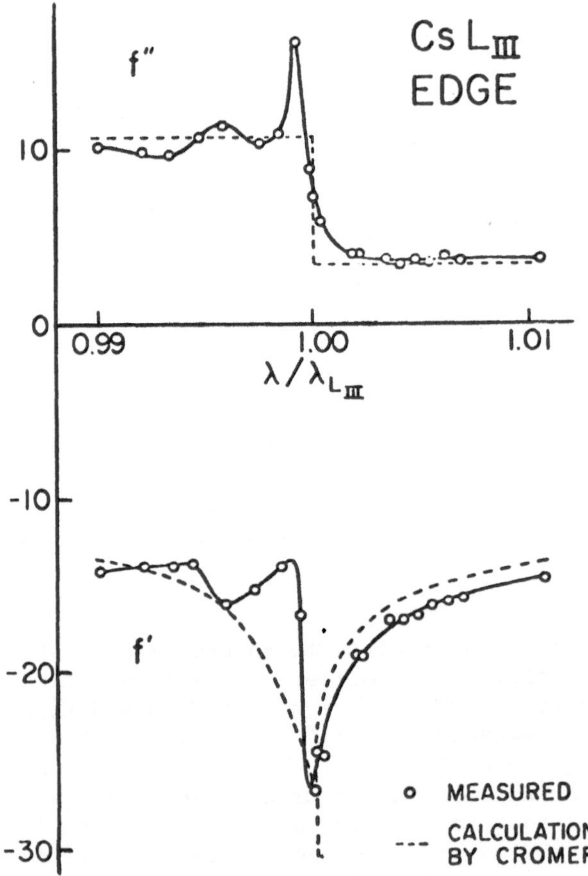

Fig. 4. Anomalous scattering terms for caesium through the L edges: the L_{III} edge.

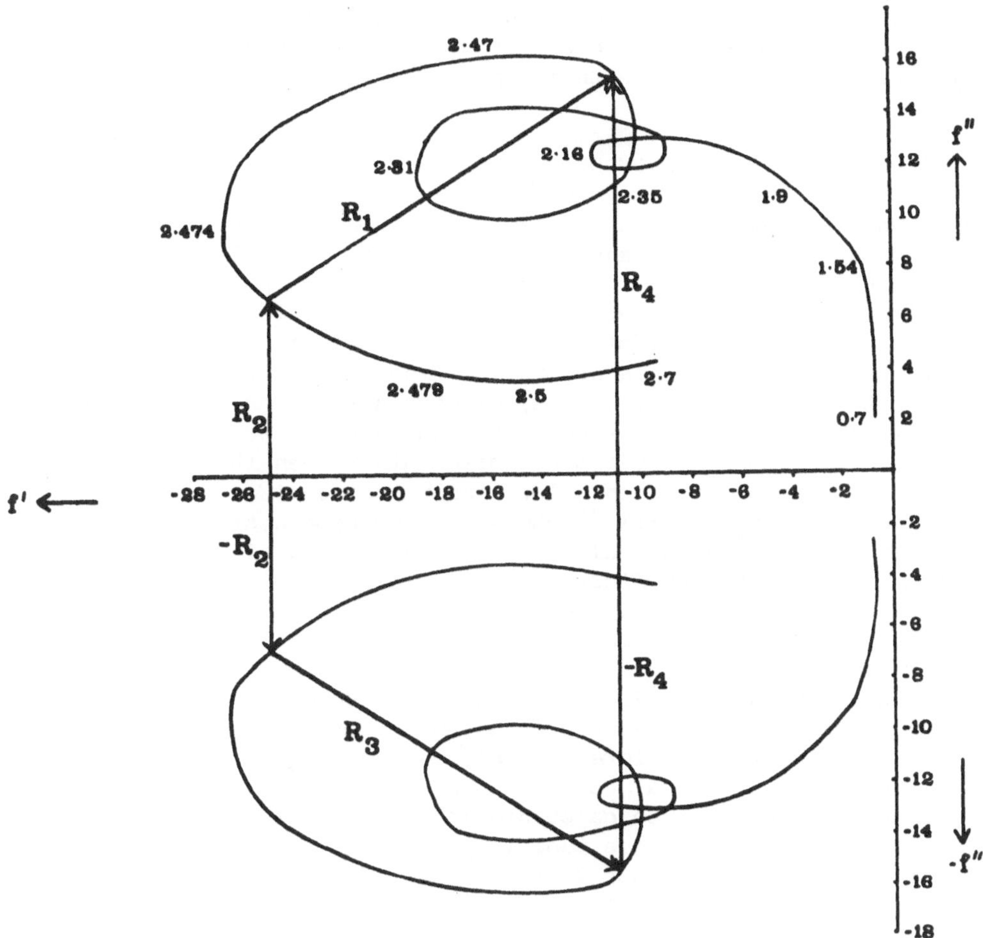

Fig. 5. Anomalous scattering terms for caesium through the L edges: the variation of f' (the dispersion component) with respect to f" (the absorption component) with variation of the wavelength (Å) indicated on the curve. (Figures 4 and 5 are derived from those of J. C. Phillips, PhD thesis, Stanford University, USA, 1978.)

mately perpendicular[18]. The best way to proceed is to first measure the EXAFS spectrum under the same experimental conditions as the diffraction[22]. The dispersion components can be calculated from the Kramers-Kronig relationship. This approach is made necessary by the dependence of the anomalous dispersion on the band width and the chemical coordination of the anomalous scatterer in the crystals studied. The method of anomalous dispersion using tunable synchrotron radiation appears to be a potentially powerful method. It may be of value either for the structure analysis of metalloproteins, especially those containing transition metals such as Fe, Co or Mo, or when a single heavy atom derivative can be prepared which is non-isomorphous with the parent crystals.

INTERPRETATION OF ELECTRON DENSITY MAPS

When a good electron density map has been calculated using isomorphous replacement, anomalous dispersion, or molecular replacement methods, the problem is to fit optimally the polypeptide chain. For structures at less than 3.5Å resolution this is not generally advisable unless a model of a known homologous structure is available. Although α-helices are well defined in low resolution (6.0-4.5Å) electron density maps, β-sheet structures are often quite unclear and the density may be dominated by clusters of closely packed side groups in the protein core rather than the polypeptide chain. An example of such a low resolution electron density is shown for γ-crystallin II[24] where the two-domain structure is clear at 5.5Å resolution, although the main chain of the polypeptide cannot be followed (see Figure 6). In this case it may be best to proceed by fitting the calculated electron density from an homologous structure, say another molecule in the γ,β-crystallin class. Dr G. Taylor (unpublished results) has devised a program, FITZ, for use on an Evans and Sutherland Picture System II - a computer graphics system - which allows the calculated and observed electron densities to be independently and interactively manipulated until an optimal fit is obtained by visual comparison on the display screen. Overlap functions may also be calculated[25]. In this kind of fitting it is not necessary to move individual amino acids independently although it is advantageous to allow individual fitting of "rigid" helices, folding units or domains.

For medium resolution electron density maps (3.5-2.0Å resolutions) interactive computer graphics techniques have proved to be very useful in protein model building. The first methods developed manipulated the model through interactive modification of the protein torsion angles[25,26]. This method minimises the number of variables and allows retention of the known geometries of the different amino acid building units. However, many protein crystallographers have found it conceptually difficult to manipulate a structure in terms of the torsion angles and a more useful approach appears to be to allow movements which one would use on a wire model. These involve isolation of fragments, movements of these by translations or rotations[27,28] followed by a geometricisation using a model-building program such as that of Hermans and McQueen[29]. Such a program, FRODO, has been developed for a Vector General computer graphics system by Jones[27] and modified for use on an Evans and Sutherland Picture System by T. A. Jones and I. J. Tickle (unpublished results). The model display can be manipulated through a tablet, tracker-ball or joy-stick by rotations and translations

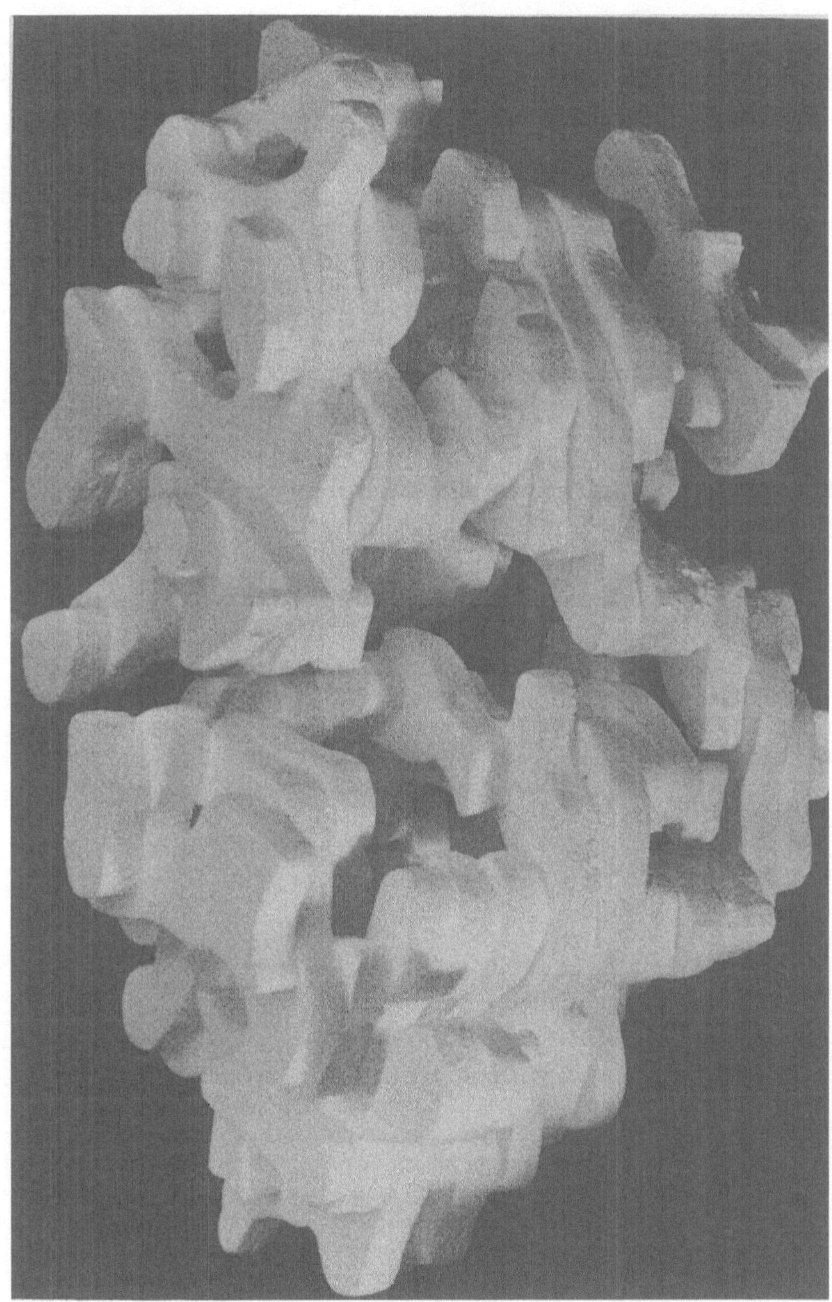

Fig. 6. A balsa wood model of the electron density at low resolution (5.5Å) of γ-crystallin II. The major axes of the molecule are 55 x 30 x 25Å. Two lobes are seen which at high resolution are each eight-stranded β-barrels. (Reproduced with permission from T. L. Blundell, P. F. Lindley, D. S. Moss, C. Slingsby, I. J. Tickle and W. G. Turnell (1978) Acta Cryst. B34 3653-3657.)

around the screen axes. By use of the pen and tablet, fragments of the structure may be isolated and manipulated to fit the electron density by rotations and translations while the interfragment distances are continuously updated on the screen. The model and density may be displayed as a cube centred on a particular atom or as a zone or mixture of zones comprising specific amino acid residues. The display may be in stereo by viewing two images with the aid of a half-silvered mirror or through lorgnettes which are synchronised with an alternating left and right stereo image. Many workers find this unnecessary and prefer to gain the three-dimensional effect by continuously rocking the image through a set angle or simply moving the model randomly with a tracker-ball in the left hand while operations are effected through the tablet with a pen in the right hand. After optimising the fits of individual fragments or by rotation of torsion angles, the structure may be "regularised"[29] and interatomic distances may be optimised by minimising the energy described by simple potential functions[30]. Computer graphics techniques have proved to be very efficient in speeding up the process of model fitting, in optimising the fit and in allowing easy storage and retrieval of the atomic coordinates.

At very high resolutions (∿1.5Å) the atoms are beginning to be resolved. Part of a good quality electron density map at 1.4Å resolution computed without introduction of information about the molecular structure is shown in Figure 7. In this structure analysis it was found that the atoms could be placed more or less individually and the manipulation of molecular fragments was less important. However, even in such a high quality, high resolution study, parts of the polypeptide chain may be very flexible and so will be thermally or statically disordered even in the crystal structure, leading to poorly resolved and weak electron density.

In principle electron density maps of several kinds may need to be displayed either separately or together. These may include the MIR (multiple isomorphous replacement) or anomalous dispersion phased maps or maps with phases calculated on the basis of a refined structure or one fitted by molecular replacement. The coefficients of the Fourier calculation for such electron density maps may be $m|nF_P(\text{observed}) - (n-1)F_P(\text{calculated})|\alpha_P(\text{calculated})$ in which n is often 2, or conventional difference Fouriers with coefficients $m|F_P(\text{observed} - F_P(\text{calculated})|\alpha_P(\text{calculated})$ where m is the figure of merit, F_P the structure factor amplitude for the protein crystals, and α_P is the best phase for the protein crystals (see ref. 1, page 151 for discussion). The electron densities may be displayed simultaneously in different textures, for instance by having one set of contours flashing (I. J. Tickle, unpublished results) and another constantly displayed, or optimally, in differing colours.

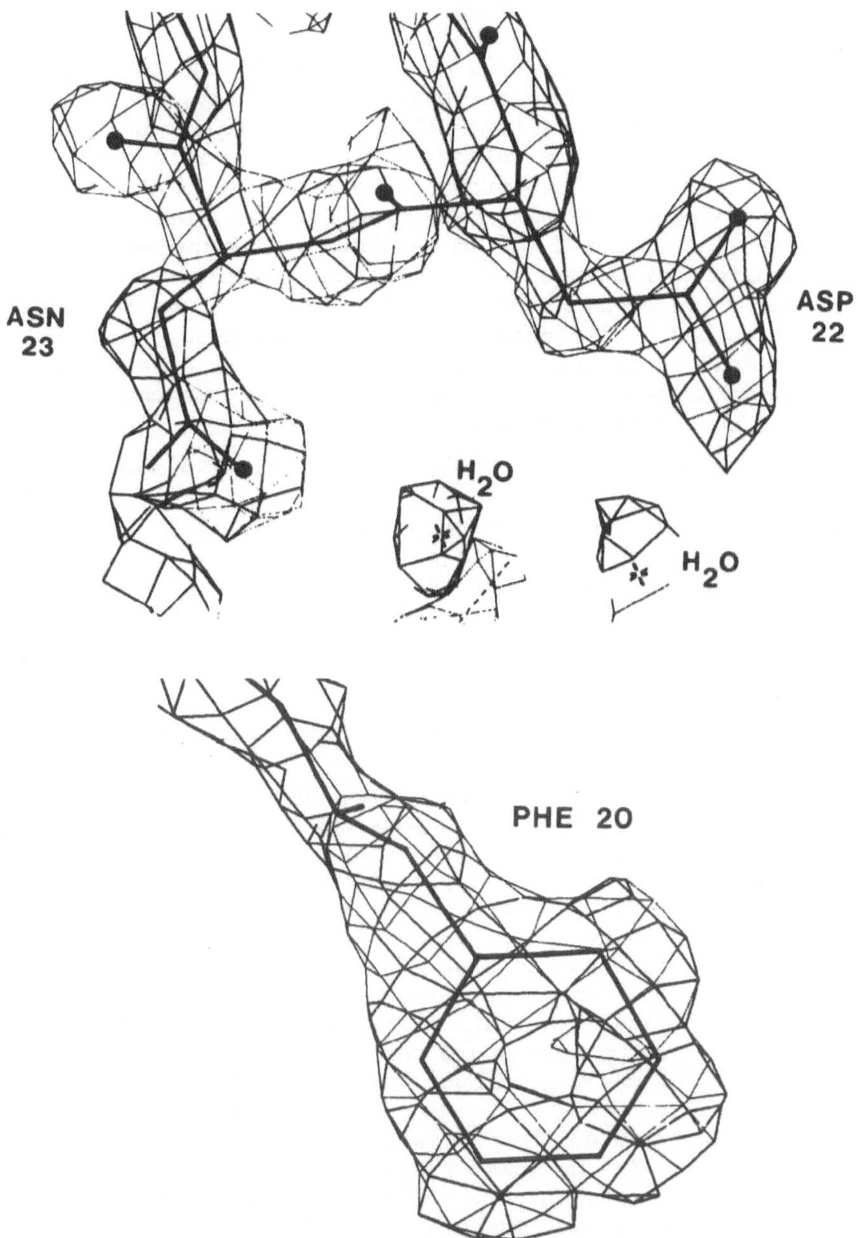

Fig. 7. Some representative parts of the 1.4Å electron density map of pancreatic polypeptide (PP) displayed on an Evans and Sutherland Picture System II using the program FRODO designed by T. A. Jones[27] and modified for the Picture System by T. A. Jones and I. J. Tickle.

REFINEMENT OF PROTEIN STRUCTURES

Although an optimisation of the model fit of the polypeptide to the electron density is achieved interactively, least-squares refinement techniques are invaluable for the improvement of the model with respect to the values of F_P (the observations). In fact, only at atomic resolution (probably better than 1.2Å) can the protein be refined using the methods developed for small molecular crystal structures. At low and medium resolutions the number of parameters (three atomic coordinates and one thermal parameter per atom) exceeds the number of independent reflections. This problem has been overcome in a number of different ways all of which depend on using "observations" about the molecular geometry to increase the ratio of observations to parameters. These methods may be characterised in terms of a number of different factors :

(i) whether they use fast Fourier[31] or block diagonal least-squares techniques;
(ii) whether they are in reciprocal or real space;
(iii) whether they optimise the protein geometries and minimise energies of non-bonded interactions as well as minimising differences between observed and calculated structure factors in the same operation;
(iv) whether they are "constrained", ie, the model is rigid, or whether they are "restrained", ie, the model is brought as close as possible to the optimal geometry.

The accumulated experience of many protein refinements is beginning to indicate that real space refinements[32] while easily accommodated in small computers as they can deal with protein in limited zones, step by step, may lead to false minima especially at medium resolutions (∿3Å) where calculated electron density maps are used. Alternative cycles of fast Fourier least-squares and model fitting[31] are powerful at high resolution but the lack of restraints can lead to dangerous false minima at medium resolutions. For most refinements a restrained least-squares process seems to be optimal. Two such programs are currently available on a wide range of computers from the VAX through the IBM range to the vector processors such as the CRAY 1. The programs of Konnert and Hendrickson[33] and Moss and Morffew[34] are essentially similar. They each seek to minimise by least-squares the differences between the observed and calculated structure factors as well as differences from perfect geometries. The program of Konnert and Hendrickson[33] uses the method of conjugate gradients and uses target positions for underdetermined cases. The program of Moss and Morffew[34] uses the Gauss-Seidel method for solving the normal equations and handles underdetermined cases by Marquardt's method. The terms minimised are listed in Table 2. An unresolved problem is the proper relative weighting of the structure factor terms and the geometric terms, which may be expressed as energies, when the refinement is far from convergence; at convergence it is advisable to allow the terms to

Table 2. Restrained Least-Squares: simultaneous minimisation of residuals

(i) Structure factor amplitudes*†

$$\sum\omega(|F_o| - |F_c|)^2$$

(ii) Structure factor phases†

$$\sum\omega(\phi_{iso} - \phi_c)^2$$

(iii) Interatomic distances between bonded atoms*†

$$\sum\omega(d_o - d_c)^2$$

(iv) Non-bonded contacts*†

$$\sum\omega(d_o - d_{min})^4 \qquad d_o < d_{min}$$

(v) Chirality*†

$$\sum\omega(V_o - V_c)^2$$

(vi) Torsion angles*

$$\sum\omega(\chi_o - \chi_c)^2$$

(vii) Planarity*

$$\sum_{k}^{\text{planes}} \sum_{i}^{\text{atoms in plane}} \omega(\underline{m}_k \cdot \underline{r}_{ik} - d_k)^2$$

Methods of : *Konnert and Hendrickson (1980)
†Moss and Morffew (1981)

contribute equally. In addition to the bond length, bond angle, and torsion angle terms, groups may be restrained to planarity (aromatic rings), the chirality of the amino acids may be preserved and intermolecular interactions may be optimised. These methods when programmed to be space-group-specific can be very fast (a protein of 173 amino acids may be refined at 2.4Å resolution, 6500 reflections, 30 minutes per cycle on an IBM370/165 and about two minutes per cycle on a CRAY 1).

All successful refinements must allow electron density maps to be calculated at regular intervals so that major model rebuilding may be carried out in parts of the structure. For this purpose it is very useful to have the mainframe computer used for the least-squares refinement linked directly into the host computer of the computer graphics. Systematic difference maps which omit, say, one eighth of the atomic coordinates in the structure factor calculation have been found to be very effective. However, it appears that difference Fouriers often reflect the evolution of the refinement; parts of the structure compensate for others incorrectly included and so the incorrect structures often reappear in the difference Fourier. This imposes a question mark on all protein structures refined at medium resolution where there is no proper method for estimating reliably the standard deviation of the coordinates obtained.

ACKNOWLEDGEMENTS

I wish to thank my colleagues at Birkbeck, especially Drs P. F. Lindley, D. S. Moss, J. E. Pitts, C. Slingsby, G. L. Taylor, I. J. Tickle and S. P. Wood, for their helpful suggestions and discussion of protein crystallographic methods which have made it easier for me to write this review.

REFERENCES

1. Blundell, T.L. and Johnson, L.N. (1976) "Protein Crystallography" Academic Press, London
2. Hendrickson, W.A. and Teeter, M.M. (1981) Nature (London) 290: 107-113
3. Frauenfelder, H., Petsko, G.A. and Tsernoglou, D. (1979) Nature (London) 280: 588-592
4. Harrison, S.C., Olson, A.J., Schutt, C.E., Winkler, F.K. and Bircogne, G. (1978) Nature 276: 368-374
5. Unge, T., Liljas, L., Strandberg, B., Vaara, I., Kannan, K.K., Fridborg, K., Nordman, C.E. and Lentz, P.J. (1980) Nature 285: 373-377
6. Tickle, I.J. (1981) in "Proceedings of symposium on refinement of proteins", Daresbury Laboratory, Science and Engineering Research Council, UK
7. McPherson, A. (1976) J. Biol. Chem. 251: 6300-6303
8. Henderson, R. (1980) Nature (London) 287: 490
9. Ozawa, T., Suzuki, H. and Tanaka, M. (1980) Proc. Natl. Acad. Sci. (USA) 77: 928-930
10. Michel, H. and Oesterhelt, D. (1980) Proc. Natl. Acad. Sci. (USA) 77: 1283-1285
11. Garavito, M. and Rosenbusch, J. (1980) J. Cell. Biol. 86: 327-332
12. Cundall, R.B. and Munro, I.H. (1979) "Applications of Synchrotron radiation to the study of large molecules of chemical and biological interest", Daresbury Laboratory, Science and Engineering Research Council, UK
13. Fourme, R. (1980) Manuscript in preparation
14. Johnson, L.N., Wilson, K., Jenkins, J.A. and Fourme, R. (1980), Unpublished results

15. Green, D.W., Ingram, V.M. and Perutz, M.F. (1954) Proc. Roy. Soc. A225: 287-291
16. Dale, D., Hodgkin, D.C. and Venkatesan, K. (1963) in "Crystallography and Crystal Perfection", 237-245 (ed. Ramachandran, G.N.) Academic Press, London
17. Hendrickson, W.A. and Lattman, E.E. (1970) Acta Cryst. B26: 136-141
18. Hoppe, W. and Jakubowski, U. (1975) in "Anomalous Scattering", eds. Ramaseshan, S. and Abrahams, S.C., International Union of Crystallography, Munksgaard, 437-447
19. Phillips, J.C., Wlodawer, A., Yevitz, M.M. and Hodgson, K.O. (1976) Proc. Natl. Acad. Sci. USA 73: 128-133
20. Karle, J. (1981) Int. J. Quant. Chem., in Press
21. Lye, R.C., Phillips, J.C., Kaplan, D., Domach, S. and Hodgson, K.O. (1980) Proc. Natl. Acad. Sci. USA, 78: 5884-5888
22. Ramaseshan, S. and Narayan, R. (1981) in "Structural studies on molecules of biological interest" ed. Dodson, G., Glusker, J.P. and Sayre, D., Oxford University Press, 233-245
23. Helliwell, J.R. (1979) in "Applications of synchrotron radiation to the study of large molecules" ed. Cundall, R.B. and Munro, I.H., Daresbury Laboratory, Science and Engineering Research Council, UK
24. Blundell, T.L., Lindley, P.F., Miller, L., Moss, D.S., Slingsby, C., Tickle, I.J., Turnell, W.G. and Wistow, G. (1981) Nature 289: 771-777
25. Barry, C.D. and North, A.C.T. (1971) Cold Spr. Harb. Symp. Quant. Biol. 36: 577-590
26. Diamond, R. (1978/79) Proceedings of the International Symposium on biomolecular structure, conformation, function and evolution, Madras, ed. Srinivasan, R., Pergamon Press
27. Jones, T.A. (1978) J. Appl. Cryst. 11: 268-272
28. Tsernoglou, D., Petsko, G.A., McQueen, J.E. and Hermans, J. (1977) Science 197: 1378-1381
29. Hermans, J. and McQueen, J.E. (1974) Acta Cryst. A30: 730-739
30. Levitt, M. (1974) J. Mol. Biol. 82: 393-420
31. Argawal, R.C. (1978) Acta Cryst. A34: 791-809
32. Diamond, R. (1971) Acta Cryst. A27: 435-452
33. Konnert, J.H. and Hendrickson, W.A. (1980) Acta Cryst. A36: 344-350
34. Moss, D.S. and Morffew, A.J. (1981) "Computers and Chemistry", in Press

X-RAY AND NEUTRON SMALL ANGLE SCATTERING

Bernard Jacrot

European Molecular Biology Laboratory
c/o C.E.N.G., L.M.A., 85X,
F-38041 Grenoble Cédex

INTRODUCTION

When radiation (light, electrons,x-rays, neutrons) passes through matter the inhomogeneities of "density" induce its scattering. This is the phenomenon which makes objects visible through different optical properties from that of the surrounding medium. This observation implies for quantitative analysis the precise definition of density, optical properties and inhomogeneities.

All the above mentioned radiations are characterised by their wave length (1-10Å for x-rays and neutrons) and by the nature of their interaction with atoms. X-rays interact with the atomic electrons and the amplitude of the scattered wave is proportional to the number of electrons of the atom. For neutrons, scattering is by the atomic nucleus and there are no simple correlations between the atomic number and the scattering amplitude. Isotopes of the same atom have different scattering amplitudes. This difference is very important for hydrogen and deuterium, and plays an essential role in the use of neutrons to study biological objects. The difference of scattering power of various atoms creates in the sample an inhomogeneity on atomic scale. X-rays and neutrons have wave-lengths comparable with inter-atomic distances; the scattered waves interfere giving rise to diffraction pattern. This is the basis for solving structure at atomic level as explained in the chapters by Marvin and Nave and by Blundell.

Now let us consider a molecule in solution, for instance a protein or a virus and forget about its atomic structure, but simply consider it as a uniform body of scattering density ρ. This density can be calculated from the sum Σb_i of all the scattering amplitudes of all the atoms, divided by the volume V occupied by those atoms. In general this "density" ρ will be different from that ρ_s of the solvent (calculated in the same way). This inhomogeneity in the solution will give rise to a scattering which is confined to an angular range λ/D where D is the size of the particle in solution. For radiation of 1Å and a protein of 30Å, this angular range is 1/30 or 2°. So this phenomenon is termed small angle scattering. In reality a protein and even more a virus does not have a uniform scattering density. The density $\rho(\vec{r})$ varies somewhat through the object; those internal inhomogeneities also give rise to scattering at small angles. The method of small angle scattering is the measurement and the analysis of the scattering due to the contrast between the object and the solvent and its large scale inhomogeneities of scattering densities. It will never give atomic resolution, but only rough information on the shape of the particle and the distribution of matter inside.

THEORY

The theory is identical for x-rays and neutrons, and it is only practical considerations which make one or the other method more useful. So here we consider only radiation of wavelengths (1-10Å) and a particle with a scattering density $\rho(\vec{r})$ as defined above. The scattering by a single particle would be too small and the sample will always be a solution of a large number N of particles. The following assumptions are then made

- All particles are identical
- Their orientation is at random
- The concentration is small enough so that interference between waves scattered by different particles can be neglected.
- The sample can be considered as a two component system : a solvent with a scattering density and a macromolecular solute . We shall come back later to this last approximation.

In this lecture only the general outline of the theory will be given. For a comprehensive review see previous articles on x-ray (1,2,3,4) and neutron (5,6) small angle scattering.

The experimental quantity is the intensity I(Q), where Q is related to the scattering angle θ and the wavelength λ by

$$Q = \frac{4\pi}{\lambda} \sin \theta/2 \cong 2\pi\theta/\lambda \quad \text{(for small } \theta)$$

This intensity is related to the distribution $\rho(\vec{r})$ of scattering density in the particle by a Fourier transformation

$$I(Q) = \left\langle \left| \int_V (\rho(\vec{r}) - \rho_s) \exp(i\vec{Q}\vec{r}) d^3r \right|^2 \right\rangle \quad (1)$$

The bracket indicates an averaging over all orientations of the particles. This averaging makes the inverse transform impossible and it is impossible to get $\rho(\vec{r})$ from the angular distribution I(Q). The relation (1) applies both to x-ray and neutron, but the scattering amplitudes of various atoms are different for the two types of radiations. Some of them are listed below. They are in units 10^{-12}cm

	X rays	Neutrons
Hydrogen	0.28	-0.3742
Deuterium	0.28	0.6671
Carbon	1.67	0.6651
Nitrogen	1.97	0.940
Oxygen	2.25	0.580
Sodium	3.10	0.36
Magnesium	3.38	0.52
Phosphorus	4.23	0.52
Sulphur	4.51	0.28
Chlorine	4.79	0.96
Potassium	5.36	0.37
Calcium	5.64	0.47

In the case of x-rays the scattering is by the electrons, and it increases proportionally to the atomic number. The neutrons are scattered by atomic nuclei, and there is no simple relation between the atomic number and the scattering amplitude. Note that hydrogen and deuterium scatter neutrons very differently.

The $\rho(\vec{r})$ are then calculated as explained above. We give now the average scattering densities of the main biological components. They are given in units of 10^{-14}cm/Å^3

	X-rays	Neutrons
H_2O	9.40	-0.562
5MNaBr in H_2O	13.3	
D_2O	9.40	6.40
Protein in H_2O	12	1.8
Protein in D_2O	12	3.1
RNA in H_2O	15.7	3.54
RNA in D_2O	15.7	4.55
DNA in H_2O	15	3.54
DNA in D_2O	15	4.29
Lipids in H_2O	8	~ 0

These values are only approximate, as they are dependent on the exact composition and on the specific volume. We note that the quantity $\rho(\vec{r})-\rho_s$, can be made equal to zero for each component in neutron scattering by using an appropriate H_2O-D_2O mixture for the solvent. For x-ray scattering this cancellation can be obtained for proteins using a solvent with a high content of sucrose or salt.

The Intensity at the origin

The intensity extrapolated to zero scattering angle (see below how this extrapolation is done) is for one particle

$$I(0) = (\Sigma b - \rho_s V)^2 \tag{2}$$

To first approximation Σb and V are proportional to the molecular weight (M) of the particle. For a sample with a concentration expressed in mg/ml, the number N of particles is inversely proportional to M. So the I(0) obtained from a solution of known concentration (C) is

$$I(0) \sim (M_w)^2 \times \frac{1}{M_w} \times C \sim C \times M_w$$

This provides a very useful method for molecular weight determination both with x-rays (1) and with neutrons (7). The main limitation is the accuracy in concentration determination. In the case of neutrons with a sample in H_2O, the term $\rho_s V$ is small compared to Σb (see table of densities above), and the uncertainties in V are not important. This is not the case for x-rays or neutrons in a solvent with a density close to that of the molecule.

From equation (2) one sees that $\sqrt{I(0)}$ must vary linearly with ρ_s, and vanishes eventually when

$$\rho_s = \Sigma b / V$$

This linear variation is true even if the scattering density of the particle varies linearly with the density of the solvent. This is certainly the case with neutrons. The scattering density of the molecule increases as a consequence of deuteration of the labile protons, and this deuteration is proportional to the amount of D_2O in the solvent. In the x-ray case the variation of ρ_s is achieved by addition to the solvent of large quantities of salts or of sucrose. Then the assumption that one is dealing with a two components (solvent and solute) system does not a priori hold . It is likely that there is a redistribution of the ions around the molecule in solution, and that one should consider in reality at least a three component system : solute, solvent far from the solute and solvent close to the solute. This problem has been dealt with in two different ways : The thermodynamical approach, and the invariant volume hypothesis. In the <u>thermodynamical approach</u> (8)(9) the intensity at the origin is expressed

$$I(0) = C \frac{M}{N_A} \left(\frac{\partial \rho_m}{\partial c} \right)_\mu^2$$

in which N_A is the Avogadro number and $\left(\frac{\partial \rho_m}{\partial c} \right)_\mu$ is the increment in scattering density as a function of concentration at constant chemical potential. The latter quantity can be calculated from experimental quantities (essentially specific volume as a function of the salt concentration). This approach is absolutely necessary in the case of nucleic acids, even to deal with neutron scattering in presence of salts. The weakness of this approach is that the thermodynamical parameters are not easy to use for the analysis of the scattering curve in structural terms.

The <u>invariant volume hypothesis</u> (4, 10) postulates that a volume is associated with each molecule in solution inside of which the scattering density distribution is invariant with the density of the solvent. In other terms the molecule is replaced by the molecule plus whatever part of the solvent around is different from the bulk solvent. This "pseudo" particle is the one which matters in all the scattering process. The argument for its invariance is based on linearity of specific volume (which is also the specific volume of that pseudo-particle) with salt concentration.

From the experimental point of view the linearity of the variation of $I(0)^{1/2}$ with solvent density is always observed, both with x-rays and neutrons and provides the basis for contrast variation, the contrast being the difference between the scattering densities of the solvent and of the particle (or the pseudo-particle).

The intensity at small angles and contrast variation

We have seen that the scattering arises from the differences in densities between the solvent and the solute and from the long range internal fluctuation of densities inside the particle. It has been shown that, to some extent, these two components can be separated by varying the density of the solvent. This is the contrast variation method (11, 12, 3); $\rho(\vec{r})$ is split in two parts

$$\rho(\vec{r}) = \bar{\rho}_V + \rho_F(\vec{r})$$

ρ_V is the average density over the volume of the particle, and ρ_F the fluctuation around this mean value. If $\rho_F(\vec{r})$ is independent of the contrast then the equation (1) can be rewritten

$$I(Q) = (\rho_V - \rho_S)^2 I_V(Q) + (\rho_V - \rho_S) I_{VF}(Q) + I_F(Q) \qquad (3)$$

So if this hypothesis is correct, experiments at various contrasts allow a separation of I(Q) into three functions; one depending only on the shape of the particle ($I_V(Q)$), an other one only on its internal structure (I_F) and a third one which is an interference term. In practice there are limitations to this separation. In the x-ray case the invariant volume hypothesis must be correct. If so, the method will give information on the shape and internal structure of the "pseudo" particle and not of the real particle. In the neutron case, the contrast variation is done by adding D_2O to the solvent. This has the consequence of modifying the internal distribution of $\rho(\vec{r})$, as the labile protons are not uniformly distributed over the volume of the particle. So both x-rays and neutrons have their limitation in this decomposition into shape and fluctuation terms. It has been assumed in the above discussion that the physical shape of the particle is unchanged upon deuteration or salt addition. This hypothesis must always be verified and a combination of x rays and neutrons is a good way to check it.

The limit of very small Q. The radius of gyration.

For very small values of Q one can expand (1)

$$I(Q) = \left\langle \left| \int_V (\rho(\vec{r}) - \rho_s)\left(1 + i\,\vec{Q}\cdot\vec{r} - \frac{1}{2}(\vec{Q}\cdot\vec{r})^2 + \cdots\right) d^3r \right|^2 \right\rangle$$

At this stage we shall use the centre of gravity of the volume as the origin of the vector $\vec{r}$. We distinguish two cases.

(a) The centre of gravity of $\rho(\vec{r})$ coincides with the centre of gravity of the volume. Then the term in $\vec{r}$ will vanish and we get, to first order,

$$I(Q) = (\bar{\rho}V)^2\left(1 - \frac{1}{3}\frac{Q^2}{\bar{\rho}V}\int_V r^2(\rho(\vec{r}) - \rho_s)\, d^3r\right)$$

In this expression we find the second moment of the distribution of the scattering density. This moment is usually (13, 11) expressed in terms of a radius of gyration R_G :

$$R_G^2 = \frac{1}{\bar{\rho}V}\int_V r^2(\rho(\vec{r}) - \rho_s)\, d^3r$$

This can be written

$$R_G^2 = R_{GV}^2 + \frac{1}{\bar{\rho}V}\int_V \rho_F(\vec{r})\, r^2\, d^3r$$

with

$$R_{GV}^2 = \frac{1}{V}\int_V r^2\, d^3r$$

R_{GV}^2 is the usual mechanical radius of gyration for a particle of the same shape as the particle under investigation, but with a uniform distribution of scattering density. Now we can rewrite (1) as

$$I(Q) = I(0)\exp\left(-\frac{R_G^2 Q^2}{3}\right) \qquad (4)$$

Expression (4) is called the Guinier law. Experiments carried out at small Q for an assembly of identical particles will always give data which can be fitted by (4) in some range of Q, which depends on the shape of the object, but is usually such that $R_G Q \lesssim 1$. The expression for R_G^2 shows that, for a given particle, it will vary with the contrast. This variation is a measure of the importance of ρ_F, that is to say, of the inhomogeneity of the scattering density in the object. The Guinier law provides a good method for extrapolation at zero angle.

(b) The centre of gravity of $\rho(\vec{r})$ does not coincide with the centre of gravity of the volume. All the above considerations about I(0) are still valid, but for the radius of gyration, one must take into account the term in $\vec{r}$. The new expression is

$$R_G^2 = R_{GV}^2 + \frac{1}{\bar{\rho} V}\int_V \rho_F(\vec{r})\, r^2 d^3r - \frac{1}{(\bar{\rho} V)^2}\left(\int_V \vec{r}\, \rho_F(\vec{r}) d^3r\right)^2$$

In this case R_G^2 is smaller by a quantity which is the square of the distance d between the centre of gravity of $\rho(\vec{r}) - \rho_S$ and that of the volume (note that the centre of gravity of $\rho(\vec{r}) - \rho_S$ is contrast-dependent and is infinity for zero contrast). For a protein this term is small. But for a complex, for instance between a detergent and a protein which for neutrons have very different scattering density, it may be very large and it then provide a method for the determination of the distance between the center of gravity of the two components (14).

The analysis of the angular distribution I(Q)

We have shown that, within limitations,this distribution can be decomposed into terms arising from the shape and the internal structure of the particle. But this does not yet provide structural information other than the radius of gyration. As relation (1) cannot be inverted the only possibility is to use some method of model fitting. Two methods have been used

a) The method of Kratky. Extensively used by KRATKY and his colleagues (1)(2) it is a search method which assumes that the shape of the particle can be approximated by simple geometrical objects (ellipsoids, cylinders or combination of them). It is specially useful for oligomeric proteins. Each monomer is assimilated to an ellipsoid and the analysis of the scattering curve gives the organization of the monomer which gives the best fit.

b) The method of STUHRMANN (15). The method is based on decomposition of $\rho(\vec{r})$ into spherical harmonics, which allows for shapes more complex than ellipsoids or cylinders.

In both cases one gets a solution which gives the best fit found with the data. It certainly does not establish the uniqueness of the solution.

So far we have neglected, in considering the analysis of the data, the fact that they are smeared by instrumental resolution. How to deal with that problem has been considered in detail by GLATTER (16, 17) who also considers seriously the effect of the limited statistical accuracy on data analysis. This important theoretical progress cannot be dealt with in the lecture, but should be familiar to those who wish to use small angle scattering.

In all model fitting the final model is defined by a set of parameters which in some cases is fairly large. It is an important question to know the number of parameters which can be deduced from a set of experimental curves. This has been approached by LUZZATI (18), but no easily usable criterion is yet available. This would be of major importance to avoid over interpretation in analysing small angle scattering data.

The case of spherical particles

In this case the averaging over all orientations disappears from equation (1) which can now be inverted. Then $\rho(r)$ (where r is now a scalar) can be obtained from the data. In a contrast variation method, $(\rho(r) - \rho_s)$ for various solvent compositions will give a hint on the chemical composition at radius r which has provided a good method to investigate spherical viruses. Neutron, due to its broader range of contrast is more suitable. Also in the case of x-rays the scattering object must include the hydration shell, specially important around nucleic acids. Then for x-rays $(\rho(r) - \rho_s)$ may be very similar for nucleic acids and proteins making the interpretation difficult (see a review in 19).

EXPERIMENTAL CONSIDERATIONS

Although the theory of x-ray and neutron small angle scattering is the same, the experimental procedure is quite different for the two types of radiation.

X-ray sources and cameras

The most common sources are x-ray generators with a copper anode delivering an intense K_α line at 1.54Å. Rotating anodes are used by most of the groups. The detection now is usually done with a position sensitive detector, which allows a recording of a useful angular distribution with a static set up. The main differences between the cameras are in the collimation of the incident beam and the methods used to eliminate the K_β line. Typical set up can be one of the following

- Collimation with slits and use of a Nickel filter (a 10 micron filter will transmit about 65% of the K_α but only 8% of the K_β).

- Collimation with slits and a curved mirror which is set up to focuss the beam on the detector.

- Utilization of a curved crystal monochromator which will at the same time eliminate the K_β line and focus the beam. This method gives a much higher background than the previous one.

All these methods are used to measure the scattering curves when they are rather smooth. In the case of a spherical or quasi-spherical object like a virus the scattering curve passes through a series of zero (see experimental data in the next section) and collimation in only one direction, as provided by the previous camera, is not good enough. A focusing in two directions is required. This is achieved by two curved mirrors or the combination of a mirror and a monochromator. For further details see for instance the review in (2) and (20).

During the past years synchrotron radiation has been widely used, as it provides intensities 100 to 1000 times larger than a rotating anode generator. Such a source delivers a white beam and a monochromator must be used which selects a wavelength band of relative width around 10^{-3}. The two important aspects of synchrotron radiation are the following : the high intensity permits kinetic experiments; experiments can be done at variable wavelengths making possible the use of anomalous scattering by specific atoms in the molecule. We shall come back to these two points. Experimentally, they require for kinetics elaborate systems of data acquisition and for the use of anomalous scattering a camera with continuously variable wavelengths.

Neutron sources and cameras

The intrinsic brilliance of a neutron source is much lower than that of an x-ray source. Neutron cameras are designed to compensate for that weakness. The best example is the one (called D11) set up on the high flux reactor at Grenoble (21). The monochromation is achieved by a mechanical device which is possible, because neutrons delivered by a reactor and used for small angle scattering have a low speed (1000m/sec) and they can be sorted on the basis of the time required to go from one slit to another movable slit. This mechanical monochromation gives a relative width of 10%, which is good enough, but compensates by two orders of magnitude the relative weakness of the intensity. The next factor is the use of a two dimensional detector. Finally the camera D11 uses a large sample area. In a x-ray camera the sample irradiated is typically 10μl, whereas with D11 it is 100μl. All these factors together make a high flux neutron source comparable for small angle scattering to synchrotron radiation.

Samples

X-ray samples are put in a capillary. In some cases the sample is circulated through the capillary to minimize radiation damage. The contrast variation, if used, is achieved by previous dialysis of the sample against the appropriate high salt medium, and the medium is used as a reference to correct for the background due to the solvent.

The neutron samples are put in standard quartz cells used on UV spectrophotometers(1 or 2 mm path usually). This allows a direct determination of the concentration of the sample. Samples for contrast variation are prepared by dialysis against the appropriate H_2O/D_2O mixture which here also is used to measure the scattering given by the solvent. We must at this stage point out an important aspect in neutron scattering which is a consequence of the values given for the scattering densities of H and D (see tables above) and of the fact that in addition to the scattering giving rise to interferences, H gives also the so-called "incoherent scattering". The waves which are incoherently scattered do not give interferences and the neutrons are sent with about equal probability in all directions. So this incoherent scattering gives a background. The point is that with H atoms this incoherent scattering is nearly two orders of magnitude more probable than coherent scattering. With D atoms coherent and incoherent scattering are of the same order of magnitude. This has several practical consequences. The first one is that the signal to background is much higher for a sample in D_2O. The second one is that H_2O scatters essentially isotropically. This scattering turns out to be a very convenient method for absolute calibration of a neutron beam (22). The last point on neutron samples is that the scattering by the salts, at usual molarities, can always be neglected compared with that given by the water (H_2O and D_2O).

Absolute calibration

Calibration is necessary for molecular weight determination, but also to put the scattering densities in the analysis on an abosolute scale, allowing an interpretation of the model in terms of chemical composition.

We have already said how this is done in the neutron case. For x-rays one either measures the intensities of the beam with the help of appropriate attenuators, or one uses a calibrated sample. (The KRATKI school uses LUPOLEN, a platelet of polyethylene.)

DATA REDUCTION

This important point cannot be dealt with in this lecture and the reader should consult previous references (2,3,4,5,16,17).

Some typical results

The first figure represent results obtained from an x-ray small angle scattering experiment with contrast variation. The data are from an experiment on rhodopsin complexed with a detergent (23). The various curves are the angular distributions (the abcissa s is the parameter Q divided by 2π) obtained with various amounts of sucrose (from 0 curve A to 49.5% curve H) in the solvent. These

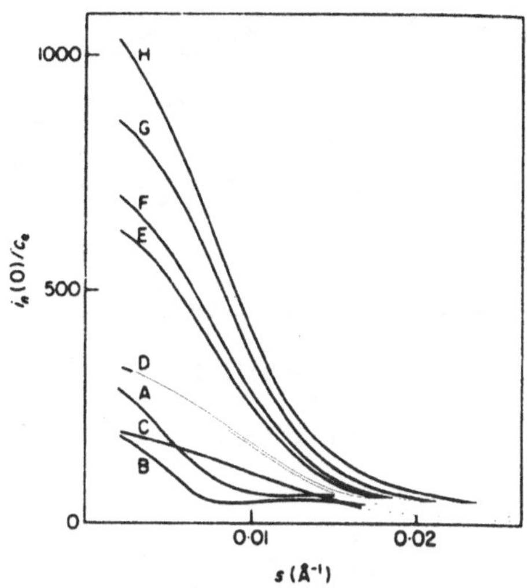

FIGURE 1

curves have been corrected for background. They have the typical aspect of small angle angular distribution : a very fast decrease, eventually modulated by some oscillation. It has been shown that for a particle without internal structure (24) or if that structure can be described by step functions(3) this decrease in intensity follows assymptotically a Q^{-4} law. One sees that the intensities do vary very strongly with contrast. The situation in this experiment is rather exceptional for x-rays : the scattering density of the solvent is higher than that of the particle, except for the two low sucrose percentages (curve A and B). This is due to the very low scattering density of the detergent.

The figure 2 represents the same data at the smallest angles,with the intensities plotted in a logarithmic scale versus the square of the scattering vector. This representation shows clearly the domain of validity of the Guinier approximation.

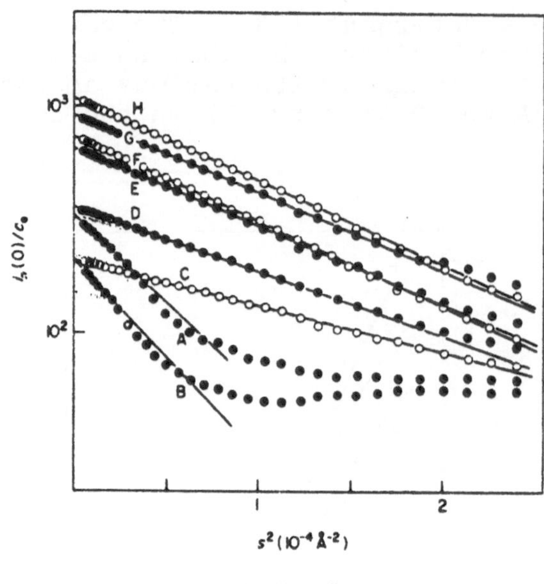

FIGURE 2

The slopes of the straight lines give the radius of gyration. The figure illustrates well the variation of this radius with contrast. From these data the authors have been able to show that the square of the radius of gyration varies linearly with the inverse of the contrast establishing the coincidence (within the accuracy of the method) of the centres of gravity of the rhodopsin and of the detergent bound to it Some information on the shape of the protein and of its complex were then deduced from the values of the radii of gyration and of the volume of the particle. The volume was deduced from the solvent composition at which the contrast vanishes (see theoretical part).[23]

With neutrons the range of contrast which can be covered by H_2O/D_2O mixture is much larger than that which can be covered with x-rays using sucrose or salt. This is illustrated in figure 3 by data obtained with neutron scattering on influenza virus (S. CUSAK and J. MELLEMA private communication). This virus has a core which is made of nucleic acid and protein surrounded by a lipid bilayer from which are emerging spikes. It appears by electron microscopy as a nearly spherical object. The curves reflect this spherical character by a succession of maxima and minima in an overall decreasing curve. For a perfect sphere and with an instrument with perfect resolution the minima would be zero. There are three

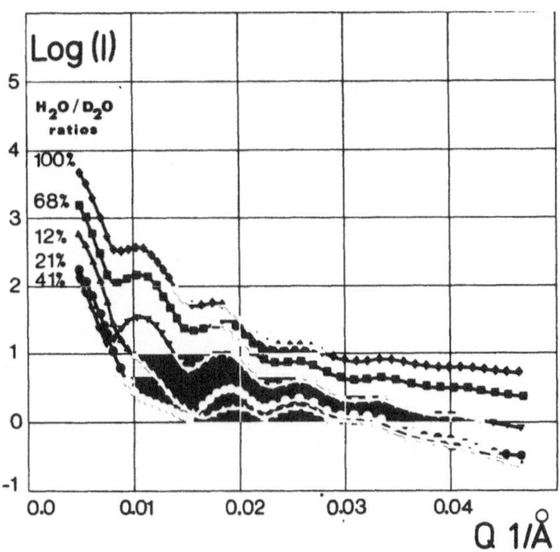

FIGURE 3

specially interesting contrasts in this set of data : 12% D_2O, the lipid contribution is very small; 41%, the protein contribution is very small; 68%, the RNA contribution is minimum. One notices the very important differences between the three corresponding scattering curves. Analysis of the data showed that the lipid bilayer is located at a radius of 425 ± 5Å. An important result has been the determination of the molecular weight which was found to be 200×10^6 daltons, a value much lower than previous estimates. A complete structural analysis is feasible in this case within the approximation of considering the virus as a sphere. The analysis is best done by fitting the set of data to a model made of concentric shells. The fit will give the radii and the scattering densities of each shell. These densities at each contrast can be unambiguously interpreted in terms of chemical composition of the shell. Such an analysis has been carried out in several cases (see a review in (19)), but cannot be done with x-rays, as the scattering densities of a protein and, for instance, of a hydrated nucleic acid may be similar. This point is well illustrated by the work done on PM2 which is a quasi-spherical virus made of a core (DNA + protein) surrounded by lipid bilayer and a protein shell. Very good data were collected from x-ray small angle scattering. As noted above these data could be Fourier inverted to give a distribution $\rho(r)$. This distribution is shown in figure 4.

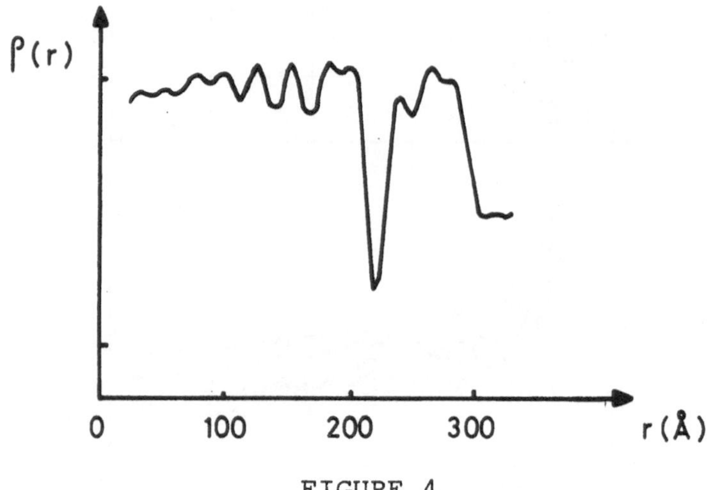

FIGURE 4

This distribution shows very clearly the lipid layer, which appears as a region of very low density (see table 2), but outside of that layer the density is uniform within experimental errors. Note that such a distribution based on FOURIER transformation requires data at relatively large angles and cannot be obtained from neutron data. On the other hand a neutron experiment with contrast variation will give information on the distribution of the DNA and the protein. The experiment was done (29). This distribution is shown in figure 5. Note that the protein and DNA are not uniformly distributed.

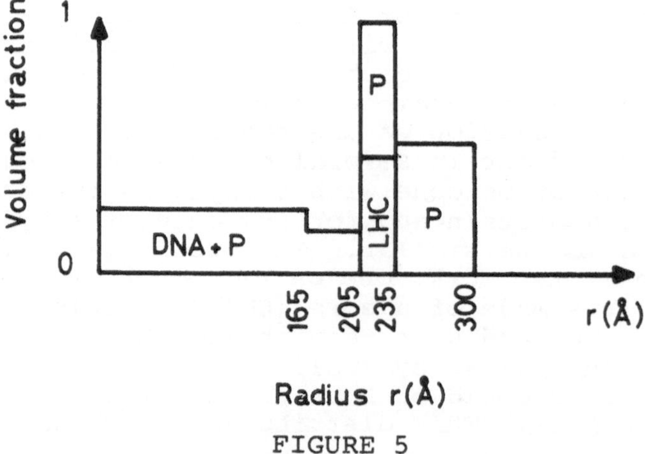

FIGURE 5

Shape analysis

A large experimental activity has been shape analysis of enzymes or enzyme complexes using either x-ray or neutron data. So far the quality of the x-ray data is, in the angular range which extends beyond the zone of GUINIER approximation, of much higher quality than the neutron data in a water solvent. This is due to the large background given by the incoherent scattering from H atoms. For a sample in D_2O the neutron data are of equivalent quality to the x-ray data.

Very often shape analysis is done with a very modest aim : for instance assimilating the particle to an ellipsoid and find its axial ratio. This is done simply by comparing the experimental curves to a set of reference curves. In this case the separation into shape and fluctuation terms by contrast variation is unnecessary. But the difficulties encountered in crystallizing complex particles (e.g. ribosome or multimeric enzymes) has stimulated various people to use small angle scattering to obtain shape information as detailed as possible. These shape analyses will not be described in this lecture. Detailed analysis based on x-ray data or neutron data can be found on multimeric enzymes (2) or ribosomes (25). In the case of ribosomes which cannot have a uniform scattering density, the use of contrast variation with neutron or x-ray is necessary.

Certainly a useful application of good data from small angle scattering is to test shapes proposed from electron microscopy. This has been done (26) to test the various models proposed from electron microscopy for the ribosomal subunits of E.coli. A practical conclusion for future work on particles not amenable to high resolution x-ray crystallography is the complementarity between small angle scattering and electron microscopy. A natural extension of small angle scattering is the low resolution crystallography with neutron which uses contrast variation to interpret low resolution maps (27). There is much less ambiguity in this interpretation than in that for the very best small angle scattering from solution.

Triangulation experiments

This is a method which replaces the scattering from a particle with that from two points in that particle. The method, originally proposed to measure the distance between heavy atom labels with x-ray scattering (30), has been successfully applied to the measurement of deuterium labels with neutron scattering. An example of successful application (see for instance 31) has been the determination of the distance between ribosomal proteins. The method consists in having in the particle two proteins which are replaced by identical but deuterated proteins. (One can as well use a deuterated particle and replace two deuterated proteins by two protonated ones). Let us call this particle PDD, and PDH the particle in which only one protein is substituted. If now one compares the scattering by an equimolar mixture of PDD and PHH with that given by a mixture PDH and PHD, all contributions other than that given by the pair of proteins will vanish. The residual scattering will have the form

$$I(Q) \sim \frac{\sin QR12}{QR12}$$

where R12 is the distance between the two proteins which can be determined from the data.

We shall not discuss here the obvious weaknesses of the method (in particular the fact that a protein is not a point), but insist on the fact that this method provides a very reliable way of establishing the architecture of oligomeric enzymes.

The use of anomalous scattering

In a molecule which has one atom with an absorption edge in the usable wavelength range of synchrotron radiation, one can isolate the scattering from that single atom, which is the only one which is wavelength dependent. The feasibility has been demonstrated by STUHRMANN (32). In the case of hemoglobin, one can separate the scattering due to the four iron atoms. The method may be useful for instance to measure the distance between two iron or two copper atoms in an enzyme. It becomes then quite similar to the triangulation method explained above.

Study of enzymatic reactions

In a reaction

$$A + B \longrightarrow AB$$

the intensity at the origin will vary following

$$\left(\Sigma_A b - b_s V_A\right)^2 + \left(\Sigma_B b - b_s V_B\right)^2 \longrightarrow \left(\Sigma_A b + \Sigma_B b - b_s (V_A + V_B)\right)^2$$

The formation of AB is accompanied by modification of the intensity at the origin. For proteins or nucleic acid the formation of a complex gives an increase of I(0) both for x-ray scattering or neutron scattering for a sample in H_2O. This method has been used to study the formation of complexes between tRNA and aminoacyl tRNA synthetase both with x-rays (33) and neutrons (34). The figure 6 shows typical neutron results.

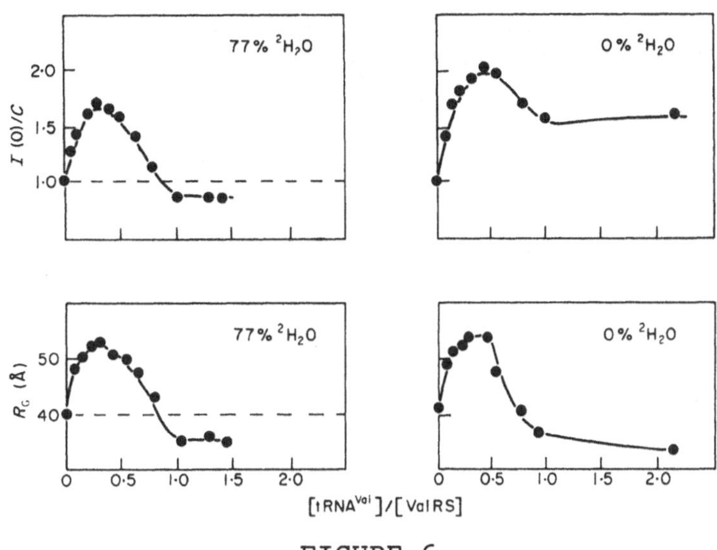

FIGURE 6

The curve indicates what happens when tRNA is added to the enzyme. In H_2O the addition of one tRNA per enzyme increases the intensity by a factor of 1.6, as expected for the formation of a one to one complex. For lower tRNA concentrations there is a large increase followed by a decrease. This corresponds to formation of enzyme aggregates which is confirmed from the data in 77% D_2O. With such a solvent the tRNA practically does not contribute to the scattering, and only the formation and dissociation of the aggregates is observed. In addition one finally has for the complex an intensity 10% lower than the initial intensity, this corresponds to a decrease of 1.3% of the specific volume during complex formation. The bottom part of the figure shows the corresponding radii of gyration.

For such studies neutrons have an advantage over x-rays in being able to distinguish what happens to the nucleic acid and to the protein. On the other hand the effect of D_2O on the reaction may create artefacts.

Kinetic experiments

The relatively poor structural information given by small angle scattering are most valuable when looking to transient phenomena. The simplest case is that of the dissociation or the association of protein complexes, such as oligomeric enzymes, viral capsids, microtubules etc. These phenomena are associated with very large differences in small angle scattering and these modifications can be followed using the high intensities provided by synchrotron radiation.

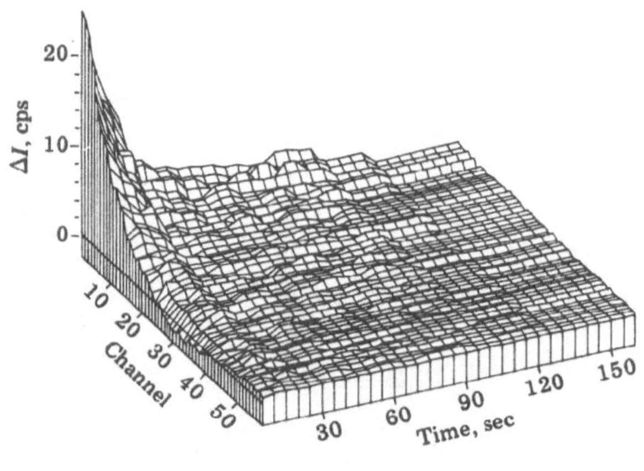

FIGURE 7

The figure 7 shows the time evolution of the dissociation of aspartate transcarbamylase, a multimeric enzyme, in presence of mercurials (35). The phenomena is followed on a time scale of a few seconds.

The formation of a viral capsid has been followed using neutron scattering (36) on an equivalent time scale, as shown in figure 8. The figure shows the intensity at the origin, which to the first approximation is proportional to the number of capsids as a function of time after a pH jump.

Technical developments, using the best equipment on synchrotron radiation,will allow work with a time-scale of a few milliseconds in the near future.

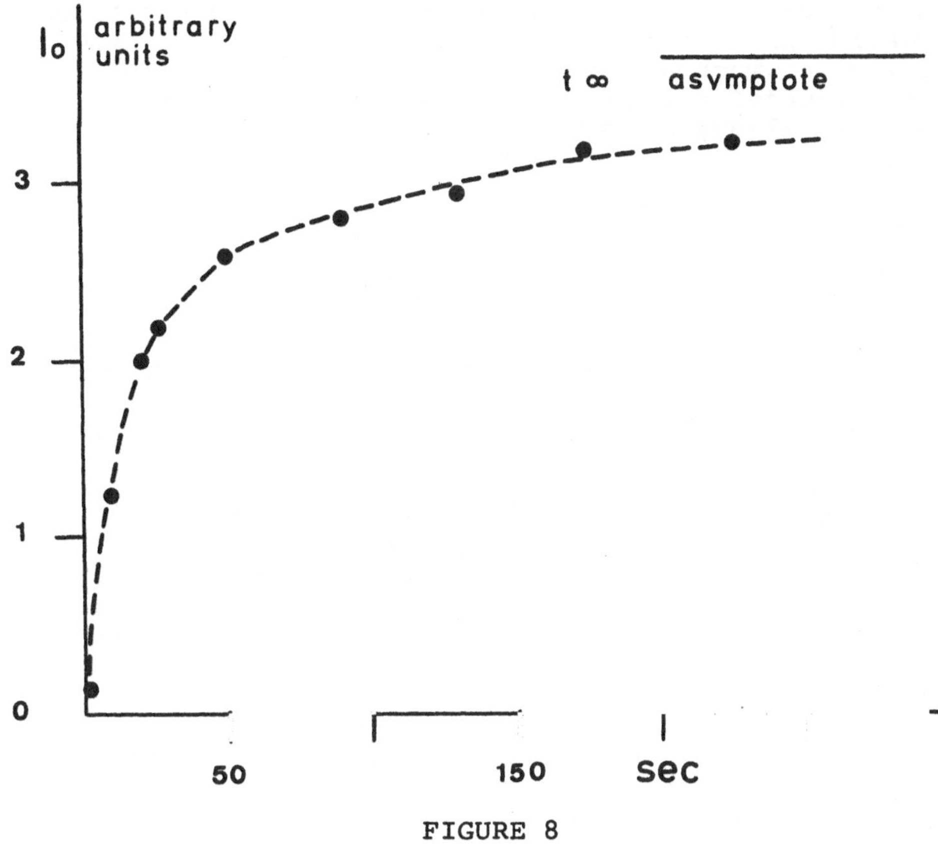

FIGURE 8

CONCLUSIONS

Small angle scattering is a technique which looks simple. In reality it is rather difficult to do reliable experiments. The method lacks internal checks which are important in crystallography. In the analysis there is nothing so far like an R factor, and over interpretation of results is a permanent danger.

However the method is useful, and becomes essential when crystallography is not possible. The availability of synchrotron radiation which permits kinetic experiments offers new scope in this field. Neutron scattering gives a unique opportunity to study protein-lipid, protein-nucleic acid or protein carbohydrate complexes. So the two types of radiation are quite complementary and should be used in parallel as much as possible. They should also be used in conjunction with other physical or biochemical methods, and always on well purified and characterized samples.

REFERENCES

1. O. Kratky and I. Pilz, Q. Rev. Biophys. 5:481 (1972)
2. I. Pilz, O. Glatter and O. Kratky, Methods in Enzymology 61:148-249 (1979)
3. V. Luzzati, A. Tardieu, L. Mateu and H. B. Stuhrmann, J. Mol. Biol. 101:115-127 (1976)
4. V. Luzzati and A. Tardieu, Ann. Rev. Biophys. Bioeng. 9:1-29 (1980)
5. B. Jacrot, Rep. Prog. Phys. 39:911-953 (1976)
6. G. G. Kneale, J. P. Baldwin and E. M. Bradbury, Q. Rev. Biophys. 10:485 (1977)
7. B. Jacrot and G. Zaccai, Biopolymers (in press) (1981)
8. H. Eisenberg, "Biological macromolecules and polyelectrolytes in solution" Oxford University Press, London (1976)
9. H. Eisenberg, Q. Rev. Biophys. (in press) (1981)
10. A. Tardieu, Eur.J.Biochem. 96:621-624 (1979)
11. H. B. Stuhrmann and R. G. Kirste, J. Phys. Chem. 46:247 (1965)
12. H. B. Stuhrmann, J. Appl. Cryst. 7:173 (1974)
13. A. Guinier and G. Fournet, "Small angle scattering of x-rays" (New York-Wiley) (1955)
14. M. Charles, M. Semeriva and M. Chabre, J. Mol. Biol. 139:297 (1980)
15. H. B. Stuhrmann, Acta Crystallogr. A26:297 (1970)
16. O. Glatter, J. Appl. Cryst. 10:415 (1977)
17. O. Glatter, J. Appl. Cryst. 12:166-175 (1979)
J. Appl. Cryst. 13:7-11 (1979)
J. Appl. Cryst. 14:101-108 (1979)
18. V. Luzzati, "Imaging processes and coherence in physics" Edited by Schlenker et al. Springer-Verlag Berlin (1979)
19. B. Jacrot, Comprehensive Virology 17:129-181 (1981)
20. J. Schelten and R. W. Hendricks, J. Appl. Cryst. 11:297-324 (1978)
21. K. Ibel, J. Appl. Cryst. 9:296-309 (1976)
22. K. Ibel and R. May, J. Appl. Cryst. (in press)
23. C. Sardet, A. Tardieu and V. Luzzati, J. Mol. Biol. 105:383-407 (1976)

24. G. Porod, Kolloid Z. 2:83 (1951)
25. M. Koch and H.B. Stuhrmann, Methods in Enzymology 59: 670-705 (1979)
26. A. Tardieu and P. Vachette, private communication
27. G. A. Bentley, J. T. Finch and A. Lewit-Bentley, J. Mol. Biol. 145:771-784 (1981)
28. S. C. Harrison, D. L. D. Caspar, R. D. Camerini-Otero and R. M. Franklin, Nature, London, New Biol. 229: 197-199 (1971)
29. D. Schneider, M. Zulauf, R. Schäfer and R. M. Franklin, J. Mol. Biol. 124:97-112 (1978)
30. W. Hoppe, Israel J. Chem. 10:321 (1972)
31. D. Engelman, Methods in Enzymology 59:656-669 (1979)
32. H. B. Stuhrmann, Acta Cryst. A36:996-1001 (1980) and private communication
33. R. Osterberg, B. Sjoberg, L. Rymo and U. Lagerkvist, J. Mol. Biol. 99:383-400 (1975)
34. G. Zaccai, P. Morin, B. Jacrot, D. Moras, J.C. Thierry and R. Giege, J. Mol. Biol. 129:483-500 (1979)
35. P. Moody, P. Vachette, A. M. Foote, A. Tardieu, M. H. J. Koch and J. Bordas, Proc. Natl. Acad. Sci USA 77:4040-4043 (1980)
36. M. Cuillel and B. Jacrot, private communication.

IMAGE RECONSTRUCTION FROM ELECTRON MICROGRAPHS OF MACROMOLECULAR STRUCTURES

R.A. CROWTHER

Medical Research Council
Laboratory of Molecular Biology
Hills Road
Cambridge CB2 2QH, England

1. INTRODUCTION
   1.1 Contrast mechanisms
   1.2 Specimen preservation and radiation damage
2. IMAGE ANALYSIS AND RECONSTRUCTION
   2.1 Two dimensional reconstruction
   2.2 Three dimensional reconstruction
3. BACTERIOPHAGE T4
   3.1 Polyheads: translational filtering
   3.2 Baseplates: rotational filtering
   3.3 Tailsheath: three dimensional reconstruction
4. SPHERICAL VIRUSES: BACTERIOPHAGE f2
5. NUCLEOSOME CORES: COMBINATION OF VARIOUS DIFFRACTION TECHNIQUES
6. RIBOSOMES: DIFFERENTIAL CONTRASTING OF PROTEIN AND NUCLEIC ACID
7. PURPLE MEMBRANE: DETAILED INTERNAL STRUCTURE
8. CONCLUSION

1. INTRODUCTION

Knowledge of the three dimensional spatial organisation of a complex biological system is generally essential, if an understanding of its working is to be achieved. The assembly of viral capsids, the functioning of a ribosome in synthesising a protein, the repetitive molecular movements generating force in muscular contraction or the operation of the various cellular organelles involved in cell division are all critically dependent on the specific structures involved. Electron microscopy provides one approach to investigating such structures and this article outlines the procedures used to reconstruct reliable two and three dimensional images from the resulting micrographs. The main emphasis however is on the sort of structural information that can be derived.

The transmission electron microscope is an indispensable tool in structural biology, providing the most direct way of visualising structure at a molecular level. Modern instruments are easy to use and, under ideal conditions, have a resolving power of about 3 Å. However, other limitations at present restrict the reliable resolution obtainable for the average biological specimen to much higher values. It must be remembered that the electron image represents a possibly drastically altered form of the original biological specimen, both because of preparation artefacts during dehydration and because of subsequent radiation damage during observation. Apart from the problems of specimen preservation, the micrograph itself does not necessarily present a simple, direct image of the specimen. Firstly, the electron image formed depends on the defocussing conditions and the aberrations in the microscope. Secondly, because of the large depth of focus of several hundred nanometers, features along the direction of view are superimposed in the image, which is therefore a projection of the specimen.

For these reasons the detail in a micrograph is often unreliable and confused and not easily interpretable without methods which assess the degree of specimen preservation, which correct for the operating conditions of the microscope and which separate contributions to the image from different levels in a specimen. These procedures for image processing of electron micrographs have been developed and applied over the last fifteen years and aim to extract from the information recorded in the electron micrographs the maximum amount of reliable information about the two or three dimensional structures which are being examined.

1.1 Contrast mechanisms

Biological specimens are composed almost entirely of atoms of low atomic number and therefore exhibit very low contrast. It has therefore been necessary to enhance the contrast by the addition of heavy metal atoms, either by staining the specimens or by shadowing them. The most successful method has been that of "negative staining", in which the specimen is embedded in a thin amorphous layer of a heavy metal salt which simultaneously preserves and maps out the shape of the region from which it is excluded. The finest detail revealed by this method has been found experimentally to be between 10 and 20 Å, and it has proved very useful in studying arrangements of molecules in complex macromolecular assemblies such as viruses, muscle, flagella and many other systems[1,2].

The other mechanism for producing contrast is phase contrast, and it is essential for dealing with unstained specimens, made up of light atoms only. It is also useful for enhancing higher resolution detail in stained specimens. As is well known in the light microscopy of transparent specimens it is necessary to shift the phases of the scattered beams relative to the unscattered beam and then let scattered and unscattered beams interfere in the image plane. An efficient and convenient way of producing phase contrast in the electron microscope is simply to record the image with the objective lens underfocussed, so changing the phases of the scattered beams. Defocussing however does not act as a perfect phase plate, since the phases are not all changed by the same amount, and successive bands of spatial frequencies contribute to the image alternately with positive and negative contrast. In order to produce a "true" image the electron image must be processed to correct for the phase contrast transfer function of the microscope, so that all spatial frequencies contribute with the same sign of contrast[3].

1.2 Specimen preservation and radiation damage

The main reason for loss of information in biological electron microscopy is specimen damage. Firstly there may be distortion of the structure as the specimen dries on the supporting substrate. Isolated molecules are particularly prone to flattening although the stain helps to preserve them, as do close packed arrays or crystals, in which the molecules support each other. Secondly, unfixed protein is rapidly degraded by the imaging electrons. The incident electrons deposit large amounts of energy by inelastic interactions with the specimens, leading to rupturing of chemical bonds, mass loss and changes in chemical composition. Fortunately stain is much more stable, but it tends to shrink during irradiation, leading to a significant loss of detail[4].

Various procedures have been devised to minimize the dose[5,6] but the difficulty is that with a dose low enough to avoid damage the statistical noise becomes so great as to obscure the signal. The dose normally given to a specimen in recording an image statistically significant to a resolution of 5 Å at a magnification of about 50,000 x is about 500 electrons/Å^2. However the maximum a molecule can tolerate without gross structural alteration at this resolution[6] is about 0.5 electrons/Å^2. Thus it is effectively impossible to obtain a high resolution image of an individual molecule. The only way of circumventing this difficulty is to average 1000 or more images of the molecules, all in the same orientation and each recorded at 0.5 electrons/Å^2, which in practice means having the molecules present as a crystalline array. We thus arrive at a situation closely related to the techniques of X-ray crystallography, with the main difference that in this case, the objective lens of the electron microscope preserves the phase information in the diffracted beams.

Because of these limitations in resolution, electron microscopy is not generally suitable for investigating the internal structure of small protein molecules; it cannot compete with X-ray diffraction of crystals. On the other hand electron microscopy, combined with the methods of image analysis and averaging, is ideal for determining the arrangements and shape of small protein molecules present within natural or artificial arrays, including macromolecular assemblies such as viruses and small cellular organelles, or two dimensional crystals. Wherever possible it is desirable to supplement the electron microscopy and check the state of preservation of the structures by X-ray analysis of wet specimens. It has been found that structural information obtainable by electron microscopy can be highly reliable with respect to detail down to the 20 Å level in the case of negatively stained material in spite of the effects of staining and of radiation damage. In a few special cases, when it has been possible to work with unstained specimens using the technique of Unwin and Henderson[6] described below, reliable information has been obtained to considerably greater resolution.

2. IMAGE ANALYSIS AND RECONSTRUCTION

The idea of averaging over many copies of a repeated motif is central to the most powerful techniques developed so far for producing reliable images of biological specimens, as these exploit the symmetries which occur naturally or which can be induced experimentally in aggregates of macromolecules. For the purposes of exposition it is convenient to consider separately those processing procedures which act on a two dimensional image to improve or modify it, and those which combine a set of two dimensional projected images to produce a three dimensional reconstruction of the original object. In either case the difficulties lie in interpreting rather than producing the reconstructed image, because of its generally low effective resolution[7].

2.1 Two dimensional reconstruction

Periodic features in an electron micrograph can easily be detected by treating the micrograph as an optical diffraction grating[8]. By illuminating the micrograph with coherent light from a laser on an optical diffractometer, its optical diffraction pattern is obtained. As in the case of X-ray diffraction from crystals the periodic part of the density gives rise to a set of discrete regularly spaced spots lying on a two dimensional lattice. The spots lying furthest out correspond to the highest spatial frequencies and therefore to the finest details recorded from the specimen. The non-periodic part gives rise to the background between the peaks.

Although the diffraction pattern can be manipulated optically with filters[9] to produce a corrected and averaged image of the repeating unit of the structure, it turns out to be much easier and more flexible to perform these operations numerically in a digital computer[10]. The optical density of the micrograph is measured at a regular grid of points by a computer linked densitometer and the computer can then calculate the Fourier transform of the resulting array of numbers to give the numerical equivalent of the optical diffraction pattern of the micrograph. The phase information is recovered and stored automatically, as the transformation procedure works with complex numbers. An averaged image of the unit cell of the structure is then resynthesised by setting all of the transform to zero except the peaks corresponding to the regular periodic part of the image and then computing the inverse transform. If necessary the phases of the peaks can be corrected for the effects of the transfer function of the microscope, prior to the Fourier inversion.

The great advantage of computed transforms over optical transforms for image processing is that the phases of the Fourier components, as well as the amplitudes, are readily available. For the purposes of certain types of image analysis, e.g. in the determination of helical parameters or in assessing the degree of preservation of a specimen, the phases are more important than the amplitudes. Of course, when it comes to recombining the Fourier coefficients in a Fourier synthesis, the phases are vital. Similarly compensation for defocussing and aberrations requires the ability to manipulate the phases.

Computational analysis is also the only satisfactory method for investigating the rotational components of an image in an objective way[11]. The image is decomposed into a set of cylindrical harmonics, the strengths of which can then be assessed, and finally a filtered or averaged image reconstructed from the desired harmonics.

2.2 Three dimensional reconstruction

The electron microscope has a depth of focus considerably greater than the thickness of most specimens. Provided that the specimen is thin enough so that multiple scattering can be neglected, the image obtained thus represents a projection of the specimen onto a plane normal to the direction of the electron beam, and features at different levels overlap to give rise, in general, to a complicated superposition pattern. It has been shown[3] that, for a specimen typical of a wide class of biological object, the single scattering regime prevails.

In order to get a picture of a three dimensional structure we must be able to view the specimen from many different directions. These different views are often provided by specimens lying in different orientations or else can be realised by tilting the specimen in the microscope. Originally the different views were interpreted by the building of models, either physically[12] or, later, graphically by computer[13]. It was then realised[14] that a set of transmission images taken in different views could be combined to give an objective reconstruction of a three dimensional object.

The method of reconstruction is based on the projection theorem, which states that the two-dimensional Fourier transform of a plane projection of a three dimensional density distribution is identical to the corresponding central section of the three dimensional transform normal to the direction of view. The three dimensional transform can therefore be built up section by section using transforms ofdifferent views of the object, and the three dimensional reconstruction then produced by Fourier inversion. The important feature of the method is that it tells one how many different views are needed for a required resolution and how these are to be recombined into a three-dimensional map of the object. The process is both quantitative and free from arbitrary assumptions. The approach is similar to conventional X-ray crystallography, except that the phases of the X-ray diffraction pattern cannot be measured directly, whereas here they can be computed from an image. Were it not for radiation damage, the different views could be collected from a single particle by using a tilting stage in the microscope, but more realistically one must use several particles in different but identifiable orientations. In general, it is desirable to combine data from different particles so that imperfections can be averaged out.

The Fourier method is only one way out of several for solving the sets of mathematical equations which relate the unknown three dimensional density distribution with known projections in different directions[15]. Historically the approach arose out of earlier work on two dimensional Fourier processing described above, but in fact no other reliable method has been shown to be superior. Moreover the Fourier method has the advantage that it is carried out in steps, i.e. formation of the two dimensional transforms, and then recombination in three dimensions, so that it is possible to assess, select and correct the data going into the final reconstruction.

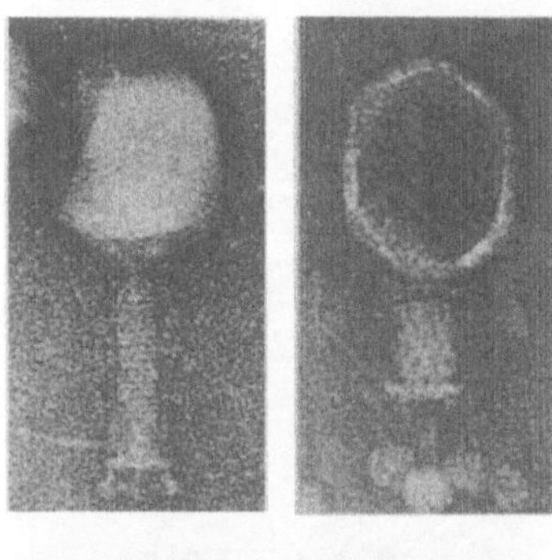

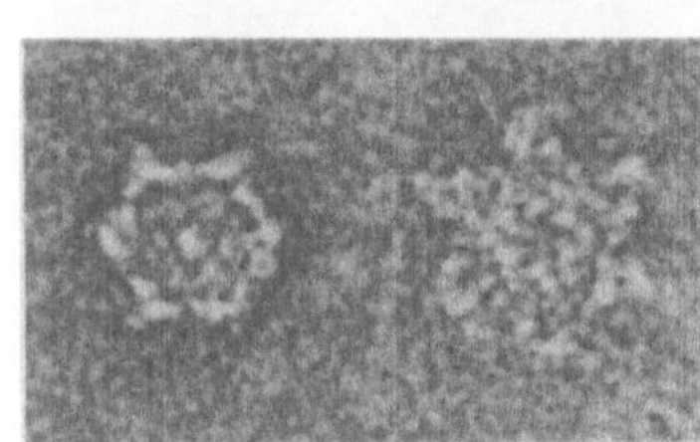

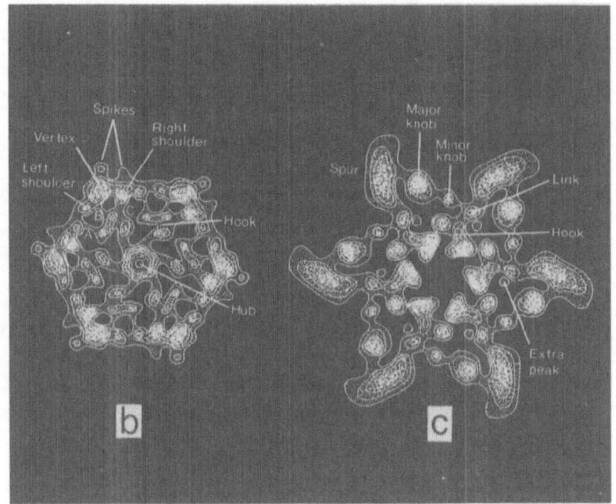

Fig. 1. Bacteriophage T4. (a) Complete particle with extended tail sheath and particle with empty head and contracted sheath. The tail is approximately 1000 Å long. Below: isolated baseplates in the hexagon and star configurations corresponding to the two states of the tail sheath above. The hexagon is approximately 350 Å across the flats. (b) and (c) show filtered images of the hexagon and star respectively, with the various prominent features named.

3. BACTERIOPHAGE T4

Bacteriophage T4 is a bacterial virus which infects the bacterium Escherischia coli. It has a DNA-containing head and a complex tail which serves to attach the phage to the bacterium during infection (Fig. 1(a)). The tail consists of a central tube surrounded by a contractile sheath and terminates in a structure called the baseplate. There are six 1600 Å long tail fibres and six 350 Å short tail fibres joined to the baseplate. Binding of the phage to the bacterium by the long and short tail fibres leads to a change in configuration of the baseplate from a hexagon to a six-pointed star (Fig. 1(a)), which in turn triggers the contraction of the tail sheath. Since the baseplate remains attached to both sheath and bacterium, this results in the tip of the tail tube penetrating the outer envelope of the bacterium. The tip of the tail tube is then unplugged and the DNA passes down the

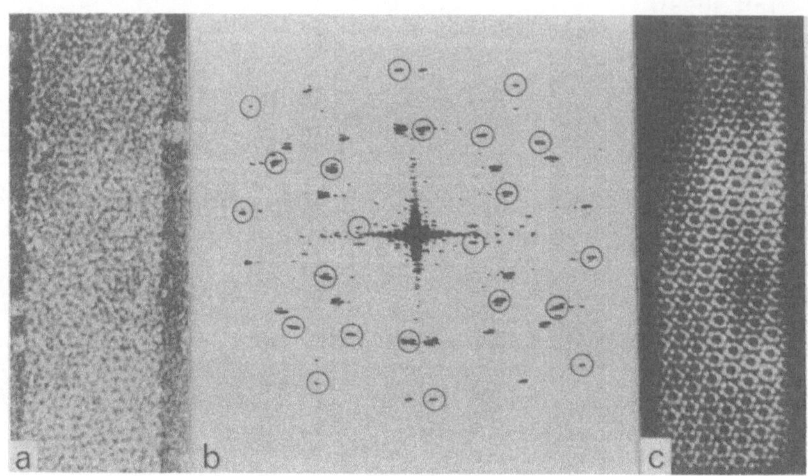

Fig. 2. Optical diffraction and filtering of a tubular coarse polyhead consisting of the major head protein of bacteriophage T4. (a) Micrograph of negatively stained flattened tube, approximate diameter 1100 Å. (b) Optical diffraction pattern of (a) with circles drawn around one set of diffraction peaks corresponding to one layer of the structure. (c) Filtered image of one layer in (a) using the diffraction mask shown in (b).

tail tube and into the target cell, in a way presently not understood, thus initiating a further round of viral replication.

3.1 Polyheads: translational filtering

The native head structure of T4 is difficult to study directly as it is rather smooth and its features are not strongly contrasted by negative staining (Fig. 1(a)). There are however a number of aberrant tubular structures, known as polyheads, which are related to the native head structure and which undergo transformations believed to mimic the complicated process of maturation associated with DNA packaging in the native head. Such tubular structures are easier to analyse than the native head.

Fig. 2(a) shows a negatively stained "coarse" polyhead[16]. The tubular structure had flattened so that the image is of two superimposed, approximately planar layers of protein molecules. This superposition plus the granularity of the support film obscures the individual subunits and their arrangement in the structure. Fig. 2(b) shows the optical diffraction pattern, which consists of spots lying on two regular lattices related by an axial mirror line and arising from the two layers in the structure. The spots from the two layers are spatially separated, so by placing in

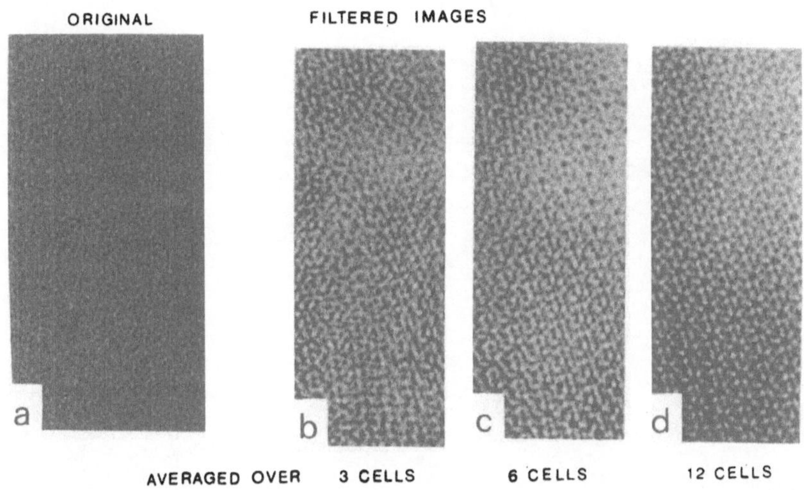

Fig. 3. Computer filtering of a fine polyhead. (a) Original micrograph. (b), (c) and (d) show filtered images of one layer, where the averaging has been carried out respectively over 3, 6 and 12 cells longitudinally.

the diffraction plane of the optical diffractometer an opaque mask with appropriate holes cut in it, as indicated by the rings in the figure, the diffracted beams from one of the layers can be selectively recombined to produce a filtered image of a single layer (Fig. 2(c)). Besides removing one of the layers, a large part of the aperiodic noise (that is the irregular part of the image, which gives rise to the speckle between the main diffraction spots) has also been removed. In the filtered image the approximately hexagonal arrangement of individual protein subunits can be clearly seen.

In cases where the diffraction spots are close together or where the signal-to-noise is particularly low it is better to perform the equivalent operations computationally[10]. An example of a "fine" polyhead is shown in Fig. 3. The filtered image is formed by setting the whole of the transform to zero, except in small regions surrounding the diffraction maxima arising from one of the two superimposed layers of the tube and then computing the inverse transform of the modified array. The range of averaging that takes place in the filtering process is controlled by the size of these "apertures"; the smaller the apertures the greater the range of averaging. If the apertures were reduced to single sample points, a perfectly periodic image would result. Fig. 3 shows a series of filtered images in which the range of averaging is 3, 6 and 12 unit cells longitudinally and half this laterally. Fig. 3(b) resembles the best that could be achieved optically and is uninterpretable. In Fig. 3(d) sufficient averaging has been performed to enable a fine hexagonal pattern of subunits to be

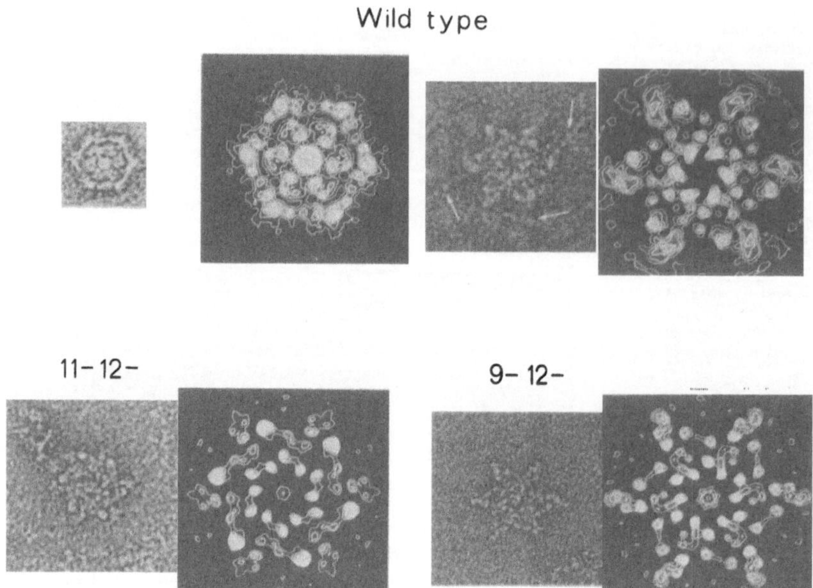

Fig. 4. Rotationally filtered images of baseplates of bacteriophage T4. Above: wild type hexagon and star, with 350 Å fibres formed by gene 12 product indicated by arrows. Below: stars lacking respectively the products of genes 11 and 12 or 9 and 12. Both are lacking the 350 Å fibres formed by gene 12 product, while the former has a much reduced "spur" and the latter a much reduced "major knob" (see Figs. 1(c) and 5).

discerned. It is believed that the protein subunits in the coarse polyhead (Fig. 2(c)) have undergone proteolytic cleavage and re-arrangement in forming the fine polyhead (Fig. 3(d)). Similar changes occur during the packaging of DNA inside the native phage head[17], which is thought to have a structure resembling that of a fine polyhead.

3.2 Baseplates: rotational filtering

The baseplate of T4 is a complex structure containing multiple copies of at least 14 different kinds of protein[18]. It is assembled sequentially and incomplete substructures accumulate if any component is missing, for example because of mutation. The whole structure in either the hexagon or the star configuration exhibits 6-fold rotational symmetry (Fig. 1), which can therefore be used for averaging. Because the number of copies of the repeated motif is small, the gain in signal-to-noise ratio is less than for extended translationally periodic specimens but is nevertheless still worthwhile.

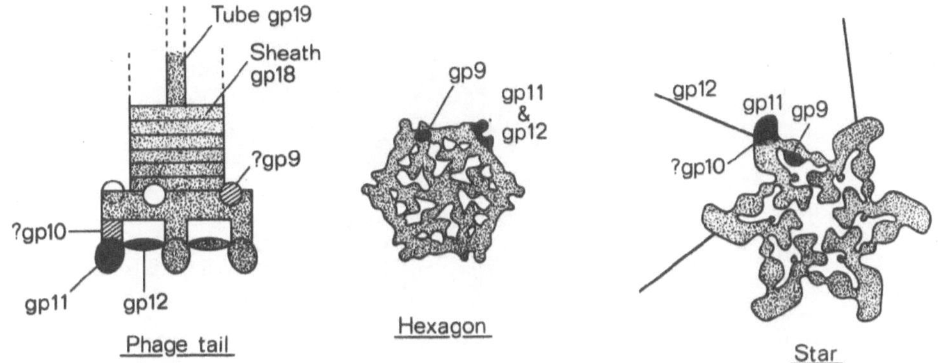

Fig. 5. Summary showing the positions of gene products 9, 11 and 12 of bacteriophage T4 in isolated hexagon and star baseplates and in intact phage tails viewed from the side.

Rotational filtering is carried out numerically by decomposing the image into a series of functions representing harmonics of increasing angular and radial frequency[11]. These are analogous to the plane sinusoidal waves that give rise to the pairs of spots in the diffraction pattern of a translationally periodic object. By plotting the power of the angular harmonics as a function of increasing angular frequency, we obtain a rotational power spectrum of the image, which is analogous to be distribution of intensities in the optical diffraction pattern of a translationally periodic image. It summarizes the results of the analysis in a convenient form and allows an objective choice of the best preserved specimens. A rotationally filtered image is produced by numerically summing just those angular harmonics which are consistent with the rotational symmetry, in this case those that are multiples of 6.

Rotationally filtered images of baseplates in the hexagon and star configurations are shown in Fig. 1(b), (c). Although such images display the anatomy of the baseplate much more clearly than the original micrographs and allow the major features to be recognised and named[19], it is impossible to take the analysis further, unless one can produce images which differ from the wild-type structure in well defined ways. This is possible here as certain of the components can be removed by mutation, while leaving a recognisable hexagon or star baseplate. It is then possible to detect consistent differences between the various images and thus ascribe certain features in the structures to particular gene products. In this way the products of genes 9, 11 and 12 have been identified in rotationally filtered images of hexagons and stars

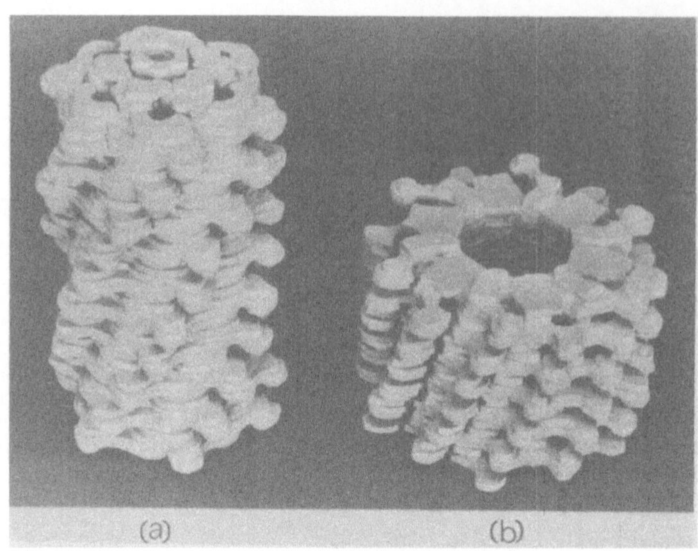

Fig. 6. Three dimensional image reconstructions of bacteriophage T4 tail. (a) Intact extended tail. (b) Polysheath, an aberrant aggregate of the sheath protein resembling contracted sheath.

and the two configurations structurally related[19].

Fig. 4 shows rotationally filtered images of various mutant baseplates and the overall results are summarised schematically in Fig. 5. Gene 9 product occupies a peripheral position in both hexagons and stars, consistent with its providing a binding site for the long tail fibres. Gene 11 product in the hexagon forms the distal part of the tail pin, seen more clearly in sideways views of the phage, which folds out to form the point of the hexagram in the star configuration. Gene 12 product is visualised as an extended 350 Å fibre in stars but appears to have a more compact structure in hexagons and intact phage. These three gene products appear to be control proteins which decorate a structural skeleton formed from all the other gene products and which control its structural transition[19,20].

3.3 Tail sheath: three dimensional reconstruction

The tail sheath of T4 was the first specimen to be reconstructed in three dimensions[14]. The sheath has 6-fold symmetry and the subunits are arranged helically in such a way that a single image of the tail provides 21 different views of the repeating subunit, tilted about the helical axis and equally spaced in angle. This is a particularly favourable geometrical arrangement and means that a three dimensional reconstruction can be made from a single view of the phage, at least to limited resolution.

The two dimensional computed Fourier transform of the image is a central section of the three dimensional transform of the three dimensional specimen. For a helical structure the two dimensional transform consists of a series of layer lines with a spacing inversely related to the helical repeat distance. The helical symmetry, established by indexing the diffraction pattern or Fourier transform, may be used to generate uniquely from this single section the complete three dimensional transform out to a limit set by the number of distinct views of the subunit. A three dimensional image of the specimen can then be computed by Fourier inversion of this filled-in transform. In practice the filling in of the three dimensional transform is implicit in the mathematical formulation as a Fourier-Bessel transform, which links the amplitude and phase of each layer line contribution to the strength and relative position of a particular helical family in the three dimensional structure. The three dimensional image is built from a superposition of a number of these helical families. Close to the meridian on each layer line there is in general only a single helical family contributing and the necessary information about it can be retrieved from a single view of the structure. Moving away from the meridian, other families start to overlap and it is only possible to sort out the various contributions by combining data from more than one view of the object. The more subunits there are in the helical repeat, the further out in the transform the overlapping starts and the higher the resolution attainable from a single view.

A reconstruction[21] of the extended tail combining data from several particles to a cut-off of about 30 Å is shown in Fig. 6(a). There is a hole of about 15 Å radius along the axis of the particle and then fairly continuous density to a radius of about 45 Å representing the tail tube. Separating this from the main bulk of the sheath is a set of six helical tunnels, with bridges between which link the sheath with the core. At outer radii the surface is divided up apparently into subunits by two strong sets of helical grooves. Although the contracted sheath itself is too short to make a good reconstruction, an aberrant structure, polysheath, made from sheath subunits assembled in a way resembling contracted sheath has been analysed (Fig. 6(b)). A comparison of the two structures, in conjunction with the geometrical analysis of the path of contraction[22], allows the subunit to be tentatively dissected out and its change in configuration followed[21].

The shape of the tracery "hooks" in the inner part of the hexagonal baseplate (Fig. 1(b)) bears a strong resemblence to the apparent subunit shape seen in the three dimensional reconstruction of the extended tail sheath. The hexagonal form of the baseplate may therefore provide a template for the assembly of the tail sheath

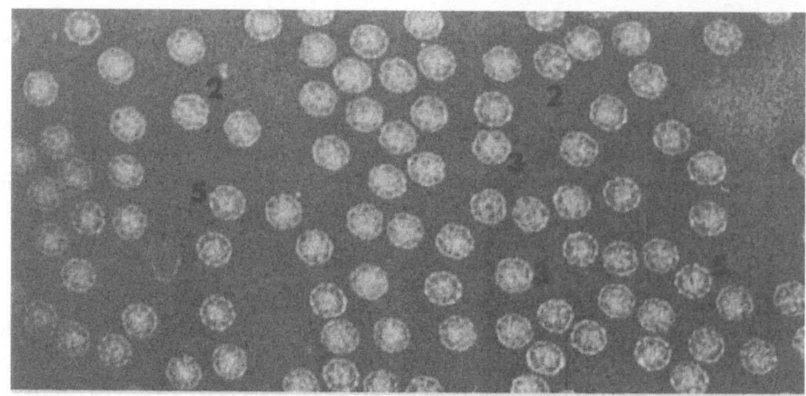

Fig. 7. Negatively stained field of bacteriophage f2. The diameter of the particles is about 250 Å. Particles marked 2, 3 and 5 show features characteristic of views close to 2-fold, 3-fold and 5-fold axes of symmetry.

in its extended form. The changes in the tracery hooks when the baseplate goes from hexagon to star configuration[19] lead to a breaking of the bonds between the tracery and the central hub of the baseplate and may also change the configuration of the first annulus of sheath subunits from an extended to a contracted state. This change of state is then propagated annulus by annulus up the sheath leading to its overall contraction.

4. SPHERICAL VIRUSES: BACTERIOPHAGE f2

The protein coats of all small spherical viruses so far investigated have icosahedral symmetry, point group 532. They contain multiple copies of one or more proteins. The largest number of identical subunits that can be arranged in identical environments in a spherical shell is 60. However, by relaxing the requirement for strict equivalence of all subunits, it is possible to build larger shells containing 60T subunits, in which the subunits are now only quasi-equivalent[23]. The triangulation number, T, can take only certain integer values, of which the smallest are 1, 3, 4, and 7. The subunits may lie in special positions in the surface lattice, giving rise to dimer, trimer, or hexamer-pentamer clustering. The resulting morphological units give rise to characteristic features in negatively stained preparations, even though the individual subunits may not be distinguishable if the stain does not penetrate between them.

For helical structures, as already described, three-dimensional reconstruction can often be performed in a fairly straightforward way from a single view of the structure. For other types of symmetry, such as icosahedral, the situation is more complicated and more than one view is necessary[15]. The two-dimensional Fourier

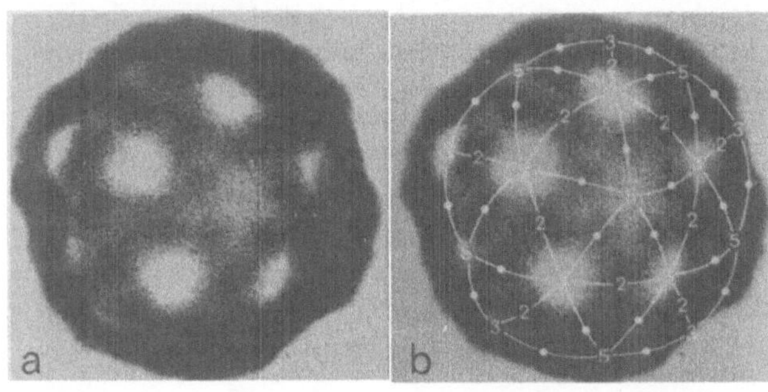

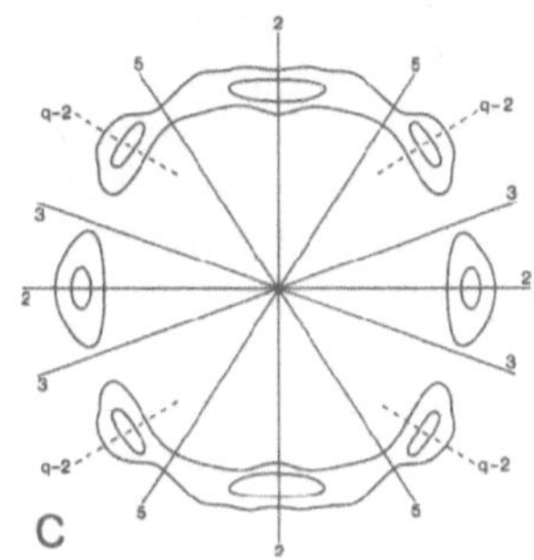

Fig. 8(a). Density plot of a reconstruction of f2 phage, in which high density represents exclusion of stain. (b) as (a) but with the T=3 icosahedral surface lattice superimposed. (c) Equatorial section of the reconstruction, showing the strict symmetry axes passing through the centre of the particle but the quasi 2-fold axes (q-2) tilted away from the 5-fold axes.

transform of each image represents a central section through the three dimensional transform of the object. The different views give different sections and so the three dimensional transform can be filled in plane by plane. When the specimen possesses symmetry, each view gives not only one plane but a whole set of equivalent planes (60 in the case of icosahedral symmetry) generated by the appropriate symmetry operations. When a sufficient number of different views has been included to fill in the three dimensional transform out to the limiting spatial frequency set by the preservation of the specimen, we perform a three-dimensional Fourier inversion to give a three dimensional image of the specimen. Since the particles lie in arbitrary orientations, the three dimensional transform is not filled in uniformly and it is therefore necessary to perform interpolation in the transform prior to inversion. This interpolation involves the solution of sets of linear equations, which will be solvable only if sufficient views have been included. Tests for the solvability of the equations provide a check that for a particular specimen the selected views do uniquely determine the three-dimensional reconstruction to a fineness of detail set by the specimen preservation. Typically for small spherical viruses three or four views are sufficient.

Before including any particular view, its orientation relative to the icosahedral symmetry axes must be found and its preservation assessed. This can be done by searching its two-dimensional transform for the position of the best set of pairs of so-called common lines[24]. These are pairs of lines along which the transform should have equal values because of the symmetry of the particle. With ideal data the values along the common lines would agree exactly if the particles were icosahedral, so the symmetry can be tested. Because of distortions of the specimen, nonuniform staining, and contributions from the supporting grid, the values will not agree exactly and the resulting discrepancy may be used as a measure of the quality of the image. It can moreover be computed as a function of increasing spatial frequency, so that it is possible to tell at what scale of detail features in the image cease to be icosahedrally correlated. Typically for small spherical viruses in negatively stained preparations, some degree of icosahedral correlation extends to spatial periodicities of about 25 Å.

Image reconstructions have been made of various spherical viruses. Here we consider bacteriophage f2 [25]. Fig. 7 shows a negatively stained field of virus particles in which certain characteristic views down symmetry axes have been marked. These symmetrical views indicate that the virus has a T=3 structure with the 180 subunits avoiding the 3-fold and 5-fold positions of the surface lattice. The particles appear roughly circular with a diameter of about 240 Å and are bounded by a heavy white line, which indicates that the protein is concentrated into a relatively thin shell about 30 Å thick. The image reconstruction (Fig. 8), made from five particles and containing data to a cut-off of 30 Å, shows that the most strongly stain-excluding features lie on the strict and quasi-2-fold axes. However the stain excluding material also extends towards the quasi-3-fold positions, which lie at the centres of the triangles of the T=3 surface lattice superimposed on Fig. 8(b). This arrangement of protein produces the characteristic pattern of holes at the 5-fold and 3-fold (quasi-6-fold) positions. The equatorial section of the reconstruction (Fig. 8(c)) shows that the subunits on the quasi-2-fold axes extend to a greater radius than do those on the strict 2-fold axes, the difference being about 7 Å. It is also clear that the quasi-2-fold axes do not pass through the centre of the particle, where the various strict symmetry elements must intersect. This popping out and tilting of the quasi-dimers is similar to that observed in the reconstruction of tomato bushy stunt virus[24] and confirmed by high resolution X-ray crystallography[26]. In f2 as in tomato bushy stunt virus this disposition of the quasi-dimers may account for the much greater stain penetration on the 5-fold positions than on the quasi-6-fold positions.

5. NUCLEOSOME CORES: COMBINATION OF VARIOUS DIFFRACTION TECHNIQUES

The fundamental unit of chromatin structure is the nucleosome[27], which consists of a length of DNA wound around a histone octamer core of composition $(H3)_2$ $(H4)_2$ $(H2A)_2$ $(H2B)_2$. Nuclease digestion of chromatin produces a nucleosome core particle, containing about 140 base pairs of DNA, which has been crystallized[28]. The unit cell of those original crystals was large, so a combined approach was made using both X-ray crystallography and electron microscopy.

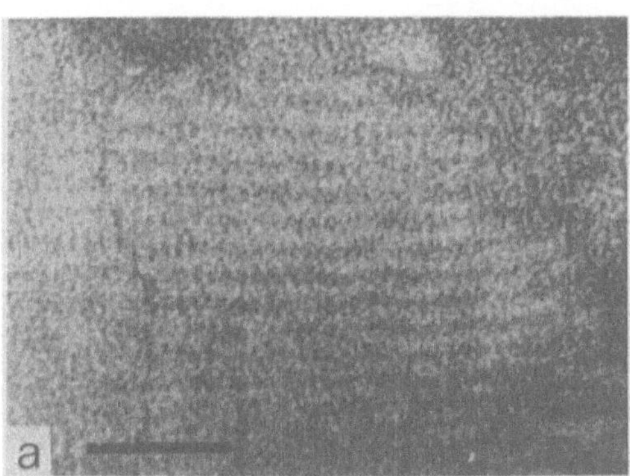

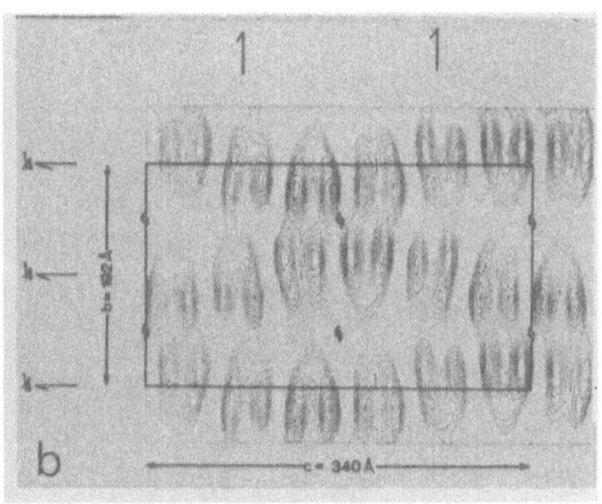

Fig. 9(a). Negatively-stained thin fragment of a nucleosome core crystal. The scale bar represents 500 Å. (b) Electron density map obtained by Fourier synthesis using amplitudes from X-ray diffraction and phases from the computed transform of (a).

One of the three centro-symmetric projections of the crystal structure gives a view looking through the width of only one particle, so a fairly clear picture of the overall structure of the nucleosome core has emerged[28].

Fig. 9(a) shows a negatively stained thin fragment of such a crystal. By computing its Fourier transform a set of phases were obtained which, when combined with the amplitudes of the corresponding X-ray reflections to 25 Å spacings, gave the map shown in Fig.9(b). The nucleosome cores are arranged in sinusoidally distorted rows and each particle presents a bipartite wedge-shaped appearance in projection.

Under certain conditions histone octamers without DNA form hollow tubular structures about 300 Å in diameter (Fig. 10)[29]. The tubes consist of stacked rings of octamers, of thickness 65 Å indicated by the first meridional reflection in the optical diffraction pattern (Fig. 10(d)), with the density in each ring divided into two layers as indicated by the strong second meridional reflection. Analysis of the computed transforms of images of the tubes indicated that each histone octamer has a 2-fold axis of symmetry normal to the axis of the tube. A three dimensional image reconstruction (Fig. 11(a)) shows the histone octamer as a wedge shaped particle formed from a roughly helical ramp of density with the same bipartite character observed previously in the nucleosome core crystals (Fig. 9(b)).

A neutron diffraction study of nucleosome core crystals using contrast variation[30], showed the $1^{3}/4$ turns of DNA in the core particle to be wrapped around the outside of the histone octamer

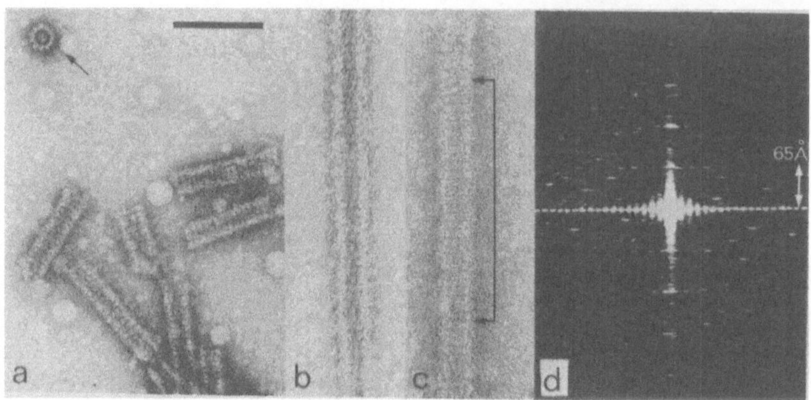

Fig. 10(a). Tubes of histone octamers negatively stained on the grid. The arrow indicates a small fragment viewed end-on. Scale bar 1000 Å. (b) Tube stained before being applied to grid, in which oblique helical lattice lines are clearly visible, as contrast of one side of particle dominates. (c) Another tube from which the region indicated gave the optical diffraction pattern (d).

ramp, accentuating its wedge-shaped appearance (Fig. 11(b)). By combining the results of histone-DNA and histone-histone chemical cross-linking studies, a tentative assignment was made of the different histones in the octamer to the different regions in the three dimensional map[29].

6. RIBOSOMES: DIFFERENTIAL CONTRASTING OF PROTEIN AND NUCLEIC ACID

Crystalline sheets of ribosomes develop on the endoplasmic reticulum membrane in oocytes during hibernation of the Italian lizard, Lacerta sicula. Each sheet is composed of two layers of ribosomes and each layer represents a two dimensional crystal of ribosome tetramers arranged on a square lattice[31]. Negatively stained crystals show characteristic moiré patterns arising from the superposition of the two layers of ribosomes (Fig. 12(a)), though occasionally one finds small areas in which there is only a single layer. By collecting data from tilted sheets it is possible to make a three dimensional reconstruction of a single layer of ribosomes (Fig. 13(a). This allows the overall shape of the ribosome to be mapped out and the large and small subunits to be identified. One can see how the ribosomes pack in the tetramer and visualise a thin stalk of density which links each ribosome to the membrane (Fig. 14(a)).

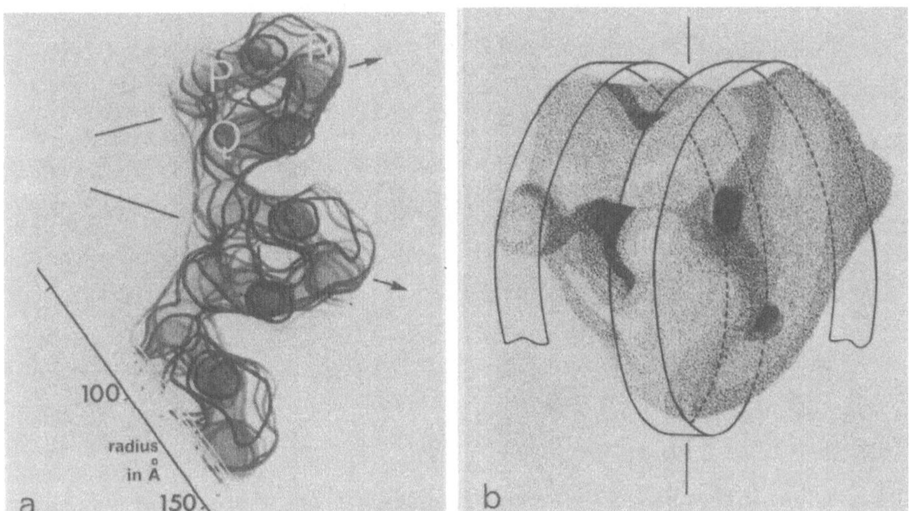

Fig. 11. Three dimensional reconstruction of the histone octamer. (a) A superimposed series of sections through the map, showing 2½ octamers. The arrows indicate the 2-fold axes of symmetry through the octamers. (b) A drawing of the histone octamer with about 1 3/4 turns of DNA, shown as a ribbon, wrapped around the histone core.

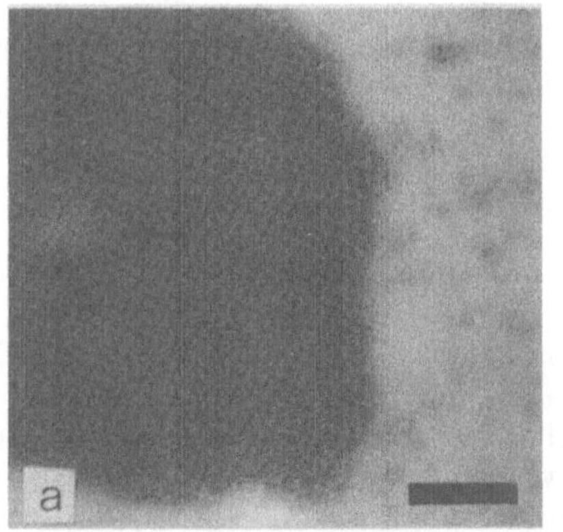

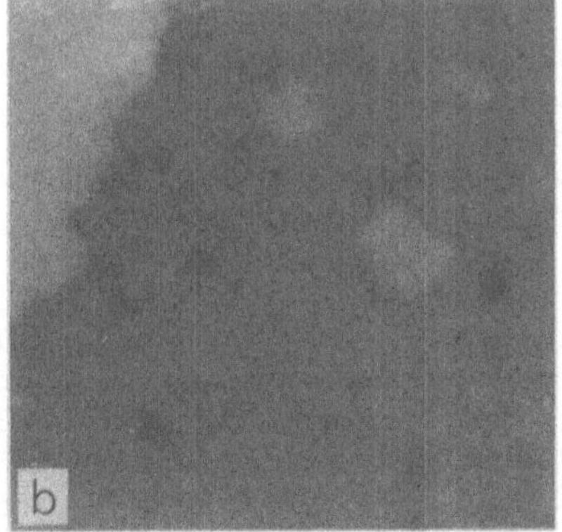

Fig. 12. Crystalline arrays of ribosomes. (a) A crystal negatively stained with gold-thio-glucose. Most of the area shows a moiré pattern arising from superposition of two layers of tetramers of ribosomes but in the small area indicated by the arrow a single layer is seen. Scale bar 5000 Å. (b) A similar crystal embedded in glucose and therefore showing much lower contrast.

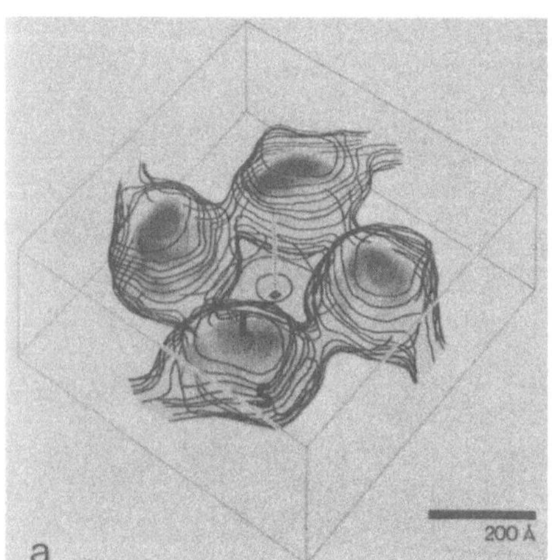

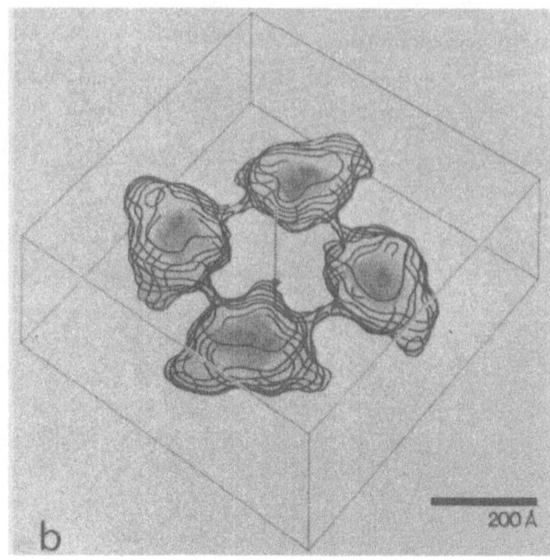

Fig. 13. Three dimensional reconstructions from ribosome crystals. (a) Reconstruction from a negatively stained crystal showing a tetramer of ribosomes. The overall shape of the ribosome can be discerned. L and S denote the large and small ribosomal subunits respectively. (b) Similar view of a reconstruction of a glucose embedded crystal of ribosomes. In this case the protein should be matched out by the glucose and only the RNA should be seen.

The eukaryotic ribosome contains roughly equal weights of RNA and protein. It is possible to get an idea of the relative distribution of the two species by using differential contrasting[32,33]. This can be achieved by embedding isolated crystalline sheets in glucose (Fig. 12(b)), which has a density very close to that of protein and thus matches out the protein, allowing the denser RNA to be seen. The three dimensional reconstruction of the glucose embedded sheet is shown in Fig. 13(b) and a sideways view of a single ribosome in Fig. 14(b). It is clear that the glucose embedded structure appears much more compact than the negatively stained one, as expected since only about 50% of the mass should be visualised. Comparing different views it is clear that the RNA constitutes a core within the ribosome, with most of the protein located around the outside. The maximum density in the RNA region occurs near the centre of the ribosome in the region thought to be the interface between the ribosomal subunits, suggesting that RNA-RNA interactions may be important in holding them together.

7. PURPLE MEMBRANE: DETAILED INTERNAL STRUCTURE

Embedding the specimen in glucose rather than negative staining was also the key to visualising the internal structure of the purple membrane protein[6]. The purple membrane is a specialised

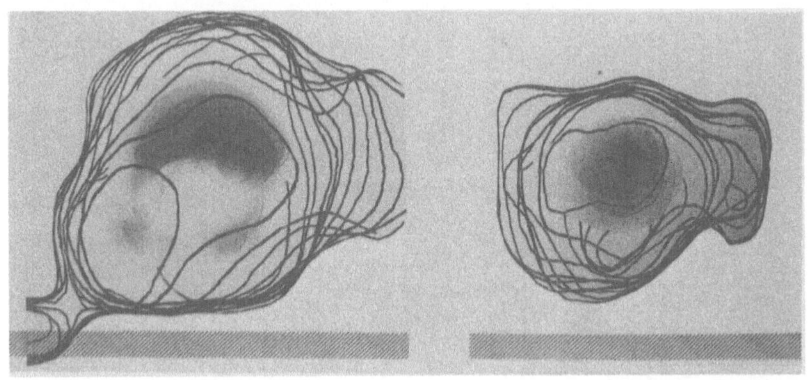

Fig. 14. Maps of single ribosomes viewed from the side, with the shaded band at the bottom indicating the membrane to which they are bound. (a) Negatively stained preparation showing link to membrane. (b) Glucose embedded preparation showing the smaller RNA core now visualised.

part of the outer membrane of a halophilic bacterium, Halobacterium halobium, which functions in vivo as a light-driven hydrogen ion pump involved in photosynthesis. The glucose replaces the aqueous environment of the membrane and allows drying of the specimen without distortion. However the resulting specimen is highly radiation sensitive and exhibits very low contrast. Pictures must therefore be taken with extremely low electron doses, not exceeding 0.5 electrons/Å^2, using a considerable amount of underfocussing to give phase contrast. The resulting image appears featureless and is dominated by statistical fluctuations in the number of electrons forming each picture element. However when such an image is optically diffracted, a regular array of diffraction spots is seen (Fig. 15(a)). This demonstrates that provided the image is averaged over enough unit cells of the purple membrane crystal, a signal well above the noise level may be retrieved.

Because the image is taken with a large amount of underfocus it is not sufficient simply to filter the diffraction pattern to produce an average image, as successive bands of spatial frequencies are contributing with opposing signs of contrast. The contrast transfer function of the microscope can be mapped out by taking a second high dose image under identical conditions to the initial low dose image. When this is optically diffracted (Fig. 15(b)), there are no longer any sharp spots as the regular crystalline structure has been destroyed but the diffraction pattern of the resulting amorphous carbon film displays the transfer function as a series of rings. It can be seen that the intensities of spots in

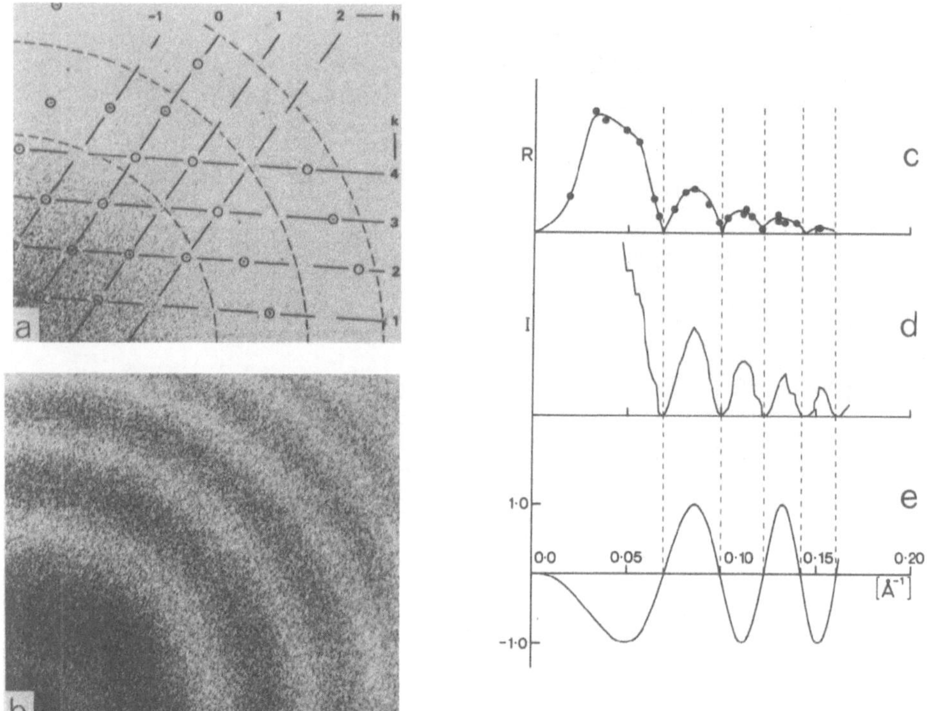

Fig. 15. Optical diffraction patterns of micrographs of purple membrane. (a) Low dose image showing diffraction spots lying on a regular lattice. (b) High dose image taken under identical conditions showing contrast transfer function as set of rings. (c) Structure factor amplitudes calculated from low dose image, plotted as ratios (R) of their values in electron diffraction patterns. (d) Densitometer tracing across the optical diffraction pattern of the high dose image shown in (b). (e) The calculated phase contrast transfer function appropriate to (c) and (d), with underfocus 5750 Å and spherical aberration coefficient 1.6 nm, illustrating how the sign corrections are made.

the optical diffraction pattern are modulated by the transfer function (Fig. 15(c)), which is displayed experimentally in Fig. 15(d) as a densitometer tracing of the rings shown in Fig. 15(b). The positions of maxima and minima line up almost exactly and enable a phase contrast transfer function to be computed (Fig.15(e)), incorporating the appropriate values of defocus and spherical aberration. After the phases have been corrected an average image of the structure can be computed (Fig. 16(a)). In practice the amplitudes of the diffraction spots were obtained from the intensities

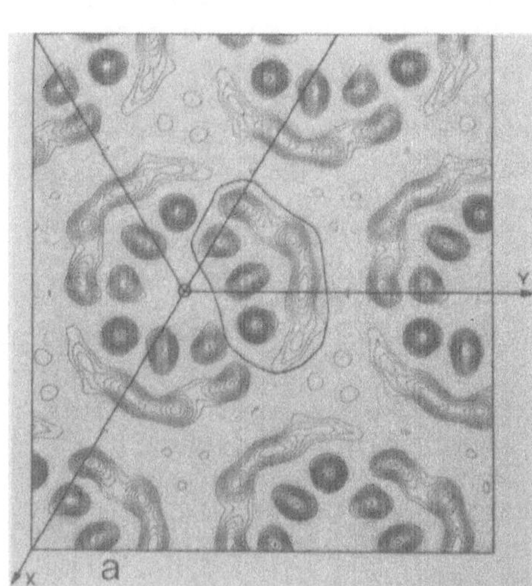

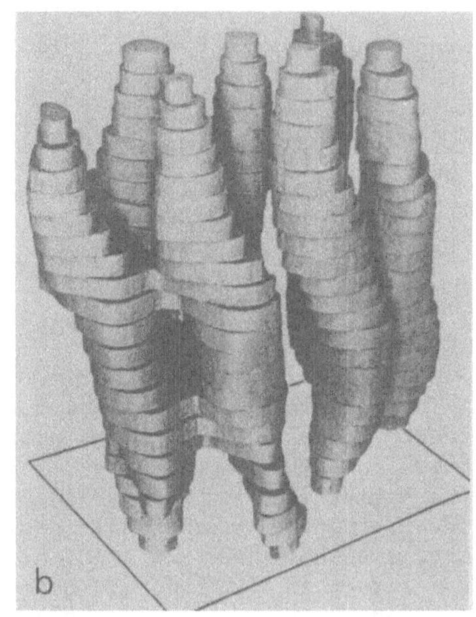

Fig. 16. Structure of the purple membrane protein. (a) Contour map of the projected structure at 7 Å resolution. The unit cell side is 62 Å and the envelope of a single protein molecule is indicated. (b) Three dimensional model showing seven rod-shaped segments, interpreted as α-helices, running approximately normal to the membrane.

in electron diffraction patterns, which are not affected by the transfer function, while the phases were averages obtained from differently defocussed micrographs, for which the zeros in the transfer function occur at different spatial frequencies. Although the effects of the contrast transfer function cannot be seen directly in the images of purple membrane, they can be seen in high resolution images of radiation insensitive inorganic crystals[34].

For three dimensional reconstruction[35] a series of two dimensional images were collected by tilting specimens through different angles. The Fourier transform of the crystalline layer is a two dimensional lattice of continuous lines of intensity pointing in a direction perpendicular to the plane. Each tilted image when Fourier transformed gives a plane section through this lattice of lines, so enough tilted views must be included to sample the lines sufficiently densely out to the required Fourier cut-off. By using 18 membranes, selected as the best preserved from several hundred, tilted by angles up to 57°, a map was computed to about 7 Å (Fig. 16(b)). For this particular specimen the missing cone of data did not present a serious problem, as X-ray diffraction showed the transform to be weak in this region.

The map shows strong rod-like features, identified as α-helices, running roughly normal to the plane of the membrane. The likely molecular boundary is indicated in Fig. 16(a) and the molecule shown in Fig. 16(b) has been cut out accordingly. In the molecular packing around the 3-fold axis there is an inner ring of nine helices about 10 Å apart and an outer ring of twelve helices which surround the inner nine. The outer helices are slightly more inclined than the inner ones and the direction of tilting is consistent with that expected for interlocking of amino acid side chain from adjacent helices.

8. CONCLUSION

The examples discussed above illustrate the power of electron microscopy combined with image processing in solving a range of problems in structural biology. They emphasise the importance of having symmetrical or crystalline specimens both for assessing and for averaging the resulting images. They also show that it is important to have specimens differing in defined ways, so that features seen in the low resolution reconstructed images can be identified. The study of purple membrane holds promise of being able to see the high resolution internal details of molecules, if suitable specimens can be prepared. This to date most sophisticated application of "electron crystallography" is a far cry indeed from the simple visual interpretation of images of stained or shadowed material, which characterised the early applications of electron microscopy in biology.

References

1. R.A. Crowther and A. Klug, Structural analysis of macromolecular assemblies by image reconstruction from electron micrographs, Ann. Rev. Biochem. 44:161 (1975).
2. J.E. Mellema, Computer reconstruction of regular biological objects, in: "Computer processing of electron microscope images", P.W. Hawkes, ed., Springer-Verlag, Berlin (1980).
3. H.P. Erickson and A. Klug, Measurement and compensation of defocussing and aberrations by Fourier processing of electron micrographs, Phil. Trans. Roy. Soc. Ser. B. 261: 105 (1971).
4. P.N.T. Unwin, Electron microscopy of the stacked disc aggregate of tobacco mosaic virus. II. The influence of electron irradiation on the stain distribution, J. Mol. Biol. 87:657 (1974).
5. R.C. Williams and H.W. Fisher, Electron microscopy of tobacco mosaic virus under conditions of minimal beam exposure, J. Mol. Biol. 52:121 (1970).
6. P.N.T. Unwin and R. Henderson, Molecular structure determination by electron microscopy of unstained crystalline specimens,

J. Mol. Biol. 94:425 (1975).
7. R.A. Crowther, The interpretation of images reconstructed from electron micrographs of biological particles, Proc. 3rd John Innes Symp. p.15 (1976).
8. A. Klug and J.E. Berger, An optical method for the analysis of periodicities in electron micrographs, and some observations on the mechanism of negative staining, J. Mol. Biol. 10:565 (1964).
9. A. Klug and D.J. DeRosier, Optical filtering of electron micrographs: reconstruction of one sided images, Nature (Lond.) 212:29 (1966).
10. L.A. Amos and A. Klug, Image filtering by computer, Proc. 5th European Congr. on Electron Microscopy p. 580 (1972).
11. R.A. Crowther and L.A. Amos, Harmonic analysis of electron microscope images with rotational symmetry, J. Mol. Biol. 60:123 (1971).
12. A. Klug and J.T. Finch, Structure of viruses of Papilloma-Polyoma type. I. Human wart virus, J. Mol. Biol. 11:403 (1965).
13. A. Klug and J.T. Finch, Structure of viruses of the Papilloma-Polyoma type. IV. Analysis of tilting experiments in the electron microscope, J. Mol. Biol. 31:1 (1968).
14. D.J. DeRosier and A. Klug, Reconstruction of three dimensional structures from electron micrographs, Nature (Lond.) 217:130 (1968).
15. R.A. Crowther, D.J. DeRosier and A. Klug, The reconstruction of a three dimensional structure from projections and its application to electron microscopy, Proc. Roy. Soc. Lond. A. 317:319 (1970).
16. D.J. DeRosier and A. Klug, Structure of the tubular variants of the head of bacteriophage T4 (Polyheads). I. Arrangement of subunits in some classes of polyheads, J. Mol. Biol. 65:469 (1972).
17. U.K. Laemmli, L.A. Amos and A. Klug, Correlation between structural transformation and cleavage of the major head protein of T4 bacteriophage, Cell 7:191 (1976).
18. Y. Kikuchi and J. King, Genetic control of bacteriophage T4 baseplate morphogenesis. III. Formation of central plug and overall assembly pathway, J. Mol. Biol. 99:695 (1975).
19. R.A. Crowther, E.V. Lenk, Y. Kikuchi and J. King, Molecular reorganisation in the hexagon to star transition of the baseplate of bacteriophage T4, J. Mol. Biol. 116:489 (1977).
20. R.A. Crowther, Mutants of bacteriophage T4 that produce infective fibreless particles, J. Mol. Biol. 137:159 (1980).
21. L.A. Amos and A. Klug, Three dimensional image reconstructions of the contractile tail of T4 bacteriophage, J. Mol. Biol. 99:51 (1975).
22. M.F. Moody, Sheath of bacteriophage T4. III. Contraction mechanism deduced from partially contracted sheaths, J. Mol. Biol. 80:613 (1973).

23. D.L.D. Caspar and A. Klug, Physical principles in the construction of regular viruses, Cold Spring Harbor Symp. Quant. Biol. 37:1 (1962).
24. R.A. Crowther, Procedures for three dimensional reconstruction of spherical viruses by Fourier synthesis from electron micrographs, Phil. Trans. Roy. Soc. Ser. B. 261:221 (1971).
25. R.A. Crowther, L.A. Amos and J.T. Finch, Three dimensional image reconstructions of bacteriophages R17 and f2, J. Mol. Biol. 98:631 (1975).
26. S.C. Harrison, A.J. Olson, C.E. Schutt, F.K. Winkler and G. Bricogne, Tomato bushy stunt virus at 2.9 Å resolution, Nature (Lond.) 276:368 (1978).
27. R.D. Kornberg, Structure of chromatin, Ann. Rev. Biochem. 46: 931 (1977).
28. J.T. Finch, L.C. Lutter, D. Rhodes, R.S. Brown, B. Rushton, M. Levitt and A. Klug, Structure of nucleosome core particles of chromatin, Nature (Lond.) 269:29 (1977).
29. A. Klug, D. Rhodes, J. Smith, J.T. Finch and J.O. Thomas, A low resolution structure for the histone core of the nucleosome, Nature (Lond.) 287:509 (1980).
30. G.A. Bentley, J.T. Finch and A. Lewit-Bentley, Neutron diffraction studies on crystals of nucleosome cores using contrast variation, J. Mol. Biol. 145:771 (1981).
31. P.N.T. Unwin, Attachment of ribosome crystals to intracellular membranes, J. Mol. Biol. 132:69 (1979).
32. W. Kühlbrandt and P.N.T. Unwin, Structural analysis of stained and unstained two dimensional ribosome crystals, in "Electron microscopy at molecular dimensions", W. Baumeister and W. Vogell, eds, Springer-Verlag, Berlin (1980).
33. W. Kühlbrandt, Structural studies on crystalline eukaryotic ribosomes. Ph.D. Thesis, University of Cambridge (1981).
34. A. Klug, Image analysis and reconstruction in the electron microscopy of biological macromolecules, Chemica Scripta 14:245 (1978-9).
35. R. Henderson and P.N.T. Unwin, Three dimensional model of purple membrane obtained by electron microscopy, Nature (Lond.) 257:28 (1975).

MAGNETIC RESONANCE PROBES OF CELLS

Ian C.P. Smith and Roxanne Deslauriers

Division of Biological Sciences
National Research Council
Ottawa, Canada K1A OR6

1. INTRODUCTION

Over the past decade there has been a revolution in magnetic resonance. At present it is possible to study a wide variety of NMR parameters of intact, live cells, tissue, and in some cases, organs or specific regions of whole bodies. These studies are possible on a time scale that is biologically reasonable. The aim of this chapter is to present an overview of the methods, illustrated by selective examples, and a reasonably thorough, but by no means comprehensive, bibliography.

2. PROPERTIES OF MEMBRANES

No organism can survive without at least one membrane. This essential membrane, which permits distinction between the inside and outside worlds, is the plasma membrane. More complex organisms may have a variety of more specialized membranes, such as those of mitochrondria, endoplasmic reticulum, or nuclei. All membranes have two major components, lipid and protein. In this section we shall deal mainly with structural aspects of the lipid components, or the influence of protein on them. Magnetic resonance studies on the protein components in membranes are in an earlier stage of development, and there will be increased activity in this area using techniques similar to those described below.

The properties of membrane lipids have been vaguely defined in terms of their "fluidity". Figure 1 attempts a decomposition of this mixed property into its fundamental components, molecular ordering and molecular mobility. On the left side of the figure, the fatty acyl chains of a lipid bilayer are represented in the all trans configuration. Next to this we see the effect of populating a single gauche conformer by rotation about a C-C single bond. The all trans chains are compact and fit tightly together in a highly-organized fashion. The presence of gauche conformers perturbs this efficient packing, making the average acyl chain separation greater, and decreasing the average thickness of the bilayer. A cooperative tilting of the entire chain has a similar effect, but it influences all positions on the chain equally. A coupled triad conformation, or kink, provides a means of minimizing the disruptive effects of gauche conformers. If cis double bonds are present in the chain, a strong local disruption of packing will be felt in the bilayer.

The packing properties of the acyl chains will obviously determine the ease with which a compound can traverse the bilayer. It is therefore important to quantitate this property. This can be done in terms of the average angle made between a specific direction in the acyl chain, say the C-D bond indicated in Fig. 1, and the average direction about which any disordering takes place, in this case the normal to the bilayer plane. The order parameter, S_{CD}, provides a convenient formalism for this purpose. It can vary from +0.125 to -0.5. An even more convenient scale, that of S_{mol}, can be used by transforming S_{CD} onto the direction of the fatty acyl long axis via the equivalent geometric function of the angle between the C-D bond and the long axis, S_{GEO}. On this scale, $S_{mol} = 1$ for a pure trans conformer, and 0 for equal populations of the trans and two gauche conformers. Strictly speaking, S_{CD} is a tensor with three principal components, but in the case of effective axial symmetry, such as pertains if rapid axial motion is present, only one component need be specified. This is

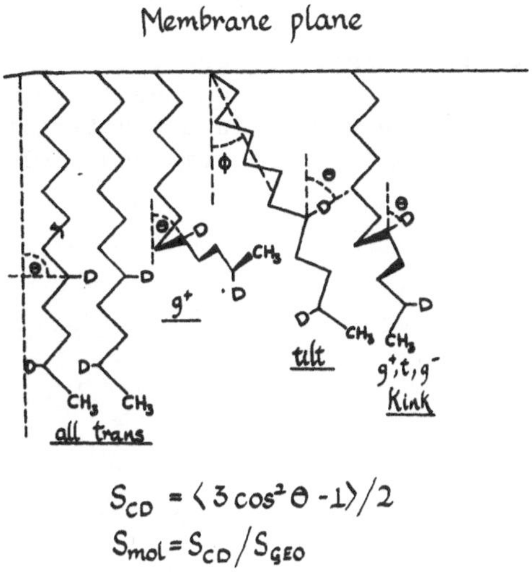

$$S_{CD} = \langle 3\cos^2\theta - 1\rangle / 2$$

$$S_{mol} = S_{CD} / S_{GEO}$$

Fig. 1. Some possible conformations of the fatty acyl chains in a membrane.

not necessarily true for lipids in the gel state, where motions about the acyl chain long axes may be slow.

A second, and essentially independent, property of the fatty acyl chains is their mobility. This may comprise axial motions, rigid body motions of the chains between different tilt angles, and interconversion between gauche and trans conformers. Clearly these rates will also determine the ease with which solutes can traverse the bilayer, since the probability of short-lived defects in the structure will increase with increasing rates of molecular motion.

Most discussions of lipids in membranes have presumed the dominance of a bilayer structure, in either the highly ordered and immobile gel phase (Fig. 2, left), or the less ordered and mobile liquid crystalline phase (Fig. 2, upper right). Recent studies have suggested the possible existence of two other morphological states of the lipids, an inverted hexagonal phase (Fig. 2, lower right) and an isotropic phase. The exact nature of the isotropic phase is still under investigation, but the hexagonal phase has been characterized by X-ray studies on model systems. Nuclear magnetic resonance has been useful in demonstrating the presence of these phases in complex lipid systems and biological membranes.

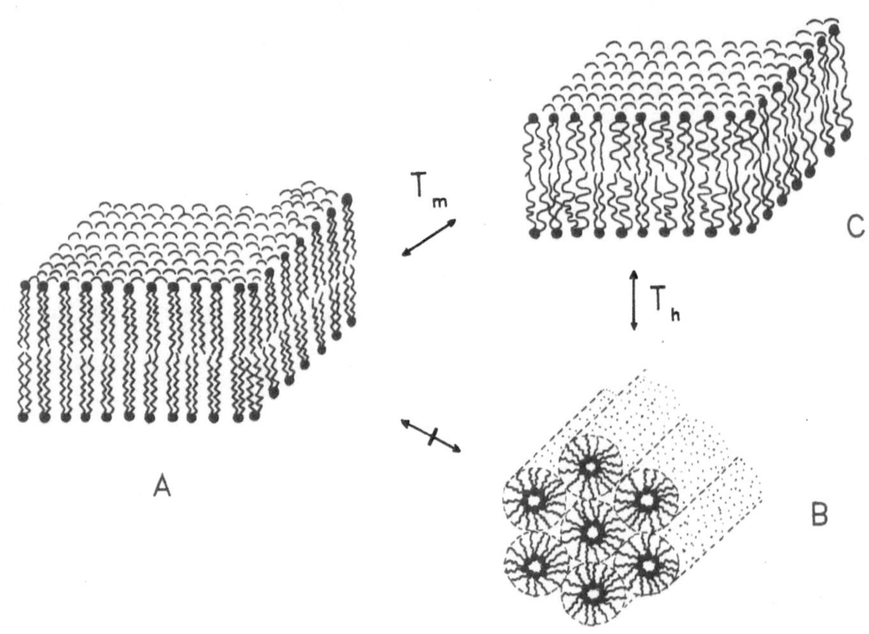

Fig. 2. Three possible morphological states of membrane lipid: bilayer gel (left); bilayer liquid crystalline; hexagonal (H_{II}). The indicated temperatures are those of the transition between the two states. A gel to hexagonal transition has not yet been observed.

3. CYTOPLASMIC PROCESSES

Biochemistry has traditionally dealt with the chemical processes which occur in living organisms. In most cases such studies involve breaking down cells in order to isolate various components, and following the chemical reactions which can occur in the subcellular constituents. A major point of such studies is to follow the means by which the potential energy of nutrients is used to provide the energy for various cellular processes. Physiology, on the other hand, is the study of function in living organisms. At the cellular level, physiology attempts to understand the functions of whole cells such as uptake and use of nutrients, response to environmental stress, growth and reproduction, cell division and differentiation. Unfortunately many of the techniques employed to study processes in living cells do not yield direct information on the chemical nature of the species involved in a given physiological response. Similarly, biochemical investigations are often difficult to carry out entirely in living cells. Physical biochemical techniques such as nuclear magnetic resonance (NMR) spectroscopy are now helping to bridge the gap between the biochemical and physiological disciplines. In many cases the responses of organisms can be followed over time courses ranging from milliseconds to days. The main advantage of NMR spectroscopy is that it provides direct information regarding the chemical nature of the species involved in a given process or response, as well as evidence of chemical and conformational changes which occur as a function of time. High-resolution NMR spectroscopy is responsive mainly to the highly mobile constituents of the cytoplasm, therefore biochemical processes which occur in solution can readily be followed in the presence of the entire complex of subcellular organelles. Conversely by appropriate manipulation of NMR excitation techniques, one can observe preferentially only the rigid subcellular constituents. Thus, NMR spectroscopy can potentially be used to monitor specific subcellular compartments within an intact cell.

4. MAGNETIC RESONANCE PROBES OF MEMBRANES

4.1 Electron spin resonance of spin labels

The magnetic resonance technique with the earliest success in membrane studies is electron spin resonance (ESR) of spin-labelled lipids or proteins[1,2]. The method takes advantage of the high detection sensitivity of ESR, and the strong dependence of ESR parameters on the angle between the applied magnetic field and the principal directions of the paramagnetic species. Since almost all membrane components are not paramagnetic, an ESR-detectable label must be added to a probe molecule. Two of the lipid probes used in the early studies are shown below. They are nitroxide-containing derivatives of stearic acid and cholesterol, both of which occur widely in biological membranes. For protein studies, spin-labelled derivatives of N-ethylmaleimide or iodoacetamide were used. The early studies are summarized in a comprehensive review[3].

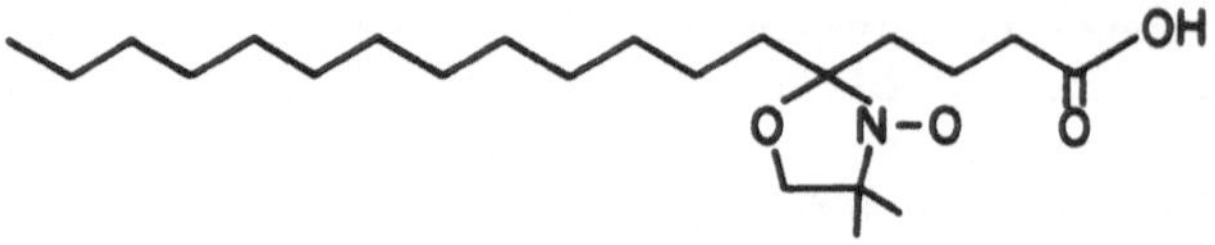

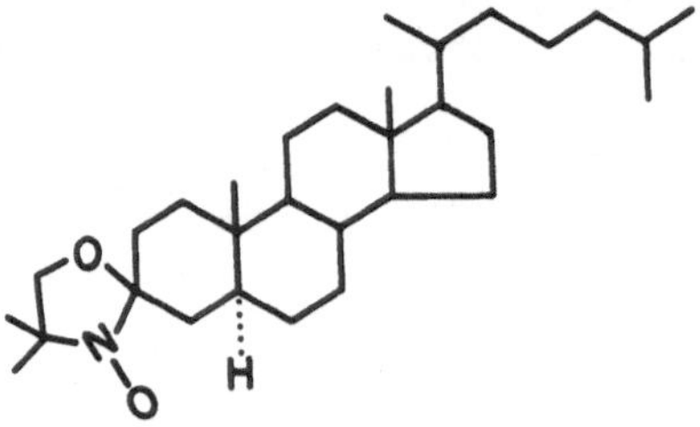

A typical ESR spectrum of a fatty acid spin label in a liquid crystalline bilayer membrane is shown in Fig. 3. The components marked $2T'_{\parallel}$ and $2T'_{\perp}$ are the partially-averaged components of the hyperfine splitting tensor for the interaction between the unpaired electron and the ^{14}N of the nitroxide moiety. The presence of these splittings indicates rapid motion within an ordered environment in the membrane. The magnitude of the order parameter may be estimated by $S_{mol} = (T'_{\parallel}-T'_{\perp})/(T_{\parallel}-T_{\perp})$, where $T_{\parallel}$ and $T_{\perp}$ are the values obtained in rigid systems with the applied magnetic field parallel or perpendicular to the π-orbital of the nitroxide. Several reviews on the use of this type of spin label in membranes contain the details of the technique[4-9] and the problems associated with it[10].

The nitroxide probes report on their local environments, and these may be strongly affected by the presence of the bulky, polar nitroxide group. A recent study has demonstrated that for spin-labelled fatty acids the absolute values of the order parameters, and even their variation due to changes in membrane structure, may be in considerable error,[11] whereas the sterol spin probe CSL gives relatively reliable information[12]. Despite these reservations, the ESR spin labels provide convenient and relatively inexpensive previews of membrane problems, thus allowing decisions to be made as to the utility of proceeding further with more difficult or expensive techniques. Their detection sensitivity is exceeded only by that of fluorescence spectroscopy, which suffers from many of the same limitations as to fidelity of response due to the need for labelling with chromophores or the use of extrinsic probes.

4.3 Carbon-13 magnetic resonance

For the study of biological membranes ^{13}C has several advantages over ^{1}H: its large range of chemical shifts (ca. 200 ppm) makes distinction of separate components easier; its low natural abundance allows the possibility of enrichment of specific components without interference from the spectra of other components; its simple relaxation mechanism (usually ^{1}H-^{13}C dipole-dipole)

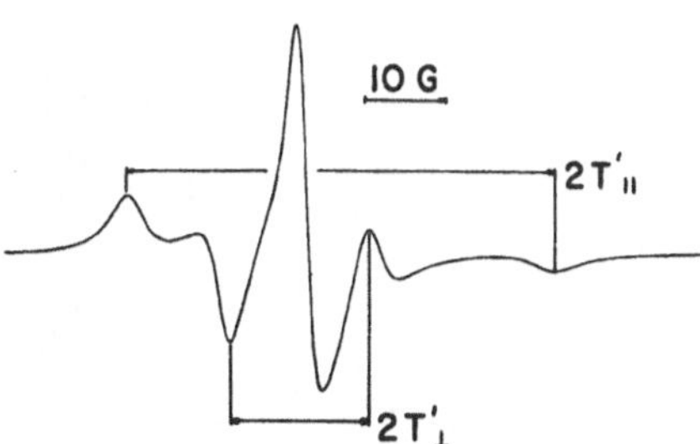

Fig. 3. The ESR spectrum (9 GHz) of 5-doxyl stearic acid in multilamellar dispersions of egg phosphatidylcholine at 25°C. The separations indicated are those usually used in estimates of the order parameter for the spin label.

makes possible the interpretation of relaxation times in terms of rates of individual motions[22].

Lipids specifically-enriched in ^{13}C have been used to simplify the spectra of membranes, particularly in the region of the methylene resonances where differences in chemical shift for many positions of the fatty acyl chains are very similar[23]. However, this is a time-consuming and expensive process. An alternative procedure is the growth of microorganisms on acetate enriched at either the carboxyl or methyl carbon atoms[24-26]. Figure 4 shows the ^{13}C spectrum of the yeast-like fungus Aureobasidium pullulans grown on acetate-$^{13}C_2$. Despite the relatively low frequency of the spectrometer, excellent resolution of many of the even-numbered carbons of the membrane fatty acyl chains is obtained. A corresponding spectrum, with the odd carbon atoms enriched, was obtained from cells grown on acetate-$^{13}C_1$[25]. The narrow resonances and excellent resolution are uncharacteristic of biological membranes, and are due to the high degree of unsaturation of the fatty acids which leads to a high fluidity (low degree of order and high mobility). They allow measurement

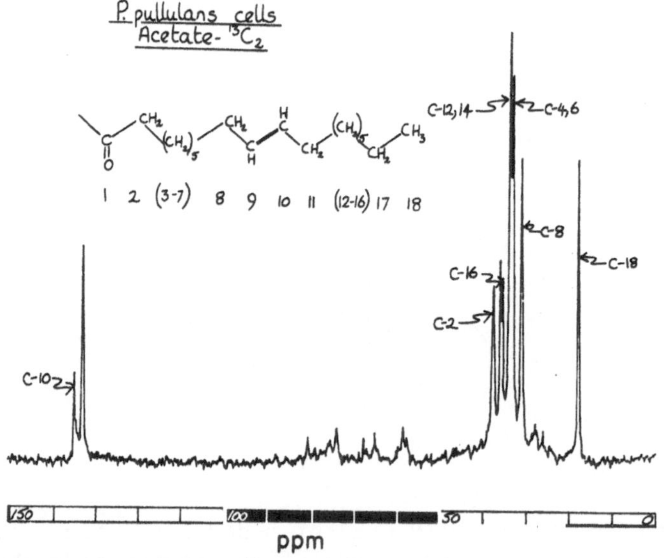

Fig. 4. ^{13}C NMR spectrum (20 MHz) of packed whole cells of A. pullulans in D_2O at 30°C. The cells were grown on media enriched in sodium acetate-$^{13}C_2$.

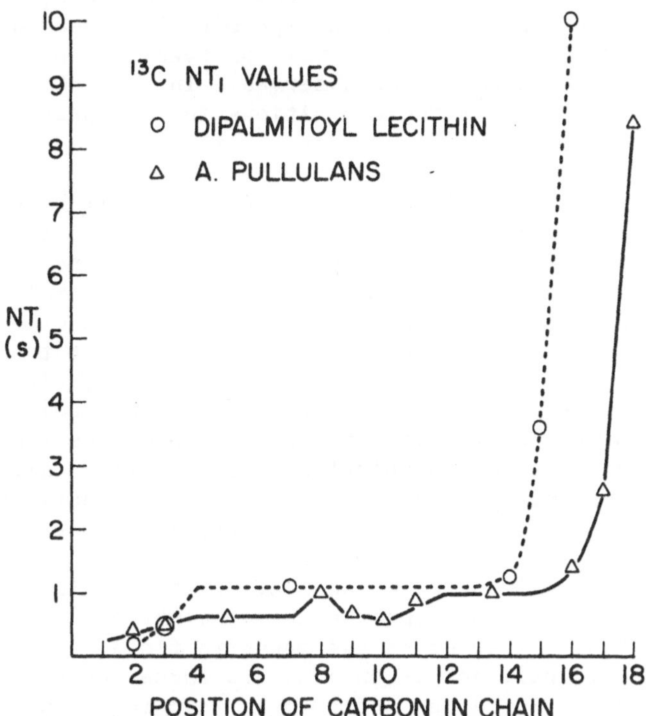

Fig. 5. Variation of NT_1, where T_1 is the spin-lattice relaxation time of ^{13}C and N is the number of directly-attached protons, with position in the oleoyl chains of _A. pullulans_ (30°C)[25] and dipalmitoyl phosphatidylcholine (52°C)[23]. For the latter, only one measurement was possible for positions 4-13 due to overlapping of resonances.

of spin-lattice relaxation times, T_1, of most of the carbon atoms of the acyl chains. In Fig. 5 these times, multiplied by N, the number of protons on the particular carbon atom, are compared with those for sonicated vesicles of dipalmitoyl phosphatidylcholine, obtained from ^{13}C at natural abundance. To a reasonable level of approximation, the NT_1 values can be taken as proportional to the mobility of the carbon atoms of the oleoyl chains. Note that this mobility is relatively constant for most positions, increasing significantly only for the last two positions of the chains. A similar mobility profile has been obtained from the T_1 values of deuterium in specifically-enriched dipalmitoylphosphatidylcholine, although a gradual increase in T_1 was observed from positions 10 to 15[27]. This increase was not observable in the ^{13}C study, due to lack of resolution for these positions in the spectra of ^{13}C at natural abundance[23].

The ^{13}C spectra of Micrococcus freudenreicheii[26], grown on acetate enriched in ^{13}C at either C_1 or C_2, are more typical of the spectra of biological membranes.[26,28,29] Despite the high resolution available in the spectra of aqueous dispersions of the isolated lipids, and the specific enrichment of odd or even carbon atoms in the fatty acyl chains, it is very difficult to assign resonances in the membrane spectra with confidence, Fig. 6. Even less useful spectra were obtained from Escherichia coli[26]. A major source of the difficulty is the breadth of the individual resonances - this is due to relatively slow motion. Although most of the broadening due to ^{13}C-^{1}H dipolar interactions can be removed by high power (dipolar) decoupling, it appears that some contributions from residual chemical shift anisotropy persist. These can only be eliminated by sonicating the sample to form small vesicles, or by spinning the intact membranes at the "magic" angle of 54.7° with respect to the magnetic field. The former technique is evidently not of much use to those interested in intact membranes. While the magic-angle spinning technique has been used with considerable success for solid synthetic polymers[30,31] and proteins[32], it is difficult to apply to aqueous suspensions of membranes. One such study, on model membranes containing only 50% by weight water, has been reported[33].

Thus, although it may appear that there is much literature on ^{13}C NMR of membranes, very few are on intact systems, and only vague conclusions have been drawn from these. It is hoped that the spinning instability problems with membrane samples will soon be overcome, as the cross-polarization, magic-angle spinning technique appears to be the only way to achieve meaningful high resolution ^{13}C spectra of intact membranes.

4.4 Nitrogen magnetic resonance

There are two magnetic isotopes of nitrogen: ^{14}N is 99.63% abundant, has a detection sensitivity of 1×10^{-3} with respect to ^{1}H, and is spin 1 with a large quadrupole moment; ^{15}N is 0.37% abundant, has a detection sensitivity of 1×10^{-3} with respect to ^{1}H, and is spin 1/2 with a negative magnetogyric ratio. Until 1980, virtually no use had been made of these nuclei in the study of membranes.

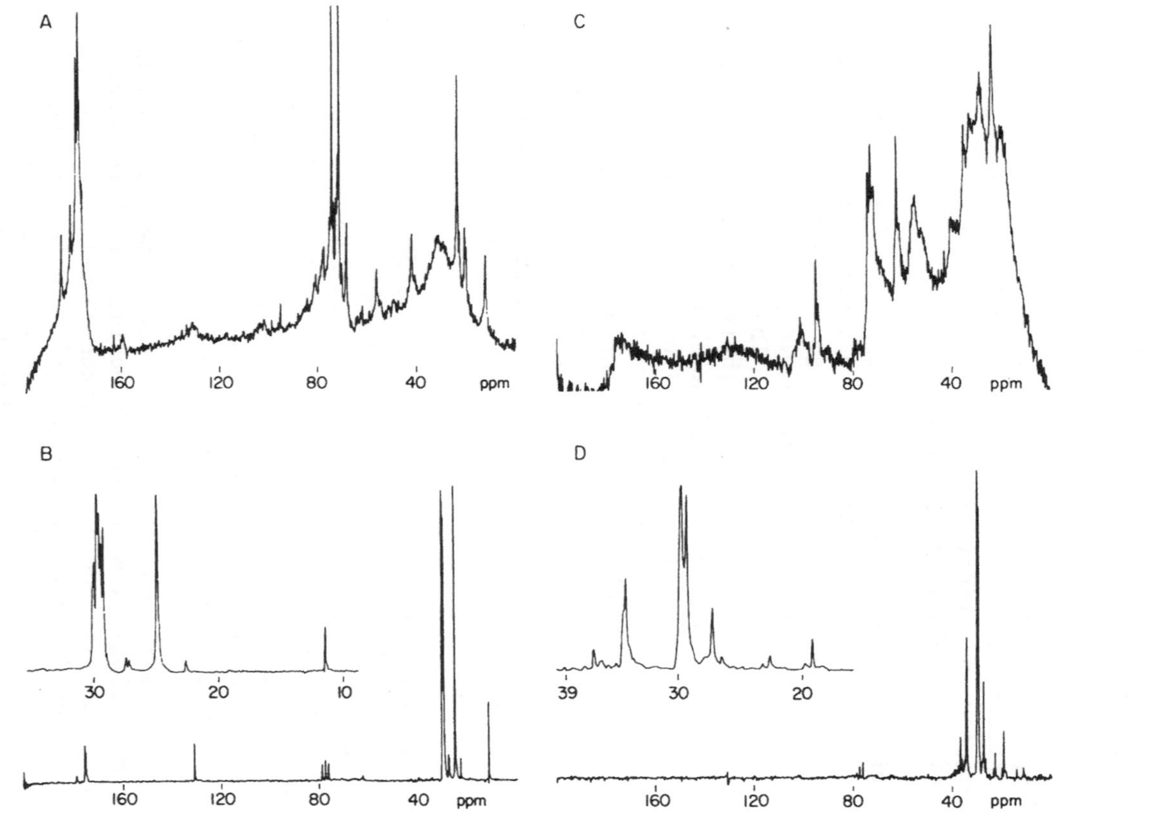

Fig. 6. ^{13}C NMR spectra (25 MHz) of packed cells (A and C), or dispersions of extracted lipids (B and D), of M. freudenreicheii grown on media enriched in sodium acetate-$^{13}C_2$ (A and B) or -$^{13}C_1$ (C and D)[26]. Samples were studied in D_2O at 30°

The principal problem with ^{14}N is its large quadrupole moment, which is 26 times greater than that of ^{2}H (vide infra). Thus, although it is readily detected for small molecules in solution, its resonances are usually very broad. This problem is minimized if the ^{14}N is in a highly symmetrical environment, such as the head group quaternary nitrogen of phosphatidylcholine. Recently a high resolution ^{14}N NMR study of several unilamellar lecithins dispersions was reported[34]. The relaxation times, T_1 and T_2, gave indications of the degree of mobility of the head-group segment, and an estimate could be made of the diffusion coefficient of the lipid in the bilayer.

The partially-averaged quadrupole splitting of the ^{14}N of phosphatidylcholine has been utilized as a monitor of the degree of ordering of the choline moiety, although no estimate of the order parameter could be made without knowledge of the quadrupole coupling constant in this environment[35]. The quadrupole splittings, Fig. 7, were of the order of 10–12 kHz; their temperature dependence for dipalmitoyl phosphatidylcholine showed a small discontinuity at the temperature of the gel to liquid crystal phase transition. Thus far no spectra have been reported for biological membranes - they would presumably be rather complex due to the high population of protein, and to the various nitrogen-containing lipid headgroups. Nitrogen species with lower degrees of symmetry than that of choline will yield very large quadrupole splittings, and will have very short transverse relaxation times, T_2, making them very difficult to detect.

Nitrogen-15 suffers from a low natural abundance in addition to its low detection sensitivity. Using enrichment in ^{15}N, it has been possible to observe a resonance from E. coli cells, which was attributed to phosphatidylethanolamine[36]. Recently we (R.A. Byrd and I.C.P. Smith, unpublished) have observed the ^{15}N resonance of egg phosphatidylcholine in unsonicated multibilayers, using the techniques of ^{1}H-^{15}N cross-polarization and magic angle spinning. The spectrum required about twelve hours of signal averaging.

Overall it appears that ^{14}N NMR will have some applications in the study of biological membranes, especially if techniques can be developed to visualize only the nitrogen-containing compounds of interest e.g. quadrupole echo sequence[35] with varying delay between the 90° pulses, ^{1}H-^{14}N cross-polarization with varying contact times. Nitrogen-15 can be used with enrichment, and it is possible that the observed residual chemical shift anisotropy will yield information related to the ordering of the head group, to complement that obtained from the residual quadrupole splittings observed for ^{14}N.

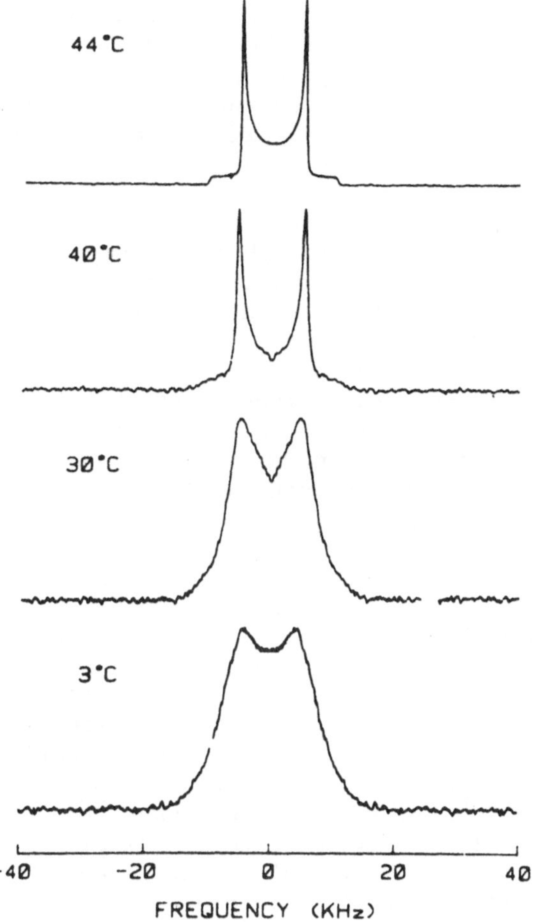

Fig. 7. ^{14}N NMR spectra (19.4 MHz) of aqueous dispersions of dipalmitoyl phosphatidylcholine at the temperatures indicated[35].

4.5 Phosphorus-31 magnetic resonance

A great deal of insight into the phase behaviour of lipids in membranes, and into the degree of organization of lipid headgroups, has been gained by ^{31}P NMR. Many of the details of the method, and examples of its use, are given in two recent reviews[37,38]. The basis of the method is the anisotropy of the chemical shift of ^{31}P; the relatively high detection sensitivity of this nuclide (6%

of that of ^{1}H) is an added advantage. Figure 8 shows the three principal components of the ^{31}P chemical shift tensor (σ_{11}, σ_{22}, σ_{33}) of a phosphodiester group, as well as the influence of a rapid axial rotation. The latter serves to average two of the components to yield an effective axial symmetry. The hypothetical ^{31}P spectrum of such a species, where the only available motion is the rapid axial rotation is shown at the bottom of the figure. Note that the pattern is right-handed, and has a total width of about 125 ppm. In membranes the phosphodiester moiety of the lipid headgroups is usually somewhat disordered, with rapid interconversions between the allowed conformations. This results in further, axially symmetric, averaging of tensor components, so that a similar pattern, but with decreased width, is obtained. The width of the pattern is indicative of the state of disorder of the phosphodiester species. In the limit of total disorder with retention of rapid motion, a single resonance would be obtained at the value of the isotropic chemical shift (σ_{iso} = ($\sigma_{11} + \sigma_{22} + \sigma_{33})/3$.

The hexagonal (H_{11}) phase of lipids yields a characteristic ^{31}P power pattern, Figure 9. Although the same axial averaging processes discussed above for the lamellar phase take place, leading to effective components of the chemical shift tensor parallel and perpendicular to the long axes of the fatty acyl chains ($\sigma^{L}_{\parallel}$ and $\sigma^{L}_{\perp}$), another averaging mechanism is possible. Rapid translational diffusion of the lipids around the cylindrical axes of the structures leads to an effective component perpendicular to the cylinders, $\sigma^{H} = (\sigma^{L}_{\parallel} + \sigma^{L}_{\perp})/2$. The component along the cylinder axis is identical to $\sigma^{L}_{\perp}$. Thus, although axial symmetry still exists, the magnitude and sign of the axial component is changed, leading to the left-handed pattern shown in Fig. 9. The overall width of the pattern is one-half of that for the corresponding lamellar phase. For both types of system, if the rates of the various averaging processes become slow on the scale of the chemical shift anisotropies, more complex spectra will result.

Lamellar and hexagonal ^{31}P NMR patterns have been observed in a wide variety of model and biological membranes[37,38]. Due to the intrinsic asymmetry of the phosphodiester chemical shift tensor, two order paramters are required to specify the average orientation of this region. Even with the use of order parameters for the neighbouring carbon segments, obtained via ^{2}H NMR, this calculation is ambiguous,[39,40] and the data from neutron and X-ray diffraction studies must be used to construct a suitable model. Nonetheless, valuable insight into the nature of the lipid phase, and its response to temperature or other perturbations, is obtainable directly from the spectra. Sophisticated spectral simulations can also yield estimates of the rates of the various motional processes[41].

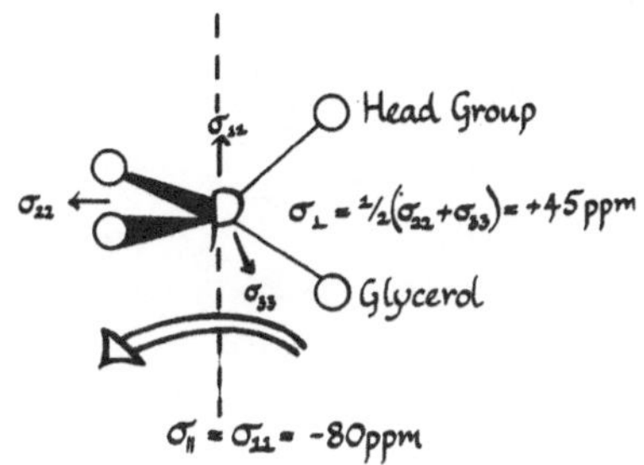

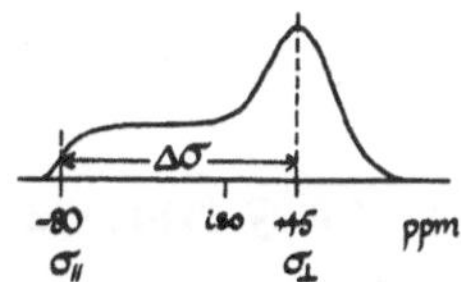

Fig. 8. Representation of the principal axes of the chemical shift tensor for the phosphodiester group of a lipid σ_{11}, σ_{22} and σ_{33} are the values of the tensor for the static situation; in the presence of rapid axial rotation as shown, averaging of σ_{22} and σ_{33} to an effective, axially symmetric tensor occurs. The expected ^{31}P NMR spectrum is shown at the bottom of the figure; $\Delta\sigma$ is the measured effective chemical shift anisotropy.

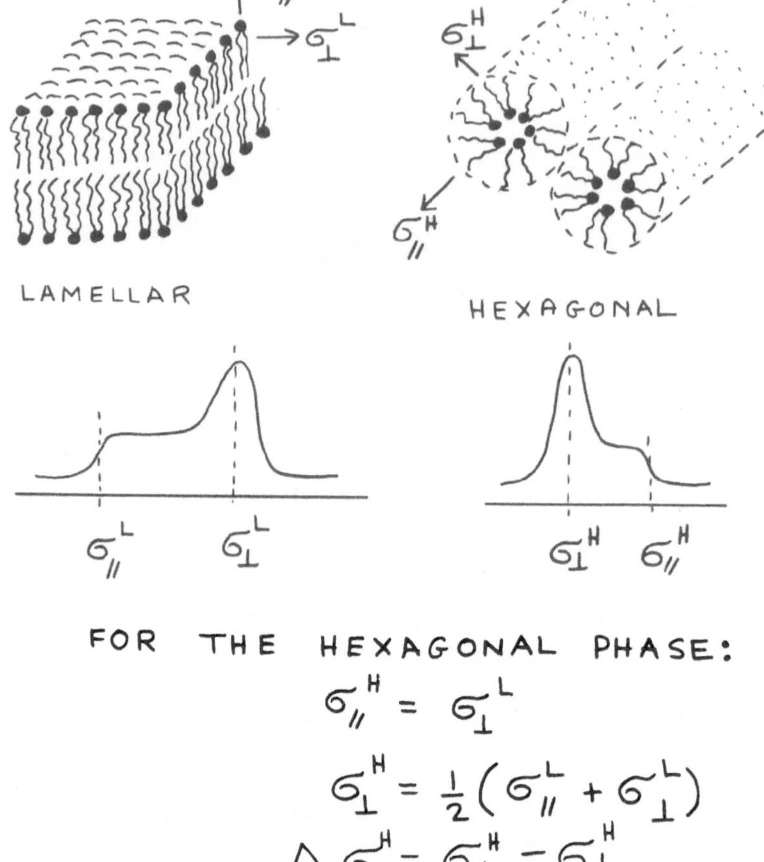

$$\sigma_{\parallel}^{H} = \sigma_{\perp}^{L}$$

$$\sigma_{\perp}^{H} = \frac{1}{2}(\sigma_{\parallel}^{L} + \sigma_{\perp}^{L})$$

$$\Delta\sigma^{H} = \sigma_{\parallel}^{H} - \sigma_{\perp}^{H} = \frac{1}{2}(\sigma_{\perp}^{L} - \sigma_{\parallel}^{L})$$

$$\Delta\sigma^{H} = -\frac{1}{2}\Delta\sigma^{L}$$

Fig. 9. ^{31}P NMR spectrum expected for the lamellar (bilayer) and hexagonal phases of lipids. $\sigma_{\parallel}$ and $\sigma_{\perp}$ refer to the chemical shifts due to the magnetic field parallel or perpendicular, respectively, to the long axes of the phospholipids.

Natural membranes often lead to ^{31}P NMR spectra more complex than those discussed so far. Such is the case with the purple membrane of the extremely halophilic bacterium _H. cutirubrum_. Although simple in containing a single protein, bacteriorhodopsin, the membrane comprises several phosphorus-containing lipids, the major one of which (85%) is phosphatidylglycerophosphate (PGP). PGP contains a phosphomonoester and a phosphodiester moiety. There have been conflicting views in the literature regarding the presence of a lipid phase transition over the range 5-60°C for these purple membranes.[42,43]

The ^{31}P NMR spectrum of the purple membrane of _H. cutirubrum_ is shown in Fig. 10.[44] Note that there are broad components to either side of the intense central peak. At lower signal-to-noise ratios one might conclude that this was a left-handed spectrum due to the presence of lipid in a hexagonal phase. However, the peculiar shape is due to different lamellar powder patterns from the phosphomonoester and diester species, the latter being three times as wide as the former. This was confirmed by application of specific saturation to the component on the right side of the spectrum, using the DANTE technique as applied to membranes by De Kruijff et al.[45]. If the saturating field is maintained long enough for all species responsible for a particular subspectrum to rotate through the angle corresponding to the spectral component saturated, the entire spectrum due to that species will be saturated and removed from the composite spectrum. In this case, the remaining spectrum resembled the dashed component in the lower part of Fig. 10. Subtraction of this component spectrum from the composite spectrum, gave the second component shown dotted in Fig. 10. This very useful approach to complex spectra has also been used recently to distinguish components due to phospholipid and phosphonolipid in the membranes of _Tetrahymena sp._[46].

Measurement of the effective chemical shift aniostropy of the individual components in complex spectra such as that of Fig. 10 can only be done by spectral simulation, as in the dashed and dotted components shown in the lower portion of the spectrum. A complete analysis would also involve the rates of motion[41], which when sufficiently slow can affect the widths of the patterns. Plots of the temperature dependences of the effective chemical shift anisotropies for the two phosphate-containing moieties showed no evidence for discontinuities due to lipid phase transitions over the range 5-60°C, in agreement with differential scanning calorimetry data[42,44]. Thus, the discontinuities seen in temperature dependences of ^{1}H and ^{13}C NMR data[43] were due to other sources, such as a variable activation energy for rotational motion of the chains. This apparent inconsistency underlines the need for more than one physical approach to a problem, and for the most practicably rigorous analysis of data.

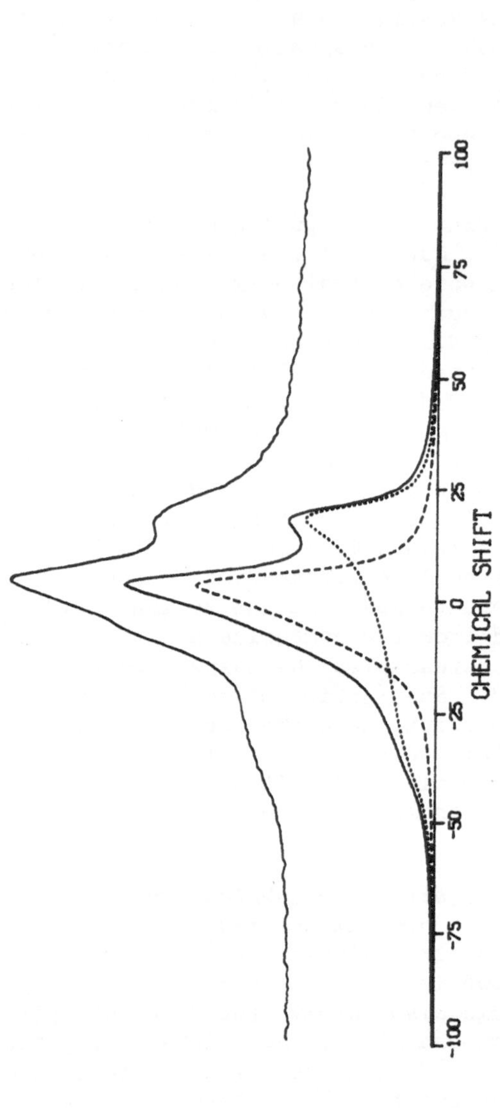

Fig. 10. ^{31}P NMR spectrum (121.5 MHz) of the purple membrane from H. cutirubrum at 15°C (upper spectrum), and simulation (lower solid spectrum) in terms of contributions from the phosphomonoester (dashed spectrum) and diester (dotted spectrum) moieties of PGP.[44]

4.6 Deuterium magnetic resonance

Deuterium NMR of specifically-enriched lipids has provided a wealth of information on membranes during the past eight years[39,47-65]. It suffered in the early days from the low detection sensitivity of ^2H (1% of that of ^1H) and the difficulty in observing broad spectra with short transverse relaxation times. The increasing use of high field instruments, improved pulse sequences for the observation of broad powder spectra, and the greater availability of instruments with high radiofrequency power and rapid data acquisition have largely overcome the earlier difficulties. ^2H NMR spectra can now be observed for the membranes of suspensions of live cells in several minutes of signal accumulation.

The usefulness of ^2H in membrane studies derives from its quadrupole moment. In ordered systems two transitions of unequal energy are possible, and their frequencies depend upon the angle between the C-D bond and the applied magnetic field, Fig. 11. In a slowly tumbling membrane, regardless of how ordered the lipids are, all angles between the C-D bond and the magnetic field will be populated. The resulting ^2H NMR spectrum will be a superposition of the many different spectra, with different values of the splitting between the two transitions. The result, shown at the bottom right of Fig. 11, is a powder spectrum with well-defined peaks and shoulders. The peaks represent the splitting for the 90° angle, and the separation between the shoulders, twice the former splitting, is that for the 0° angle. If the molecular order parameter at the labelled position were 1, or if no rapid motion within a partially-ordered environment took place, the separation between the peaks, Dq, would be roughly 128 kHz. However if, as in lipid systems, rapid motion within a disordered environment occurs, partial averaging of the quadrupole splitting results. If the rapid motion is axially symmetric, we can describe the ordering by a single order parameter,

$$S_{CD} = \frac{3}{4}\frac{e^2qQ}{h}\,Dq, \text{ where } \frac{e^2qQ}{h}$$

is the quadrupole coupling constant for deuterium in that chemical environment (ca. 170 kHz for attachment to an sp^3 carbon). As the rate of motion decreases, spectra with more complex shapes are observed[66].

Deuterium NMR studies have now been performed on a variety of biological membranes; those of <u>Acholeplasma laidlawii</u>[50,56,59,62,65,72-75], <u>E. coli</u>[62,63,67,68,71], sarcoplasmic reticulum[70], red cells[69], <u>Lactobacillus plantarum</u> (R.A. Byrd, A. Joyce, and I.C.P. Smith, unpublished), and the purple membrane of <u>Halobacterium cutirubrum</u>

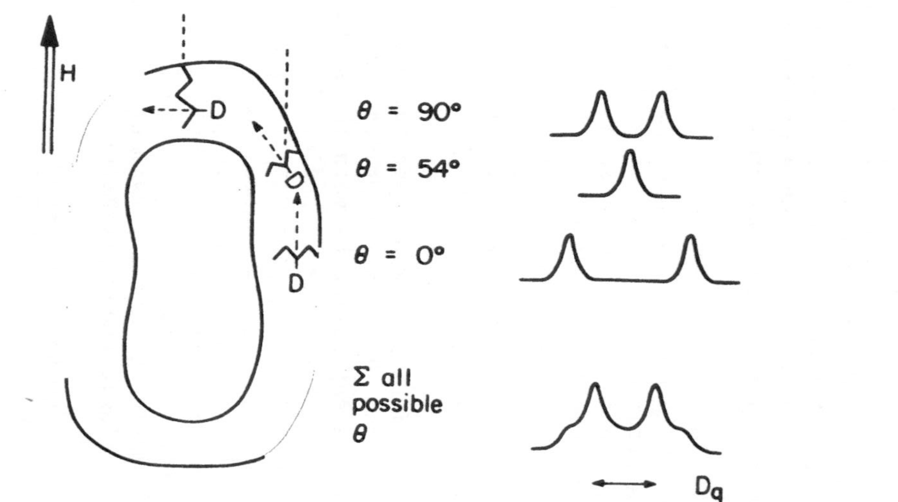

Fig. 11. Schematic of a large membrane showing some of the possible angles (Θ) between the applied magnetic field (H) and the carbon-deuterium bond. The resultant powder spectrum for the system is the sum of those for all possible values of Θ.

(I.C.P. Smith, I. Ekiel, and A. Lewis, unpublished). The most detailed results have been obtained from A. laidlawii, so we shall use this system to outline the method.

Acholeplasma laidlawii B is an ideal organism for the study of membrane lipids by ^{2}H NMR. It has only one membrane, the plasma membrane, and when grown on a medium of controlled fatty acid content, will incorporate mainly the exogenous fatty acid with little change in length, degree of unsaturation, or position of label. Addition of the egg white protein avidin serves to complex all the coenzyme biotin required for de novo fatty acid synthesis, and incorporation of a given fatty acid to levels near 100% can be achieved[75]. Ghost membranes can be easily prepared by hypotonic lysis, and freeze-dried for storage. The ^{2}H NMR spectra of fresh cells, lysed membranes, and rehydrated freeze-dried cells are very similar, a convenience when one is subject to the vagaries of complex electronic equipment.

Some ^{2}H NMR spectra of A. laidlawii membranes, enriched to 75% in 13-d_2-palmitic acid, are shown in Fig. 12. At 45°C the lipids are in the liquid crystalline state, and a typical powder pattern is observed[72]. Measurement of Dq for this and other positions leads to the variation of the order parameter with position along the chain[56]. S_{mol} is of intermediate magnitude (ca. 0.4) and constant for the first ten positions from the carboxyl group, decreasing rapidly thereafter to a very small value (ca. 0.05) for the terminal methyl group. This behaviour parallels that found earlier for the model membranes of egg lecithin[52] and dipalmitoyl-[58] and dimyristoylphophatidylcholine[77]. Decreasing the temperature through the region of the calorimetric liquid crystal-gel phase transition (43-27°C) results in the appearance of a second spectrum on the extremities. By 25°C essentially all the spectrum due to the liquid crystalline component has disappeared. Those spectra demonstrated the advantage of high power NMR spectrometers, and the use of the quadrupole echo technique[78]. Previously the onset of the gel phase was noted in ^{1}H, ^{2}H and ^{13}C spectra by the disappearance of a measurable spectrum. The first significant conclusion of this study is that the gel and liquid crystalline states of the lipid coexist throughout the transition, i.e. rather than a state with intermediate properties, the two extreme states coexist in slow exchange. This indicates that during the transition the membrane is extremely heterogeneous. The second conclusion concerns the properties of the gel state lipid - it is not rigid. The approximate splitting of 60 kHz is that expected for a highly ordered chain ($S_{mol} \simeq 1$) with rapid motion about the long axes of the fatty acyl chains. Decreasing the temperature further results in a decrease in the rate of this motion, until a spectrum of width ca. 120 kHz is observed at 3°C. Accurate measurement of this latter type of spectrum is difficult

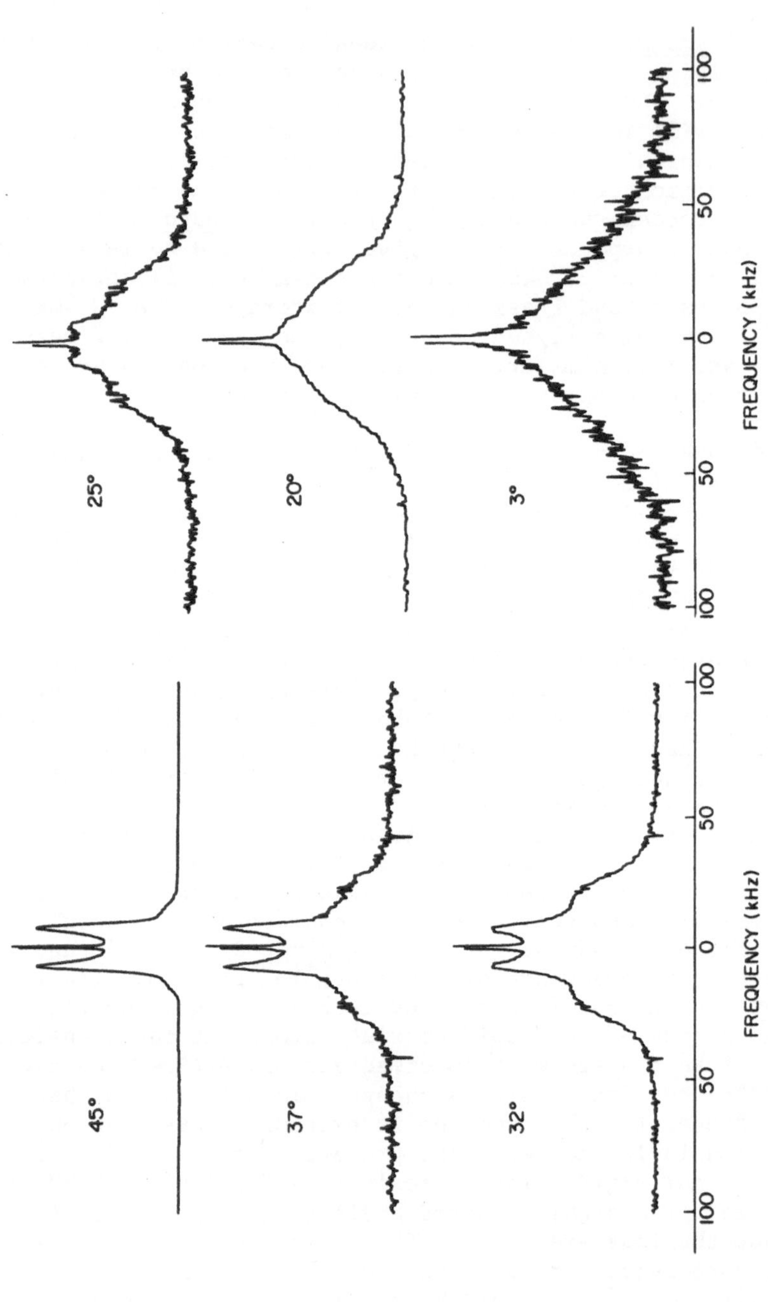

Fig. 12. ^{2}H NMR spectra (34.4 MHz) of A. laidlawii membranes, enriched in 13,13-d_2-palmitic acid, at the indicated temperatures[72].

due to insufficiently uniform distribution of radiofrequency power across the spectrum, but the situation can be improved by the use of pulses of less than 90° with a concomitant loss in signal-to-noise ratio. Measurement of the second moments of the spectra (vide infra) confirmed the interpretation in terms of decreasing rates of axial motion with decreasing temperature.

Extension of this type of study to A. laidlawii membranes containing 90% or more of myristic acid (C14:0), using avidin in the medium, have been made (H.C. Jarrell, R. Deslauriers, K.W. Butler, and I.C.P. Smith, unpublished). They yielded similar order-position profiles, and a somewhat narrower range for the phase transition. The various moments of the spectra were utilized to calculate the fraction of lipid in the liquid crystalline state, according to the formalism of Jarrell et al.[79]. The variation of this fraction, and the moment-derived Δ_2 parameter which is a measure of heterogeneity in lipid order[68], is shown in Fig. 13. The phase transition temperature estimated calorimetrically is indicated by T_c. Note that the heterogeneity in lipid order does not necessarily reach a maximum at either T_c or at the temperature where equal amounts of liquid crystalline and gel state lipid coexist; the position of the maximum depends upon the properties of both types of lipid.

Davis et al. used perdeuterated palmitic acid to explore the influence of cholesterol incorporated into the membranes of A. laidlawii[73]. Due to the breadth of the powder patterns for different positions, not all quadrupole splittings could be observed. A novel use was made of the spectral moments to calculate the order-position profile and its response to cholesterol. The order parameters increased for most positions on incorporation of cholesterol, with the increases being greatest and constant for the first 10-12 carbon segments, in agreement with earlier conclusions for the egg lecithin-cholesterol system[54].

The influence of unsaturated fatty acids on lipid ordering in A. laidlawii was studied recently by Rance et al.[74]. They incorporated specifically-deuterated oleic acid, with a 9,10 cis double bond, and measured quadrupole splittings, spectral moments, and longitudinal relaxation times, T_1. The positional variation of the quadrupole splitting, Dq, and of the relaxation time, T_1, are shown in Fig. 14. Note the large dip in Dq at carbon-10; this is mainly due to the geometry of the cis double bond. Transformation to account for this shows that the double bond is in fact highly-ordered. Some diminution of order occurs for carbons adjacent to this position. The three splittings for carbon-2 arise from two sources - the inequivalence of the sn-1 and sn-2 chains, and the inequivalence of the two deuterons at position-2 of the sn-2 chain. This is similar to behaviour observed for dipalmitoyl phosphatidylcholine[51], which was readily explained by

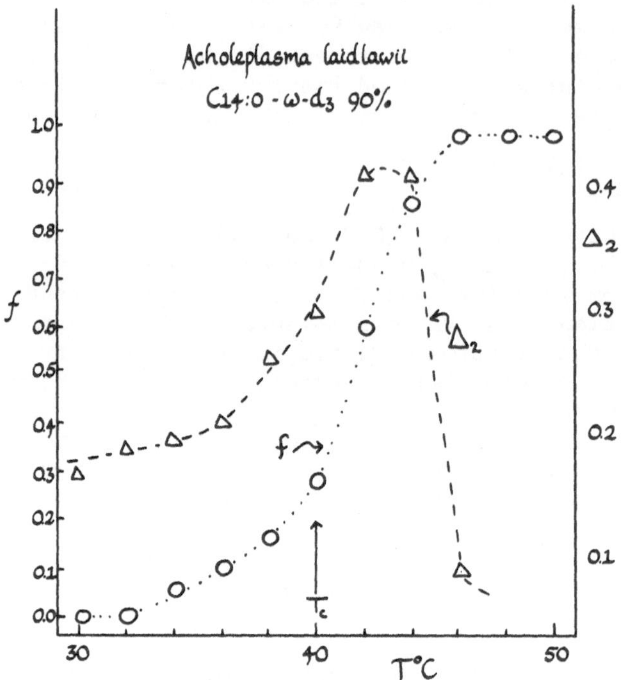

Fig. 13. Temperature-dependence of the fraction of lipids in the liquid crystalline state (f) and of the moment parameter (Δ^2) derived from the ^2H NMR spectra of A. laidlawii enriched in 14,14,14-d_3-myristic acid. The Δ^2 parameter is a measure of the mean square deviation of the order parameter.

the subsequent neutron diffraction data on this system[60], and for A. laidlawii enriched in palmitic acid[56]. The dependence of T_1 on position is mostly an indication of the degree of mobility of the chains, since the influence of order on this parameter is slight[80]. The mobility is quite constant along the chains, increasing steeply only for the last few carbon segments, in a fashion very similar to that discussed earlier for the ^{13}C relaxation in the oleate-rich membranes of A. pullulans (section 4.3, Fig. 5). The influence of cholesterol on these parameters was slight, causing an increase in ordering for carbons in the early portion of the chain, very minor increases elsewhere, and negligible changes in T_1 (M. Rance, K.R. Jeffrey, and I.C.P. Smith, unpublished).

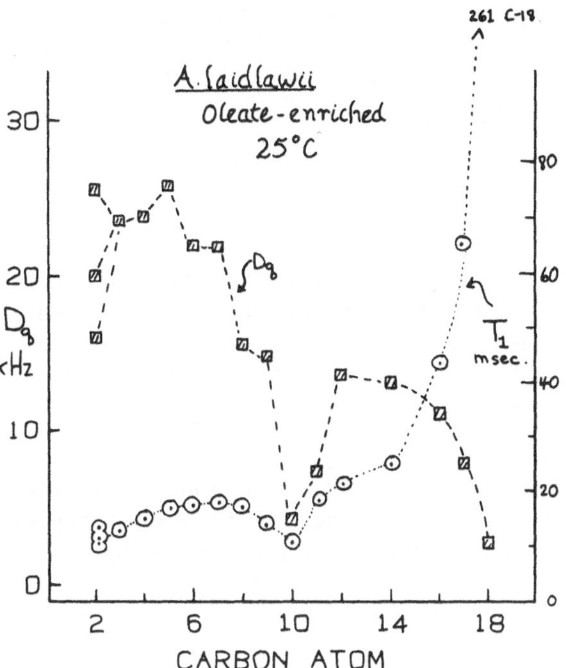

Fig. 14. Dependence on position of the quadrupole splitting (Dq) and the spin-lattice relaxation time (T_1) for specifically-deuterated oleic acid in membranes of A. laidlawii at 25°C.[74]

^{2}H NMR of membranes is now established as the most reliable monitor of the lipid properties. The subject has matured from the simple measurement of quadrupole splittings in liquid crystalline systems, to analysis of lineshapes[66,81] and relaxation rates[74,80] for the estimation of correlation times, measurement of complex gel state spectra[72,81] and the use of spectral moments[68,72,74,79] to obtain a wide variety of parameters. The synthetic pathways to most of the fatty acids[82] and head group components of interest have been established. Excellent spectrometers are now available which overcome most of the instrument problems encountered earlier. The challenge is now to do the biologically significant experiments.

5. MAGNETIC RESONANCE PROBES OF THE CYTOPLASM

5.1 Introduction

Technological advances in both NMR instrumentation and related data processing techniques in the last 5 years have allowed physical biochemists to probe the structural and metabolic characteristics of whole cells, tissues and organs as well as intact organisms.

^{31}P NMR has been most widely exploited in the study of living tissues, followed by ^{13}C and ^{1}H. ^{31}P spectra of living tissue are spectroscopically the easiest to obtain and interpret. Although ^{31}P spectra may take longer to acquire than proton spectra, the limited number of phosphorus-containing compounds in the cytoplasm renders spectral interpretation simpler. The use of complete proton decoupling further simplifies the spectra. The chemical shift range of the most common biological phosphates is 30 ppm, the chemical shift spread increases to 60 ppm if biologically-occuring phosphonates are included.

The chemical shift range of ^{13}C is over 200 pm, allowing greater dispersion of resonances than for ^{1}H. As with ^{31}P, complete proton decoupling yields one resonance for each carbon atom. As the natural abundance of ^{13}C is only 1%, carbon-carbon spin-spin couplings are generally not observed. The low natural abundance of ^{13}C allows the reseacher to use ^{13}C-enriched substrates as a probes of specific metabolic rates and pathways.

Among the less commonly employed nuclei for studies of live biological systems are ^{15}N, ^{23}Na, ^{17}O, ^{2}H and ^{39}K.

^{23}Na is a major constituent of biological systems; like ^{14}N, it has a quadrupole moment (spin = 3/2) which can lead to broad resonances. ^{17}O, which has a 5/2 spin, has been used for studies of water structure in biological systems after specific enrichment in the isotope. Similarly ^{2}H has been employed as a probe of water structure in tissues.

5.2 Proton magnetic resonance

The great potential advantages of ^{1}H NMR spectroscopy for studies of living systems lie in the sensitivity of ^{1}H as well as in the large abundance of protons in all metabolically important compounds. However, these attractive features of ^{1}H NMR are largely negated by the presence of large amounts of relatively mobile water molecules which create dynamic range problems in conventional data acquisition techniques. Various solvent-suppression techniques have been devised to circumvent the dynamic range problem created by the presence of water, thus eliminating the necessity of working with deuterated solvents. Another potential difficulty in using ^{1}H NMR for metabolic studies, resides in the fact that ^{1}H is 100% naturally abundant. It is difficult if not impossible, to follow particular substrates undergoing metabolic conversion. One approach to circumvent the problem is to use ^{13}C-enriched substrates. The natural abundance of ^{13}C being only 1%, ^{1}H NMR spectra do not usually show ^{1}H-^{13}C spin-spin coupling; the abundant isotope of carbon, ^{12}C, has 0 spin. By using a specifically ^{13}C-enriched substrate, all protons directly bonded to ^{13}C will show spin-spin couplings of approximately 145 Hz. Thus, all one needs to follow in a metabolic study is the time evolution of the coupled protons. An example of the above approach has been provided by Ogino and co-workers,[83] who followed glucose metabolism and pyruvate transport in anaerobic suspensions of *Escherichia coli*. The labelled substrate was [1-^{13}C] glucose. They determined quantitatively the relative importance of the Embden-Meyerhof pathway versus that of the pentose shunt (22%). It was further possible to evaluate the effect of pH on the glycolytic rates. In these studies dynamic range problems were circumvented by using proton correlation spectroscopy.

^{1}H NMR has been also applied to whole organs, in this case, conventional high-resolution Fourier transform methods were used. Such spectra will reveal only the highly mobile cellular components. The major organs studied were those where large quantities of given molecular species might be stored. Adrenal gland for instance, was subdivided into cortex and medulla. The former yielded spectra corresponding to the fatty acyl chains of sterol esters; no cholesterol was observed and it was concluded that it was immobilized. The medulla on the other hand, gave a spectrum corresponding to ATP, adrenalin and a protein. The spectra of ATP and adrenalin were different from those of free compounds at low concentration. Isolation of the hormone storage vesicles did not perturb the spectrum, thus demonstrating that the vesicles are not damaged by isolation and represent accurately the state of the hormones in the cell. The major soluble protein, chromogranin A, was shown to be in a random coil state both *in vitro* and *in vivo*. High resolution spectra were obtained under all conditions for this 80,000 molecular weight protein; T_1 and T_2 relaxation times supported the conclusion. Further studies, using paramagnetic Mn^{++} ions, were performed and led to the model of the storage vesicle in which mutual association between protein:metal ion or adrenalin:ATP helps lower the osmotic pressure of the vesicle. The random-coil protein was proposed to be the major factor in preventing precipitation of the vesicle contents by stabilizing the system as a loose jelly.

Other storage systems which yielded ^{1}H NMR spectra were whole seeds, zooplankton, octopus salivary gland, splenic nerve, and blood platelets.[84]

One of the goals of biological NMR spectroscopy is to monitor living systems and obtain chemical information in a non-invasive fashion. One approach involves mapping of internal structures in two and three dimensions. In such cases, the NMR signals which are monitored arise from the highly abundant fluid water within the organism. In NMR imaging, or zeugmatography, the sample is placed in a non-uniform magnetic field in order to label different parts of the sample with different Zeeman field strengths. The nuclei in the various parts of the sample respond with different NMR frequencies, and spatial information can be obtained which allows image reconstruction in two or three dimensions[85]. A number of experimental approaches for the formation of images in human subjects are being pursued. The main purpose of such endeavours is to make NMR an effective medical diagnostic tool, particularly for the diagnosis of malignancy in tissues.[88-90]

5.3 Carbon-13 magnetic resonance

^{13}C NMR spectroscopy was one of the first magnetic resonance techniques applied to metabolic studies, mainly due to the usefulness of ^{13}C-enriched material in tracing metabolic pathways in cell-free systems.

The increased sensitivity of high-field spectrometers (67-90 MHz for ^{13}C) has allowed study of metabolic pathways in cell suspensions and whole tissues. These studies normally involve ^{13}C-enriched material both for increased spectral sensitivity and as a necessary tracer of metabolic transformations. At the cellular level, ^{13}C-labelled glycerols, [2-^{13}C]glycerol and [1-^{13}C] glycerol, were used to follow gluconeogenic pathways in isolated rat hepatocytes.[91] Treatment of rats with triiodothyronine allowed comparisons of glucose formation and glycerol consumption in normal and hyperthyroid rats. In the latter animals, a 2-fold increase in glucose formation and glycerol consumption was observed and differences in flux through various pathways were noted. Using both labelling patterns and spin-spin couplings it was possible to draw conclusions regarding the enzyme reversibility.

Yeast cells have been used extensively as model eukaryotic systems. Anaerobic glycolysis has been studied in suspensions of yeast cells using [1-^{13}C]glucose and [6-^{13}C]glucose.[94] Among the prokaryotic cells, _Escherichia coli_ has been studied using [1-^{13}C] glucose as a tracer of glycolysis.[95,96]

^{13}C NMR studies of metabolism have been carried out in tissues as well as isolated cells. [3-^{13}C]alanine and [^{13}C]glycerol labelled at C1,3 or C2 have been used to follow gluconeogenesis in perfused mouse liver and to investigate the competition between ethanol and alanine in the tricarboxylic acid cycle.[92,93]

Natural abundance ^{13}C NMR spectroscopy is proving useful in detection of unusual metabolites. For example, studies of the marine mollusc *Tapes watlingi* have revealed high concentrations of free taurine, betaine and glycine.[97] It has been suggested that these metabolites may play an osmoregulatory role in some molluscs.

In the course of our studies on the differentiation process in eukaryotic cells we have used the soil amoeba *Acanthamoeba castellanii* as a model system. This free-living amoeba can be readily cultured in liquid medium yielding up to 5 x 10^6 cells/ml. *Acanthamoeba* can exist in either of two states: the mobile vegetative state known as the trophozoite (Figure 15), or as the dormant cyst (Figure 15). ^{13}C NMR spectra obtained on whole cysts have revealed a previously unknown metabolite in this organism: the non-reducing disaccharide α,α-trehalose[98]. Only two major classes of resonances are observed in the ^{13}C spectrum of the whole packed cysts of *Acanthamoeba* (Fig. 16). The most intense arise from α,α-trehalose, and the minor resonances are from lipids. Identification of the disaccharide was provided by comparison with spectral characteristics of the synthetic material. The minor peaks were identified in lipid extracts of the cysts and correspond well with the major fatty acids found in *Acanthamoeba*. No resonances could be attributed to protein or to cell wall cellulose in the high resolution spectra, indicating restricted motional freedom for the major proteins and polysaccharides. The narrow resonances observed for α,α-trehalose and lipid indicate that at least some compartments of the cysts are very fluid. ^{13}C NMR spectra of vegetative cells revealed mostly free glucose with minor amounts of maltose, a precursor of the storage polysaccharide of the vegetative cell, glycogen, and only traces of α,α-trehalose. The presence of large amounts of free glucose is easily understood because the vegetative cells ingest nutrients from glucose-rich

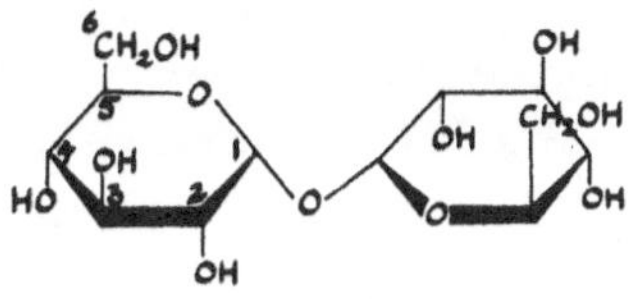

α,α TREHALOSE

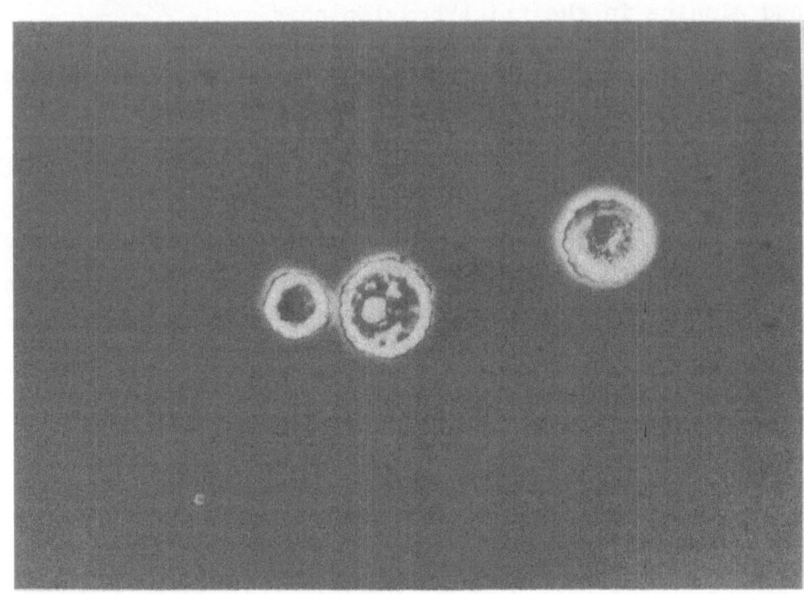

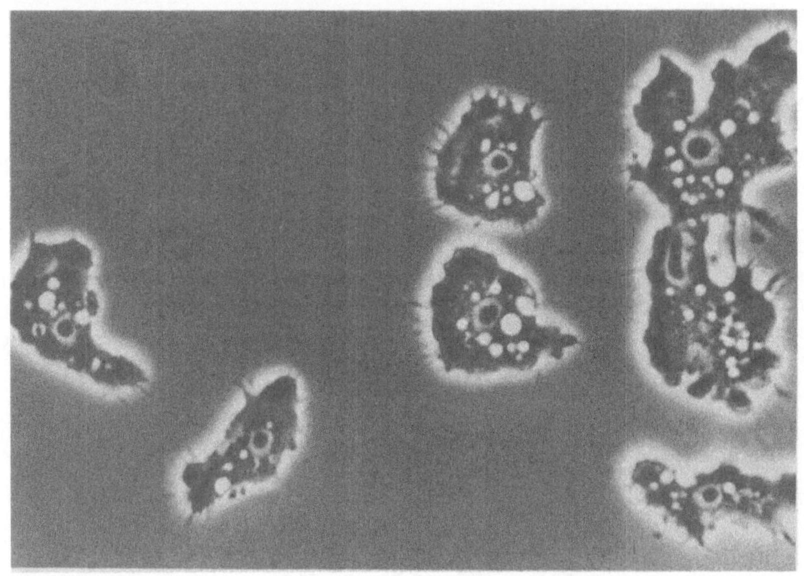

Fig. 15. Light microscope views of _A. castellanii_ at 900x magnification: left, vegetative cells; right, cysts.

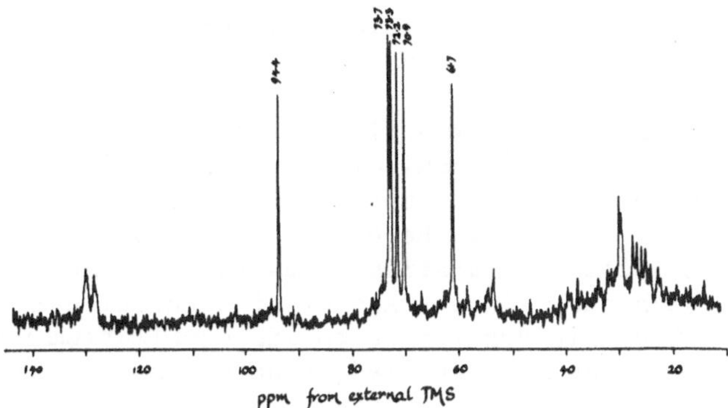

Fig. 16. ^{13}C NMR spectrum (75 MHz) of encysted A. castellanii.

liquid medium by phagocytosis and pinocytosis. The role of trehalose in the cyst may be at least twofold: it can serve as an energy source upon excystment but, more importantly, the non-reducing nature of the sugar may prevent the irreversible interaction which often occurs between protein side chains and reducing sugars upon partial dehydration, and which appears to occur upon encystment.

5.3 Phosphorus-31 magnetic resonance

^{31}P NMR is the most widely employed NMR technique for metabolic and kinetic studies. A number of articles have appeared which survey the progress accomplished in this field over the last 10 years.[96,99-107]

One of the earliest applications of ^{31}P NMR to intact cellular systems was in determining the intracellular pH of erythrocytes.[108] It was found that the chemical shift of intracellular inorganic phosphate, as well as the chemical shift difference between the phosphates in 2,3-diphosphoglycerate, reflected the intracellular pH of erythrocytes whose hemoglobin had been liganded with carbon monoxide.[108] Since then, ^{31}P chemical shifts have been used to detect heterogeneous erythrocyte populations in patients with congenital hemolytic anemia, as well as to quantify intracellular levels of 2,3-diphosphoglycerates.[109] Intracellular pH differences between sickle cell blood and normal blood have been monitored by ^{31}P NMR. In contrast, the pH of whole blood, as measured using a standard pH meter, was similar for both samples.[110]

^{31}P NMR is now in wide use for determining intracellular pH values as well as in detecting pH in subcellular compartments[111-116]. Such studies have provided direct support for the chemiosmotic hypothesis of ATP generation.[117] Measurements of intracellular pH using ^{31}P NMR have also been possible in bacterial spores[118], yeast spores[119] and cysts of amoeba.[120]

Kinetic data on rates of reactions have been followed _in vivo_ using saturation transfer ^{31}P NMR. The method is based on transfer of saturation from one molecular species to another when the two species are chemical exchange. However the method requires knowledge of the spin-lattice relaxation times (T_1) of the species involved in the exchange, since transfer of saturation must occur before the saturated spin can return to its equilibrium magnetization. The technique has been applied to determine the unidirectional rate of ATPase-catalyzed synthesis of ATP from ADP and P_i in _Escherichia coli_[96,121]. Saturation transfer studies on human red blood cells have helped determine exchange times of phosphorus in ATP, ADP and AMP, and shown a slow exchange rate of ATP with 2,3-diphosphoglycerate and with P_i[122]. Such studies have also been applied to whole tissues, in particular to measuring enzyme-catalyzed fluxes between phosphocreatine and ATP in isolated perfused rat hearts.[123]

Although the 100% natural abundance of ^{31}P does not permit its use as a tracer in NMR studies, in the manner of ^{13}C, studies of metabolic pathway are among the most prevalent. Fructose metabolism has been studied in perfused livers from 48-hour starved rats[124]. The time course of changes in ATP, P_i, fructose-1-phosphate, and intracellular pH were followed after infusions of fructose. Similarly glucose metabolism has been followed upon incubation of Ehrlich ascites cells in the sugar for 15 minutes[125]. The metabolite resonances were assigned from chemical shifts, pH titration of chemical shifts in percloric acid extracts, and spin-spin coupling patterns.

Physiological studies have been carried out on a variety of organs and whole tissues, mainly heart[123,126-128], muscle[113,115,129], kidney[116] and liver[124]. The response of kidney and heart to ischemia and the effect of contraction on muscle can all be monitored by variations in such parameters as pH, ATP levels, and changes in glycolytic intermediates and inorganic phosphate. The use of gated ^{31}P NMR has made it possible to acquire spectra at various times _during_ the cardiac cycle. Thus, changes in levels of ATP, creatine phosphate, and P_i can be monitored within each cardiac cycle[128].

Most recently, the use of radiofrequency coils applied to the surface of whole animals has yielded high resolution ^{31}P NMR spectra, and has been used to detect regional disturbances in metabolism, such as localized ischaemia[106].

The possibility of obtaining three-dimensional images of phosphorus-containing metabolites in whole organisms is being actively explored using zeugmatographic imaging techniques[130]. At present, simulated samples containing creatine phosphate, ATP and inorganic phosphate in separate compartments have yielded reasonably well resolved two-dimensional images[130].

Another area of research in which ^{31}P NMR has proven useful is in following changes which occur as a function of development of a complete organism. The concentrations of nucleoside triphosphate, inorganic phosphate and yolk proteins have been monitored in living embryos of the frog *Xenopus laevis*[131]. Intracellular pH measurements were made using the chemical shift of inorganic phosphate, and the complexation of nucleoside triphosphate to divalent cations was detected.

The differentiation process in *Acanthamoeba castellanii* has been followed using ^{13}C and ^{31}P high-resolution and high-power NMR spectroscopy[120]. Figure 17 shows the life cycle of *Acanthamoeba castellanii*[132]. Studies of vegetative cells have revealed stable levels of phosphonic acids which previously had not been reported in the cytoplasm: 2-aminoethylphosphonic acid (AEP) and 1-hydroxyl-2-aminoethylphosphonic acid[133]. These phosphonic acids are found incorporated into the membrane lipophosphonoglycan as well as into the exterior wall of the organism upon encystment[120]. *Acanthamoeba* lends itself well to NMR studies, 30% suspensions of washed cells will maintain constant levels of nucleoside triphosphates and intracellular pH when bubbled overnight with O_2 (Fig. 18). Under anaerobic conditions the internal pH of the suspensions drops from ≃6.5 to 5.8 with a significant decrease in nucleoside triphosphate (Fig. 18).

Acanthamoeba castellanii can be induced to encyst by transfer of cells to a non-nutrient salt medium buffered to pH 9.0. The morphological changes[132] which occur (Fig. 19) have been correlated with the spectroscopically observable changes taking place (Fig. 20). After 3 hours in encystment medium the formation of polyphosphate, a polymer previously undetected in this organism, is evident. The major portion of the polyphosphate is excreted from the cyst[134]. The role of polyphosphates may be to provide a stable phosphorus source to meet the requirements of excystment. A further role for polyphosphate may be in osmoregulation: polymerization of the inorganic phosphorus released from catabolism of cellular components in the course of cyst formation will decrease the number of osmotically active particles, thus helping maintain cell volume and integrity. Phosphonoprotein, which yields a prominent peak at -22 ppm (downfield from H_3PO_4), is a major component of the spectra of cells which have been in encystment medium for 24 hours. The phosphonoprotein is the major constituent of the external wall of the cyst. Nucleoside di- and triphosphates are visible 50 hours after placing cells in encyst-

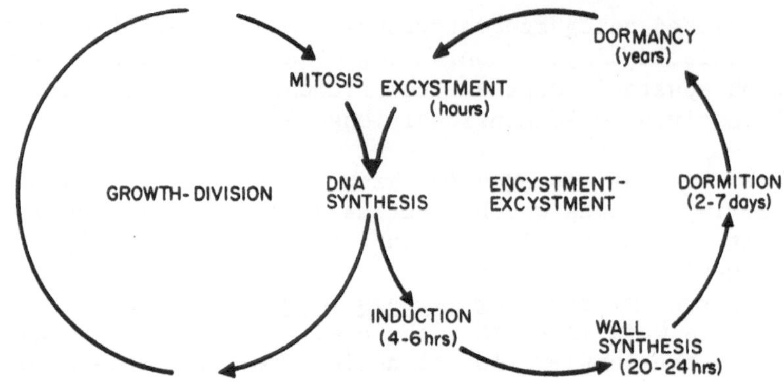

Fig. 17. Life cycle of A. castellanii (redrawn from ref. 121).

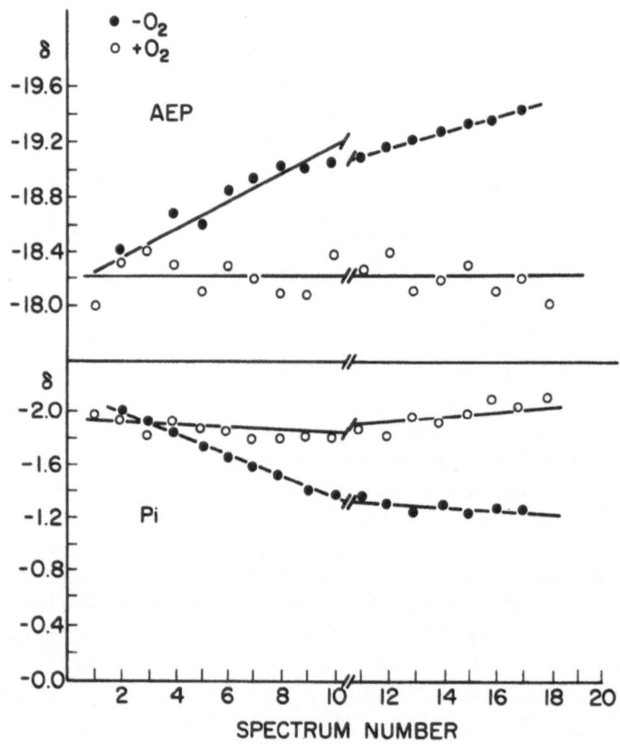

Fig. 18. Changes in intracellular pH of A. castellanii as monitored by the ^{31}P NMR chemical shifts of inorganic phosphate and 2-aminoethylphosphonic acid: O aerobic conditions; ● anaerobic conditions. Spectra 1-10 were acquired in 10 min., spectra 11-20 in 1.5 h. each.

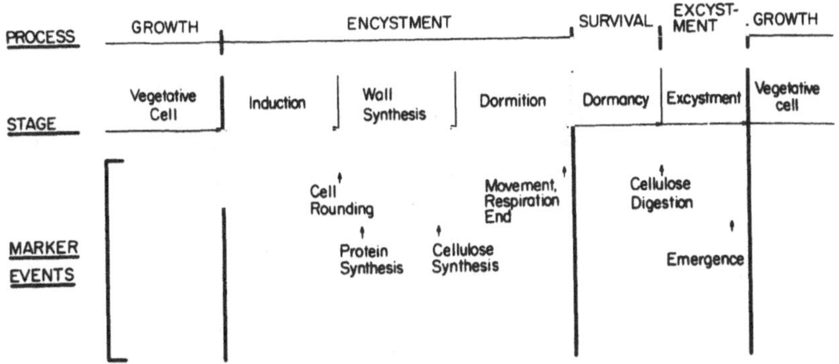

Fig. 19. Marker events in the encystment-excystment cycle of *Acanthamoeba* (redrawn from ref. 121).

ment medium. A large number of intermediates of glycolysis became visible during the encystment process. The peaks observed in the whole cells at various times during the encystment process were identified by titration of perchloric acid extracts of the cells. Fully mature cysts, which occur after weeks in encystment medium, show no signs of nucleoside phosphates. Washed mature cysts show simple ^{31}P NMR spectra, consisting of two major phosphonate resonances (-22, -17 ppm), inorganic phosphate (-3.4 ppm) and a phosphodiester peak (≈1.0 ppm). The peak at -22 ppm (phosphonoprotein) sediments with the empty cyst upon disruption of the cyst, the inorganic phosphate remains in the supernatant, and the peaks at -17 and +1 ppm sediment with the microsomes. Identification of the latter two peaks are underway.

Vegetative cells of *Acanthamoeba* in normal growth medium have an intracellular pH of 6.5; in encystment medium at pH 9.0 the pH of the cell increases slightly. Acidification of the encystment medium occurs as encystment proceeds, reaching a pH of 8.4 when cyst formation is complete. However, the cyst pH remains ≈7.5.

High-resolution NMR spectra reveal only the fluid constituents of the cell cytoplasm. The disappearance of peaks may result from depletion of a given metabolite, or from sequestration in insoluble granules or other structures where rotational mobility is restricted. In order to discriminate between these alternatives, high-power solid-state NMR approaches have been employed to visualize solid-like components of the cells. For example, membrane-bound polyphosphates might be visualized using such techniques. Preliminary studies on whole cysts of *Acanthamoeba* using magic-angle spinning and dipolar decoupling to reduce linewidths have yielded high resolution spectra of previously unobservable polyphosphates in washed cysts (R.A. Byrd, R. Deslauriers, I.C.P. Smith,

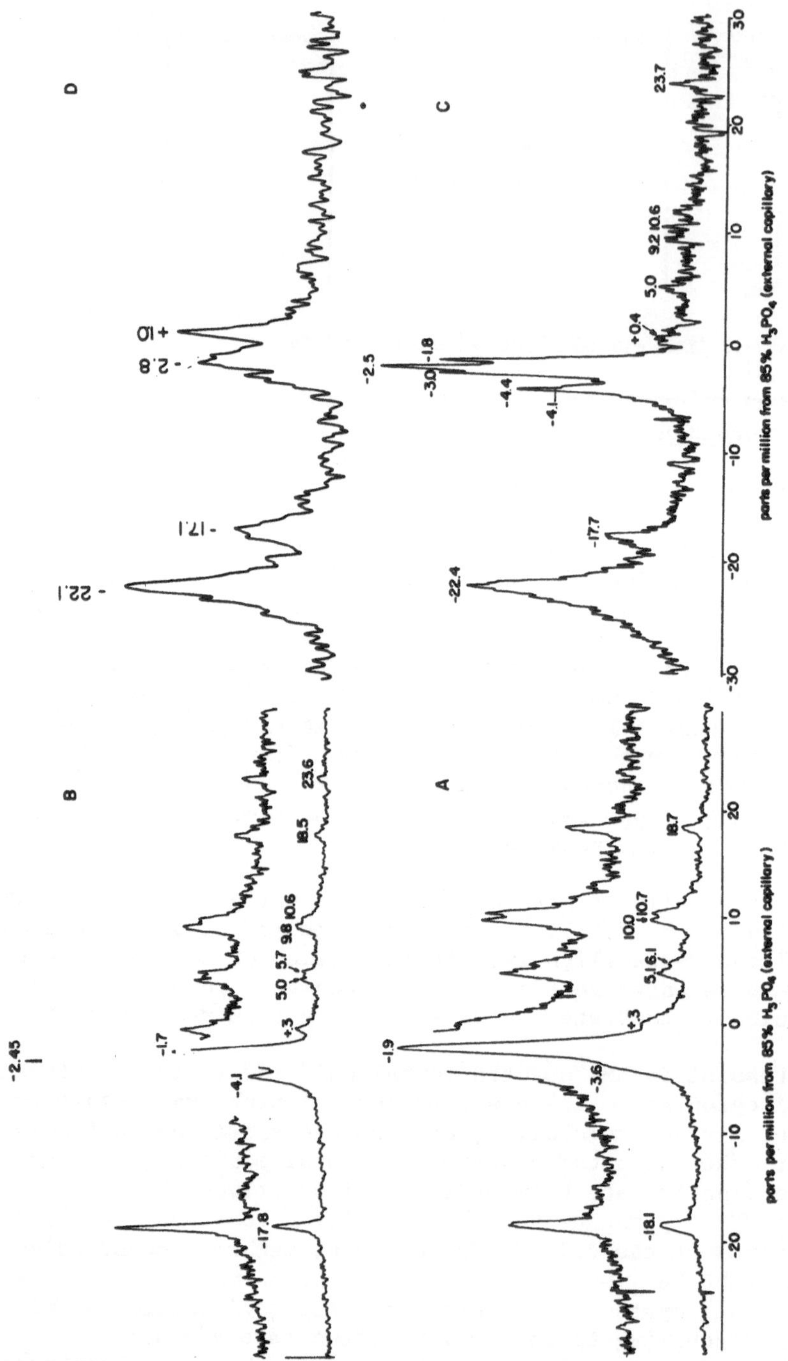

Fig. 20. ^{31}P NMR spectra (121 MHz) of A. castellanii cells as a function of time after transfer to non-nutrient salt medium; (A) 0 h; (B) 3 h; (C) 22 h; (D) washed mature cysts, 3 weeks in encystment medium.

unpublished). Such approaches appear promising for yielding a more complete description of cytoplasmic contents and processes which occur therein.

^{31}P NMR has been proven helpful in studying developmental processes. Applications of these techniques to areas of drug therapy and pathologically-altered tissues appears promising.

5.4 Other nuclei

Nitrogen has not been used extensively to study cytoplasmic constituents or processes *in vivo*. One recent application of ^{15}N to metabolic studies has traced the incorporation of ^{15}N into protein and amino acids in *Neurospora crassa*[135]. Incorporation of label into soluble (amino acid) and solid (protein) components was followed in intact lyophilized mycelia of *N. crassa* by using the solid state techniques of magic-angle spinning and cross-polarization to reduce linewidths and increase sensitivity to solid components of the system.

NMR studies of ^{23}Na have concentrated mainly on determining relaxation times in systems such as frog muscle[136,137] and halotolerant[138] bacteria in order to determine the fraction of ions bound to macromolecular structures where molecular tumbling is restricted compared to true solutions. Similarly, ^{17}O and ^{2}H have been employed to study the state of water in frog muscle[139,140] and human erythrocytes[141]. Relaxation time measurements of ^{39}K have been used to argue that the ion is associated with charged macromolecular sites in muscle, brain and *E. coli*[142].

6. REFERENCES

1. W. L. Hubbell and H.M. McConnell, Orientation and motion of amphiphilic spin labels in membranes, *Proc. Natl. Acad. Sci. U.S.* 64:20 (1969).
2. H. Schneider and I. C. P. Smith, A study of the structural integrity of spin-labelled proteins in some fractions of human erythrocyte ghosts, *Biochim. Biophys. Acta* 219:73 (1970).
3. I. C. P. Smith, The spin label method, *in* "Biological Applications of Electron Spin Resonance", H. M. Swartz, J. R. Bolton and D. C. Borg, eds., John Wiley and Sons, New York (1972).
4. L. J. Berliner, ed., "Spin Labelling Theory and Applications", Academic Press, New York (1976).
5. J. Seelig, Anisotropic motion in liquid crystalline structures, Chapt. 10 in ref. 4, 1976.
6. I. C. P. Smith and K. W. Butler, Oriented lipid systems as model membranes, Chapt. 11 in ref. 4, 1976.

7. O. H. Griffith and P. C. Jost, Lipid spin labels in biological membranes, Chapt. 12 in ref. 4, 1976.
8. H. M. McConnell, Molecular motion in biological membranes, Chapt. 13 in ref. 4, 1976.
9. A. D. Keith, M. Sharnoff and G. E. Cohn, A summary and evaluation of spin labels as probes for biological membrane structures, Biochim. Biophys. Acta 300:379 (1973).
10. S. Schreier, C. F. Polnaszek and I. C. P. Smith, Spin labels in membranes: problems in practice, Biochim. Biophys. Acta 515:375 (1978).
11. M. G. Taylor and I. C. P. Smith, The fidelity of response by nitroxide spin probes to changes in membrane organization: the condensing effect of cholesterol, Biochim. Biophys. Acta 599:140 (1980).
12. M. G. Taylor and I. C. P. Smith, The reliability of nitroxide spin probes in reporting membrane properties: a comparison of nitroxide- and deuterium-labelled steroids, Biochemistry (in press, 1981).
13. D. Chapman, Nuclear magnetic resonance spectroscopic studies of biological membranes, Ann. N. Y. Acad. Sci. 195:175 (1972).
14. A. L. Y. Lau and S. I. Chan, Voltage-induced formation of alamethecin pores in lecithin bilayer vesicles, Biochemistry 15:2551 (1976).
15. N. O. Petersen and S. I. Chan, More on the motional state of lipid bilayer membranes: interpretation of order parameters obtained from nuclear magnetic resonance experiments, Biochemistry 16:2657 (1977).
16. A. G. Lee, N. J. M. Birdsall and J. C. Metcalfe, Measurement of fast lateral diffusion of lipids in vesicles and in biological membranes by ^{1}H nuclear magnetic resonance, Biochemistry 12:1650 (1973).
17. G. Lindblom, H. Wennerström, G. Arvidson and B. Lindman, Lecithin translational diffusion studied by pulsed nuclear magnetic resonance, Biophys. J. 16:1287 (1976).
18. H. L. Kantor and J. H. Prestegard, Fusion of phosphatidylcholine bilayer vesicles: role of free fatty acid, Biochemistry 17: 3592 (1978).
19. A. L. MacKay, E. E. Burnell, C. P. Nichols, G. Weeks, M. Bloom and M. I. Valic, Effect of viscosity on the width of the methylene proton magnetic resonance line in sonicated phospholipid bilayer vesicles, FEBS Lett. 88:97 (1978).
20. T. H. Fischer and G. C. Levy, Electron and proton magnetic resonance studies of the effect of rhodopsin incorporation on molecular motion in dimyristoylphosphatidylcholine bilayers, Chem. Phys. Lipids 28:7 (1981).
21. G. W. Feigenson and P. R. Meers, ^{1}H NMR study of valinomycin conformation in a phospholipid bilayer, Nature 283:313 (1980).

22. A. G. Lee, N. J. M. Birdsall and J. C. Metcalfe, Nuclear magnetic relaxation and the biological membrane, Methods in Membrane Biology 2:1 (1974).
23. A. G. Lee, N. J. M. Birdsall, J. C. Metcalfe, G. B. Warren and G. C. K. Roberts, A determination of the mobility gradient in lipid bilayers by ^{13}C magnetic resonance, Proc. R. Soc. Lond. B. 193:253 (1976).
24. N. J. M. Birdsall, D. J. Ellar, A. G. Lee, J. C. Metcalfe and G. B. Warren, ^{13}C-enriched phosphatidylethanolamines from Escherichia coli, Biochim. Biophys. Acta 380:344 (1975).
25. I. C. P. Smith, A. P. Tulloch, G. W. Stockton, S. Schreier, A. Joyce, K. W. Butler, Y. Boulanger, B. Blackwell and L. Bennett, Determination of membrane properties at the molecular level by carbon-13 and deuterium magnetic resonance, Ann. N. Y. Acad. Sci. 308:8 (1978).
26. A. Joyce and I. C. P. Smith, Carbon-13 nuclear magnetic resonance studies of bacterial membranes enriched via biosynthetic pathways, Microbiology 1979, p. 5.
27. M. F. Brown, J. Seelig and U. Häberlen, Structural dynamics in phospholipid bilayers from deuterium spin-lattice relaxation time measurements, J. Chem. Phys. 70:5045 (1979).
28. J. D. Robinson, N. J. M. Birdsall, A. G. Lee and J. C. Metcalfe, ^{13}C and ^{1}H nuclear magnetic relaxation measurements of the lipids of sarcoplasmic reticulum membranes, Biochemistry 11:2903 (1972).
29. R. E. London, V. H. Kollman and N. A. Matwiyoff, ^{13}C Fourier transform nuclear magnetic resonance studies of fractionated Candida utilis membranes, Biochemistry 14:5492 (1975).
30. E. O. Stejskal, J. Schaefer and R. A. McKay, High-resolution, slow-spinning magic-angle carbon-13 NMR, J. Magn. Res. 25: 569 (1977).
31. C. A. Fyfe, J. R. Lyerla, W. Volksen and C. S. Yannoni, High resolution carbon-13 nuclear magnetic resonance studies of polymers in the solid state. Aromatic polyesters, Macromolecules 12:757 (1979).
32. S. J. Opella, M. H. Frey and T. A. Cross, Detection of individual carbon resonances in solid proteins, J. Amer. Chem. Soc. 101:5856 (1979).
33. R. A. Haberkorn, J. Herzfeld and R. G. Griffin, High resolution ^{31}P and ^{13}C nuclear magnetic resonance spectra of unsonicated model membranes, J. Amer. Chem. Soc. 100:1296 (1978).
34. K. Koga and Y. Kanezawa, Dynamical structure of phosphatidylcholine molecules in single bilayer vesicles observed by nitrogen-14 nuclear magnetic resonance, Biochemistry 19: 2779 (1980).
35. D. J. Siminovitch, M. Rance and K. R. Jeffrey, The use of wide line ^{14}N nitrogen NMR as a probe in model membranes, FEBS Lett. 112:79 (1980).

36. C. S. Irving and A. Lapidot, The dynamic structures of the Escherichia coli cell envelope as probed by ^{15}N nuclear magnetic resonance spectroscopy, Biochim. Biophys. Acta 470:251 (1977).
37. J. Seelig, ^{31}P Nuclear magnetic resonance and the head group structure of phospholipids in membranes, Biochim. Biophys. Acta 515:105 (1978).
38. P. R. Cullis and B. De Kruijff, Lipid polymorphism and the functional roles of lipids in biological membranes, Biochim. Biophys. Acta 559:399 (1979).
39. H. U. Galley, W. Niederberger and J. Seelig, Conformation and motion of the choline head group in bilayers of dipalmitoyl-3-sn-phosphatidylcholine, Biochemistry 14:3647 (1975).
40. R. Skarjune and E. Oldfield, Physical studies of cell surface and cell membrane structure. Determination of phospholipid head group organization by deuterium and phosphorus nuclear magnetic resonance spectroscopy, Biochemistry 18:5903 (1979).
41. R. F. Campbell, E. Meirovitch and J. H. Freed, Slow-motional NMR line shapes for very anisotropic rotational diffusion. Phosphorus-31 NMR of phospholipids, J. Phys. Chem. 83:525
42. M. B. Jackson and J. M. Sturtevant, Phase transitions of the purple membranes of Halobacterium halobium, Biochemistry 17:911 (1978).
43. H. Degani, D. Bach, A. Danon, H. Garty, M. Eisenbach and S. R. Caplan, Phase transition of the lipids of Halobacterium halobium, in "Energetics and Structure of Halophilic Organisms", S. R. Caplan and N. Ginzburg, eds., Elsevier-North Holland, Amsterdam, pp. 225-232 (1978),
44. I. Ekiel, D. Marsh, B. W. Smallbone, M. Kates and I. C. P. Smith, The state of the lipids in the purple membrane of Halobacterium cutirubrum as seen by ^{31}P NMR, Biochem. Biophys. Res. Commun. (1981, in press).
45. B. De Kruijff, G. A. Morris and P. R. Cullis, Application of ^{31}P-NMR saturation transfer techniques to investigate phospholipid motion and organization in model and biological membranes, Biochim. Biophys. Acta 598:206 (1980).
46. H. C. Jarrell, R. A. Byrd, R. Deslauriers, I. Ekiel and I. C. P. Smith, Characterization of the phase behaviour of phosphonolipids in model and biological membranes by ^{31}P NMR, Biochim. Biophys. Acta (1981, in press).
47. H. Saitô, S. Schreier-Muccillo and I. C. P. Smith, High resolution deuterium magnetic resonance - an approach to the study of molecular organization in biological membranes and model systems, FEBS Lett. 33:281 (1973).
48. J. Seelig and A. Seelig, Deuterium magnetic resonance studies of phospholipid bilayers, Biochem. Biophys. Res. Commun. 57:406 (1974).

49. G. W. Stockton, C. F. Polnaszek, L. C. Leitch, A. P. Tulloch and I. C. P. Smith, A study of mobility and order in model membranes using ^{2}H NMR relaxation rates and quadrupole splittings of specifically-deuterated lipids, Biochem. Biophys. Res. Commun. 60:844 (1974).
50. G. W. Stockton, K. G. Johnson, K. W. Butler, C. F. Polnaszek, R. Cyr and I. C. P. Smith, Molecular order in *Acholeplasma laidlawii* membranes as determined by deuterium magnetic resonance of biosynthetically-incorporated specifically-labelled lipids, Biochim. Biophys. Acta 401:535 (1975).
51. A. Seelig and J. Seelig, Bilayers of dipalmitoyl-3-sn-phosphatidylcholine. Conformational differences between the fatty acyl chains, Biochim. Biophys. Acta 406:1 (1975).
52. G. W. Stockton, C. F. Polnaszek, A. P. Tulloch, F. Hasan and I. C. P. Smith, Molecular motion and order in single-bilayer vesicles and multilamellar dispersions of egg lecithin and lecithin-cholesterol mixtures. A deuterium magnetic resonance study of specifically labelled lipids, Biochemistry 15:954 (1976).
53. J. Seelig and H. U. Gally, Investigation of phosphatidylethanolamine bilayers by deuterium and phosphorus-31 nuclear magnetic resonance, Biochemistry 15:5199 (1976).
54. G. W. Stockton and I. C. P. Smith, A deuterium magnetic resonance study of the condensing effect of cholesterol on egg phosphatidylcholine bilayer membranes. I. Perdeuterated fatty acid probes, Chem. Phys. Lipids 17:251 (1976).
55. A. Seelig and J. Seelig, Effect of a single cis double bond on the structure of a phospholipid bilayer, Biochemistry 16:45 (1977).
56. G. W. Stockton, K. G. Johnson, K. W. Butler, A. P. Tulloch, Y. Boulanger, I. C. P. Smith, J. H. Davis and M. Bloom, Deuterium NMR study of lipid organization in *Acholeplasma laidlawii* membranes, Nature 269:267 (1977).
57. H. H. Mantsch, H. Saitô and I. C. P. Smith, Deuterium magnetic resonance, applications in chemistry, physics and biology, Prog. NMR Spec. 11:211 (1977).
58. J. Seelig, Deuterium magnetic resonance: theory and applications to lipid membranes, Quart. Rev. Biophys. 10:353 (1977).
59. I. C. P. Smith, Organization and dynamics of membrane lipids as determined by magnetic resonance spectroscopy, Can. J. Biochem. 57:1 (1979).
60. J. Seelig and A. Seelig, Lipid conformation in model membranes and biological membranes, Quart. Rev. Biophys. 13:19 (1980).
61. D. M. Rice, J. C. Hsung, T. E. King and E. Oldfield, Protein-lipid interactions. High field deuterium and phosphorus nuclear magnetic resonance spectroscopic investigations of the cytochrome oxidase-phospholipid interaction and the effects of cholate, Biochemistry 18:5885 (1979).

62. S. Y. Kang, R. A. Kinsey, S. Rajan, H. S. Gutowsky, M. C. Gabridge and E. Oldfield, Protein-lipid interactions in biological and model membrane systems, J. Biol. Chem. 256:1155 (1981).
63. C. P. Nichol, J. H. Davis, G. Weeks and M. Bloom, Quantitative study of the fluidity of Escherichia coli membranes using deuterium magnetic resonance, Biochemistry 19:451 (1980).
64. M. I. Valic, H. Gorrissen, R. J. Cushley and M. Bloom, Deuterium magnetic resonance study of cholesteryl esters in membranes, Biochemistry 18:854 (1979).
65. I. C. P. Smith, The states of the lipids in biological membranes as visualized by deuterium NMR, Bull. Magn. Res. (1981, in press).
66. H. W. Spiess, Rotation of molecules and nuclear spin relaxation in "NMR Basic Prinicples and Progress", P. Diehl, E. Fluck and R. Kosfeld, eds., Springer Verlag, Berlin, pp. 55-214 (1978).
67. S. Y. Kang, H. S. Gutowsky and E. Oldfield, Spectroscopic studies of specifically deuterium labelled membrane systems. Nuclear magnetic resonance investigation of protein-lipid interaction in Escherichia coli membranes, Biochemistry 18:3268 (1979).
68. J. H. Davis, C. P. Nichol, G. Weeks and M. Bloom, Study of the cytoplasmic and outer membranes of Escherichia coli by deuterium magnetic resonance, Biochemistry 18:2103 (1979).
69. J. H. Davis, B. Maraviglia, G. Weeks and D. V. Godin, Bilayer rigidity of the erythrocyte membrane, ^{2}H NMR of a perdeuterated palmitic acid probe, Biochim. Biophys. Acta 550:362 (1979).
70. J. Seelig, L. Tamm, L. Hymel and S. Fleischer, Deuterium and phosphorus NMR and fluorescence depolarization studies of functional reconstituted sarcoplasmic reticulum membrane vesicles, Biochemistry (1981, in press).
71. H. U. Gally, G. Pluschke, P. Overath and J. Seelig, Structure of Escherichia coli membranes. Fatty acyl chain order parameters of inner and outer membranes and derived liposomes, Biochemistry 19:1638 (1980).
72. I. C. P. Smith, K. W. Butler, A. P. Tulloch, J. H. Davis and M. Bloom, The properties of gel state lipid in membranes of Acholeplasma laidlawii as observed by deuterium magnetic resonance, FEBS Lett. 100:57 (1979).
73. J. H. Davis, M. Bloom, K. W. Butler and I. C. P. Smith, The temperature dependence of molecular order and the influence of cholesterol in Acholeplasma laidlawii membranes, Biochim. Biophys. Acta 597:477 (1980).
74. M. Rance, K. R. Jeffrey, A. P. Tulloch, K. W. Butler and I. C. P. Smith, Orientational order of unsaturated lipids in the membranes of Acholeplasma laidlawii as observed by ^{2}H NMR, Biochim. Biophys. Acta 600:245 (1980).

75. A. Wieslander, J. Ulmius, G. Lindblom and K. Fontell, Water binding and phase structures for different Acholeplasma laidlawii membrane lipids studied by deuteron nuclear magnetic resonance and X-ray diffraction, Biochim. Biophys. Acta 512:241 (1978).
76. J. R. Silvius, N. Mak and R. N. McElhaney, Lipid and protein composition and thermotropic lipid phase transitions in fatty acid-homogeneous membranes of Acholeplasma laidlawii Biochim. Biophys. Acta 597:199 (1980).
77. D. Rice and E. Oldfield, Deuterium nuclear magnetic resonance studies of the interaction between dimyristoylphosphatidylcholine and gramicidin A', Biochemistry 18:3272 (1979).
78. J. H. Davis, K. R. Jeffrey, M. Bloom, M. I. Valic and T. P. Higgs, Quadrupole echo deuteron magnetic resonance spectroscopy in ordered hydrocarbon chains. Chem. Phys. Lett. 42:390 (1976).
79. H. C. Jarrell, R. A. Byrd and I. C. P. Smith, Analysis of the composition of mixed lipid phases by the moments of ^{2}H NMR spectra, Biophys. J. (1981, in press).
80. M. F. Brown, Deuterium relaxation and molecular dynamics in lipid bilayers, J. Magn. Res. 351:203 (1979).
81. T. H. Oldfield, Restricted rotational isomerization in polymethylene chains, J. Amer. Chem. Soc. 102:7377 (1980).
82. A. P. Tulloch, Synthesis of deuterium and carbon-13 labelled lipids, Chem. Phys. Lipids 24:391 (1979).
83. T. Ogino, Y. Arata and S. Fujiwara, Proton correlation nuclear magnetic resonance study of metabolic regulation and pyruvate transport in anaerobic Escherichia coli cells, Biochemistry 19:3684 (1980).
84. A. J. Daniels, J. Krebs, B. A. Levine, P. E. Wright and R. J. P. Williams, The proton NMR spectra of whole organs, in "NMR in Biology", R. A. Dwek, I. D. Campbell, R. E. Richards and R. J. P. Williams, eds., Acad. Press, London, pp. 277-287.
85. E. R. Andrew, NMR imaging of intact biological systems, Phil. Trans. R. Soc. Lond. B. 289:471 (1980).
86. R. Damadian, Field focusing NMR (FONAR) and the formation of chemical images in man, Phil. Trans. R. Soc. Lond. B 289:489 (1980).
87. W. S. Moore and G. N. Holland, Experimental considerations in implementing a whole body multiple sensitive point nuclear magnetic resonance imaging system, Phil. Trans. R. Soc. Lond. B 289:511 (1980).
88. J. Mallard, J. M. S. Hutchison, W. A. Edelstein, C. R. Ling, M. A. Foster and G. Johnson, In vivo NMR imaging in medicine; the Aberdeen approach both physical and biological, Phil. Trans. R. Soc. B 289:519 (1980).

89. P. Mansfield, P. G. Morris, R. J. Ordidge, I. L. Pykett, V. Bangert and R. E. Coupland, Human whole body imaging and detection of breast tumors by NMR, Phil. Trans. R. Soc. Lond. B 289:503 (1980).
90. W. M. M. J. Bovee, J. H. N. Creyghton, K. W. Getreuer, D. Korbee, S. Lobregt, J. Smidt, R. A. Wind, J. Lindeman, L. Smid and H. Posthuma, NMR relaxation and images of human breast tumors in vitro, Phil. Trans. R. Soc. Lond. B. 289:535 (1980).
91. S. M. Cohen, S. Ogawa and R. G. Shulman, ^{13}C NMR studies of gluconeogenesis in rat liver cells: utilization of labelled glycerol by cells from euthyroid and hyperthyroid rats, Proc. Natl. Acad. Sci. U.S.A. 76:1663 (1979).
92. S. M. Cohen, R. G. Shulman and A. C. McLaughlin, Effects of ethanol on alanine metabolism in perfused mouse liver studied by ^{13}C NMR, Proc. Natl. Acad. Sci. U.S.A. 76:4808 (1979).
93. S. M. Cohen and R. G. Shulman, ^{13}C NMR studies of gluconeogenesis in rat liver suspensions and perfused mouse liver, Phil. Trans. R. Soc. Lond. B 289:407 (1980).
94. J. A. den Hollander, T. R. Brown, K. Ugurbil and R. G. Shulman, ^{13}C nuclear magnetic resonance studies of anaerobic glycolysis in suspensions of yeast cells, Proc. Natl. Acad. Sci. U.S.A. 76:6096 (1979).
95. T. R. Brown, J. A. den Hollander, R. G. Shulman and K. Ugurbil, ^{13}C NMR studies of glycolysis in suspensions of Escherichia coli cells, in "Frontiers of Biological Energetics, Vol. II: Electrons to Tissues", P. L. Dutton, J. S. Leigh and A. Scarpa, eds., Academic Press, New York, pp. 1365-1370 (1978).
96. K. Ugurbil, R. G. Shulman and T. R. Brown, High-resolution ^{31}P and ^{13}C nuclear magnetic resonance studies of Escherichia coli cells in vivo, in "Biological Applications of Magnetic Resonance", R. G. Shulman, ed., Acad. Press, New York, pp. 537-589 (1977).
97. R. S. Norton, Identification of mollusc metabolites by natural-abundance ^{13}C NMR studies of whole tissue and tissue homogenates, Comp. Biochem. Physiol. 63B:67 (1979).
98. R. Deslauriers, H. Jarrell, R. A. Byrd and I. C. P. Smith, Observation by ^{13}C NMR of metabolites in differentiating amoeba - Trehalose storage in encysted Acanthamoeba castellanii, FEBS Lett. 118:185 (1980).
99. D. G. Gadian, G. K. Radda, R. E. Richards and P. J. Stanley, ^{31}P NMR in living tissue: the road from a promising to an important tool in biology, in "Biological Applications of Magnetic Resonance", R. G. Shulman, ed., Acad. Press, New York, pp. 463-535 (1977).
100. D. P. Hollis, Phosphorus NMR of cells, tissues and organelles, in "Biological Magnetic Resonance, Vol. 2", L. J. Berliner and J. Reuben, eds., Plenum Press, New York, pp. 1-44 (1980).

101. P. J. Seeley, P. A. Sehr, D. G. Gadian, P. B. Garlick and G. K. Radda, Phosphorus NMR in living tissue, in "NMR in Biology", R. A. Dwek, I. D. Campbell, R. E. Richards and R. J. P. Williams, eds., Acad Press, London, pp. 247-275 (1975).
102. J. Dawson, D. G. Gadian and D. R. Wilkie, Studies of living contracting muscle by ^{31}P nuclear magnetic resonance, in "NMR in Biology", R. A. Dwek, I. D. Campbell, R. E. Richards and R. J. P. Williams, eds., Acad. Press, London, pp. 289-322 (1977).
103. C. T. Burt, S. M. Cohen and M. Barany, Analysis of intact tissue with ^{31}P NMR, Ann. Rev. Biophys. Bioeng. 8:1 (1979).
104. M. K. Battersby, P. M. Garlick, P. J. Seeley, P. A. Sehr and G. K. Radda, Phosphorus nuclear magnetic resonance studies of living tissue, in "Biomolecular Structure and Function", P. F. Agris, ed., Acad. Press, New York, pp. 175-193 (1978).
105. R. G. Shulman, T. R. Brown, K. Ugurbil, S. Ogawa, S. M. Cohen, and J. A. den Hollander, Cellular applications of ^{31}P and ^{13}C nuclear magnetic resonance, Science 205:160 (1979).
106. J. J. H. Ackerman, T. H. Grove, G. C. Wong, D. G. Gadian and G. K. Radda, Mapping of metabolites in whole animals by ^{31}P NMR using surface coils, Nature 283:167 (1980).
107. D. I. Hoult, S. J. W. Busby, D. G. Gadian, G. K. Radda, R. E. Richards and P. J. Seeley, Observation of tissue metabolites using ^{31}P nuclear magnetic resonance, Nature 252:285 (1974).
108. R. B. Moon and J. H. Richards, Determination of intracellular pH by ^{31}P magnetic resonance, J. Biol. Chem. 248:7276 (1973).
109. R. J. Labotka and G. R. Honig, ^{31}P NMR spectroscopy of erythrocytes in congenital hemolytic anemias: detection of heterogeneous erythrocyte populations and quantification of intracellular 2,3-diphosphoglycerate, Amer. J. Hematology 9:55 (1980).
110. Y. F. Lam, A. K. L. C. Lin and C. Ho, A phosphorus-31 nuclear magnetic resonance investigation of intracellular environment in human normal and sickle cell blood, Blood 54:196 (1979).
111. H. B. Pollard, H. Shindo, C. E. Creutz, C. J. Pazoles and J. S. Cohen, Internal pH and state of ATP in adrenergic chromaffin granules determined by ^{31}P nuclear magnetic resonance spectroscopy, J. Biol. Chem. 254:1170 (1979).
112. D. Njus, P. A. Sehr, G. K. Radda, G. A. Ritchie and P. J. Seeley, Phosphorus-31 nuclear magnetic resonance studies of active translocation in chromaffin granules, Biochemistry 17:4337 (1978).
113. S. J. W. Busby, D. G. Gadian, G. K. Radda, R. E. Richards and P. J. Seeley, Phosphorus nuclear magnetic resonance studies of compartmentation in muscle, Biochem. J. 170:103 (1978).

114. S. M. Cohen, S. Ogawa, H. Rottenberg, P. Glynn, T. Yamane, T. R. Brown, R. G. Shulman and J. R. Williamson, ^{31}P nuclear magnetic resonance studies of isolated rat liver cells, Nature 273:554 (1978).

115. K. Yoshizaki, H. Nishikawa, S. Yamada, T. Morimoto and H. Watari, Intracellular pH measurement in frog muscle by means of ^{31}P NMR, Jap. J. Physiol. 29:211 (1979).

116. P. A. Sehr, P. J. Bore, J. Papatheofanis and G. K. Radda, Non-destructive measurement of metabolites and tissue pH in the kidney by ^{31}P nuclear magnetic resonance, Br. J. Exp. Path. 60:632 (1979).

117. G. Navon, S. Ogawa, R. G. Shulman and T. Yamane, High-resolution ^{31}P nuclear magnetic resonance studies of metabolism in aerobic Escherichia coli cells, Proc. Natl. Acad. Sci. U.S.A. 74:888 (1977).

118. B. Setlow and P. Stelow, Measurement of the pH within dormant and germinated bacterial spores, Proc. Natl. Acad. Sci. U.S.A. 77:2474 (1980).

119. J. K. Barton, J. A. den Hollander, T. M. Lee, A. MacLaughlin and R. G. Shulman, Measurement of the internal pH of yeast spores by ^{31}P nuclear magnetic resonance, Proc. Natl. Acad. Sci. U.S.A. 77:2470 (1980).

120. R. Deslauriers, R. A. Byrd, H. C. Jarrell and I. C. P. Smith, NMR studies of differentiation in Acanthamoeba castellanii, in "Non-Invasive Probes of Tissue Metabolism", J. S. Cohen, ed., John Wiley, New York (1981, in press).

121. T. R. Brown, K. Ugurbil and R. G. Shulman, ^{31}P nuclear magnetic resonance measurements of ATPase kinetics in aerobic Escherichia coli cells, Proc. Natl. Acad. Sci. U.S.A. 74:5551 (1977).

122. R. K. Gupta, Saturation transfer ^{31}P NMR studies of the intact human red blood cell, Biochim. Biophys. Acta 586:189 (1979).

123. J. J. H. Ackerman, P. J. Bore, D. G. Gadian, T. H. Grove and G. K. Radda, NMR studies of metabolism in perfused organs, Phil. Trans. R. Soc. Lond. B 289:425 (1980).

124. R. A. Iles, J. R. Griffiths, A. N. Stevens, D. G. Gadian and R. Proteous, Effects of fructose on the energy metabolism and acid-base status of the perfused starved-rat liver, Biochem. J. 192:191 (1980).

125. G. Navon, S. Ogawa, R. G. Shulman and T. Yamane, ^{31}P nuclear magnetic resonance studies of Ehrlich ascites tumor cells, Proc. Natl. Acad. Sci. U.S.A. 74:87 (1977).

126. T. H. Grove, J. J. H. Ackerman, G. K. Radda and P. J. Bore, Analysis of rat heart in vivo by phosphorus nuclear magnetic resonance, Proc. Natl. Acad. Sci. U.S.A. 77:299 (1980).

127. D. P. Hollis, Nuclear magnetic resonance of phosphorus in the perfused heart, IEEE Transactions on Nuclear Science NS-27:1250 (1980).

128. E. T. Fossel, H. E. Morgan and J. S. Ingwall, Measurement of changes in high-energy phosphates in the cardiac cycle using gated ^{31}P nuclear magnetic resonance, Proc. Natl. Acad. Sci. U.S.A. 77:3654 (1980).
129. M. J. Dawson, D. G. Gadian and D. R. Wilkie, Studies of the biochemistry of contracting and relaxing muscle by the use of ^{31}P NMR in conjunction with other techniques, Phil. Trans. R. Soc. Lond. B 289:445 (1980).
130. P. C. Lauterbur, Progress in NMR zeugmatographic imaging, Phil. Trans. R. Soc. Lond. B 289:483 (1980).
131. A. Coleman and D. G. Gadian, ^{31}P nuclear magnetic resonance studies on the developing embryos of Xenopus laevis, Eur. J. Biochem. 61:387 (1976).
132. R. J. Neff and R. H. Neff, The biochemistry of amoebic encystment, in "Dormancy and Survival", Symp. Soc. Exptl. Biol., Cambridge Univ. Press, pp. 51-81 (1969).
133. R. Deslauriers, R. A. Byrd, H. C. Jarrell and I. C. P. Smith, ^{31}P NMR studies of vegetative and encysted cells of Acanthamoeba castellanii - Observation of phosphonic acids in live cells, Eur. J. Biochem. 111:369 (1980).
134. R. Deslauriers, H. C. Jarrell, R. A. Byrd and I. C. P. Smith, ^{31}P NMR studies of metabolism in Acanthamoeba castellanii: polyphosphate release from encysted cells, Biochem. Biophys. Res. Commun. 95:1211 (1980).
135. G. S. Jacob, J. Schaefer, E. O. Stejskal and R. A. McKay, Magic-angle ^{15}N NMR study of nitrate metabolism of Neurospora crassa, Biochem. Biophys. Res. Commun. 97:1176 (1980).
136. M. Shporer and M. M. Civan, Effects of temperature and field strength on the NMR relaxation times of ^{23}Na in frog striated muscle, Biochim. Biophys. Acta 354:291 (1974).
137. H. Monoi, Nuclear magnetic resonance of tissue ^{23}Na correlation time, Biochim. Biophys. Acta 451:604 (1976).
138. M. Goldberg and H. Gilboa, Sodium exchange between two sites. The binding of sodium to halotolerant bacteria, Biochim. Biophys. Acta 538:268 (1978).
139. M. M. Civan and M. Shporer, Pulsed NMR studies of ^{17}O from $H_2{}^{17}O$ in frog striated muscle, Biochim. Biophys. Acta (1974).
140. M. M. Civan and M. Shporer, Pulsed nuclear magnetic resonance study of ^{17}O, ^{2}D and ^{1}H of water in frog striated muscle, Biophys. J. 15:299 (1975).
141. M. Shporer and M. M. Civan, NMR study of ^{17}O from $H_2{}^{17}O$ in human erythrocytes, Biochim. Biophys. Acta 385:81 (1975).
142. R. Damadian and F. W. Cope, Potassium nuclear magnetic resonance relaxations in muscle and brain, and in normal E. coli and a potassium transport mutant, Physiol. Chem. Phys. 5:511 (1973).

HIGH RESOLUTION NMR STUDIES OF NUCLEIC ACIDS

Cornelis Altona

Gorlaeus Laboratory of the University,
P.O. Box 9502, 2300 RA Leiden, The Netherlands

INTRODUCTION

Nuclear Magnetic Resonance (NMR) spectroscopy is by far the most powerful physical technique available today for the study of conformation, structure and other properties of (bio)molecules in solution. NMR spectroscopy on the one hand stands as a mature branch of science, used by scores of research groups throughout the world as a tool to solve structural, conformational and kinetic problems in almost all fields of chemistry; on the other hand it is a field in which astounding new developments still take place on a great scale. These new developments pertain both to "hardware", (e.g. the most powerful superconducting systems from the mid-seventies are already superseded) as well as to "software". Computer-controlled pulse-sequences now allow the execution of experiments that were unheard of a few years ago.

Nucleic acids constitute a very interesting class of compounds, not only from the standpoint of biology and biochemistry but also as seen through the eyes of the crystallographer, the conformational analyst or the NMR spectroscopist, to name but a few. Nucleic acids contain several types of atomic nuclei that have attracted the attention of NMR spectroscopists: ^{1}H, ^{13}C, ^{31}P, ^{15}N, and some others that for obvious reasons have not enjoyed popularity to date (^{2}H, ^{14}N, ^{17}O, ^{33}S). Some NMR properties of these nuclei are collected in Table 1.

By far the largest portion of NMR work on nucleic acids during the last two decades has been carried out by means of proton spectroscopy. ^{1}H NMR has the advantages of great sensitivity compared to NMR of other nuclei, widespread availability of instrumentation,

Table 1. Properties of Atomic Nuclei of Main Interest for NMR Spectroscopy of Nucleic Acids.[a]

	^{1}H	^{13}C	^{15}N	^{31}P
% Natural Isotopic Abundance	99.985	1.108	0.37	100
Rel. sensitivity for equal no. of nuclei	1.00	1.59×10^{-2}	1.04×10^{-3}	6.63×10^{-2}
Rel. sensitivity in natural abundance	1.00	1.76×10^{-4}	3.85×10^{-6}	6.63×10^{-2}
Rel. NMR frequency at constant field	100	25.14	10.13	40.48
Nuclear spin	1/2	1/2	1/2	1/2

[a] Data from Bruker NMR/NQR Periodic Table.

and versatility. The parameters obtained (chemical shifts, spin-spin coupling constants, relaxation times, nuclear Overhauser intensity enhancements) often allow insights into intimate geometrical and thermodynamical details of conformational behaviour. However, one should not neglect the possibilities offered by ^{13}C, ^{15}N and ^{31}P spectroscopy, each of which can yield additional information not available from ^{1}H NMR. As matters stand today, no ^{17}O studies appear to have been carried out on nucleic acid constituents, whereas the number of ^{13}C and ^{31}P studies appears to be on the increase and it can b foreseen that the advent of two-dimensional (2D) NMR will stimulate much research in the near future.

The present paper is not intended as a general review covering all main aspects of NMR as applied to nucleic acid conformation and chemistry. The subject has already become much too vast for such a treatment. Instead, a selection of topics is presented. As the critical description of each topic will be oriented more toward aspects of problem solving than toward the various technical and theoretical aspects of NMR spectroscopy or, for that matter, toward the details of results obtained. Emphasis will be laid on scope and limitations of the different possible approaches. Key references provide entries into the more technical details. Discussions will emphasize structural and conformational problems, therefore, reference to other powerful techniques besides NMR, e.g. single crystal X-ray analysis and circular dichroism (CD), will be made in order to correct for a too onesided view. This reflects the author's contention that continuous efforts to aim at the judicious combination of information obtained on a given compound or class of compounds by means of several (bio)chemical or physical techniques, aided by theoretical calculations, are highly rewarding in the long run.

Principles underlying various approaches to NMR spectroscopy will not be treated. Familiarity with standard Fourier transform (FT) techniques is assumed. Background and details are plentiful in some of the references cited, especially those in recent authoritative books or chapters in books.

CONFORMATIONAL ANALYSIS [1]

A few words on the basic tenets of conformational analysis are now in order. The rationale behind this section is the fact that the literature abounds with examples of erroneous use of conformational principles and sometimes flagrant violations of the ground rules of this branch of science are noted.

Let us start from the primary element that the discipline of conformational analysis concerns itself with the various possible three-dimensional shapes that a given molecule can assume by rotation around one or more (nominally single) chemical bonds. Covalent bonds should remain essentially intact during these rotations. Each of the infinite number of geometries, described as points on the potential energy surface, that are met during rotation can be called a conformation. The stable molecular geometries that correspond to (local) energy minima in the total conformational (potential energy) space are called conformers or rotamers, as the case may be. In practice these definitions are arbitrarily narrowed to exclude molecular species that are separated by rotational energy barriers of such a magnitude that they can be isolated and studied as pure chemical entities at ambient temperatures. Hindered biphenyls are a classical example. At the same time a lower limit must be placed on the rotational barrier. Although in open chain compounds so-called "free-rotation" does not exist (barriers range from about 1 kcal.mole^{-1} up to about 15 kcal.mole^{-1} or more), one encounters cases, e.g. in the field of five-membered rings, where torsional and bond-angle strain compensate in such a way that the molecular shape cannot be described in terms of a thermal equilibrium between discrete conformers. A quantum-mechanical description in terms of torsional energy levels that encompass a broad range of geometries then becomes more appropriate. Cases in point are molecules like cyclopentane and tetrahydrofuran. In short, for a proper conformational analysis one presupposes that the interconversion barriers lie between about 1.0 and 20-22 kcal.mole^{-1}.

The following questions constitute the central themes of conformational analysis.

1) What is the geometry of each conformer present in the equilibrium blend? Of course, if the blend is strongly biased toward a single (dominant) species attention is necessarily limited to this species. The term "geometry" can be interpreted on various levels of sophistication. If we take a look at chlorocyclohexane, for example, for many chemical purposes it may be sufficient to state that an equilibrium exists between the axial and equatorial conformers in ideal chair geometries. Similarly, the 5-membered sugar ring in nucleic acids is most often described in terms of an equilibrium between the C3'-endo and C2'-endo forms.

On a higher level one is concerned with more precise geometrical details, i.e. bond distances, bond angles and torsion angles. Although in simple rigid molecules bond distance and -angle information can be gleaned from NMR studies in liquid crystal solvents, this approach is less practical in conformational work. Instead, one takes bond length information from other sources (X-ray diffraction, electron diffraction, microwave techniques) and assumes constant length, independent of molecular surroundings (solid, solution or gas). For lack of better means, bond angles usually are obtained in the same way. However, it should be remembered that in cyclic systems endocyclic bond angles and endocyclic torsion angles are linked by geometrical relationships.[2] Remains the determination of the torsion angles ϕ. A thorough analysis of three-bond spin-spin coupling constants $^3J_{HH}$, corrected for the effect of electronegativity and orientation of substituents[3] provides this information along CH-CH bonds. More approximate $\cos^2\phi$ methods serve to delineate torsion angles with the aid of $^3J_{CH}$ and $^3J_{PH}$.

Returning to the example of chlorocyclohexane, certain physical data require that the axial carbon-halogen bond does not occupy an ideal staggered position but is splayed outward. In the course of the 1960's it was established that cyclohexanes and heterocyclic 6-membered rings do not conform to the ideal tetrahedral shape ($\phi = 60^o$) postulated by Sachse[4a] and by Mohr[4b] but are either flattened or super-puckered. The actual geometry assumed depends mainly on the requirements imposed on the endocyclic bond angles by the atoms that make up the ring skeleton. Steric repulsions between axial substituents also play a role, leading to extra flattening.[5] Cyclohexanes and e.g. 6-membered ring sugars are flattened by at least 4° owing to the fact that the "natural" C-C-C bond angle $>109.5^o$. This flattening automatically brings about a splaying out of the axial valencies.

More subtle is the reaction of the 5-membered ring toward endocyclic or exocyclic substitution. Next to simple overall flattening or super puckering the 5-membered ring adapts itself to torsional and steric strain by adjusting its phase angle of pseudorotation. Therefore, classifications like C3'-endo or C2'-endo forms for the sugar ring in nucleosides and poly(nucleotides) must be seen as rough (first order) approximations of the true situation, useful, but not necessarily accurate.[3d,6,7] The point that I wish to make here is the following: just as in the case of cyclohexanes, the exact conformational preference of the sugar ring in nucleic acids is unimportant from the standpoint of the synthetic chemist. It is of crucial importance, however, for the correct interpretation (and prediction) of physico-chemical properties as well as, in part, for the understanding of the difference in behaviour between DNA and RNA.

2) What is the magnitude of a given physical property associated with each of the pure conformers present in the equilibrium blend? Given these magnitudes question 3), vide infra, can be answered easily. However, experimental isolation of the properties of pure conformers often is an impossible task. In cyclohexane chemistry recourse can be taken to the introduction of so-called "holding-groups", but this approach is not possible in the 5-membered ring series.[1,2a] Conformational studies of open-chain compounds, including exocyclic groups, are similarly hampered. Again, 3-bond NMR coupling constants, in conjunction with statistical data from X-ray studies, may provide a useful starting point for a detailed conformational analysis.[3c,d] In the case of dinucleotides and trinucleotides, extrapolation to the properties (CD and NMR) associated with the fully stacked single helical form in aqueous solution can sometimes be carried out with success.[7]

3) What are the relative populations of the equilibrating conformers at a given temperature? In conformational analysis of nucleic acid oligomers it is imperative to answer this question: What is the relative enthalpy and entropy content of each conformer?

$$\Delta G = \Delta H - T\Delta S \qquad (1)$$

Knowledge of ΔH and ΔS is more essential than knowledge of ΔG at a single temperature because the random coil $\rightleftharpoons$ single-helical stack equilibrium in DNA and RNA fragments is characterized by a balance between large and opposing factors: ΔH and $T\Delta S$. Typical ΔH values range about -6000 to -8000 $cal.mole^{-1}$, ΔS values between about -18 and -25 $cal.mole^{-1}.deg^{-1}$ appear most common. At the "transition temperature" T_m the free energy difference ΔG by definition is zero:

$$\Delta G = \Delta H - T\Delta S = 0 \qquad (2)$$

Let us consider the thermodynamics of ApA and dApdA as an example:

	$-\Delta H$	$-\Delta S$	T_m	Ref.
ApA	7.2	24.5	295	8
dApdA	7.3	22.7	322	9

(ΔH in $kcal.mole^{-1}$, ΔS in $cal.mole^{-1}.deg^{-1}$, T_m in K)

It is of interest to note that the "heat of melting" of the vertical base-base stacked form is the same in the ribo and in the deoxy derivative. The appreciable difference in T_m of 27° appears to stem solely from a small (7 percent) difference in the entropy factor, perhaps reflecting the greater conformational freedom in stacked dApdA.[9] Obviously, comparisons of "percentage of stack" obtained at ambient temperature for a series of dinucleotides could easily lead to erroneous interpretations in terms of "stronger base-base interactions".

4) Which are the preferred geometrical pathways (saddle points) that connect the deepest minima on the multi-dimensional potential energy surface? In other words, what are the distortions the molecule has to suffer when proceeding from one conformer to the next?

5) What are the activation energies associated with the conformational interconversions, i.e. the heights of the energy barriers that separate the various stable species in terms of $\Delta H^{\ddagger}$ and $\Delta S^{\ddagger}$, or expressed in interconversion rates at a given temperature?

6) Finally, a question can be posed that resides on the borderline between conformational analysis and vibrational analysis: How large are the (torsional) motions described by the molecule or part(s) of the molecule <u>within</u> a given conformational energy well?

It cannot be emphasized strongly enough that experimental knowledge of the details of a given conformational equilibrium (i.e. an experimental answer to questions 1-3) <u>does not allow any conclusions</u> regarding answers to questions 4-6. In fact, the actual <u>interconversion pathways taken by the molecule are immaterial to equilibrium conformational analysis</u> as long as the barriers are between the limits discussed above. Question 6 has special relevance for NMR spectroscopy of nucleic acids because it has been recently discovered[10,11] that relatively large amplitude internal motions of base, sugar and phosphate backbone occur within the intact double helix of DNA molecules, 150-300 base pairs long. In fact, without the existence of these motions ^{1}H, ^{13}C and ^{31}P NMR of such stiff rod--like molecules would not be practical. The rates deduced for the internal motions (about $10^9 s^{-1}$) correspond to lattice vibrations with frequencies of the order of 0.1 - 0.01 cm^{-1}.

NITROGEN - 15 NMR [12,13]

The ^{14}N isotope is by far more abundant than ^{15}N. However, ^{14}N NMR is difficult because the absorption lines suffer from quadrupolar broadening and are usually 50 - 1000 Hz wide and often wider still. In contrast, ^{15}N nuclei yield spectra characterized by lines as sharp as those displayed by ^{13}C and ^{31}P. For this reason ^{15}N NMR is greatly preferred over ^{14}N spectroscopy. The ^{15}N and ^{14}N chemical shift data may be used interchangeably, i.e. there is no observable isotope effect on the screening constant. The range of N-shifts in organic compounds is comfortably large, about 800 ppm. Several shift reference compounds are in popular use, the most common ones are ^{15}N labelled NO_3^- ion (or CH_3NO_2) and ^{15}N labelled Me_4N^+ (or NH_4^+) ion. The chemical shift difference between these groups of standards is about 330-340 ppm, with ammonia upfield from nitrate. Several authors denote upfield shifts as positive, whereas others follow the common convention for ^1H NMR and take downfield shifts (from R_4N^+) to be positive.

Unfortunately, several factors combine to make ^{15}N studies more expensive than ^{13}C or ^{31}P NMR, either in terms of machine-hours or in terms of cost of ^{15}N isotopic enrichment. These factors are the small natural abundance (0.37%), the low relative sensitivity (roughly 15 times less than that of ^{13}C) and the negative sign of the magnetogyric ratio γ of the ^{15}N nucleus. The latter property implies that proton noise decoupling may give rise to various phenomena. The nuclear Overhauser effect[14] (NOE, redistribution of spin population among the spin levels under the action of strong decoupling conditions) on broadband irradiation of the proton frequencies often leads to a strongly enhanced but inverted signal in cases where dipole-dipole interactions are dominant. If these are not dominant, the NOE may lead to small positive or negative peaks or even result in a complete disappearance of the signal.

Natural abundance ^{15}N NMR can be carried out on conventional spectrometers (e.g. ^1H 90 or 100 MHz, 10 mm sample tube) provided the molar concentration can be made sufficiently high. Concentrated solutions, 5-10 M, require ca 20,000 pulses with a delay between pulses of a few seconds. However, for many reasons much lower concentrations are desired in the study of nucleic acids. An experiment on a 0.05 M solution on a wide bore instrument (^1H 200 MHz, 25 mm sample tube) again would require the accumulation of ca 20,000 transients or more, i.e. a running time between 12 and 24 h. It takes little imagination to conclude that an extensive study of pH or temperature dependence of ^{15}N spectra of nucleic acid polymers without isotopic enrichment presently appears rather impractical and expensive. Nevertheless, a study on the natural abundance ^{15}N NMR spectrum of tRNA has appeared.[15]

^{15}N NMR of isotopically enriched nucleosides and nucleotides appears quite useful to detect favoured interaction sites; problems such as preferred hydrogen-bonding schemes, protonation sites and cation binding can be solved. A pioneering study on the interaction between Zn^{2+} (and Mg^{2+}) and 5' ATP was published as early as 1966.[16] The assignment is shown in Table 2. Note that the original[16] assignment of N3 and N1 signals had to be interchanged by later workers.[17] The new assignment is used in Table 2. It was found[16] that Zn^{2+} produces shifts, indicative of binding, in the 6-NH_2, N7 and N9 resonances of 5' ATP. Interestingly, no shifts were observed on addition of Mg^{2+} ions.

Table 2. Nitrogen Shifts of 5' ATP in H_2O.[12,16,17] Shifts are Expressed Relative to External Me_4N^+, Downfield Shifts are taken Positive (ppm).

NH_2	39.5	(s)	
N9	133.7	(s)	
N3	176.6	(d)	$^2J_{NH}$ = 16 Hz
N1	185.7	(d)	$^2J_{NH}$ = 16 Hz
N7	191.8	(d)	$^2J_{NH}$ = 10 Hz

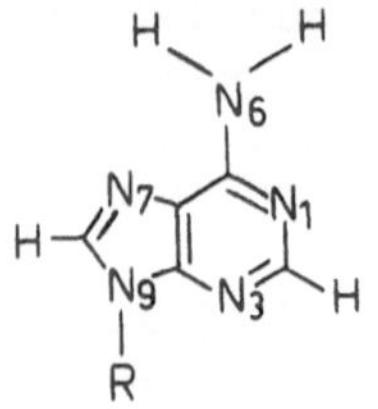

Natural abundance ^{15}N studies of the common ribonucleosides and of deoxyadenosine (0.5-1 M in DMSO) and also of the corresponding 5'-phosphates (in water) have been conducted.[17a] The addition of excess trifluoroacetic acid caused N1 of Ado and N7 of Guo to shift upfield by 71.7 and 66.3 ppm, respectively. Protonation shifts in water were somewhat smaller. It was also found that N1, N3 and N7 of 5' AMP in water shift 7-12 ppm upfield compared to that observed in nonprotic solvents; the resonance of N9 remains approximately constant. These shifts were attributed to hydrogen bond formation with the solvent. It would be of interest to repeat the experiments at much lower concentration.

The common nucleoside 3'-phosphates, biologically enriched at all nitrogen positions, have been studied in water over a large pH range.[17b,c] Very large shifts were observed on protonation (or deprotonation) of the nitrogen bases. The deprotonated nitrogen invariably resonates at much lower field compared to its protonated counterpart. Examples are N3 of 3' UMP and 3' CMP (40 and 60 ppm, respectively) and N1 of 3' AMP and 3' GMP (about 70 ppm) in the appropriate pH ranges. The other N resonances move relatively little and then mostly in opposite direction. At pH values below 3 an upfield shift of the N7 resonance of 3' GMP (not of N7 in 3' AMP) occurs which indicates preferred protonation at N7 under these conditions. Both groups of workers[17] concluded that protonation at N3 of guanosines is much less favourable.

Interbase hydrogen bonding is also expected to reveal itself in the nitrogen chemical shifts. An attempt to observe such shifts in an equimolar mixture of uridine and adenosine in DMSO was unsuccessful.[18] However, DMSO is long known to be unsuitable for the study of A.U base pairing.[19] More indicative proved a study of the synthetically N3-labelled tribenzoate of uridine in the presence of various amounts of 5'-acetylated 2',3'-O-isopropylidene adenosine in deuteriochloroform.[20] The N3 doublet moved downfield to a limiting value ($\Delta\delta$ = 4.7 ppm) in the presence of a three molar excess of the purine nucleoside. One notes that the hydrogen-bonding shift is about ten percent of the full deprotonation shift at N3 of uridine.

Whether this behaviour is general for nucleic acid bases remains to be investigated.

With the widespread advent of superconducting spectrometer systems equipped with multinuclear probes one may expect a rapid increase in application of ^{15}N NMR to specific problems of nucleic acid chemistry in the near future. New techniques are actively being explored. For example, the enhancement of protonated ^{15}N signals by magnetization transfer via J_{NH} couplings by as much as a factor of seven greater than the NOE has been reported recently.[21] Unfortunately, the method is not, in general, applicable to macromolecules.

PHOSPHORUS - 31 NMR

The presence of the ^{31}P nucleus in the backbone of RNA and DNA and their constituents makes ^{31}P magnetic resonance an attractive method for structural studies. The sensitivity of ^{31}P is about 7% of that of protons. However, this is hardly a shortcoming in view of its 100% natural abundance and the possibilities of modern NMR spectrometers. The chemical shift range of this nucleus extends over about 500 ppm.[22] The actual shifts for the phosphate mono- and diesters observed in nucleic acids comprise only a small fraction of total ^{31}P shift range. For each residue only one phosphorus atom is present against many ^{1}H and ^{13}C nuclei and far fewer resonance lines are observed. This is not much of an advantage for detailed assignment purposes, as we will see, but at the same time it offers unique possibilities to study chemical and conformational properties that are not easily probed otherwise. Still, one should best view ^{31}P magnetic spectroscopy of nucleic acids as supplementary to ^{13}C and ^{1}H NMR. The theory of ^{31}P chemical shifts was originally developed by Letcher and van Wazer[23], who suggested that ^{31}P shifts are affected by changes in bond angles, P - X ligand electronegativity and π-bonding. In general, the factors that have a more or less important influence on the ^{31}P chemical shift in nucleic acids are: (i) the ionization state of the phosphate, of particular relevance for terminal phosphates; (ii) changes in the O-P-O diester bond angle;[23,24] (iii) changes in the phosphate torsional angles (which are coupled to O-P-O diester angles);[24,25] (iv) complexation with divalent cations such as Mg^{2+};[26] (v) the substitution pattern of (at least) the α-and β-carbons attached to the phosphate ester oxygens;[27] (vi) the temperature, as exemplified in cases where the phosphate group is incorporated in a rigid 6-membered ring;[27] (vii) changes in solvation and/or hydrogen bonding;[28–30] (viii) the nature of the sugar ring for a 3'-bonded phosphate (ribo vs deoxy, perhaps this effect reflects a combination of (v) and (vii) above);[29] (ix) the nature of the base at the 3' and 5' ends.[27,29,31] The latter influence appears relatively minor compared to the dominating effect of ring current shifts on proton spectra of nucleotides, but certainly cannot be neglected, as is often done.

Some remarks on technical aspects of ^{31}P NMR are now in order. There is the old question of selection of the chemical shift reference compound and also that of definition of upfield or of downfield shifts being positive. External H_3PO_4 (85% or, in modern work, 15-20%) is a very popular reference standard in ^{31}P NMR but the values of the line positions will depend on the bulk susceptibility of the sample. This is usually not a problem in work carried out in a given solvent at a constant temperature. However, in shift _vs_ temperature studies, where accuracy is of prime importance, recourse must be taken to an internal standard. For this reason several groups have used internal trimethylphosphate (TMEP) as chemical shift reference.[30,32-34a] Still, this choice may not represent the best one possible. It was found[27] that the ^{31}P shift position of TMEP moves non-linearly downfield by more than 0.5 ppm between 0 and 100 ^{o}C. This temperature profile virtually parallels that displayed by 3',5'-cAMP (A>P). For this reason, and in analogy to tetramethylammonium chloride (TMA), a well-known and satisfactory reference for nitrogen and proton NMR, a new standard was proposed:[27] tetramethylphosphonium bromide (TMPB). TMPB has the following characteristics: (i) its resonance position is 22.9 ppm downfield of external 85% H_3PO_4 at 25 ^{o}C, (ii) the compound is water-soluble and the ^{31}P shift is independent of pH and nucleotide concentration, (iii) the compound is magnetically isotropic and no hydrogen bonding is expected. The chemical shift of A>P can be used as a secondary standard when corrections are necessary for temperature-induced shifts. Throughout the present paper shifts are denoted using the δ convention, i.e. positive values represent shifts to _lower_ field.

Rigorous removal of paramagnetic contaminants from ^{31}P NMR samples proved to be absolutely necessary;[29] signal line widths of 0.3 Hz or less are then routinely obtained for simple mono- and dinucleotides on spectrometers operating at 40.5 MHz. Much broader lines are observed for large nucleic acid molecules, e.g. in the case of tRNA (MW $\sim$27,000) about 3 Hz line width at 40 MHz and 14 Hz at 109 MHz.[32] This broadening is attributed to the anisotropic chemical shift relaxation mechanism, which should give rise to a width proportional to the square of field (but see also ref. 34b). Considering resolution and sensitivity, Guéron and Shulman[32] showed that sensitivity increases with field in the same way for protons and phosphorus, however, resolution actually diminished at high field since the width of the ^{31}P signals increases as ω^2 whereas the chemical shift splittings go as ω only. Another problem in high-field ^{31}P (and ^{13}C) NMR is the substantial heating of the sample under proton noise decoupling conditions; without special measures the sample temperature may easily increase by as much as 21 ^{o}C at 146 MHz (8.46 T, 360 MHz ^{1}H).[24e] Design improvements have largely alleviated the latter problem on present-day spectrometers.

An interesting application of ^{31}P NMR is the study of the interaction of Mg^{2+} ions with the nucleoside triphosphates ATP, GTP, CTP and UTP in aqueous solution.[26] These four nucleotides behave similarly. Purine and pyrimidine bases have little effect on the phosphate groups even in the N1 and N7 pK region of ATP and GTP, respectively. The Mg^{2+} ion binds exclusively to the β-phosphate.

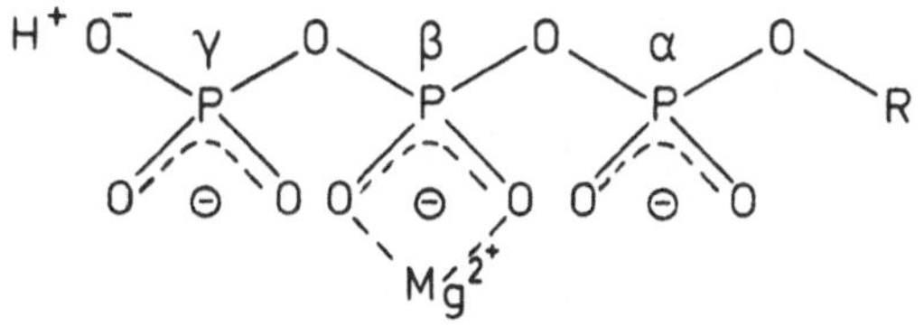

Terminal phosphate resonances (monophosphate) can be spotted easily, even in large molecules, by their pH titration behaviour.[29,32,35] Similarly, a good check on the integrity of tRNA samples is afforded by ^{31}P NMR. Traces of nucleases may introduce nicks in the sugar-phosphate backbone of the molecule with the concomitant formation of a 2',3' cyclic phosphate,[32,35,36] this is then revealed by a characteristic signal about 20 ppm downfield from the main cluster of phosphate peaks. The ^{31}P spectrum of tRNAs has been studied by several groups.[24e,11,35,36] The most complete study to date is that of Salemink et al.[36] who used not only intact $tRNA^{Phe}$ but also combined the principles of chemical modification, specific fragmentation, "and melting" vide infra. The ^{31}P spectra of native tRNAs studied so far are characterized by one main cluster of resonances (approximately 60 phosphates, including the about 42 phosphates in cloverleaf duplex regions) and a number of well resolved peaks at both high and low field from the main cluster. About 17 phosphate resonances are resolved from the main resonance in yeast $tRNA^{Phe}$.[36] It was concluded that these resolved resonances are generated by the specific folding and/or hydrogen bonding of the phosphate backbone, imposed by the tertiary structure and the interactions within the loops.[36] Gorenstein and coworkers[24] interpret the various ^{31}P resonance positions observed in oligonucleotides and in tRNA almost entirely in terms of the conformation of the phosphodiester group, i.e. O-P-O bond angle distortions in the tertiary structures. The scattered downfield peaks would represent gauche-trans (gt) or trans-gauche (tg) conformers about the phosphodiester bonds with bond angles several degrees smaller than normal for gauche-gauche (gg). Of course, a regular righthanded stack implies a g^-g^- phosphate conformation. The scattered upfield peaks in this view represent phosphates with bond angles several degrees larger than normal. Salemink et al.[36] infer that these upfield shifts come from phosphates hydrogen-bonded to adenine NH_2 groups. On the other hand, Evans and Kaplan[37] indicate that hydrogen bond formation in 8-amino--AMP produces upfield shifts of only 0.4 ppm. Finally, Lerner and Kearns[30] recently issued a warning to the effect that the change in solvation, measured by using mixed organic solvent/water solutions, can cause upfield shifts greater than 3 ppm. Therefore, steric constraints which prevent full hydration of a given phosphate by water molecules might be responsible for these upfield shifted resonances in tRNA. Evidently, this matter needs further attention.

Other recent applications of ^{31}P NMR include the study of DNA duplexes,[11,34] deoxyoligonucleotides,[31] DNA unwinding,[38] 5S RNA,[39] and codon-anticodon interaction in tRNA.[40] The reader is referred to the papers cited and the references contained therein in order to appreciate the many excellent possibilities offered by phosphorus NMR in the field of DNA and RNA structure and dynamics.

Finally, it should be mentioned that ^{31}P NMR <u>cannot</u> be used to monitor the quantitative thermodynamics of the unstack ⇌ stack equilibrium of dinucleoside monophosphates in a straightforward way.[27] This is so because the geometry about the P-O ester bonds must be g^-g^- in the righthanded stacked conformation, whereas the unstacked molecules exist in a large number of states characterized by conformational freedom of the sugar and of the

C3'-O3'-P-O5'-C5'-C4' (ε-ζ-α-β-γ)

backbone sequence.[3a,c,7-9] The conformational freedom about the (ζ,α) P-O bonds appears to include most, if not all, of the eight possible combinations of g and t (excepting the fully extended tt form), although not necessarily in equal amounts. Therefore, under conditions where conformational transitions are fast on the NMR time scale, the recorded ^{31}P chemical shift is a weighted time-average of the gg and gt conformations. In short, ^{31}P shifts thus should be analyzed in terms of the following model:

(i) a stacked conformer having a g^-g^- conformation,
(ii) a blend of unstacked forms having gg orientations (various combinations of + and -),
(iii) a blend of unstacked forms having gt and/or tg orientations.

Under the assumption that all + and - combinations of gg resonate at approximately the same frequency and all gt/tg combinations together at another frequency, located downfield from gg,[24] the problem is reduced to a three-state analysis, Figure 1. This analysis[27] was carried out on a model compound, m_2^6ApU. The thermodynamic parameters of stacking of this compound had been determined[41] from both NMR and circular dichroism measurements; it shows a large shift of the destack-stack equilibrium over the accessible temperature range (15% stack → 78% stack going from 100 °C to 0 °C) and should be eminently suited for the purpose. A satisfactory least-squares fit to the ^{31}P shift <u>vs</u> temperature profile was obtained with the gt resonance approximately 2.35 ppm downfield from the gg signal and the unstacked gt blend having a slightly higher overall population compared to the unstacked gg blend,[27] see Fig. 1.

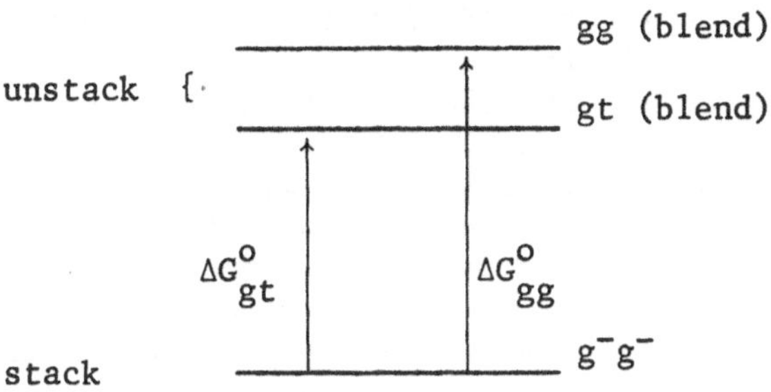

Fig. 1. Energy diagram for a three-state unstack-stack equilibrium.

CARBON - 13 NMR

^{13}C NMR spectroscopy[42-45] has come to stay. This nucleus exhibits a 200 ppm range of chemical shifts and during the past decade a tremendous rise in its popularity for the structural analysis of organic molecules has followed the commercial introduction of FT techniques. The methyl peak of tetramethylsilane (TMS) has become the standard reference signal and chemical shifts are taken as positive in downfield direction. Information may be obtained directly from carbon atoms constituting the molecular backbone. With complete proton decoupling each magnetically distinct carbon appears as a sharp single line. Also, the sensitivity is increased up to a factor of about three in the AX case as a consequence of NOE enhancement. A superior enhancement technique by magnetization transfer via C-H couplings was recently published.[21] Still, the application of ^{13}C NMR to conformational analysis of nucleic acid constituents is beset with difficulties. A similar remark can be made concerning the study of high-molecular weight RNA and DNA.

Anet[42b] has discussed the advantages and disadvantages of high-field (superconducting systems) ^{13}C NMR. Although higher fields should lead to higher sensitivities, several problems can occur to negate this advantage. This is because in large molecules the NOE[14] enhancement becomes frequency dependent and can be much smaller at high fields than at low fields for a certain narrow range of tumbling rates. Small and medium-sized molecules, e.g. steroids, dissolved in mobile organic solvents at room temperature and above have correlation times $\tau_c \leqslant 10^{-10}$ s and NOE is at a maximum even at fields corresponding to 360 MHz proton NMR. Under these conditions the higher magnetic field will yield spectra of considerably better signal-to-noise ratio than will lower fields, if other things are equal. This is no longer true when $\tau_c \sim 10^{-9} - 10^{-8}$ s, i.e. values typical for DNA, tRNA and poly (A). The reader is referred to recent papers by Bolton and James[34] and by Hogan and Jardetzky[11] for appreciation of this point. In short, natural abundance ^{13}C NMR studies of high-molecular weight nucleic acids are best carried out on (wide-bore) instruments of the

200 MHz class. The experimental ^{13}C investigations on DNA mentioned above[11,34] (see also 46) in fact were not aimed at elucidating structural details, but made use of line-broadening effects to learn more about the direction and amplitude of internal motion in duplexes 140-260 base pairs long. Typical ^{13}C line widths for these polymers are: ribose carbons, 150-220 Hz; non-protonated aromatic carbons, 290-350 Hz; methyl carbon, 70 Hz.[11] Some interesting discrepancies are noted between the conclusions reached by different investigators. According to Hogan and Jadetzky[11] the base planes at positions C6 and C8 and the deoxyribose carbons C1', C2' and C3' all experience $\pm$ 20° fluctuations in geometry which occur with a time constant near 10^{-9} s. On the other hand, Bolton and James[34c] conclude that the internal motion of the bases is restricted relative to that of the ribose for DNA and tRNA by at least an order of magnitude.

An entirely different approach to the use of ^{13}C NMR for the study of ^{13}C isotopically enriched tRNAs is exemplified by the work of Agris and coworkers,[47-49] Schweizer et al.,[50] and Yokoyama et al.[51] E.coli or Salmonella was grown under conditions where bases labelled at a specific position are *in vivo* incorporated into tRNAs by blocking the normal biosynthetic pathways.[47-51] It has been stressed[47] that these carbon probes do not affect the native conformation for they are simply isotopic replacements at natural sites achieved *in vivo*. From here on different lines of investigation have been followed. Schweizer et al.[50] chose the carbonyl C4 position as a site for labelling because (1) it is a quaternary carbon having a fairly narrow resonance and (2) this site is involved in secondary and tertiary structure hydrogen bonding and thus may be expected to be sensitive to its surroundings.[50b] Purified $tRNA^{Val}$ (Salmonella) showed assignable DHU, S^4U and uridine 5-oxyacetic acid (V) ^{13}C signals. The uridine (U, Ψ, rT) C4 resonances were spread out over 3-4 ppm and the shifts of individual lines could be followed as function of temperature, but assignments could not be made. Schmidt et al.[49] used unfractionated tRNAs (coli) enriched in either position 2 of adenine (60 atom %) or in position 2 of uracil (82%) and cytosine (63%). The ^{13}C NMR spectra were recorded in H_2O/D_2O (90:10). Adenine C2 with a directly bonded proton has resonances of about 40 Hz line width (judging from T_2 values) but in folded tRNA the C2 peak was about six times wider (5 ppm); it narrowed as the molecule unfolded. This was ascribed to unresolved chemical shift non-equivalence, which is progressively lost due to loss of ordered structure. It seems unlikely that A C2's will be of much help in future work. The C2 of cytosine appeared quite insensitive to surroundings judging from its intrinsically narrow peak at both low and high temperatures. In contrast, C2 of uracil may turn out to be a resonably good candidate for ^{13}C NMR of purified tRNAs.

The labelling discussed above is at locations all around the cloverleaf structure of tRNA and specific assignments of non-modified bases cannot be made. Instead, ^{13}C enrichment of the methyl groups of modified nucleosides offers interesting perspectives and ^{13}C NMR investigation is expected to supplement the proton NMR research of certain tRNA regions. In early experiments Chang and Lee[51b] identified several methyl carbon resonances in a spectrum of yeast tRNA with the use of a ^{13}C enriched methylating agent. Tompson et al.[47] studied _in vivo_ ^{13}C-methylated (57%) nucleoside resonances in unfractionated tRNA (coli) and reported assignments for methyl carbons of rT, ms^2i^6A, m^2A, m^6A, m^1G, mam^5s^2U, m^7G, methyl ester of V base (mo^5U) and 2'-O-methylribose nucleosides (Gm, Cm, Um). It was pointed out[47b] that these modifications represent minor bases of singular locations in tRNA and therefore can serve as intrinsic probes of local molecular conformation. Moreover, their resonances occur in spectral regions free from ribose and major base interference. The methyl carbons were classified in three categories: (i) bonded directly to ring carbon or to sulfur (10-20 ppm); (ii) bonded to nitrogen (20-40 ppm); (iii) bonded to oxygen (40-60 ppm). The detailed assignments[47] were confirmed by Yokoyama et al.[51a] who studied various purified E. coli tRNAs. The CH_2 group of the V base was also assigned to a specific resonance.[51a]

Indeed, the full strength of the enrichment method is realized only with the ability to study resonances from individual carbons within purified species. This objective was recently explored by Agris and Schmidt.[48] E. coli tRNA species specific for phenylalanine, tyrosine and cysteine (57 atom % enriched) were studied at 67.9 MHz. In typical experiments the spectra of $tRNA^{Phe}$ (8 mg in 1.1 ml D_2O, 10 mm tube) were obtained by block averaging 36,000 - 54,000 transients, requiring about 9 hrs of data accumulation for each spectrum. The $tRNA^{Phe}$ spectrum is especially interesting in that it shows two peaks for the rT methyl, 12.5 ppm (major) and 11.2 ppm (minor) under "Mg-free" conditions at 30 °C (Fig. 2).[48] Raising the temperature of the sample to 45 °C leaves only a single resonance at 12.3 ppm (fast exchange on the NMR time scale). In contrast, addition of 10 mM Mg^{2+} to the $tRNA^{Phe}$ sample led to coalescence of the rT resonances to a single upfield peak at 11.1 ppm (30 °C). It should be noted that the rT methyl carbon resonates at 11.1 - 11.2 ppm[51a] in seven other tRNA species in the presence of Mg^{2+}. These findings imply that the rT nucleoside exists in at least two different environments that are slowly interconverting at 30 °C when Mg^{2+} content is low. The ^{13}C spectrum cannot aid us to specify particulars about these environments (conformations of the TΨC loop?), but perhaps this finding helps to clear up some problems connected with the low-field proton spectrum of tRNAs, see below. The m^7G methyl perhaps also samples different environments (10 mM Mg^{2+}, 30 °C).[48] Summing up, it appears that methyl carbon NMR may in the near future serve in probing tRNA conformational equilibria and tRNA-protein and tRNA-tRNA interactions.

In the past, a large number of studies have appeared in which simple pyrimidine and purine nucleosides and nucleotides were

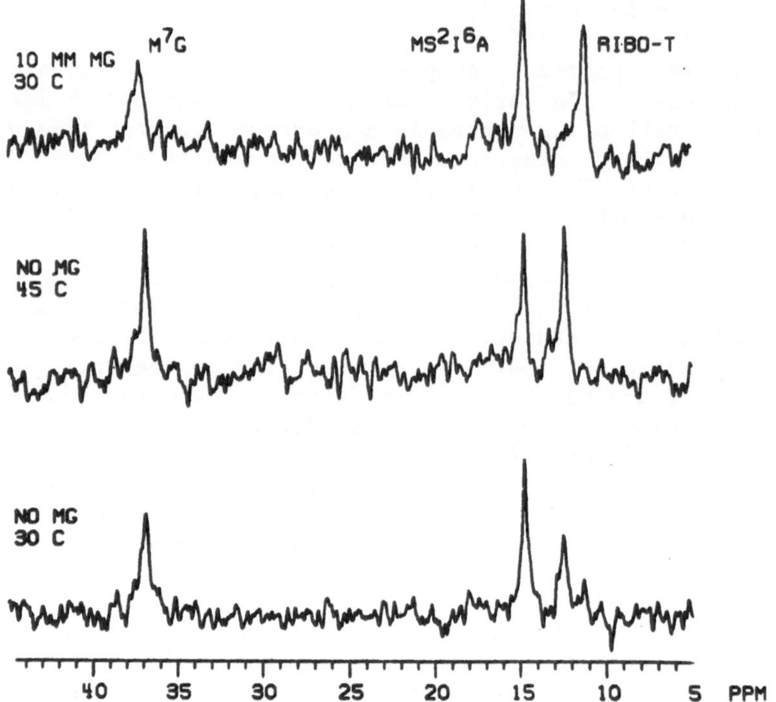

Fig. 2. ^{13}C NMR spectra of tRNA^{Phe} (coli) under different solution conditions. The methyl carbon region is shown. A total of 54,000 transients were averaged for each of the two spectra at 30 °C; 36,000 scans were sufficient at 45 °C (from Ref. *48*)

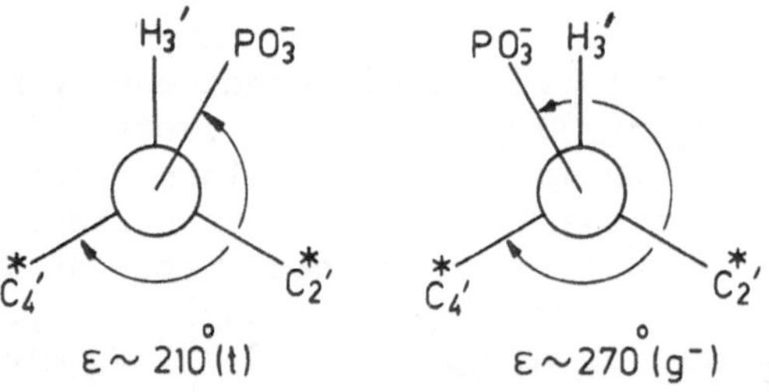

Fig. 3. Newman projection along C3'-O3' (ε) showing the non-classical ε^t and ε^- conformations. Note that the phosphate is *gauche* with respect to H3' in both forms, whereas it is either *anti* or *gauche* with respect to C2' and C4'.

examined by ^{13}C NMR.[52] In general, these spectra have been assigned in a straightforward manner. An exception must be made for the ribose carbons C2' and C3'. For example, the original assignment[42,52] placed C2' at about 71 ppm and C3' at 75 ppm in virtually all ribonucleosides and nucleotides. Following Mantsch and Smith,[53] this assignment was reversed by various authors for both 5' and 3' nucleotides as well as for nucleosides.[52] The correct assignment is of some importance because the three-bond carbon-phosphorus spin-spin coupling, $^3J_{(CP)}$ in principle can serve as a useful probe for the angular preference about the C3'-O3' bond (ε). The couplings $^3J_{(C2'P)}$ and $^3J_{(C4'P)}$ should, in conjunction with the proton-phosphorus coupling $^3J_{(H3'P)}$, give us a clear picture of the rotamer distribution about ε. Alderfer and Ts'o[54] found that the chemical shifts of the 2' and 3' carbons of 3' AMP are pH dependent. Their titration shows that in going from pH 5 (monoanion) to pH 8 (dianion) the respective chemical shifts actually cross over. The original assignment (C3' at lowest field) appeared to be correct for the monoanionic state and for the central phosphates in ribo-oligomers, whereas the Mantsch and Smith[53] reversed assignment holds for the dianion, i.e. terminal 3' phosphate esters at neutral pH. The conformation-sensitive coupling constants $^3J_{(C2'P)}$ and $^3J_{(C4'P)}$ of 3'-AMP and 3'-UMP were also pH dependent; $^3J_{(C2'P)}$ strongly decreases and the C4'P coupling increases in going from the mono- to the dianion, while maintaining a constant sum. With the aid of an approximate Karplus-type equation:[55]

$$^3J_{(CP)} = 9.5 \cos^2\phi - 0.6 \cos\phi$$

the data were interpreted in terms of a shifting population distribution of two nonclassical rotamers ($\varepsilon = 210^o$ and $\varepsilon = 270^o$), Fig. 3. Similarly, temperature-dependent variations of $^3J_{(CP)}$ values of ApA were taken to indicate a shift of the conformational equilibrium towards ε(210) when the bases stack, i.e. at lower temperatures. Recently, accurate parameters for the description of the thermodynamics of the unstack-stack conformational equilibrium as well as reliable $^3J_{(H3'P)}$ coupling constants for some dinucleoside monophosphates have become available[7-9] and a quantitative reassessment of the carbon-phosphorus couplings may well be worthwile.

HYDROGEN - 1 NMR

Proton NMR is an enormously powerful technique for the study of the structure and dynamics of nucleic acids in solution. The chemical shift range of protons bound to carbon, nitrogen and oxygen, i.e. the usual "organic" protons, spans about 15 ppm. This range appears rather smaller than that displayed by the other nuclei discussed above but this apparent drawback is easily offset by a number of advantages inherent to proton NMR. A few of these advantages can be listed as follows:

(i) The natural abundance of 99.9$^+$ %, highest sensitivity compared to other nuclei, and top position on the magnetic resonance frequency scale (only ^3H scores higher on the latter two aspects but, of course, scores zero for natural abundance) combine to minimize instrumental limits of detection. Nevertheless, even ^1H NMR remains a relatively insensitive assay compared to other modern chemical and biochemical analysis techniques. For example, a few milligrams of tRNA must be isolated and purified in order to observe a useful proton spectrum.

(ii) In the sometimes complex spectra of small molecules proton chemical shifts can be measured easily with an accuracy of 0.5 Hz or better relative to a chosen standard. On instruments of the 360-500 MHz class this corresponds to 0.001 ppm. In polymers, where resonances are much broader, one is usually content with shifts accurate to $\leqslant$ 0.01 ppm.

(iii) Proton chemical shifts, unlike shifts displayed by ^{13}C, ^{15}N and ^{31}P, are rather sensitive toward subtle conformational changes along bonds that are far removed in the primary structure but spatially relatively close to the proton in question. A good example is the sensitivity of proton shifts toward the exact orientation of the nucleic acid bases in the vicinity. The ring current effects exerted by these bases can be calculated with some confidence.[55,56] Unfortunately, ^1H chemical shift perturbations caused by neighbouring, non-aromatic but strongly anisotropic, structural elements in the molecule (C=O, C-O, C-N, C=C, to name a few) are hard to calculate with the required accuracy from first principles, although trends can be established.[55c-d] Usually one resorts to the enormous body of available experimental material to work by analogy. However, the flexibility inherent to small nucleic acid molecules makes the latter approach more hazardous than, for example, in the field of rigid 6-membered ring sugars.

(iv) Spin-spin couplings between protons, or between a proton and another nucleus, are a rich source of information. First of all, double or triple resonance experiments are indispensable for the unambiguous identification of each signal in a complex pattern. In its most elementary form one irradiates a given proton signal, thereby effectively removing the spin-spin coupling of this proton to the other nuclei from the spectrum, and observes changes in the multiplet patterns exhibited by these other nuclei. After assignment is completed the multiplets themselves are analyzed by computer simulation of the observed spectrum, for example, by means of program LAME (LAOCOON with Magnetic Equivalence).[57] Especially in cases where a complete set of coupling constants J can be extracted from the spectra, one has access to detailed structural information. References to the use of vicinal (or three-bond) couplings, $^3J_{(HH)}$, for this purpose have already been given earlier, _vide supra_.[3] The correct assignment and analysis of the ribose region of molecules as "small" as a trinucleoside diphosphate is by no means trivial and requires NMR instruments in the 300-400 MHz class, Fig. 4. Recently, full assignments of ribo- and deoxyribotetranucleoside triphosphates have been carried out on the most powerful instrument available today, the Bruker WM-500 (this Workshop, _vide infra_).

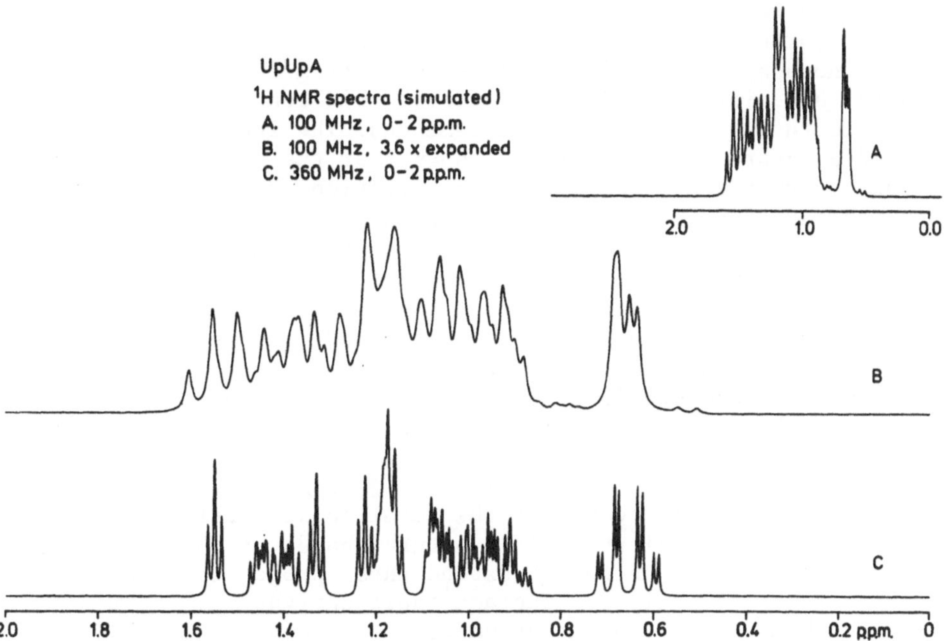

Fig. 4. Computer-simulated NMR spectrum of the ribose protons (H2'-H5" region) of the trimer UUA. (A) 100 MHz spectrum, 0-2 ppm. (B) As (A), 3.6 times expanded. (C) 360 MHz spectrum. Chemical shift reference is TMA, see text (H.P.M. de Leeuw and C. Altona, unpublished).

In the field of biological polymers, proton shifts are usually reported relative to the methyl signal of internal DSS (sodium 2,2--dimethyl-2-silapentane-5-sulphonate), even if actually measured with respect to water or dioxane. The DSS resonance position is essentially independent of pH[58] and occurs at the same place as the resonance of TMS (tetramethyl silane), the universal standard for ^{1}H NMR in non-aqueous solvents. In organic chemistry the δ scale is commonly adhered to, i.e. chemical shifts are denoted positive in the downfield direction, and this convention will be adhered to in this paper. Most researchers on e.g. tRNA prefer the opposite convention, but not always consistently, i.e. negative numbers are displayed in tables and figures of spectra, but the minus sign is often conveniently neglected in the discussions. It should be noted that DSS nowadays is usually replaced by tetramethylammonium chloride (TMA) in those investigations on nucleic acid constituents where highly accurate proton shifts are demanded. The introduction of TMA followed a report in which it was shown that the position of the DSS resonance varied with the purine concentration.[59] The TMA methyl resonance has other advantages. By virtue of $^2J_{HN}$ coupling the signal is a narrowly spaced multiplet and observation of this ensures optimum tuning of the spectrometer. Moreover, calibration of the chemical shift difference δHDO-δTMA against temperature by means of standard methanol and ethylene glycol samples provides a reliable temperature scale. The spectra have, in fact, a built-in temperature probe which is accurate to better than 0.5 ^{o}C.[60] For most practical purposes the TMA scale can be converted to the DSS scale by setting δTMA-δDSS = 3.18 ppm.

Several proposals to divide the total ^{1}H spectral range displayed by RNA[61] and DNA[11a] molecules into natural bands or regions have been made. It appears useful to extend these proposals to cover all natural classes of nucleic acids and their derivatives, from monomers to high-molecular weight polymers. In keeping with the δ convention we will number the regions I-VI in the downfield direction, Fig. 5.

Region I, 0-4.0 ppm, contains methyl resonances of modified bases. In DNAs a subregion, I', 1.5-3.0 ppm, contains the H2', H2" multiplets. Region II, 4.0-5.4 ppm, contains the ribose proton signals, except H1' and the H5', H5" signals of the 5' end residue, the latter occur near the upper end of region I. In region III, 5.4-6.7 ppm, the pyrimidine H5 and sugar H1' resonances are concentrated. Region IV, 6.7-9.2 ppm, is denoted the aromatic region and is made up of purine H8 and H2 and pyrimidine H6 peaks. This region also contains the hydrogen-bonded 2'OH resonance of tRNA at 6.8 ppm,[62] the amino resonances of the adenines,[63] cytidines, guanosines (at the lower end) and the imino signal of uridines.[19] Of course, OH and NH protons rapidly exchange with deuterium and the signals disappear completely when solutions in D_2O are prepared. In solutions of tRNA in water broad lines have been observed recently

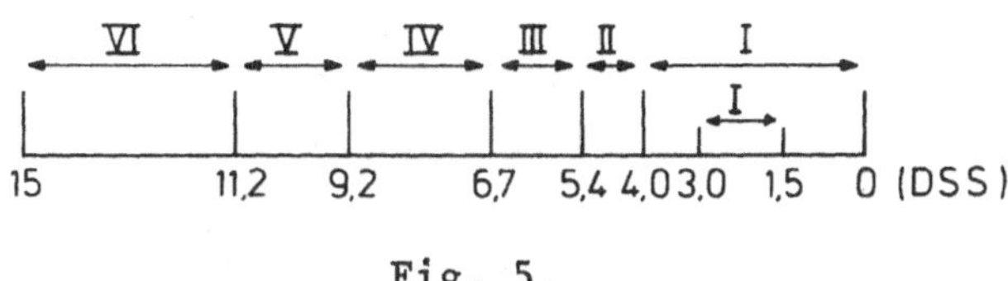

Fig. 5.

that are believed to represent amino signals.[63]

The intermediate region V, 9.2-11.2 ppm, contains peaks of ring NH protons that are easily exchanged in water and do not form part of hydrogen-bonded base pairs. For example, the "free" imino proton signals of thymidylyl residues in T-containing loops and hairpins occur between 10.0 and 11.2 ppm.[64,65] This region has received early attention,[66] but is at present still very much "terra incognita".

Finally, region VI, $\delta > 11.2$ ppm, is of utmost importance for the study of duplexes and tertiary base-base interactions because it displays peaks originating from ring NH protons (G·, U· and T·) of hydrogen-bonded base pairs in water solution. The extreme deshielding of these NH protons in non-aqueous solutions was studied by several workers in 1966[19] and first exploited by Kearns et al. in a series of pioneering studies on duplex regions in tRNA.[66,67] For practical purposes region VI of the tRNA spectrum has been subdivided into seven spectral sections labelled A to G.[68] Spectra of the 9-15 ppm regions V and VI in H_2O are usually accumulated by the use of correlation spectroscopy.[69] More recently, the interesting possibilities of the Redfield 21412 observation pulse FT NMR method[70] were demonstrated and applied to tRNA.[63,71,72] This technique allows the recording of superior proton spectra in 95% H_2O - 5% D_2O in a short time whilst maintaining the advantages offered by the nuclear Overhauser effect[71] (NOE) and by time-resolved NMR.[72]

Transfer RNA

The study of tRNA in solution by means of NMR represents one of the most fascinating illustrations of the scope and limitations of this method for the study of biological macromolecules. In terms of molecular biology tRNAs are small molecules, but from the standpoint of contemporary X-ray crystallography or NMR spectroscopy these polymers, usually consisting of some 75-80 nucleotide residues, are huge (Fig. 6).

After much effort by several competing groups the X-ray method has yielded a fairly accurate three-dimensional picture of a single

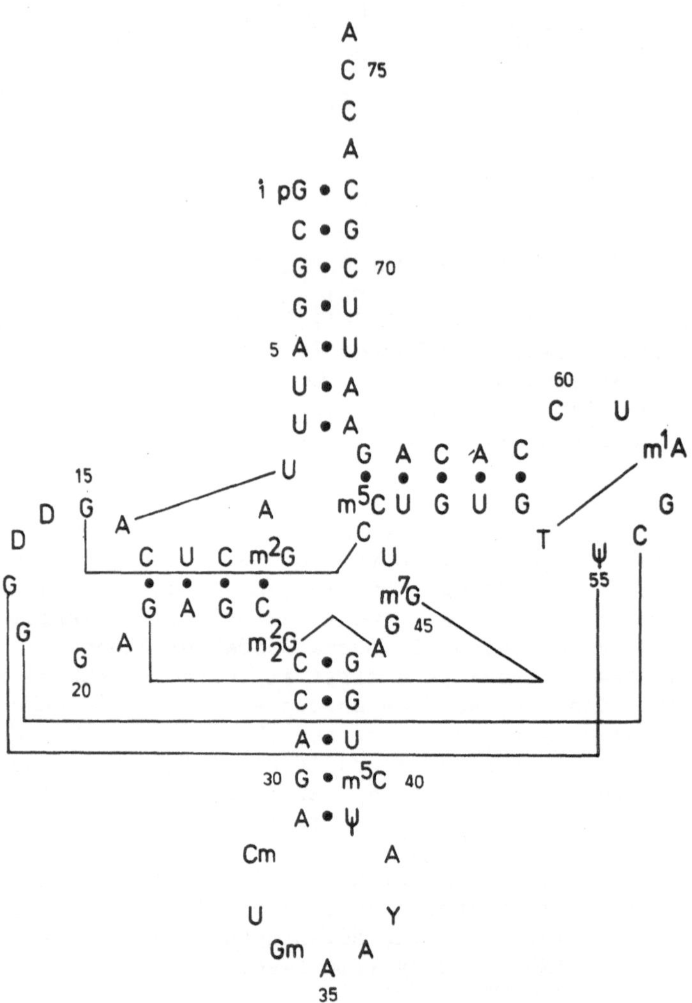

Fig. 6. (a) Cloverleaf sequence of yeast $tRNA^{Phe}$. The solid lines indicate tertiary structure interactions (A9-A23 not shown);

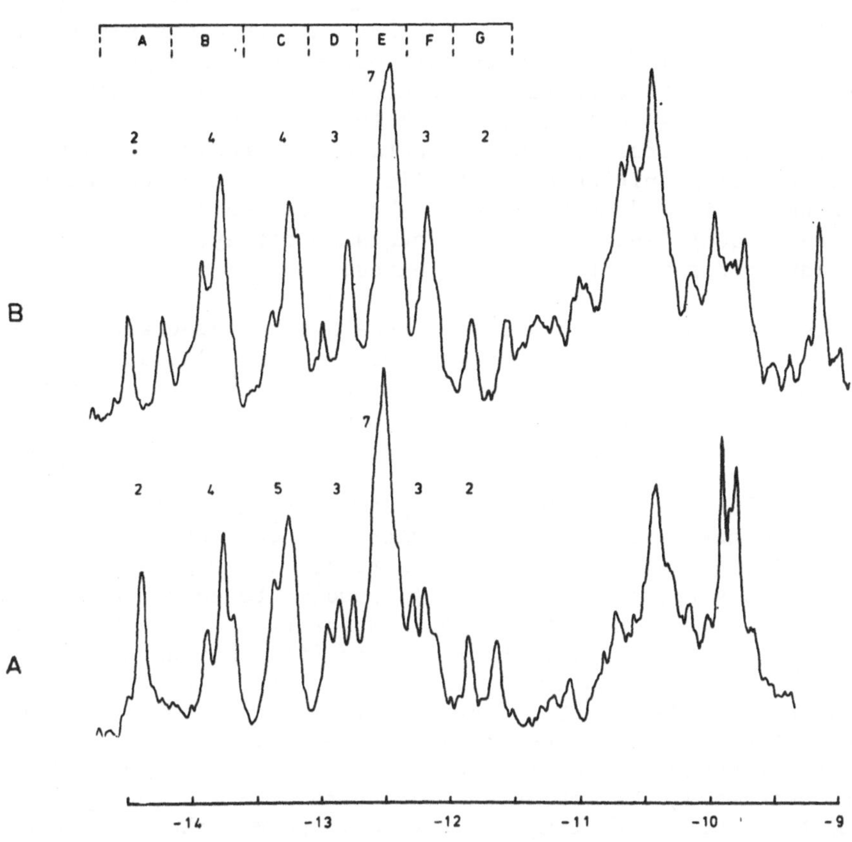

Fig. 6. (b) 360 MHz proton NMR spectra of region VI of yeast $tRNA^{Phe}$ (0.5 mM): (A) at 35 °C in the presence of 0.5 mM Mg^{2+}; (B) Mg^{2+} free (from Ref. *68*, detailed assignment omitted).

species of this class of molecules in the crystalline state, bakers yeast tRNA^{Phe}.[73] The present-day status of NMR research on tRNA would be unthinkable without this structural information. Now this stage has been reached, NMR is at some advantage because solution structures of many related tRNA species can be studied and compared in a reasonably short time.

The interested reader can spend many happy and instructive hours in his library reading up the original tRNA-NMR references, starting from 1971[67] and working his way up to the present time. Alternatively, recent reviews produced by the leading authorities in the field can be consulted.[71b,74] The present brief survey, written by a (hopefully) unbiased outsider, attempts to point out some of the difficulties and pitfalls that NMR spectroscopists have met in the past, and methods that had some success, in the hope that the present generation of students in biophysics and biochemistry may profit from these experiences.

Disagreements in the world of tRNA NMR has provoked very lively debates. Even the answer to the ostensibly simple first question as to how many proton signals contribute to the intensity in region VI proved to be far from unambiguous. Since each base pair displays only one ring NH hydrogen bond (G·, U·, T·), the number of resonances in this region should correspond to the number of "stable" base pairs in solution, i.e. those base pairs with helix lifetimes of 5 ms and longer. The number of regular secondary base pairs in the cloverleaf model of tRNA$_f^{Met}$ amounts to 20. The region VI NMR spectrum of this species was originally found[67b] to contain 23 ± 3 resonances. Integration of the signal intensity was then based on an external proton reference. Shortly thereafter, Kearns and coworkers[66,67c-e] using an internal comparison of resonances, concluded that yeast tRNA^{Phe} has 20 ± 2 base pairs in accordance with the cloverleaf model which predicts 20 ± 1 secondary base pairs. Therefore, contributions from ring NH protons in tertiary pairing situations were thought to be absent in the spectra, although it was stated that the "experimental error would permit the existence of perhaps one".[67e] These ideas were apparently strongly supported by ring current shielding calculations[67e], which led to plausible assignment of the resonances, and by the study of tRNA^{Phe} fragments (5' half, 3' half and 3' three-quarter molecule).[67c,75]

In 1974 the first detailed results of the X-ray crystallographic studies on yeast tRNA^{Phe} definitely established the presence of several extra base pairs or triples which help to maintain the amazing stability of the tertiary structure, Fig. 7. Immediately, the spectral integration problem was reinvestigated. Using highly resolved spectra of pure tRNA, Reid and coworkers[77] and, independently, Daniel and Cohn[78] came to the revolutionary conclusion that most, if not all, signals from tertiary bound ring NH protons indeed must be present in the 11-15 ppm spectrum of class I tRNAs, although

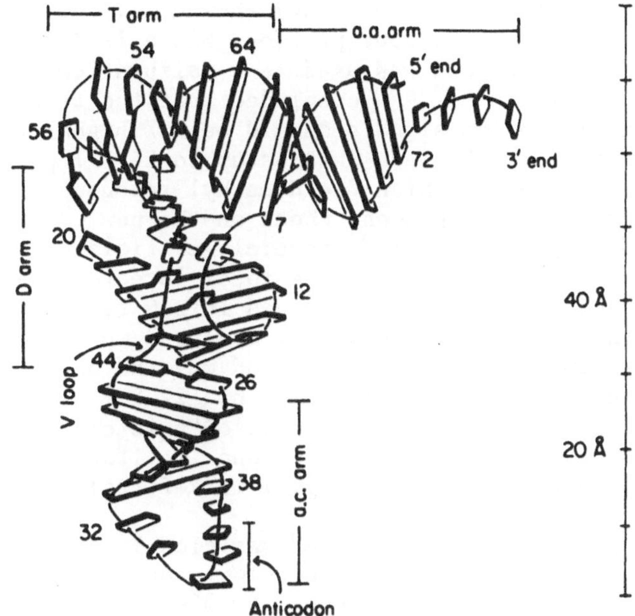

Fig. 7. Illustration of the extensive base stacking in the crystal structure of yeast $tRNA^{Phe}$. The Watson-Crick base-pairs are shown as long slabs and unpaired bases as short slabs. The tertiary base-pairs are shown either as bent slabs or slabs connected by dark rods. a.a., amino acid; a.c., anticodon (from Ref.*85*).

precise assignments of these protons could not be given at the time. E. coli $tRNA^{Val}$ and $tRNA_f^{Met}$ provided the first keys to this important breakthrough.[68,77-79]

The assessment on the total number of proton resonances visible in a certain spectral area depends on the correct choice of the unitary proton signal intensity. External standards were first used[67] but soon abandoned. A composite peak at 13.7 ppm in the yeast $tRNA^{Phe}$ spectrum was thought to represent three protons[66] and this peak became the internal standard of choice in subsequent work.[67c,d] However, a reexamination of this peak at higher resolution than was previously available (360 MHz against 270-300 MHz) showed a 1 : 2 : 1 triplet, that is, the peak represents four protons. Since it has been assumed to contain three protons and was used as a calibration standard, Reid, Robillard and coworkers[68a,77b,79] concluded that the true integrated intensity of the 11-15 ppm signals should be 33% higher than reported, bringing the total for all class I tRNAs up to 26 ± 1 protons. The integrated intensity of the high field methyl resonance in E. coli $tRNA^{Val}$ was used as an independent check on the new calibration.[77b] Shortly afterward, Bolton et al.[80] severely criticized the latter method and advocated comparison of the total integrated area of the peaks in the aromatic region IV, measured in D_2O to remove amino resonances, with the area of region VI, measured in H_2O. Reviewing matters, it was concluded that this method was the only valid one.[80] On this basis the yeast $tRNA^{Phe}$ molecule has 22.2 ± 1 ring NH proton signals in region VI. However, Kan and Ts'o[81] pointed out that, on integration involving such a wide area, one inevitably encounters serious baseline ambiguities (see Fig. 1 in Ref.80a).

In fact, the "integration debate" was closely tied up with the "assignment debate", as will be shown below. The intrinsically improved spectral resolution afforded by the 360 MHz generation of NMR instruments made it possible to base integration methods on areas displayed by <u>single</u> resolved ring NH resonances (the yeast $tRNA^{Phe}$ spectrum resolves into 14 peaks at 360 MHz, eight of which are single proton peaks, whereas only nine peaks are seen at 300 MHz), and so helped to virtually end this particular debate. It should be pointed out that the case of tRNA is a special one because the structure is so extremely stable that, in the presence of Mg^{2+} ions, spectra are usually measured at temperatures around 35 °C. In the case of short double-helical fragments one should keep in mind that base pairs could be partially "melted out" and then the integrated intensity of a single peak would correspond to less than one proton. Moreover, it is necessary to integrate over a spectral distance that equals at least six times the line width at half height in order to include 90% of the true intensity of a Lorentzian line.[80] Therefore, the "standard" single proton peaks should be well resolved to carry out the integration over a sufficient area. Modern computer curve analysis programs alleviate these problems because any complex spectrum can be simulated in terms of Lorentzian peaks.[61]

Much recent research effort on the NMR spectra of tRNAs has gone into the assignment of the ∿ 20 secondary and ∿ six tertiary ring NH protons. A correct assignment of all resonances is mandatory before the NMR spectrum is to be used with confidence as a valuable tool with which local structural perturbations and conformational dynamics can be monitored. These assignments have been, and still are, extremely difficult. Overlooking the literature, it is clear that assignments based on a single tRNA species, measured at a single temperature and medium composition, and reliance on a given set of ring current additivity rules,[81] cannot possibly yield meaningful results. On the contrary, several variational principles must be applied, as many as possible in fact, in combination with NMR techniques other than simply recording the spectrum. Each of the following methods has been used by workers in the field with greater or lesser success. The reader is referred to a series of papers by Reid and coworkers[72,79,82] for full appreciation of these points:

(i) chemical modification
(ii) "melting" behaviour
(iii) paramagnetic probes
(iv) fragment analysis
(v) prediction of ring current shifts
(vi) NMR pulse techniques

Chemical modification. Not only induced chemical changes in one or more bases, but also excision of a residue and detailed comparison of the spectra of closely related tRNA species differing in one or a few base pairs fall in this category. This approach entails the assumption, often quite reasonable, that the chemical change does not affect the detailed tertiary structure of the molecule. A classic example is the discovery that removal of the sulphur from the modified uridine S^4U8 in E. coli $tRNA^{Val}$ and $tRNA^{Arg}$ causes the loss of a resonance at 14.9 ppm and the appearance of a new signal at 14.3 ppm.[77,82b,83] This was easily interpreted to represent a non-Watson-Crick type tertiary (Hoogsteen) interaction, the change in line position in going from $S^4U8{\cdot}A14$ to $U8{\cdot}A14$ is caused by the intrinsic extra deshielding of N_3H in the thiocompound.[82b] Thus, the first unequivocal assignment of a ring NH resonance was firmly established.

A second example concerns the chemical excision of m^7G46, accomplished without chain cleavage.[82d] This base forms a triple interaction $m^7G{\cdot}G22{\cdot}C13$ in the crystal structure of yeast $tRNA^{Phe}$ and is also present in a number of E. coli tRNAs. Excision of m^7G caused the disappearance of a tertiary resonance at 13.4 ppm. E. coli $tRNA_1{}^{Gly}$ does not contain m^7G in its variable loop and, according to expectations, was found to lack the 13.4 ppm resonance.[82d] As an interesting corollary the spectrum of E. coli $tRNA_f{}^{Met}$ was considered. In this particular species the environment of the m^7G base differs from that of the homologous series related to yeast $tRNA^{Phe}$. In the crystal structure[73] of yeast $tRNA^{Phe}$ the m^7G46 base is stacked with A9 and A12. In $tRNA_f^{Met}$ the analogous environment would be G9 and C12. Because the sum of the ring currents of a guanine and a cytosine is known to be much smaller than that of the two adenines[55,56] one predicts that the m^7G ring NH resonance in $tRNA_f^{Met}$ occurs downfield from 13.4 ppm. It is actually found at 14.55 ppm. The inherent unshifted or starting position for the m^7G N1H proton in the triple interaction was deduced to be about 15.1 ppm. A large part of this extreme deshielding with respect to the starting position of a ring NH proton in a Watson-Crick G·C base pair at 13.4 ppm is caused by delocalization of the positive charge in m^7G.[82d]

Melting behaviour. In the presence of 10-15 mM Mg^{2+} ions the tRNA structure is highly stable and melts out in a narrow temperature range near 70 °C, the actual temperature varies a little with the chemical composition. In the particular case of E. coli $tRNA^{Phe}$ the acceptor stem is unusually stable in that it contains six consecutive G·C pairs. In the presence of Mg^{2+} the molecule can be partially unfolded by heating. At 68 °C, 21 of the 27 base pair resonances are eliminated from the spectrum, only the six G·C resonances remain.[79b] Magnesium does not seem to be required for the three-dimensional folding of the molecule, since at low temperatures[68] all but one (13.2 ppm) of the tertiary interactions appear to be present in the complete absence of Mg^{2+}, but the structure of all tRNAs is thermally less stable under this condition as judged from the melting behaviour. Some tertiary interactions in "Mg^{2+}-free" tRNA disappear at relatively low temperatures and several consecutive transitions have been traced in the high-field spectral region I as well as in the low-field region VI.[68] The "differential melting process" in Mg-free yeast $tRNA^{Phe}$ starts at about 18 °C. Unfolding of the least stable tertiary elements ($T_m \sim 25$ °C) is followed by breakdown of residual tertiary structure, acceptor stem and anticodon helix (T_m 35-40 °C), TΨC helix (T_m 45-50 °C) and finally the D helix (60-65 °C). A study on the saturation-recovery behaviour of the G·C rich acceptor stem in E. coli $tRNA^{Phe}$ will be discussed below.

Paramagnetic probes. In the $tRNA^{Phe}$ crystal structure there is a Co^{2+} site in which the metal ion is directly coordinated to G15.[84] There are only three hydrogen-bonded ring NH protons within a distance of 10 Å of this site, namely G15·C48, C13·G22 and the 8·14 hydrogen--bonded proton. If metal binding takes place at the same site in solution, these three resonances should be paramagnetically relaxed by low levels of Co^{2+} and perhaps by other metal ions as well. The 8·14 proton signal had been assigned earlier in chemical modification studies, vide supra, but no consensus had been reached about the position of the signals of the 15·48 and 13·22 pairs: G15·C48 had been variously relegated to peaks at 13.5, 12.9, 11.7, 10.5 and 10.0 or 9.5 ppm. Similarly, C13·G22 had been assigned to various positions between 13.1 and 11.5 ppm. This confusion will not be traced here, but is mentioned to serve as a warning against unduly optimistic dealings with complex spectra.

Hurd et al.[82c] reasoned that these resonances should be paramagnetically relaxed by low levels of Co^{2+} and Mn^{2+} and investigated the effect of these additives on the region VI spectrum of several E. coli tRNAs. Indeed, selective line broadening was affected by 0.05 Co^{2+} ions per molecule of tRNA as well as by a tenfold lower Mn^{2+} concentration in the presence of magnesium. The Mn^{2+} was found to bind at a site close to, but different from, the Co^{2+} site. As expected, the known 14.9 ppm signal from s^4U8·A14 is severely broadened by Co^{2+} and by Mn^{2+}. A comparative study of broadenings elsewhere in the spectra of a homologous series of tRNA species, under the assumption that the known crystal structure is the correct model for the tertiary folding of E. coli $tRNA_1^{Val}$, $tRNA^{Lys}$ and $tRNA_m^{Met}$, led to consistent assignments: C13·G22 occurs at 12.1-12.15 ppm in all three molecules, whereas the position of G15·C48 is found at ∿ 12.3 ppm in the $tRNA^{Val}$ species (and presumably in yeast $tRNA^{Phe}$ also) and at 12.7-12.8 ppm in the other two species. The difference between the two resonance positions was related to a difference in surroundings: in $tRNA^{Val}$ the G15 has U59 on top, in $tRNA^{Lys}$ and $tRNA_m^{Met}$ U59 is replaced by A59. Why this particular replacement of U by A should give rise to a downfield shift was not explained.[82c]

Fragment analysis. The tRNA cloverleaf can be cleaved at specific positions either by careful manipulation with enzymes or by purely chemical methods. In this way hairpin fragments like the 5' half, anticodon half, 3' half as well as an assortment of other parts can be isolated and purified, Fig. 8. Since each fragment contains only a few base pairs, the spectra are much simpler than those of the intact structure. This type of analysis was introduced at an early stage[75] (1973) in the expectation that the $tRNA^{Phe}$ fragment spectra would provide an unambiguous test of the ring current shift calculations. It was also hoped that all resonances of the intact molecule could be assigned by simply superposing the appropriate fragment spectra. Unfortunately, at the time spectral resolution and sensitivity were insufficient to take full advantage of these ideas.

Similar reasons prompted Reid et al.[82b] to investigate the 360 MHz spectra of hairpin-duplex fragments of $tRNA_1^{Val}$. Fig. 8 shows the spectrum of the 47-76 fragment (obtained by cleavage at m^7G46 by chemical means) taken at 27 °C. Besides four sharp resonances assigned to G·C and C·G base pairs 49, 51, 52 and 53 on the basis of Arter and Schmidt's predictions[56] for 11-fold RNA helices, no less than five extra peaks are observed between 10 and 15 ppm. The three peaks at the high end, 11.95, 11.35 and 10.7 ppm (two protons) were assumed to be related to the wobble pair G·U50. In a subsequent paper Hurd and Reid[82a] were able to show by NOE cross-relaxation experiments on the intact tRNA that the resonances at 11.95 and 11.35 ppm indeed belong to the same base pair, i.e.

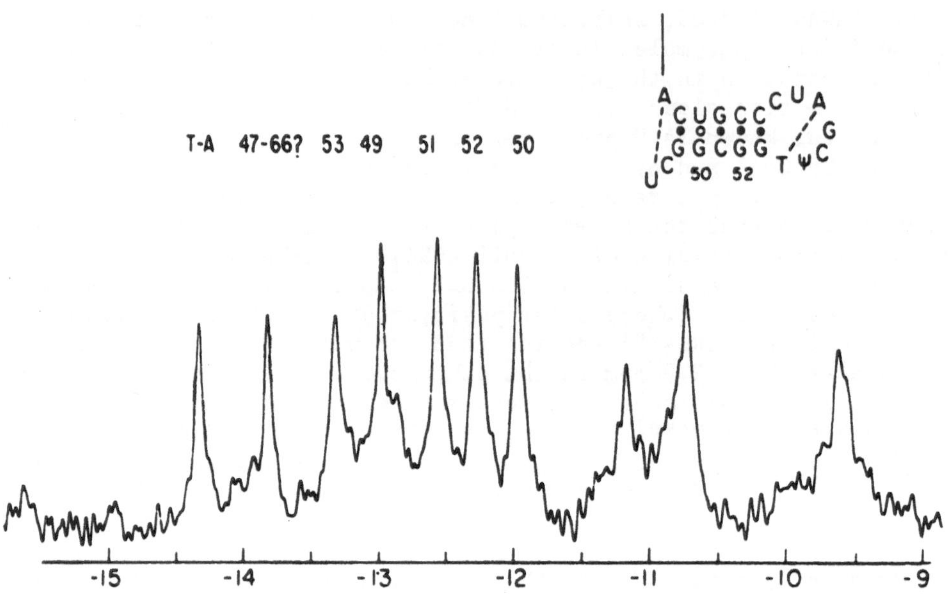

Fig. 8. The low-field part of the 360 MHz spectrum of the E. coli $tRNA_1^{Val}$ fragment 47-76 containing the TΨC loop. The fragment was dissolved in 100 mM NaCl, 5mM phosphate buffer, pH 7, and the spectrum taken at 25 °C under correlation sweep conditions (Ref. *74b*).

Fig. 9. Hoogsteen-type base-pairing between rT54 and m^1A58 in yeast $tRNA^{Phe}$. (R=ribose)

UN3H and GN1H in the G·U pair. The complete absence of a cross-relaxation effect at 10.7 ppm was not discussed.

The low-field peak at 14.3 ppm in the fragment spectrum integrated to almost a full proton and was, on the basis of a large number of considerations, assigned to the rT54·A58 reversed Hoogsteen tertiary interaction involving hydrogen bonding to N7 of A58, Fig. 9. The unexpected new peak at 13.9 ppm was tentatively attributed to base pairing between the single-stranded tails of the fragment, e.g. U47 and A66. This assignment seems rather unlikely, however, because one would expect such a tail-end pairing to be rather unstable, whereas this particular resonance melted out between 50 and 55 ^{o}C. A reversed Hoogsteen 54·58 pairing scheme (Fig. 9) would imply an observable NOE between rT54N3H (14.35 ppm) and the A58H8 proton located "somewhere" in the aromatic region. Modern difference NMR spectroscopy is sensitive enough to detect the intensity enhancement of one such proton. The assignment riddle deepens here, because Sânchez et al.[63] recently reported a clear NOE peak in the aromatic proton region IV of yeast $tRNA^{Phe}$ on irradiation of the 14.35 ppm resonance under conditions where <u>all purine C8 positions had been deuterated</u>. Sânchez et al.[63] interpreted this observation to mean that the 14.35 ppm peak comes from a secondary A·U Watson-Crick pair (U·A6?). It should be mentioned in this connection that the compound investigated, yeast $tRNA^{Phe}$, has a modified adenine at position 58 (m^1A^+) that cannot form a normal Watson-Crick type tertiary interaction, Fig. 9. However, before these far-reaching conclusions can be accepted with confidence it would seem advisable to extend the NOE studies to fragments and perhaps even to simple nucleosides in non-aqueous solvents.

Boyle et al.[85] recently studied the NMR spectra of yeast $tRNA^{Phe}$ fragments with an entirely different objective in mind. Their fragments have a common 5' end and were chosen to investigate the possibility of sequential folding of tRNA during its biosynthesis (which proceeds from the 5' end). It could be shown that, as expected, the secondary structure forms as soon as the nucleotides necessary for the formation of an intact duplex stem are available. Moreover, some correct tertiary hydrogen bonds are made at an intermediate stage. It was suggested that the D stem-anticodon domain may be stabilized even early in its synthesis. However, formation of the complete acceptor stem appeared necessary to achieve the correct 3D structure.

<u>Prediction of ring-current shifts</u>. From the beginning of NMR research on tRNA the "success" of the theoretical prediction of line positions often has played a decisive role in assignment strategies. This approach is not without dangers, however. The correct application of ring-current shifts rests upon three foundations: (i) the theoretical evaluation of the magnetic field created by the ring current of the nucleic acid bases in the vicinity of the proton under consideration; (ii) the experimental determination of the reference

(or offset) value, i.e. the shift position that the proton would exhibit in the absence of perturbing ring currents. In the case of a hydrogen-bonded base pair proton the ring currents of this base pair are included in the reference position; (iii) the known or assumed geometric factor, i.e. the spatial position of the adjacent bases or base pairs with respect to the proton studied.

As regards (i), the classical mechanical approach (Johnson and Bovey[86]), advocated by the Pullman school[55], has been most popular, although Robillard et al.[88] preferred to use the Haigh-Mallion approach.[87] After appropriate scaling the difference between the two approaches mentioned does not appear to be significant. Later refinements of the classical approach included contributions from local diamagnetic anisotropies.[55c-e] The Pople equivalent dipole model[89] appears to have been ignored, although it has been used with success in the porphyrin field.[90] In all, the theoretical calculations appear sufficiently accurate to be applied with confidence.

The choice of reference values for ring NH base-pair protons (X°) proved somewhat controversial. It is tied up with the assignment problem (vide supra). Kearns and coworkers[67c,75] originally placed the A·U^{o} reference position at 14.8-14.7 ppm in order to account for the shift of presumed Watson-Crick A·U interactions at ∿ 14.3 ppm. The calculated ring-current effects[55a] were scaled up by 20% in order to correctly reproduce the remaining resonance positions.[67e] Studies[91] on a pentamer duplex d(AACAA).d(TTGTT) showed conclusively that G·C hydrogens resonated upfield from A·T protons. The G·C^{o} reference value was set at 13.6 ± 0.1 ppm. It was stated[67c] that an equally good fit could be obtained by using AU^{o}=14.6 ppm and GC^{o}=13.6 ppm without scaling. However, when the downfield (> 14 ppm) resonances in tRNAs were assigned to tertiary interactions a reappraisal was called for. Robillard et al.[88] endeavoured to calculate the ring current shift, which each separate hydrogen-bonded proton experiences from all other aromatic rings within a radius of 10 Å, directly from the rather crude X-ray coordinates available in 1975. Empirical adjustments for each of the ring currents were made by iterative computer fit to the observed spectrum and gave AU^{o} = 14.35 ppm and GC^{o} = 13.54 ppm. The use of better coordinates did not alter the main conclusions. This brings us to the question of choice of helix geometry, point (iii) above. Shulman and coworkers[67e] selected the 12-fold duplex tRNA model (A'-RNA)[93] to represent the helical regions in tRNA. This proved to be an unfortunate choice, because in the crystal structure of yeast tRNA^{Phe} the actual number of residues per turn was found to be 11 or less (A-RNA) and all four cloverleaf helices have considerable irregularities.[94] Reid et al.[79,82] strongly advocate the use of the current values for a regular A-RNA (11-fold) helix calculated by Arter and Schmidt[56] for nearest neighbour and next-nearest neighbour stacking. From the study of fragment spectra reference values of AU^{o} = 14.35 ppm and GC^{o} = 13.4 ppm were determined, i.e. remarkably close to the values deduced by Robillard et al.[88,92] in a different manner.

Oligoribonucleotide duplexes differ in their helical geometries from their deoxy counterparts and the respective reference values need not be the same for RNA and DNA helices. The single-crystal X-ray structure of a DNA dodecamer[95] shows deviations from the previously established B-DNA model duplex[96] and irregularities as well proceeding from one base pair to the next. Further ring-current calculations based on single-crystal geometries are obviously necessary to clear up this point. It should be pointed out that Haasnoot et al.[64,65] did not find "predicted" ring NH shifts calculated from the B-helix[96] to be of much use in making detailed assignments in synthetic self--complementary DNA fragments, 12 to 17 residues long. Instead, the principle of chemical modification was adhered to. One of the most interesting aspects in the work of Haasnoot et al.[64,65] was the finding that non-base-paired thymine NH protons contained in a wobble--pair-like position or in a hairpin loop display clear resonances in the 10-11 ppm range, Fig. 10. Presumably these ring NH protons are hydrogen bonded to "inside" water which is in some measure protected against rapid exchange with "outside" solvent by the enclosing bulge[64] or hairpin.[65]The data appear to indicate[65] that, upon enlarging, the loop becomes more penetrable to outside water. However, further work, e.g. under different conditions of added salt and buffer, is necessary in order to quantify these results. We wish to point out, however, that this finding implies that similar non-base-paired ring NH signals, at least of U· and rT· , may be found in tRNA spectra taken under suitable conditions. Salemink et al.[39] also point out this possibility for the loops in 5S RNA.

Methyl resonances of tRNA

The methyl resonances of the modified bases occur generally in the upfield region I, Fig. 5. Several extensive investigations of this region and assignments made will not be treated here.[68,74,97] Instead, the work of Kastrup and Schmidt[98] is selected because of its relevance for the conformational possibilities of the TΨC loop, already mentioned in the ^{13}C NMR section. Fig. 11 shows the upfield spectrum of $tRNA_1^{Val}$ (coli). The rT methyl group resonance presents an interesting behaviour. A single peak at 1.0 ppm occurs at 27 ^{o}C. On raising temperature this peak diminishes and disappears; the transition has a T_m of 51 ± 5 ^{o}C. At 85 ^{o}C a single rT methyl peak is seen at 1.8 ppm, which is ascribed to the random coil conformational mixture of the molecule. At intermediate temperatures, however (40° - 70° C) two more resonances, at 1.25 and 1.9 ppm, first appear and then disappear. Therefore, four distinct sites are available to this methyl group. The 1.0 ppm site corresponds to a native conformation with tertiary structure intact. The peaks at 1.25 and at 1.9 ppm correspond to intermediate conformations that defy further specification at the moment, except that the low-field peak at 1.9 ppm resonates at the same position as free ribothymidine. This

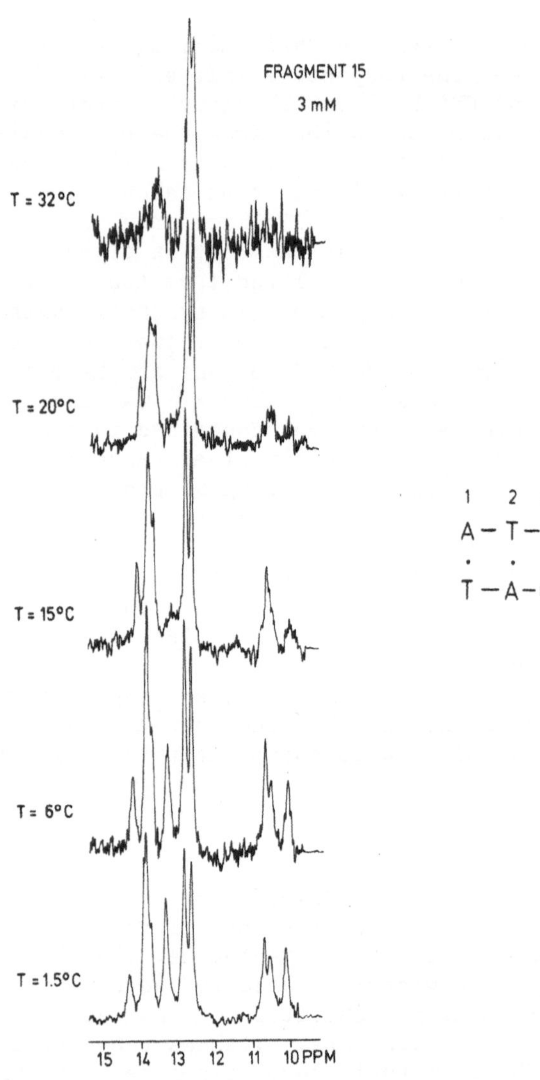

Fig. 10. The low-field 360 MHz proton NMR spectra of fragment 15 (3 mM, 500 mM NaCl, 100 mM cacodylate, pH 7) at various temperatures. Shifts are referenced to DSS. Note the upfield (10-11 ppm) signals of non-hydrogen-bonded ring NH protons of the three non-equivalent thymines in the loop (Ref. *65*).

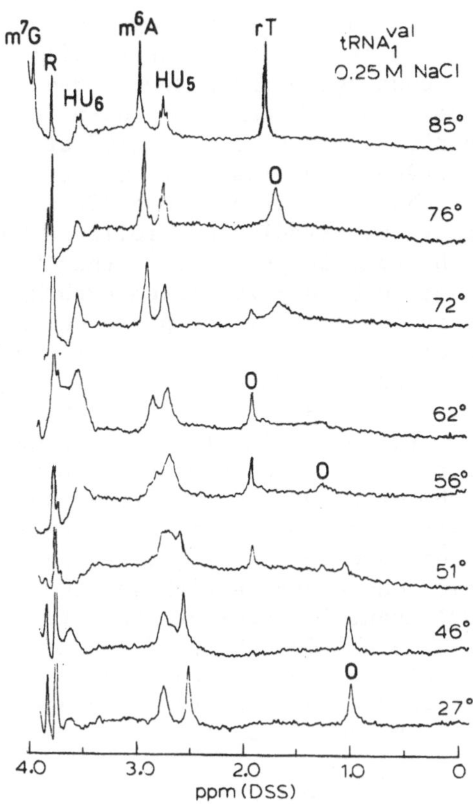

Fig. 11. The upfield 220 MHz proton NMR spectrum of $tRNA_1^{Val}$ showing the methyl resonances at a series of temperatures (250 mM NaCl, 10 mM phosphate, 0.4 Mg^{2+}/tRNA, pH 7). The four signals assigned to the rT methyl proton are indicated by O. The sharp peak R comes from dioxane added as internal reference (Ref.*98*).

appears to indicate that rT in this site is no longer stacked on G53 and that the A58·rT54 tertiary interaction probably is lost.

NMR pulse techniques. In a recent study[72] of the dynamic aspects of the acceptor stem duplex in E. coli $tRNA^{Phe}$ the differential thermal stability of the acceptor stem under magnesium-free conditions was utilized. The spectrum simplifies to the resonances of the six G·C pairs in the range from 52° - 65° C and the exchange-dominated saturation recovery behaviour of each ring NH proton in the acceptor stem could be determined at 360 MHz, a feat still impossible for the intact molecule. The good agreement between observed and predicted positions of these six G·C resonances is shown in Fig. 12. A resonance at 13.7 ppm derives from base pair A·U7 at the bottom of the acceptor stem; it broadens and disappears between 51° and 55° C because of solvent exchange at rates above 200 s^{-1}. When the water resonance was saturated by irradiation, followed by a one millisecond delay and then the observation pulse, more than 90% of the G·C resonances disappeared due to base-pair proton exchange with the saturated water protons. This indicates that the proton exchange rate is at least 5 s^{-1}. Conversely, each of the base-pair ring NH protons was selectively preirradiated to saturation and after a variable delay (1 ms to 500 ms) a 0.38 ms Redfield[70-71] observation pulse was applied. The longer the delay, the more signal intensity was recovered for the particular proton irradiated. The recovery rate is found from the slope of the logarithm of remaining fractional saturation plotted against delay time. The measured saturation recovery rates are shown in Table 3, taken from Hurd and Reid.[72] These rates represent the sum of two rates: (i) the exchange with an unsaturated water proton and (ii) the natural magnetic spin-lattice relaxation. The latter was tentatively set at 6 s^{-1} for the C·G3 resonance in order to fit the linear Arrhenius plot criterion. Subtraction of 6 s^{-1} from the values in Table 3 leads to a pure proton exchange rate of 3 s^{-1} at 52 °C and 30 s^{-1} at 62 °C for C·G3, i.e. a tenfold increase over a 10 deg. range.

The ring NH protons in the intact duplex are shielded from contact with the solvent. It is thought that proton exchange involves two separate steps: (i) a local opening of the structure, followed by (ii) chemical exchange from the now exposed position with water, Fig. 13. The chemical coil-water rate, k_{CW}, is buffer-catalyzed and can be experimentally manipulated by changing pH and buffer concentration. If step (ii) is limiting the exchange process, such manipulation should strongly affect the exchange rates and thus the observed saturation recovery rates. Furthermore, one would expect, at any given temperature, roughly equal exchange rates for G·C and C·G pairs 1-6 in the duplex. This is not found, Table 3. It was concluded[72] that the helix-coil rate, k_{HC}, is the rate-limiting step; seen in this light it is of interest to note that k_{HC} (measured saturation-recovery rate minus 6 s^{-1}) strongly decreases in going from either end of the acceptor stem towards its center. For example,

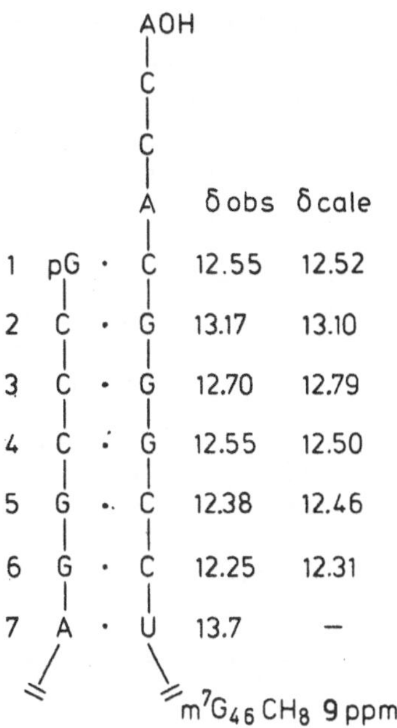

Fig. 12. G·C rich acceptor stem of tRNA^{Phe} (coli) showing observed and calculated[56] chemical shifts (data from Ref. *72*).

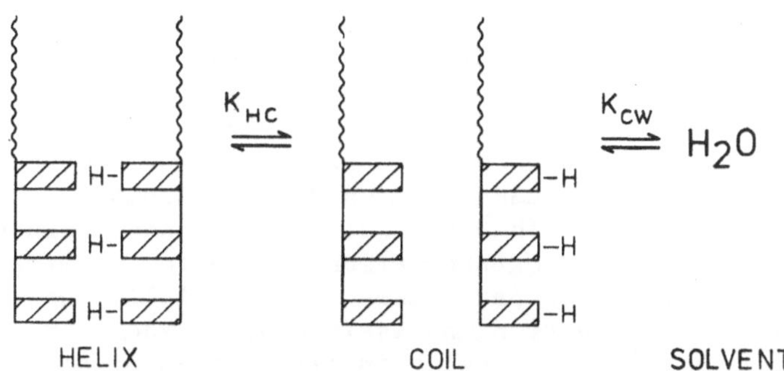

Fig. 13. Schematic representation of proton exchange from a double-helical region. The three states involved, duplex, coil and water, are shown. Note that the coil state is envisaged to consist in a mixture of substates having 1,2,3...n "free" bases, each substate also is engaged in a stack--destack equilibrium (this work). Short double helices progressively open up from both ends (including hairpin loops) and from weak spots in the center, if present (e.g. a wobble pair or mismatch).[64,65]

Table 3. Saturation-recovery rates (s^{-1}) determined for the ring NH protons of the six G·C base pairs in the acceptor stem of $tRNA^{Phe}$ (coli) under a variety of conditions[72], Fig. 12

5 mM-sodium phosphate 100 mM-sodium chloride (pH 7)				5 mM-sodium phosphate 408 mM-total Na^+ 63 °C			
Base pair	52°C	58°C	62°C	pH 5.1	pH 5.8	pH 7.2	pH 7.2 + 10 mM-Tris
1		41	90				
2	14	22	42	25	18		
3	9	18	35	16	13	10	10
4	6	11	21				
5		12	28				
6		73	115				

at 62 °C one finds for G·C pairs 1-6: 84, 36, 29, 15, 22 and 109 s^{-1}, respectively. This finding is entirely consistent with the notion that, at least in short helical fragments, the "melting process" proceeds from the 5' and 3' terminals towards the interior base pairs.[64,65,91] It would be of great interest to see the results of similar experiments on different homologous sequences as well as on mixed A·U and G·C containing duplexes.

Another interesting application of pulsed NMR to tRNA was recently given by Schmidt and Edelheit.[61] Purified tRNAs in D_2O yield a set of narrow peaks in the carbon-bound aromatic proton region IV (Fig. 5) by the application of partially relaxed spin-echo techniques. This way, peaks broader than about 20 Hz (W1/2) were essentially eliminated, leaving sharp peaks from only 15-20 aromatic protons out of a possible 89, Fig. 14. The rationale behind this approach can be summed up as follows: (i) extrapolation from a few chemical shifts to give detailed structural parameters is a "woefully underdetermined" problem,[61] and the observation and assignment of additional protons per base pair could at least assist in elimination of alternative possibilities; (ii) the single-strand loop regions, which are rather important in terms of biomolecular function, carry some probes (tertiary resonances and methyl resonances, *vide supra*) but not the CCA end. A drawback of the technique is that the exact number of protons contributing to unresolved resonances is not easily established. A closer analysis showed that of 33 possible doublets with J5,6 ∿8 Hz from C or U, there are none resolved in the $tRNA_1^{Val}$ spectrum. This implies that all H6 protons of unmodified pyrimidines have linewidths greater than 10-15 Hz; conversely, the 45 peaks with W1/2 less than 10 Hz must come from A and G protons (and perhaps H6 of T, Ψ and V).

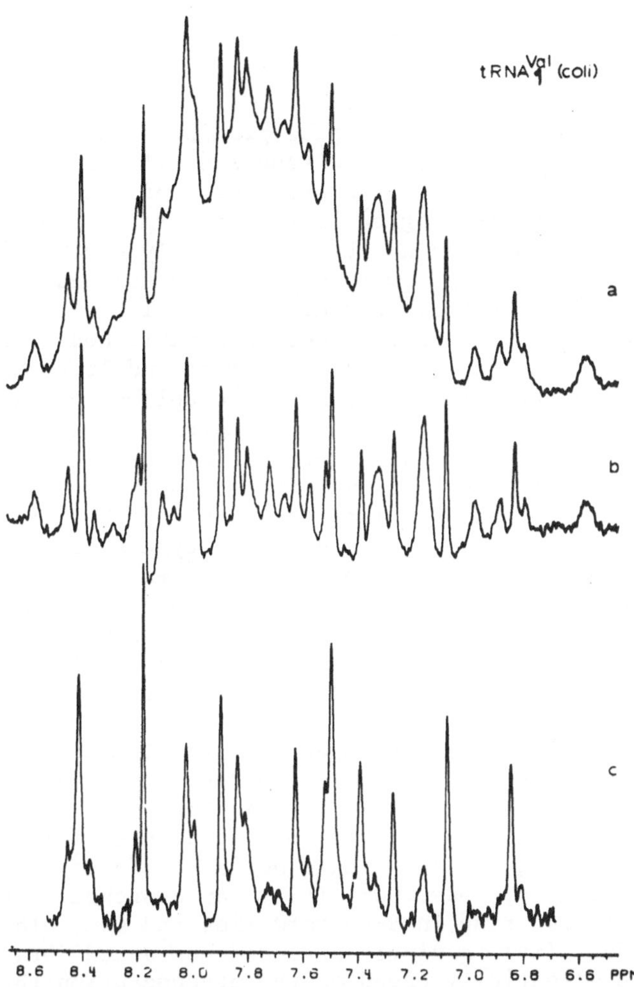

Fig. 14. 270 MHz proton spectrum of the aromatic region IV of $tRNA_1^{Val}$. (a) Normal FT spectrum, 1-Hz line broadening. (b) Convolution difference spectrum. (c) Partially relaxed spin-echo spectrum; peaks with line widths of 20 Hz or more are essentially eliminated (from Ref. *61*).

Further discrimination was made by deuterium exchange. Because H8 protons are effectively eliminated in this way, the problem reduces to the analysis of some 15 AH2 and three H6 resonances of the modified pyrimidines. The sharp peak at 8.17 ppm belongs to H2 of A76, the 3'-terminal base. Of course, it is of interest to use this signal as a probe when the tRNA molecule is bound to proteins. Further detailed assignments will undoubtedly be made in the not too distant future.

Oligonucleotides. Lack of space prohibits an adequate survey of this enormous subject and our discussion will be limited to a few examples with emphasis on the assignment of the ribose and deoxyribose proton signals (region II (I') in Fig. 5). Moreover, the literature up to about 1977 has been adequately covered in an exhaustive review by Davies[99] (with emphasis on nucleosides, mononucleotides and dinucleoside monophosphates). The first complete NMR data sets for two ribonucleotides were published by Altona et al. in 1974: m^6ApU[100] and ApU.[101] The assignment and line-shape simulation of a third dinucleotide, m^6_2ApU, and coupling constants and conformational aspects of these three compounds as well as of i^6ApU were fully discussed at the 4th Steenbock symposium in June 1974.[102] The data were interpreted in terms of a strongly predominant N3', 5'N ribose conformational combination in the stacked state (N-type sugars roughly correspond to 3' endo and S-type sugars roughly to 2' endo, but the N/S nomenclature has far wider scope than the endo/exo nomenclature[6,99,103]). The overall backbone conformation in the stacked state of these dinucleotides in solution was deduced to be: β(C5'-O5') ∿ 180°, γ (C4'-C5') ∿ 60° and ε (C3'-O3') ∿ 210° (270°). Two years later Kondo and Danyluk[104] endorsed these conclusions in an extensive discussion on the conformational properties of ApA, but at the same time they postulated the simultaneous existence of a "more loosely" base-stacked lefthanded loop structure with a ζ^+, α^+ phosphate diester conformation. A third stacked form was introduced by Lee and Tinoco,[105] who envisaged an equilibrium between: (i) a classical righthanded single helix, ζ,α/300°, 290°; (ii) a lefthanded helix, ζ,α/30°,100°; (iii) an unorthodox stacked form having the Xp-ribose ring in S-conformation and some other unusual features besides; (iv) a single extended unstacked form. These postulates are essentially based on the preconception that all observed dimerization shifts, δ(monomer)-δ(dimer) must be ascribed to ring-current shifts of the nearest-neighbour base.

Altona et al.[41] criticized these far-reaching generalizations. First of all, too little is known at present about the many factors which influence the composite shielding or deshielding effects on the various protons upon helix formation. More important, contrary to previous suggestions, intermolecular base-base association cannot be neglected, especially for purine nucleotides, even when measurements are carried out on 5-10 mM solutions. This association induces shift effects of considerable magnitude on some protons and should be corrected for by running the spectra at a series of temperatures for at least two concentrations.[41] Third, if it is true that the shift of certain protons, e.g. H5' of the Xp-residue, is due solely to the formation of the lefthanded loop and the dimerization shift of other protons to the formation of the righthanded stack one would expect different thermodynamics for the two groups of protons implicated. It is hard to imagine that an unstack-loose stack equilibrium (with parallel bases postulated at a distance of ca 4.2 Å) would be characterized by the same ΔH, ΔS and T_m as an unstack-righthanded stack equilibrium with parallel bases at ca 3.4 Å. Nevertheless it was found[41] that, after correction for intermolecular association, all stacking-sensitive protons of m_2^6ApU obeyed the same thermodynamics. The combined NMR and CD data definitely ruled out the presence of a lefthanded base-base overlapping stack in m_2^6ApU in amounts large enough to cause the observed upfield shifts of UH2' and AH3', or the downfield shifts of AH5' and AH5".

It should be realized that dinucleoside monophosphates represent at the same time 3'- and 5'-terminal fragments of RNA and DNA and their behaviour may not be characteristic of conformational properties of central residues in longer chains. Extension of these detailed researches to trimers and higher oligomers is therefore mandatory. Following this line of thought the first unambiguous complete assignment of the sugar protons of a trimer, d(T-T-A), measured at 360 MHz, was published by Altona et al. in 1976.[106] The assignment of A-A-A, proposed by Evans and Sarma[107] in the same year, was recently shown to be untenable.[108] Unfortunately, no less than six signals had been assigned incorrectly. The NMR spectrum of d(T-T-A), however interesting, did not allow quantitative conclusions regarding the stacking interaction of the d(T-T-)part of the molecule. In contrast, our recent investigation of d(A-A) and d(A-A-A) served to settle several long-standing questions.[7,9] It has long been known that double-helical DNA under conditions of high humidity assumes a B-type geometry.[96] The B-DNA genus of geometries (B,C,D-DNA) displays a single outstanding common feature, i.e. an S-type[6,103] conformation of the deoxyribose ring, whereas the sugar ring adopts an N-type conformation in the A-DNA or A(A')-RNA genus. One can thus describe the deoxyribose chain of B(C,D)-DNA in terms of an S-S-S... sequence of sugar rings and the A family in terms of an N-N-N...sequence. It was an open question whether or not the sugar ring in the single-helical (stacked) conformation of deoxyribonucleic acids has an outspoken conformational preference, either N-N-N... or S-S-S..., or assumes both N and S forms in a more or less random fashion (mixed-type stack). In other words, would the "conformational adjustment" of the deoxyribose-phosphate residues include adjustment of the sugar conformation? This is by no means a trivial question because any major conformational adjustment along one or more of the degrees of freedom of the nucleic acid backbone implies an enthalpy and entropy contribution on formation of the duplex which must be counterbalanced by the base-base hydrogen bonding energy.

According to CD[9] and 360 MHz ^{1}H studies[7] the dimer d(A-A) and the trimer d(A-A-A) display excellent stacking proclivities (about 87-90% stack at 0 °C) but it was a surprise to find that stacking occurs in two specific modes that differ in sugar conformational type: about 70% of the total stacked species in the dimer d(A-A) prefers a B-DNA-like S-S stack and 30% occurs in the form of a hitherto unsuspected mixed S-N stack. The stacked trimer d(A-A-A) consists of a major (70%) regular single-helical S-S-S sequence accompanied by a minor (30%) S-S-N component. Similarly, the stacked nonamer, $d(A)_9$, is appropriately described as a 7:3 mixture of $(S)_8$-S and $(S)_8$-N forms. The conclusion must be that completely stacked poly(dA)chains are conformationally homogeneous (B-DNA type) in aqueous solution except for residual conformational freedom of the sugar at the 3'-OH terminus. The detailed geometry of the S-type sugar rings is not invariable but undergoes a slight shift in phase angle P (from about 170° to about 152°) when another base stacks at the 5'-end. The backbone angles of the fully stacked poly(dA) single helix in aqueous solution is inferred to be as follows (starting the notation at the P atom): $\beta = 187^\circ$, $\gamma = 50^\circ$, $\delta = 138^\circ$, $\varepsilon = 186^\circ$. It should be mentioned that the above conformational analyses could not have been carried out with such great precision without the benefit of the pioneering studies by Haasnoot et al.[3] that led to the development of a new empirical generalization of the Karplus equation[3a]. For the first time the influence of Electronegativity and Orientation of Substituents (EOS) on $^3J_{HH}$ of the ribose and deoxyribose ring[3d] and its C4'-C5' side chain[3c] could be taken into account. Moreover, a parallel study by de Leeuw et al.[6b] resulted in the formulation of empirical correlations between the proton-proton torsion angles and the pseudorotational parameters,[6a,103] i.e. phase angle P and puckering amplitude Ψ_m. These correlations, combined with the EOS method, were incorporated into a least-squares computer program (PSEUROT)[109] which computes P, Ψ_m and the N/S equilibrium constant given a set of vicinal couplings.[3d]

A mystery that has intrigued many workers during the past 15 years was solved[9] by the discovery of a mixed-type stack in $d(A)_n$. Only a brief indication will be given here. The CD (and ORD) spectra of poly(dA) are long known to show anomalous behaviour. Unlike the CD of poly(A), the CD of poly(dA) could not be predicted from the CD of the corresponding dimer. Moreover, the CD amplitude of single-stranded nucleic acid dimers, oligomers and polymers usually decreases strongly with increasing temperature along with a decrease in the population of stacked species. Poly(dA) remains a glaring exception; its CD shows an anomalously small temperature dependence and even an increase of $\Delta\varepsilon$ with T in the 269 nm band. This behaviour is strongly reminiscent of premelting changes seen in the CD spectra of double-helical DNAs and therefore deserves close attention. It now turns out that the mixed-type S-N stacking mode has far larger CD amplitudes compared to the S-S stack.[9] For example, around 269 nm the CD of the S-S mode is practically zero, whereas $\Delta\varepsilon 265$ of the S-N stack appears quite large. At increasing temperatures increasing random destacking occurs along the poly(dA) chain with a concomitant release of the conformational restraint to adopt an exclusively S-type deoxyribose form for those residues that no longer "feel" a stacked base underneath. As a consequence, a redistribution of S-S and S-N stacking interactions takes place. An assumed ratio of 85:15 of -pSpSp- and -pSpNp- interactions readily reproduces the observed CD phenomena,[9] Fig. 15.

The Leiden group, in keeping with tradition, recently published[110] the first complete spectral assignment (500 MHz) of the sugar protons of a tetramer: m$_2^6$A(1)-U(2)-m$_2^6$A(3)-U(4), abbreviated $\bar{\mathrm{A}}$U$\bar{\mathrm{A}}$U, Fig. 16. In view of the well known stacking proclivity of 6-N-dimethylated $\bar{\mathrm{A}}$U[8,41] (about 80% stack at 0 °C) it was expected that the tetramer $\bar{\mathrm{A}}$U$\bar{\mathrm{A}}$U would exist in two well-stacked $\bar{\mathrm{A}}$U regions, perhaps separated by a weaker-stacked U$\bar{\mathrm{A}}$ hinge. The results proved otherwise: the 3'-terminal -$\bar{\mathrm{A}}$U part indeed retains to a large measure the stacking properties of the corresponding dimer but, in sharp contrast, the 5'-terminal $\bar{\mathrm{A}}$U- as well as the middle moiety -U$\bar{\mathrm{A}}$- bases hardly stack on each other. In fact, the involvement of U(2) in stacking with nearest-neighbour $\bar{\mathrm{A}}$ bases is strikingly small (< 10%). A next nearest neighbour interaction between purine bases $\bar{\mathrm{A}}$(1)...$\bar{\mathrm{A}}$(3) appears to be present, judging from the observed shieldings of H8 and H2 of both adenines. This leads to the conclusion that a considerable amount of conformers in which U(2) is "bulged out", and $\bar{\mathrm{A}}$(1) and $\bar{\mathrm{A}}$(3) mutually stack, must be present, next to the normal stacking of fragment $\bar{\mathrm{A}}$(3)-$\bar{\mathrm{U}}$(4), Fig. 17. A similar "bulging out" of the central pyrimidine residue in purine-pyrimidine-purine trimers was postulated by Lee and Tinoco[108] in an investigation of differential shieldings of base- and H1' protons. Noteworthy is the complete conformational freedom displayed by the A(1) and U(2) ribose rings and by the backbone β-ε angles in the AUAU underlined part of the sequence.[110] Because conformational freedom implies a significant gain in entropy of mixing it is proposed that the (1)...(3) stacking energy of the purine bases, concomitant with the gain in entropy of mixing, is sufficient to offset the loss of one AU and one UA stacking interaction. Thus, the bulged-out situation does not represent a single conformer but rather a large assembly of forms each of which is characterized by a purine-purine stacking interaction.[110] Tetramers are now definitely amenable to routine investigation of their complete sugar proton spectra. A 360 and 500 MHz spectral assignment and conformational analysis of a deoxyribotetramer, d(T-A-A-T), will be published shortly.[111]

Two-dimensional NMR. One of the principal useful attributes of 2D Fourier transformation is the separation of NMR parameters into different frequency dimensions F1 and F2.[112] For the study of biological macromolecules (proteins) 2D J-resolved[113a] and 2D NOE[113b] proton NMR experiments are rapidly becoming important. Spin-echo correlated spectroscopy (SECSY)[114a] and foldover-corrected correlated

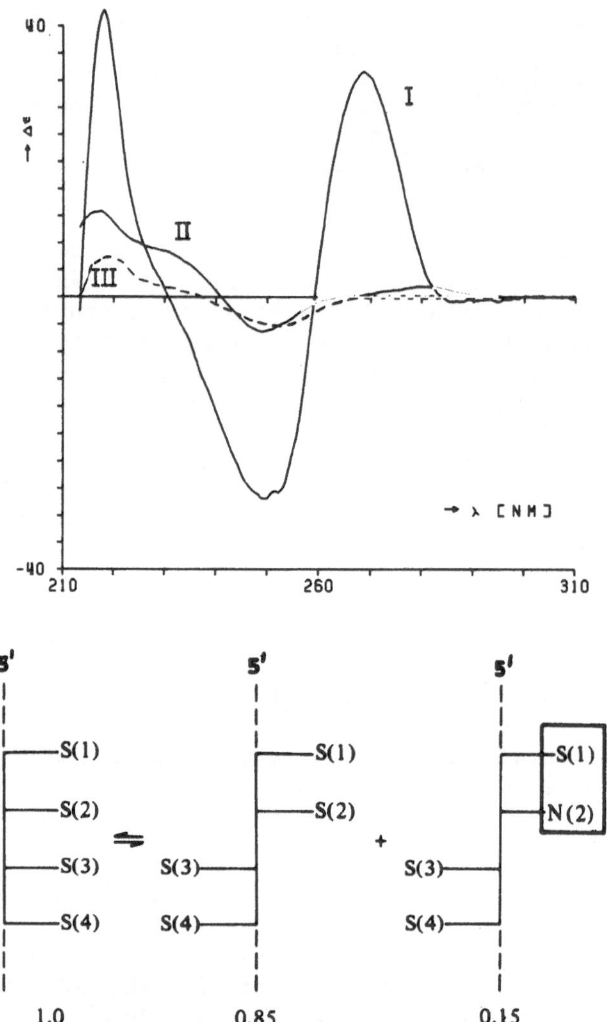

Fig. 15. Left: (I) extrapolated CD spectrum of the mixed S-N stacking mode in a dApdA sequence; (II) experimental low-temperature spectrum of poly(dA) taken as the CD of the S-S stacking mode; (III) CD spectrum of the unstacked states (Δε values given per base-base interaction). Right: Schematic representation of the redistribution of S-S and S-N stacking interactions in a $(dA)_n$ sequence when destacking occurs between non-terminal residues somewhere along the strand. The occurrence of two stacking modes with widely different CD gives rise to the curious reversed temperature-induced changes in the CD of $(dA)_n$ (from Ref. *9*).

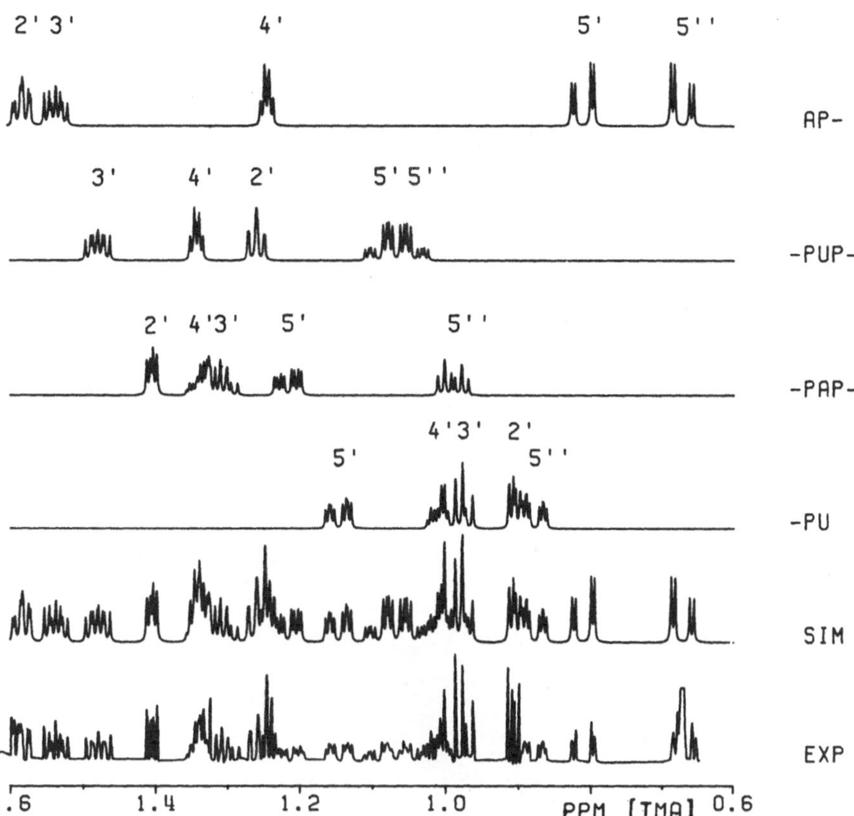

Fig. 16. High-field part of the ribose region of the resolution-enhanced 500 MHz spectrum of $\overline{A}U\overline{A}U$ at 17 °C. The experimental spectrum (Exp) was digitized by hand for illustration purposes. The four upper spectra are computer simulations of the signals for each fragment as indicated in the margin. The assignment of each signal is shown. The spectrum denoted SIM represents the complete computer simulation. Asterisk in spectrum Exp indicates signal from added EDTA (from Ref. *110*).

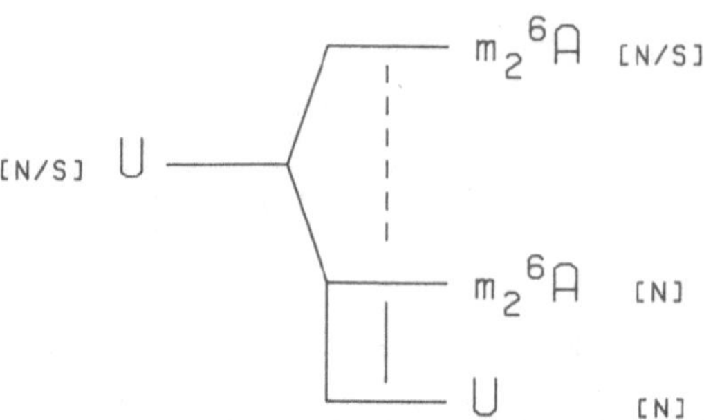

Fig. 17. Highly schematic representation (and ribose conformational properties) of favoured nearest neighbour and next-nearest neighbour stacking interactions in the tetramer $\overline{A}U\overline{A}U$ (from Ref. *110*).

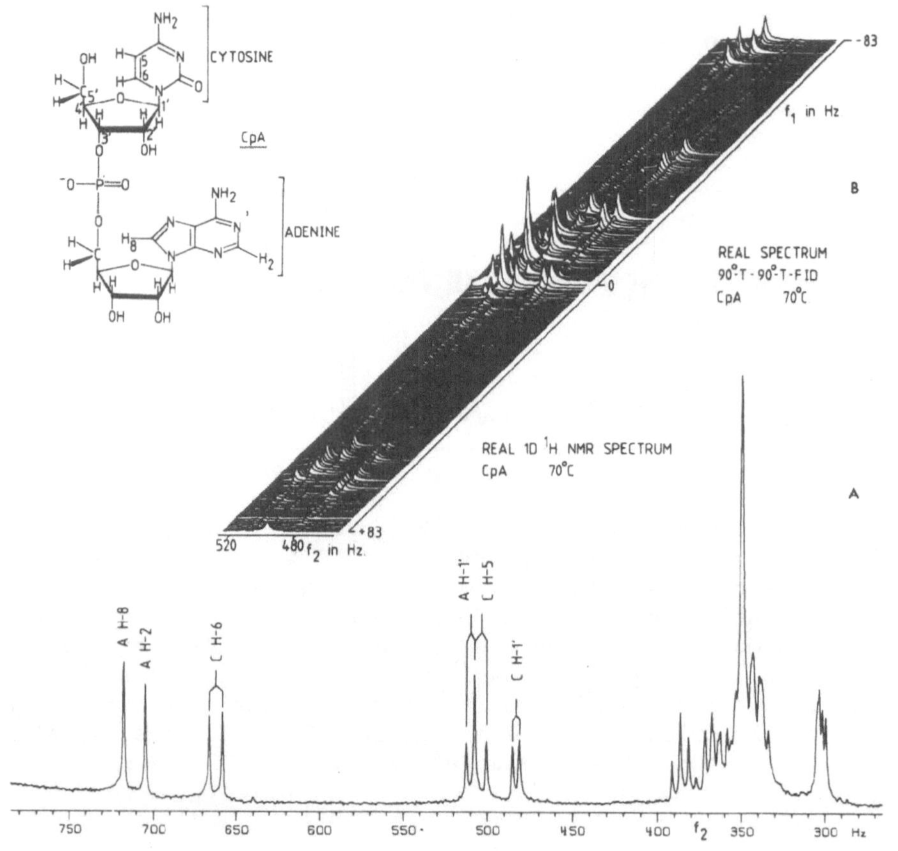

Fig. 18. (A) The 1D 90 MHz proton NMR spectrum of CpA at 70 °C.
(B) The 2D spectrum of CpA in the region of the H5 and H1' protons (from Ref. *116*).

spectroscopy (FOCSY)[114b] are new roots from this rapidly growing tree. Thus far, application of 2D NMR to the study of nucleic acids has been rather limited. Broido and Kearns[115] studied the direct dipolar connectivity between cytidine H5 and ribose H1' in poly(C) and interprete their data in terms of a surprisingly small value (< 2.9 Å) of the H5-H1' internuclear distance. A left-handed helical model for poly(C) was implicated.

Bain et al.[116] used an alternative two-dimensional pulse sequence (90° - ½t - 90° - ½t - FID) to correlate the chemical shift of the strongly spin-spin coupled protons in 5'-AMP and in CpA. Although these experiments were carried out at 90 MHz only, they show great promise. Chemical shifts of completely hidden protons could be deduced. Moreover, a connection between cytidine H5 and C1' protons via 5-bond coupling was demonstrated. Perhaps this finding necessitates a reinterpretation of the poly(C) data mentioned above. The 2D proton NMR spectrum of CpA is illustrated in Fig. 18. It is not hard to predict that further developments in this promising area, and especially 2D NMR carried out on 400-500 MHz instruments, will soon revolutionize proton spectroscopy of nucleic acids.

ACKNOWLEDGEMENTS

I am much indebted to my students A.J.Hartel, J.Doornbos, J.-R. Mellema, P.P.Lankhorst and many others for their efforts to keep track of the vast literature and to Dr. C.A.G.Haasnoot for his daring exploration of new fields. Our contributions to nucleic acid NMR would not have been possible without the pioneering synthetic work carried out by Prof. Dr. J.H. van Boom and his associates. Support by the Netherlands Foundation for Chemical Research (SON) with financial aid from the Netherlands Organization for the Advancement of Pure Research (ZWO) is gratefully acknowledged. I wish to thank Paul G.Schmidt, George T. Robillard, Russell A. Bell and Alex D. Bain for their generous cooperation and gift of illustrations from their work.

REFERENCES

1. Standard works devoted to conformational analysis in organic chemistry are:
(a) E. L. Eliel, N. L. Allinger, S. J. Angyal and G. A. Morrison, "Conformational Analysis", Wiley, New York (1965); (b) J. Dale, "Stereochemistry and Conformational Analysis", Verlag Chemie, New York (1978); (c) "Conformational Analysis, Scope and Present Limitations", G. Chiurdoglu, Ed., Organic Chemistry Monographs, Vol. 21, Academic Press (1971); (d) "The Conformational Analysis of Heterocyclic Compounds", F.G. Riddell, Ed., Academic Press, New York (1980).
2. (a) C. Romers, C. Altona, H.R. Buys and E. Havinga, in: "Topics in Stereochemistry", E. L. Eliel and N. L. Allinger, Eds., Vol. 4,

p. 39, Wiley, New York (1969); (b) L. Pauling, Proc. Natl. Acad. Sci. USA. 35:495 (1949); (c) H. J. Geise, C. Altona and C. Romers, Tetrahedron Lett. 1383 (1967).

3. (a) C. A. G. Haasnoot, F. A. A. M. de Leeuw and C. Altona, Tetrahedron 36:2783 (1980); (b) idem, Bull. Soc. Chim. Belg. 89:125 (1980); (c) C. A. G. Haasnoot, F. A. A. M. de Leeuw, H. P. M. de Leeuw and C. Altona, Recl. Trav. Chim. Pays-Bas 98:576 (1979); (d) C. A. G. Haasnoot, F. A. A. M. de Leeuw, H. P. M. de Leeuw and C. Altona, Org. Magn. Reson. 15:43 (1981).
4. (a) H. Sachse, Chem. Ber. 23:203 (1892); J. Prakt. Chem. 10:203 (1892); (b) E. Mohr, J. Prakt. Chem. 98:315 (1918).
5. C. Altona, H. R. Buys, H. J. Hageman and E. Havinga, Tetrahedron 23: 2265 (1967).
6. (a) C. Altona and M. Sundaralingam, J. Am. Chem. Soc. 94:8205 (1972); (b) H. P. M. de Leeuw, C. A. G. Haasnoot and C. Altona, Isr. J. Chem. 20:108 (1980).
7. C. S. M. Olsthoorn, L. J. Bostelaar, J. H. van Boom and C. Altona, Eur. J. Biochem. 112:95 (1980).
8. C. S. M. Olsthoorn, C. A. G. Haasnoot and C. Altona, Eur. J. Biochem. 106:85 (1980).
9. C. S. M. Olsthoorn, L. J. Bostelaar, J. F. M. de Rooij, J. H. van Boom and C. Altona, Eur. J. Biochem. 115:309 (1981)
10. T. A. Early and D. R. Kearns, Proc. Natl. Acad. Sci. USA. 76:4165 (1979).
11. (a) M. E. Hogan and O. Jardetzky, Proc. Natl. Acad. Sci. USA. 76:6341 (1979); (b) idem, Biochemistry 19:3640 (1980); (c) S. J. Opella, W. B. Wise and J. A. DiVerdi, Biochemistry 20:284 (1981).
12. (a) G.A.Webb and M.Witanowski,in: "Nitrogen NMR", M. Witanowski and G. A. Webb, eds., p. 1, Plenum, London (1973); (b) E. W. Randall, ibid. p. 41; (c) M. Witanowski, L. Stefaniak and G. A. Webb, in: "Annual Reports on NMR Spectroscopy", G. A. Webb, ed. Vol. 7, p. 118, Academic Press, New York (1977).
13. (a) G. C. Levy and R. L. Lichter, "Nitrogen-15 Nuclear Magnetic Resonance Spectroscopy", Wiley, New York (1979); (b) G. J. Martin, N. L. Martin and J.-P. Gouesnard, in: "NMR: Basic Principles and Progress", P. Diehl and E. Fluck, eds., Vol. 18, Springer, Heidelberg (1981).
14. J. H. Noggle and R. E. Schirmer, "The Nuclear Overhauser Effect", Academic Press, New York (1971).
15. D. Gust, R. B. Moon and J. D. Roberts, Proc. Natl. Acad. Sci.USA. 72:4696 (1975).
16. J. A. Happe and M. Morales, J. Am. Chem. Soc. 88:2077 (1966).
17. (a) V. Markowski, S. R. Sullivan and J. D. Roberts, J. Am. Chem. Soc. 99:714 (1977); (b) P. Büchner, W. Maurer and H. Rüterjans, J. Magn. Reson.29:45 (1978); (c) P. Büchner, F. Blomberg and H. Rüterjans, in: "Nuclear Magnetic Resonance Spectroscopy in Molecular Biology", B. Pullman, ed., p. 53, Reidel, Dordrecht, Holland (1978).

18. G. E. Hawkes, E. W. Randall and W. E. Hull, J. Chem. Soc. Perkin II, 1268 (1977).
19. (a) L. Katz and S. Penman, J. Mol. Biol. 15:220 (1966); (b) R. R. Shoup, H. T. Miles and E. D. Becker, Biochem. Biophys. Res. Commun. 23:194 (1966).
20. C. D. Poulter and C. L. Livingston, Tetrahedron Lett. 755 (1979).
21. P. H. Bolton, J. Magn. Reson. 41:287 (1980).
22. G. Mavel, "NMR Studies of Phosphorus Compounds", in: "Annual Reports on NMR Spectroscopy", E. F. Mooney, ed., Vol. 5B, Academic Press, New York (1973).
23. (a) J. H. Letcher and J. R. von Wazer, in: "Topics in Phosphorus Chemistry", M. Grayson and E. J. Griffith, eds, Vol. 4, p. 75, Interscience, New York (1966); (b) M. Mark and J. R. von Wazer, J. Org. Chem. 32:1187 (1967).
24. (a) D. G. Gorenstein, J. Am. Chem. Soc. 97:898 (1975); (b) D. G. Gorenstein and D. Kar, Biochem. Biophys. Res. Commun. 65:1073 (1975); (c) D. G. Gorenstein, J. B. Findlay, R. K. Momii, B. A. Luxon and D. Kar, Biochemistry 15:3796 (1976). (d) D. G. Gorenstein, J. Am. Chem. Soc. 99:2254 (1977); (e) D. G. Gorenstein and B. A. Luxon, Biochemistry 18:3796 (1979).
25. D. Perahia and B. Pullman, Biochim. Biophys. Acta, 475:184 (1977); 435:282 (1976).
26. T.-D. Son, M. Roux and M. Ellenberger, Nucl. Acids Res. 2:1101 (1975).
27. C. A. G. Haasnoot and C. Altona, Nucl. Acids Res. 6:1135 (1979).
28. C. G. Reinhardt and T. R. Krugh, Biochemistry 16:2890 (1977).
29. P. J. Cozzone and O. Jardetzky, Biochemistry 15:4853 (1976).
30. B. D. Lerner and D. R. Kearns, J. Am. Chem. Soc. 102:7611 (1980).
31. P. Davanloo, I. M. Armitage and D. M. Crothers, Biopolymers 18:663 (1979).
32. M. Guéron and R. G. Shulman, Proc. Natl. Acad. Sci. USA. 72:3482 (1975).
33. Y. H. Mariam and W. D. Wilson, Biochem. Biophys. Res. Commun. 88:861 (1979).
34. (a) R. T. Simpson and H. Shindo, Nucl. Acids Res. 8:2093 (1980); (b) L. Klevan, I. M. Armitage and D. M. Crothers, Nucl. Acids Res. 6:1607 (1979); (c) P. H. Bolton and T. L. James, J. Phys. Chem. 83:3359 (1979); (d) P. H. Bolton and T. L. James, J. Am. Chem. Soc. 102:25 (1980).
35. L. M. Weiner, J. M. Backer and A. I. Rezvukin, FEBS Lett. 41:40 (1974).
36. P. M. J. Salemink, T. Swarthof and C. W. Hilbers, Biochemistry 18:3477 (1979).
37. F. E. Evans and N. O. Kaplan, FEBS Lett. 105:11 (1979).
38. (a) G. J. Garssen, C. W. Hilbers, J. G. G. Schoenmakers and J. H. van Boom, Eur. J. Biochem. 81:453 (1977).
39. P. M. J. Salemink, H. A. Raué, H. Heerschap, R. J. Planta and C. W. Hilbers, Biochemistry 20:265 (1981).
40. H. A. M. Geerdes, J. H. van Boom and C. W. Hilbers, J. Mol. Biol. 142:195 (1980); 142:219 (1980).

41. C. Altona, A. J. Hartel, C. S. M. Olsthoorn, H. P. M. de Leeuw and C. A. G. Haasnoot, *in*: "Nuclear Magnetic Resonance in Molecular Biology", B. Pullman, ed., Jerus. Symp. Series Vol. 11, p. 87, Reidel, Dordrecht, Holland (1978).
42. (a) J. B. Stothers, "Carbon-13 NMR Spectroscopy", Academic Press, New York (1972); (b) F. A. L. Anet, "^{13}C NMR at High Magnetic Fields", *in*: "Topics in Carbon-13 Spectroscopy", G. C. Levy, ed., Vol. 1, p. 209, Wiley, New York (1974).
43. S. N. Rosenthal and J. H. Fendler, "^{13}C NMR Spectroscopy in Macromolecular Systems of Biochemical Interest", *in*: "Advances in Physical Organic Chemistry", V. Gold and D. Bethell, eds, Vol. 13, p. 280, Academic Press, New York (1976).
44. R. A. Komoroski, I. R. Peat and G. C. Levy, "^{13}C NMR Studies of Polymers", *in*: "Topics in Carbon-13 Spectroscopy", G.C.Levy, ed., Vol. 2, p. 180, Wiley, New York (1976).
45. J. Feeney, "The Application of ^{13}C NMR Spectroscopic Techniques to Biological Problems", *in*: "New Techniques in Biophysics and Cell Biology", R. H. Pain and B. J. Smith, eds, Vol. 2, p. 287, Wiley, New York (1975).
46. R. L. Rill, P. R. Hilliard, J. T. Bayley and G. C. Levy, *J. Am. Chem. Soc.* 102:418 (1980).
47. (a) J. G. Tompson and P. F. Agris, *Nucl. Acids Res.* 7:765 (1979) (b) J. G. Tompson, F. Hayashi, J. V. Paukstelis, R. N. Loeppky and P. F. Agris, *Biochemistry* 18:2079 (1979).
48. P. F. Agris and P. G. Schmidt, *Nucl. Acids Res.* 8:2085 (1980).
49. P. G. Schmidt, J. G. Tompson and P. F. Agris, *Nucl. Acids Res.* 8:643 (1980).
50. (a) M. P. Schweizer, W. D. Hamill, I. J. Walkiw, W. J. Horton and D. M. Grant, *Nucl. Acids Res.* 8:2075 (1980); (b) W.D.Hamill, D. M. Grant, W. J. Horton, R. Lundquist and S. Dickman, *J. Am. Chem. Soc.* 98:1276 (1976).
51. (a) S. Yokoyama, K. M. J. Usuki, Z. Yamaizumi, S. Nishimura and T. Myazawa, *FEBS Lett.* 119:77 (1980); (b) C. J. Chang and C. G. Lee, *Arch. Biochem. Biophys.* 176:801 (1976).
52. Consult Tables 7 and 8 in Ref. 44.
53. H. H. Mantsch and I. C. P. Smith, *Biochem. Biophys. Res. Commun.* 46:808 (1972).
54. J. L. Alderfer and P. O. P. Ts'o, *Biochemistry* 16:2410 (1977).
55. (a) C. Giessner-Prettre and B. Pullman, *J. Theor. Biol.* 27:87 (1970); (b) C. Giessner-Prettre, B. Pullman, P. N. Borer, L. S. Kan and P. O. P. Ts'o, *Biopolymers* 15:2277 (1976); (c) C. Giessner-Prettre and B. Pullman, *Biochem. Biophys. Res. Commun.* 70:578 (1976); (d) idem, *J. Theor. Biol.* 65:171 (1977); (e) F. Ribas-Prado and C. Giessner-Prettre, *J. Mol. Struct.* 76:81 (1981).
56. D.B. Arter and P. G. Schmidt, *Nucl. Acids Res.* 3:1437 (1976).
57. (a) C. W. Haigh, *in*: "Annual Reports on NMR Spectroscopy", E. F. Mooney, ed., Vol. 4, p. 311, Academic Press, New York (1971); (b) P. Diehl, H. Kellerhals and E. Lustig, *in*:"NMR: Basic Principles and Progress", P. Diehl, E. Fluck and

R. Kosfeld, eds, Vol. 6, p. 1, Springer, Heidelberg (1972).
58. A. de Marco, J. Magn. Reson. 26:527 (1977).
59. D. H. Live and S. I. Chan, Org. Magn. Reson. 5:275 (1973).
60. A. J. Hartel and C. Altona, unpublished.
61. P. G. Schmidt and E. B. Edelheit, Biochemistry 30:79 (1981).
62. P. H. Bolton and D. R. Kearns, Biochim. Biophys. Acta 517:329 (1978).
63. (a) V. Sãnchez, A. G. Redfield, P. D. Johnston and J. Tropp, Proc. Natl. Acad. Sci. USA. 77:5659 (1980); (b) P. D. Johnston and A. G. Redfield, Biochemistry 20:1147 (1981).
64. C. A. G. Haasnoot, J. H. J. den Hartog, J. F. M. de Rooij, J. H. van Boom and C. Altona, Nature (London) 281:235 (1979). Note that the authors' names were mutilated in this paper because of an editorial error after proofs were returned.
65. C. A. G. Haasnoot, J. H. J. den Hartog, J. F. M. de Rooij, J. H. van Boom, and C. Altona, Nucl. Acids Res. 8:169 (1980).
66. Y. P. Wong, D. R. Kearns, B. R. Reid and R. G. Shulman, J. Mol. Biol. 72:725 (1972)
67. (a) D. R. Kearns, D. Patel and R. G. Shulman, Nature (London) 229:338 (1971); (b) D. R. Kearns, D. Patel, R. G. Shulman, and T. Yamane, J. Mol. Biol. 61:265 (1971); (c) D. R. Kearns, and R. G. Shulman, Acc. Chem. Res. 7:33 (1974); (d)D.R.Kearns, in: "Progress in Nucleic Acids Research and Molecular Biology", W. Cohn, ed., Vol. 18, p. 91, Academic Press, New York (1976); (e) R. G. Shulman, C. W. Hilbers, D. R. Kearns, B. R. Reid and Y. P. Wong, J. Mol. Biol. 78:57 (1973).
68. G. T. Robillard, C. E. Tarr, F. Vosman and B. R. Reid, Biochemistry 16:5261 (1977).
69. (a) J. Dadok and R. F. Sprecher, J. Magn. Reson. 13:243 (1974); (b) R. Gupta, J. Ferretti and E. D. Becker, J. Magn. Reson. 13:275 (1974).
70. (a) A. G. Redfield, S. D. Kunz and E. K. Ralph, J. Magn. Reson. 19:114 (1975); (b) A. G. Redfield, in: "NMR: Basic Principles and Progress", P. Diehl, E. Fluck and R. Kosfeld, eds, Vol. 13, p. 137, Springer, Heidelberg; (c) A. G. Redfield and R. Gupta, J. Chem. Phys. 54:1418 (1971); Adv. Magn. Reson. 5:81 (1971).
71. (a)P.D.Johnston and A. G. Redfield, Nucl. Acids Res. 4:3599 (1977);(b) P. D. Johnston and A. G. Redfield, Nucl. Acids Res. 5:3913 (1978); (c) P. D. Johnston and A. G. Redfield, in: "Transfer RNA Monograph", J. Abelson, P. R. Schimmel and D.Soll, eds, p. 191, Cold Spring Harbor Press, Cold Spring Harbor, N.Y. (1979); (d) P. D. Johnston, N. Figuera and A. G. Redfield, Proc. Natl. Acad. Sci. USA, 76:3130 (1979).
72. R. E. Hurd and B. R. Reid, J. Mol. Biol. 142:181 (1980)
73. (a) G. Quigley, N. Seeman, A. Wang, F. Suddath and A. Rich, Nucl. Acids Res. 2:2329 (1975); (b) A. Jack, J. Ladner and A. Klug, J. Mol. Biol. 108:619 (1976); (c) J. L. Sussman and S.-H. Kim, Biochem. Biophys. Res. Commun. 68:89 (1976); (d) J. Sussman, S. Holbrook, R. W. Warrant, G. Church and S.-H. Kim, J. Mol. Biol. 123:607 (1978); (e) C. D. Stout, H. Mizuno,

S. T. Rao, P. Swaminathan, J. Rubin, T. Brennan and M. Sundaralingam, Acta Cryst. B 34:1529 (1978).

74. P. H. Bolton and D. R. Kearns, in: " Biological Magnetic Resonance", L. J. Berliner and J. Ruben, eds, Vol. 1, p. 91, Plenum, New York (1978); (b) G. T. Robillard and B. R. Reid, in: " Biological Applications of Magnetic Resonance", R. G. Shulman, ed., p. 45, Academic Press, New York (1979); (c) P. R. Schimmel and A. G. Redfield, Ann. Rev. Biophys. Bioeng. 9:181 (1980); (d) G. T. Robillard, in: "NMR in Biology", R. A. Dwek, I. D. Campbell, R. E. Richards and R. J. P. Williams, eds, p. 201, Academic Press, New York (1977).

75. (a) D. R. Lightfoot, K. L. Wong, D. R. Kearns, B. R. Reid and R. G. Shulman, J. Mol. Biol. 78:71 (1973); (b) R. G. Shulman, C. W. Hilbers, Y. P. Wong, K. L. Wong, D. R. Lightfoot, B. R. Reid and D. R. Kearns, Proc. Natl. Acad. Sci.USA, 70:2042 (1973).

76. (a) S.-H. Kim, F. L. Suddath, G. J. Quigley, A. McPherson, J. S. Sussman, H. A. J. Wang, N. C. Seeman and A. Rich, Science 185:435 (1974); (b) J. D. Robertus, J. E. Ladner, J. T. Finch, D. Rhodes, S. R. Brown, B. F. C. Clark and A. Klug, Nature (London) 250:546 (1974).

77. (a) B. R. Reid, N. S. Ribero, G. Gould, G. T. Robillard, C. W. Hilbers and R. G. Shulman, Proc. Natl. Acad. Sci. USA, 72:2049 (1975); (b) B. R. Reid and G. T. Robillard, Nature (London) 262:424 (1975).

78. (a) W. E. Daniel and M. Cohn, Proc. Natl. Acad. Sci. USA, 72:2582 (1975); (b) W. E. Daniel and M. Cohn, Biochemistry 15:3917 (1976).

79. (a) B. R. Reid, N. S. Ribero, L. McCollum, J. Abbate and R. E. Hurd, Biochemistry 16:2086 (1977); (b) B. R. Reid and R. E. Hurd, Acc. Chem. Res. 10:396 (1977).

80. (a) P. H. Bolton, C. R. Jones, D. Bastedo-Lerner, K. L. Wong and D. R. Kearns, Biochemistry 15:4370 (1976); (b) P. H. Bolton and D. R. Kearns, Nature (London) 262:423 (1976).

81. L. Kan and P. O. P. Ts'o, Nucl. Acids Res. 4:1633 (1977).

82. (a) B. R. Reid, L. McCollum, N. S. Ribero, J. Abbate and R. E. Hurd, Biochemistry 18:3996 (1979); (b) R. E. Hurd and B. R. Reid, Biochemistry 18:4005 (1979); (c) R. E. Hurd , E. Azhderian and B. R. Reid, Biochemistry 18:4012 (1979); (d) R. E. Hurd and B. R. Reid, Biochemistry 18:4017 (1979).

83. (a) K. L. Wong, P. H. Bolton and D. R. Kearns, Biochim. Biophys. Acta 383:464 (1975); (b) K. L. Wong, D. R. Kearns, W.Wintermeyer and H. Zachau, Biochim. Biophys. Acta 395:1 (1975).

84. A. Jack, J. E. Ladner, D. Rhodes, R. S. Brown and A. Klug, J. Mol. Biol. 111:315 (1977).

85. J. Boyle, G. T. Robillard and S.-H. Kim, J. Mol. Biol. 139:601 (1980).

86. C. E. Johnson and F. A. Bovey, J. Chem. Phys. 29:1012 (1958).

87. (a) C. W. Haigh and R. B. Mallion, Mol. Phys. 22:955 (1971); (b) R. B. Mallion, in: "Nuclear Magnetic Resonance in Molecular

Biology", B. Pullman, ed., p. 183, Reidel, Dordrecht, Holland, (1978).
88. G. T. Robillard, C. E. Tarr, F. Vosman and H. J. C. Berendsen, Nature (London) 262:363 (1976), and Ref. 74d.
89. J. A. Pople, J. Chem. Phys. 24:1111 (1956).
90. R. J. Abraham, in: "Nuclear Magnetic Resonance in Molecular Biology", B. Pullman, ed., p. 461, Reidel, Dordrecht, Holland, (1978).
91. D. M. Crothers, C. W. Hilbers and R. G. Shulman, Proc. Natl. Acad. Sci. USA.70:2899 (1973).
92. G. T. Robillard, C. Tarr, F. Vosman and J. Sussman, Biophys. Chem. 6:291 (1977).
93. S. Arnott, in: "Progress in Biophysics and Molecular Biology", J. A. V. Butler and D. Noble, eds, Vol. 22, p. 179, Pergamon, Oxford (1971).
94. S. R. Holbrook, J. L. Sussman, R. W. Warrant and S.-H. Kim, J. Mol. Biol. 123:631 (1978).
95. (a) R. Wing, H. R. Drew, T. Takano, C. Broka, S. Tanaka, K. Itakura and R. E. Dickerson, Nature (London) 287:755 (1980); (b) R. E. Dickerson and H. R. Drew, J. Mol. Biol. submitted, and private commun. to Prof. J. H. van Boom.
96. S. Arnott and D. W. L. Hukins, Biochem. Biophys. Res. Commun. 47:1504 (1972).
97. (a) L. S. Kan, P. O. P. Ts'o, M.Sprinzl, F. van der Haar and F. Cramer, Biochemistry 16:3143 (1977); (b) P. Davanloo, M.Sprinzl and F. Cramer, Biochemistry 18:3189 (1979).
98. R. V. Kastrup and P. G. Schmidt, Nucl. Acids Res. 5:257 (1978).
99. D. B. Davies, "Conformations of Nucleic Acids", in: "Progress in Nuclear Magnetic Resonance Spectroscopy", J. W. Emsley, J. Feeney and L. H. Sutcliffe, eds, Vol. 12, p. 135, Pergamon, Oxford (1978).
100. C. Altona, H. J. Koeners, J. R. de Jager, J. H. van Boom and G. van Binst, Recl. Trav. Chim. Pays-Bas 93:169 (1974).
101. C. Altona, J. H. van Boom, J. R. de Jager, H. J. Koeners and G. van Binst, Nature (London) 247:558 (1974).
102. C. Altona, in: "Structure and Conformation of Nucleic Acids and Protein-Nucleic Acid Interactions", M. Sundaralingam and S. T. Rao, eds, p. 163, University Park Press, Baltimore (1975). Note that the spectrum in Fig. 1 is of m_2^6ApU, as is correctly stated in the text.
103. C. Altona and M. Sundaralingam, J. Am. Chem. Soc. 95:2333 (1973).
104. N. S. Kondo and S. S. Danyluk, Biochemistry 15:756 (1976).
105. C.-H. Lee and I. Tinoco, Biochemistry 16:5403 (1977).
106. C. Altona, J. H. van Boom and C. A. G. Haasnoot, Eur. J. Biochem. 71:557 (1976).
107. F. E. Evans and R. H. Sarma, Nature (London) 263:567 (1976).
108. C.-H. Lee and I. Tinoco, Biophys. Chem. 11:283 (1980).
109. F. A. A. M. de Leeuw, Thesis Leiden, in preparation.
110. A. J. Hartel, G. Wille-Hazeleger, J. H. van Boom and C. Altona, Nucl. Acids. Res. 9:1405 (1981).

111. J.-R. Mellema, C. A. G. Haasnoot, J. H. van Boom and C. Altona, Biochim. Biophys. Acta, in press; J.-R. Mellema, this workshop.

112. (a) W. P. Aue, E. Bartholdi and R. R. Ernst, J. Chem. Phys. 64:2229 (1976); (b) J. Jeener, B. H. Meier, P. Bachman and R. R. Ernst, J. Chem. Phys. 71:4546 (1979); (c) R. Freeman and G. A. Morris, Bull. Magn. Reson. 1:5 (1979).

113. (a) K. Nagayama, K. Wüthrich, P. Bachman and R. R. Ernst, Biochem. Biophys. Res. Commun. 78:99 (1977); (b) A. Kumar, G. Wagner, R. R. Ernst and K. Wüthrich, Biochem. Biophys. Res. Commun. 96:1156 (1980).

114. (a) K. Nagayama, K. Wüthrich and R. R. Ernst, Biochem. Biophys. Res. Commun. 90:305 (1979); (b) K. Nagayama, A. Kumar, K. Wüthrich and R. R. Ernst, J. Magn. Reson. 40:321 (1980).

115. (a) K. Nagayama and K. Wüthrich, Eur. J. Biochem. 114:365 (1981); (b) G. Wagner, A. Kumar and K. Wüthrich, Eur.J.Biochem. 114:375 (1981).

116. M. S. Broido and D. R. Kearns, J. Magn. Reson. 41:496 (1980).

117. A. D. Bain, R. A. Bell, J. R. Everett and D. W. Hughes, Can. J. Chem. 58:1947 (1980).

HIGH RESOLUTION NMR EXPERIMENTS FOR

STUDIES OF PROTEIN CONFORMATIONS

Kurt Wüthrich

Institut für Molekularbiologie und Biophysik
Eidgenössische Technische Hochschule
ETH-Hönggerberg, CH-8093 Zürich, Switzerland

CONTENTS:

1. INTRODUCTION

Applications of nuclear magnetic resonance (NMR) in biological research cover a wide spectrum, including studies of structure and conformation of biological macromolecules, investigations of biomembranes, studies of intermediary metabolites in intact, live cells and organs, and imaging of macroscopic objects.[1,2] In all these different areas the use of NMR has gained much momentum from rapid progress in the development of improved instrumentation and methodology. For studies of protein conformations in solution, which is the theme of this presention, the introduction of ever higher polarizing fields and of Fourier transform techniques was of particular interest. At present, two-dimensional (2D) experiments[3,4] promise to further increase the potentialities of NMR for delineating biomacromolecular structures.[5] In Section 2 this paper starts with a brief survey of fundamental aspects of NMR spectra of proteins. The main part of the paper, Section 3, describes some more recently introduced experiments. Section 4, finally, contains some general comments on the potentialities of modern NMR experiments for studies of protein conformation.

2. NMR SPECTRA OF PROTEINS

2.1. NMR nuclei in polypeptide chains

Polypeptide chains contain three nuclei with spin I = 1/2, which are suitable for high resolution NMR experiments. These are ^{1}H, ^{13}C and ^{15}N. The relative ease of observation of the NMR signals for the different nuclei at constant field is determined by the NMR sensitivities and the isotope abundance.[6] At natural isotope abundance, relative signal intensities are 1 for ^{1}H, $1.7 \cdot 10^{-4}$ for ^{13}C and $3.8 \cdot 10^{-6}$ for ^{15}N. Hence, NMR observation of ^{13}C and ^{15}N is much more difficult than observation of ^{1}H. As a consequence, ^{1}H NMR has played a dominant role in many biological applications. However, with the improved sensitivity of modern Fourier transform (FT) spectrometers, ^{13}C and ^{15}N have recently also become attractive for NMR studies of biopolymers. Besides optimal instrumentation, isotope enrichment can greatly improve the conditions for ^{13}C and ^{15}N NMR experiments, since the natural abundance of these isotopes is only 1.11% and 0.37%, respectively.[6] As an illustration, ^{1}H, ^{13}C and ^{15}N spectra of peptides or proteins are presented in Figs. 1-3.

Fig. 1 shows two natural abundance ^{13}C NMR spectra of the basic pancreatic trypsin inhibitor (BPTI), a small globular protein of molecular weight 6500, which were recorded at two different field strengths. From comparison with the spectra of the individual amino acid residues in model peptides [6-8] the resonances between 0 and 25ppm can be attributed to methyl carbons of the aliphatic amino acid side chains, between 25 and 70ppm to methylene and methine carbons of the side chains and to the backbone α-carbons, between 110 and 160ppm to the aromatic carbons and the guanidinium group of arginine, and between 165 and 185 to the carbonyl and carboxyl carbons of the polypeptide backbone and the side chains. With the use of Fourier transform spectrometers and large sample volumes, observation of proton noise-decoupled ^{13}C NMR spectra of peptides and proteins is readily achieved even at relatively low magnetic fields, and many of the pioneering studies were done at a frequency

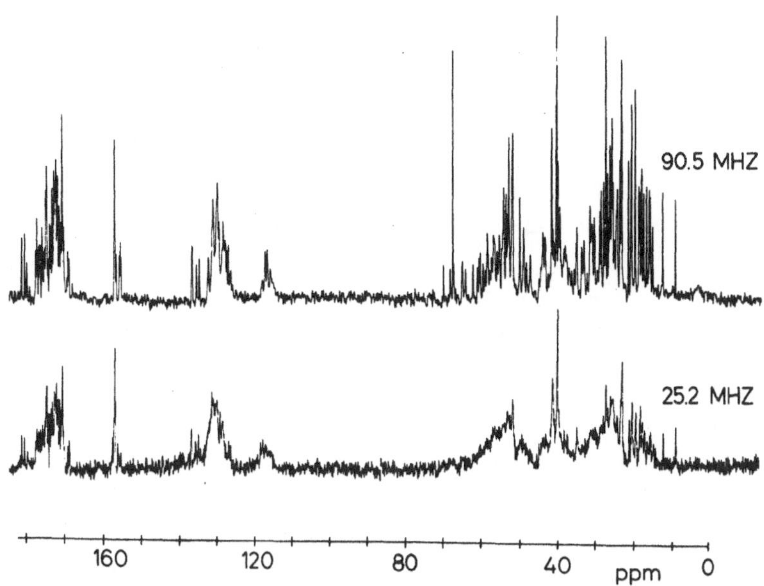

Fig. 1 Natural abundance ^{1}H noise-decoupled FT ^{13}C NMR spectra at 25.2 MHz and 90.5 MHz of a 0.025 M solution of the basic pancreatic trypsin inhibitor (BPTI) in D_2O, pD = 8.2, T = 35° C, accumulation time 12 h. At 25.2 MHz, the sample diameter was 12 mm, 54'000 transients were accumulated with a recycle time of 0.8 s, the digital resolution is 1.25 Hz/point. At 90.5 MHz, the sample size was 10 mm, 86'000 transients were accumulated with a recycle time of 0.5 s, the digital resolution is 2.5 Hz/point. At both frequencies, a digital broadening of 1 s was applied. The chemical shifts are relative to external TMS.[6] (Reproduced from ref. 9).

of 15 MHz.[10,11] Nevertheless, as is illustrated in Fig. 1, greatly improved resolution can be obtained at higher field strength, in particular for the spectral regions which contain resonances of protonated carbon atoms. The use of high fields is of particular interest for assignments of ^{13}C NMR lines by ^{1}H - ^{13}C heteronuclear double resonance techniques, which depend critically also on the resolution of the ^{1}H NMR spectrum.[9,12]

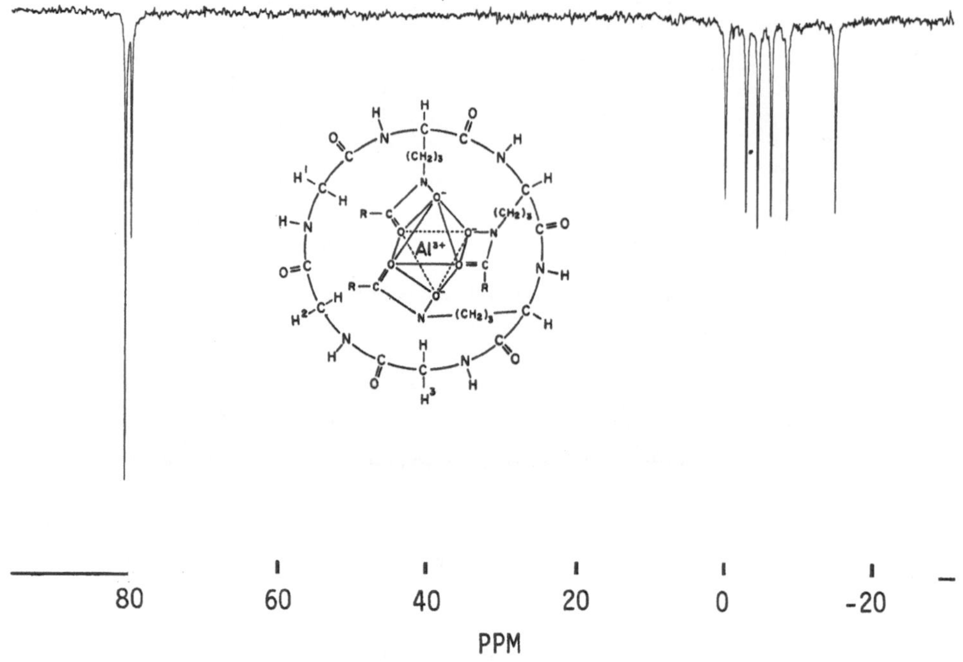

Fig. 2 ^{1}H noise-decoupled FT ^{15}N NMR spectrum at 10.1 MHz of the cyclohexapeptide alumichrome. The spectrum was recorded in ca. 6 h in a 0.07 M solution of 99.2% ^{15}N enriched peptide in deuterated dimethylsulfoxide, T = 45°. The isotope enriched peptide was obtained from a culture of Ustilago sphaerogena which was grown on a medium containing ^{15}N enriched ammonium acetate as the sole nitrogen source. The structure of alumichrome is also indicated in the figure. The backbone peptide nitrogen resonances extend from 0 to approx. -15 ppm and the three metal-coordinated hydroxamate resonances appear at approx. 80 ppm relative to the lowest field amide nitrogen line. The negative sign of the resonance lines is a consequence of the nuclear Overhauser enhancement (NOE). (Reproduced from ref. 13).

Fig. 2 shows the proton noise-decoupled ^{15}N NMR spectrum of a 99% ^{15}N enriched cyclic hexapeptide.[13] When recording proton decoupled ^{15}N spectra, a big gain in sensitivity can be attained through the large negative Overhauser enhancement (NOE).[6,14] Even though natural abundance ^{15}N NMR spectra have been recorded for a variety of peptides and proteins,[15,16] use of ^{15}N labelled molecules appears to be a more promising approach. Particularly at-

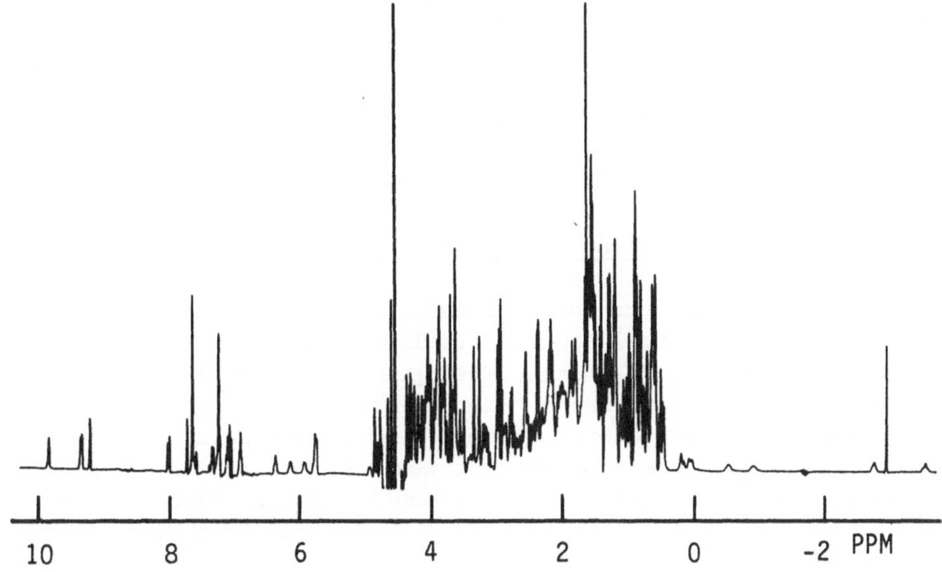

Fig. 3 ^1H NMR spectrum at 360 MHz of a 0.002 M solution of ferrocytochrome c-551 from Pseudomonas aeruginosa[21] in 0.05 M deuterated phosphate buffer, pD = 6.6, T = 40°C. This hemoprotein, which has as molecular weight of 8600, consists of one polypeptide chain with 82 amino acid residues and one heme c group. The spectral resolution was improved by multiplication of the free induction decay with a shifted sine bell, $\sin[\pi(t + t_o/t_s)]$, with t_s equal to the acquisition time and $t_o/t_s = 1/64$.

tractive experiments can be devised with selective ^{15}N or ^{13}C enrichment of amino acid residues with key roles for the structural and/or functional properties of the protein.[17-19]

A ^1H NMR spectrum of a medium size globular protein is shown in Fig. 3. From comparison with the resonances of the individual amino acid residues,[6,20] peaks between 0 and 2 ppm can be attributed primarily to methyl groups of aliphatic amino acid side chains, between 2 and 3.5 ppm to methine and methylene protons, between 3.5 and 5 ppm to backbone α-protons, and from 6 to 10 ppm to protons of aromatic amino acids and the heme group. The residual solvent protons of HDO give rise to an intense line near 4.5 ppm.[6] The lines between 0 and -4 ppm will be discussed in the following section. Even though high field was used and the resolution was further improved by digital filtering[6,22,23], the crowded region from 0 to 5 ppm of the spectrum in Fig. 3 is only partially resolved. Compared to ^{13}C and ^{15}N, relatively little can be gained from high resolution ^{1}H NMR studies of proteins at low magnetic fields and therefore advances in the use of ^{1}H NMR were closely linked with the development of high field spectrometers.[6]

2.2. Conformation-dependent ^{1}H NMR chemical shifts

Conformation-dependent chemical shifts arising from interactions of protons with the local magnetic fields of aromatic rings have played an important role in the development of general notions on protein NMR spectra.[6,24] These "ring current shifts" are used here as an illustration for conformation-dependent NMR parameters. Fig.4 shows schematically the ring current field of an aromatic ring. Protons of other segments of the polypeptide chain which are located near the aromatic amino acids in the globular form of the protein, experience the local ring current field, H_R, in addition to the external polarizing field, H_o. The resulting chemical shifts may be as large as ca. 2 ppm for protons near phenylalanine, tyrosine or tryptophan and ca. 5 ppm for protons near porphyrin rings in hemoproteins.[6] Since for a given ring size the extent of the ring current shifts depends only on the relative spatial arrangement of the ring and the observed protons (Fig. 4), conformational features can be clearly manifested in the ring current shifted lines. Thus, in ferrocytochrome c-551 (Fig. 3) the resonances at the high field end of the spectrum are shifted to their extreme positions by the ring current field of the heme group. The five lines between 0 and -4 ppm, where the spectra of diamagnetic organic molecules do not usually contain any resonances,[6] correspond to the methyl and methylene protons of the axially bound methionine side chain (Fig. 5).

On a more general level, "conformation-dependent chemical shifts" are the chemical shift differences for corresponding protons in the globular protein and the random coil form of the polypeptide chain. In a hypothetical "random coil" NMR spectrum of a polypeptide chain computed as the sum of the resonance lines of the constituent amino acid residues measured in small model peptides,[6,20] all the resonance lines are in the spectral regions from ca. 0.8 to 4.8 ppm and 6.8 to 8.2 ppm. They coincide usually closely with the corresponding lines in the experimental spectrum of the denatured protein.[6] In a globular protein, most protons are exposed to local magnetic fields of neighbouring groups in the protein, e.g. aromatic rings (Fig. 4), carbonyl double bonds, etc. Even though most of the local magnetic fields are small compared to the ring current

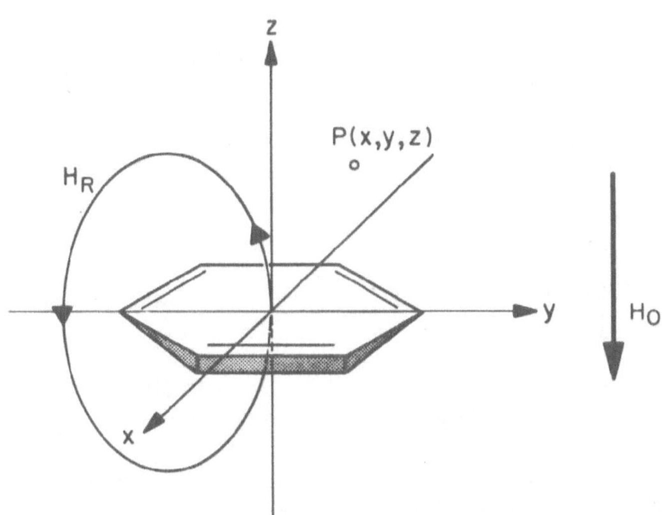

Fig. 4 The local magnetic field of an aromatic ring. H_0 is the external polarizing field. For a fixed ring size, the field strength experienced by nearby protons is determined by the position P(x,y,z) relative to the center of the aromatic ring. In a protein, the ring current shifts of nuclei located near aromatic rings are hence determined by the molecular conformation in the environment of the aromatic residue.

fields, they nevertheless cause a dispersion of the chemical shifts about the corresponding random coil values. Since there is generally no periodicity in protein tertiary structures, the local environment of each proton is in general unique, which is manifested by a unique chemical shift in the NMR spectrum. On the one hand this provides that each proton can in principle be observed individually, and hence truly a many-parameter characterization of the protein conformation can be obtained. On the other hand exceedingly complex ^1H NMR spectra are obtained even for small and medium size globular proteins, and sophisticated techniques are required to resolve and assign individual resonance lines.

2.3. NMR parameters and protein conformation

In addition to the fore-mentioned manifestations of the spatial polypeptide structure in the chemical shifts, data on protein conformations may be obtained e.g. from spin-spin coupling constants,[25] measurements of spin relaxation parameters, observation of labile

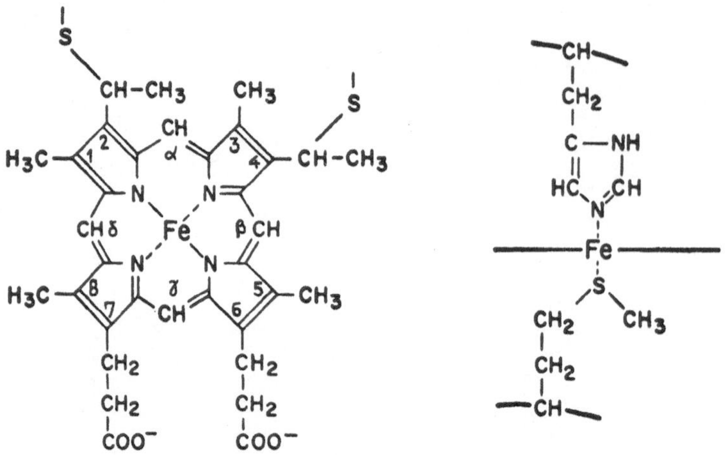

Fig. 5 Heme group (left) and axial ligands of the heme iron (right) in cytochrome c-551. The protons of the axial ligands are located in the area where the ring current field of the porphyrin ring opposes the external field H_0 (Fig. 4). As a consequence, in the ^1H NMR spectrum of Fig. 3 the methyl groups and four lines of the four methylene protons of the axial methionine are between 0 and -4 ppm.

protons in D_2O and H_2O solution, and studies of the effects of pH, temperature or shift reagents on the protein NMR spectra.[6] Overall, a wealth of interesting data on differences between protein conformations in different solvent media or between protein crystals and solution,[1,2] on various aspects of protein function[1,2] and, perhaps most important, on internal flexibility of globular proteins[1,2,6,26,27] was thus obtained.

One tackles a considerably more difficult problem when trying to determine the conformation of a polypeptide chain from the known amino acid sequence and the NMR data. The arrangement of a polypeptide chain in space can be characterized e.g. by a complete set of atomic coordinates,[28] by a complete set of torsion angles about all the bonds in the molecular structure[6,28] or by a complete set of intramolecular proton-proton distances.[52] Except in few particularly favorable circumstances, such as the ring current effects near aromatic rings (Fig. 4) or pseudocontact shifts near paramagnetic centers[6,29], the present understanding of NMR chemical

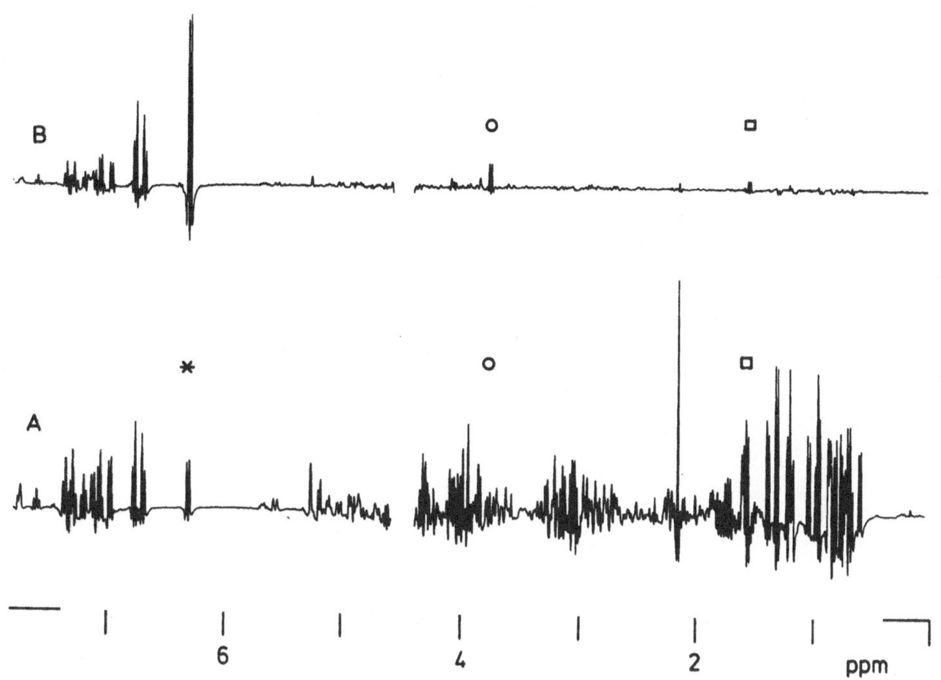

Fig. 6 Selective NOE's between protons of Tyr-23 and Ala-25 (Fig.7) in truncated driven NOE (TOE) difference spectra[34] of BPTI. The figure shows ^1H NMR spectra at 360 MHz of BPTI in D_2O, pD = 7.0, T = 35°C. The spectral resolution was improved by multiplication of the free induction decays with a sine bell.[23] (A) Normal ^1H NMR spectrum. (B) TOE difference spectrum obtained with presaturation of the doublet resonance of the ε-protons of Tyr-23 at 6.30 ppm (*). The multiplets ○ and □ come from Ala-25. (Reproduced from ref. 36).

shifts is not quite sufficient for this parameter to be used systematically for measurements of non-bonding interatomic distances. More direct distance information can be obtained from studies of the manifestations of dipolar spin-spin interactions in the spin relaxation times and in nuclear Overhauser effects (NOE),[6] as will be discussed in more detail in the following section.

3. NEW NMR EXPERIMENTS FOR BIOPOLYMERS IN SOLUTION

3.1. Selective ^1H-^1H Overhauser effects

As was indicated at the end of the foregoing section, studies of nuclear Overhauser enhancements (NOE) appear at present to be

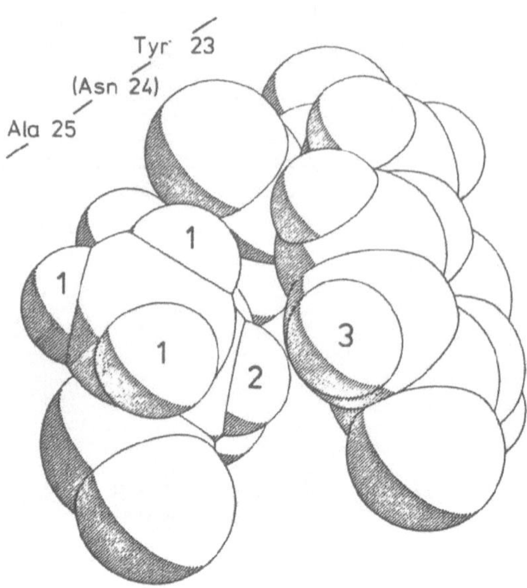

Fig. 7 Computer drawing of the peptide fragment from Tyr-23 to Ala-25 in the refined crystal structure of BPTI.[37] The side chain of Asn-24 was omitted. The size of the individual atoms corresponds to the van der Waals radii. The following atoms are identified by numbers: (1) methyl protons of Ala-25; (2) α-proton of Ala-25; (3) ε-proton of Tyr-23. (Reproduced from ref. 36).

a particularly promising approach for ^1H-^1H distance measurements in proteins, which might eventually provide a sufficient number of intramolecular distance constraints to characterize spatial polypeptide structures. The NOE is the fractional change in intensity by cross relaxation of one NMR line when another resonance is perturbed. It has long been a valuable tool for measurements of internuclear distances in small molecules.[30] In macromolecules at high magnetic fields, however, spin diffusion can become quite efficient[31-33], causing the conventional steady-state NOE's[30] to be less specific and hence less useful. Theory shows that, in contrast, the initial build-up rates of NOE's are simply related to the inverse sixth power of the distance between the observed and the presaturated proton[30-35] and that adverse effects of spin diffusion can be eliminated by suitable selection of the experimental conditions.[33,34] This is illustrated in Fig. 6 which shows selective NOE's between two nearby amino acid side chains (Fig. 7) in the globular structure of the basic pancreatic trypsin inhibitor (BPTI). The two multiplets (O, □) in the TOE difference spectrum[34] in Fig. 6B indicate close proximity between the side chains of Tyr-23 and Ala-25 in the solution conformation of BPTI. For cytochromes c similar TOE experiments resulted in individual assignments of the heme c proton resonances[21,38] and detailed descriptions of the spatial arrangement of the axial ligands of the heme iron (Fig. 5).[39]

3.2. One-dimensional and two-dimensional NMR

Recent experience has shown that some limitations of conventional, one-dimensional NMR experiments can be overcome with the use of two-dimensional (2D) NMR techniques.[3-5] The experiment of Fig. 6 may serve to illustrate two salient points. Firstly, this particular experiment was successful because the preirradiation was on the well separated line at 6.30 ppm (*), which could selectively be saturated. If instead the preirradiation had been somewhere between 1 and 4 ppm, several resonance lines would have been perturbed simultaneously (Fig. 6), and even without adverse effects of spin diffusion an ambiguous result would have been obtained. Secondly, practical applications of one-dimensional NOE experiments are often discouraged by the low sensitivity, which requires accumulation times of several hours for each individual experiment. Both these difficulties are largely eliminated by 2D NOE spectroscopy (NOESY), which does not require selective preirradiation of individual lines and which yields with a single instrument setting a complete set of selective ^1H-^1H NOE's in a protein.[40] Furthermore, since the resonance peaks are spread out in two dimensions, the spectral resolution at a given field strength H_o is considerably improved in 2D NMR spectra,[3-5] and it is an important advantage for biological studies that many 2D NMR experiments can be performed nearly as easily in H_2O as in deuterated solvents.[41]

The general principles of 2D spectroscopy[3] provide room for a large number of different 2D NMR experiments.[42] The experiments which are described in the following have so far been particularly useful for work with proteins.

3.3. Two-dimensional correlated spectroscopy

Correlated spectroscopy (COSY)[3], spin echo correlated spectroscopy (SECSY)[43,44] and foldover-corrected correlated spectroscopy (FOCSY)[44] are three 2D NMR experiments for delineating connectivities between J-coupled nuclei. Compared to conventional, one-dimensional spin decoupling experiments, these techniques have the

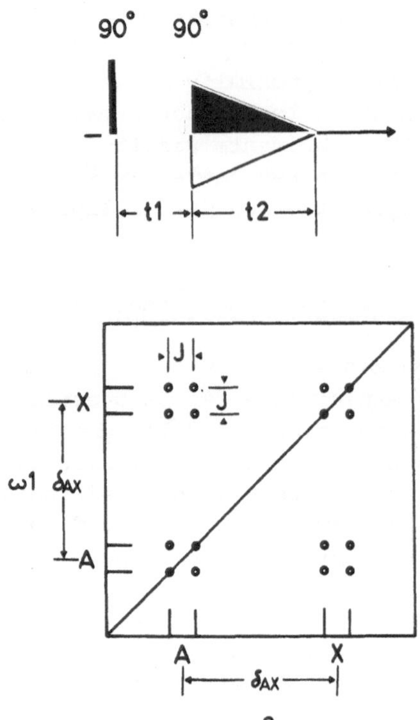

Fig. 8 (A) Experimental scheme for correlated spectroscopy (COSY). The experiment uses two non-selective 90° pulses, which are separated by the "evolution period", t_1. The "detection period", t_2, follows immediately after the second pulse. As in all 2D experiments, the measurement is repeated for a set of equidistant t_1-values.[3] (B) Schematic COSY spectrum for an AX spin system. ω_1 and ω_2 represent the chemical shift on the horizontal and vertical axes. Peaks corresponding to the resonances in the normal, one-dimensional ^{1}H NMR spectrum are on the diagonal. J-connectivities are manifested by cross peaks between the diagonal peaks.

advantage that they do not require selective irradiation of individual multiplets and that a complete set of J-coupling connectivities in a macromolecule can be obtained with a single instrument setting.[3,43,44] The experimental scheme for COSY and a schematic COSY spectrum for an AX spin system are shown in Fig. 8.

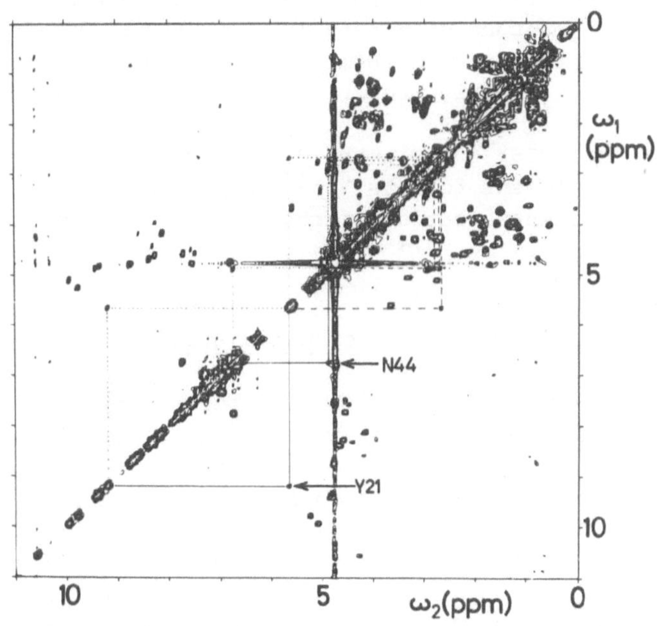

Fig. 9 Contour plot of a 2D correlated (COSY) ^1H NMR spectrum at 360 MHz recorded in a 0.02M solution of BPTI in D_2O, pD = 4.6, T = 24°C. An absolute value plot is shown. The residual water protons are observed at 4.8 ppm and cause the appearence of a strong vertical and a weaker horizontal noise band at this position. The peaks on the diagonal correspond to the normal, one-dimensional ^1H NMR spectrum.[45] Cross peaks manifesting proton-proton J-connectivities appear in symmetrical locations with respect to the diagonal. The J-connectivities for Tyr-21 and Asn-44 are indicated as follows: ——— Connectivities between amide proton and C^{α}-proton in the lower triangle. - - - Connectivities between $C^{\alpha}H$ and $C^{\beta}H_2$. The two β-protons of Tyr-21 have identical chemical shifts, whereas two separate peaks prevail for $C^{\beta}H_2$ of Asn-44. ····· Symmetrical connectivities in the upper triangle. (Reproduced from ref. 46).

Fig. 9 shows a COSY spectrum recorded in a solution of BPTI and illustrates how the J-connectivities between protons of the polypeptide backbone and the amino acid side chains can be delineated (for details see figure captions 8 and 9). Note that in addition to the types of protons discussed in Fig. 3, ca. 30 slowly exchanging amide protons with chemical shifts between 6 and 11 ppm are observed in a freshly prepared D_2O-solution of BPTI.[45]

In SECSY and FOCSY spectra the J-connectivity information is presented in different formats.[43,44] SECSY has been used extensively for the identification of the spin systems of amino acid side chains in proteins.[43,44,47]

3.4. Two-dimensional nuclear Overhauser spectroscopy

The experimental scheme for 2D NOE spectroscopy (NOESY)[40,48] is shown in Fig. 10A. It includes three non-selective 90° pulses. After frequency labelling of the various magnetization components during t_1, cross relaxation leads to exchange of magnetization between nearby protons during the mixing time, τ_m. The interval τ_m is kept fixed and the signal recorded immediately after the third pulse as a function of t_2. In the frequency domain spectrum obtained after 2D Fourier transformation of the data set $s(t_1, t_2; \tau_m)$, the diagonal peaks correspond to the resonance positions in the normal, one-dimensional spectrum and NOE's are manifested by pairs of cross peaks in symmetrical locations with respect to the diagonal (Figs. 10B). Since the build-up rates of the NOE's are related to the inverse sixth power of the proton-proton distances[30,33-35], different data sets are generally obtained with different mixing times τ_m.[40] With short mixing times of up to ca. 100msec, only the shortest proton-proton distances likely to occur in a protein will be seen. In NOESY spectra recorded with longer mixing times, say 300msec, additional correlation peaks manifesting longer distances, possibly also via spin diffusion,are likely to occur.[40]

In the NOESY spectrum of BPTI in Fig. 11 a large number of cross peaks can be seen which indicate selective NOE's. The extent of the structural information contained in such a spectrum is indicated by the few cross peaks which are identified in the figure. Firstly, there are NOE's between covalently linked protons of the same amino acid residue, exemplified by the NOE connectivities between $C^\alpha H$, $C^\beta H$ and $C^\gamma H_3$ of Thr-32. Secondly, there are NOE's between the backbone protons of neighbouring residues in the amino acid sequence, illustrated by the cross peaks between $C^\alpha H$ and amide protons of residues 30 and 31, and 32 and 33, respectively. Finally there are

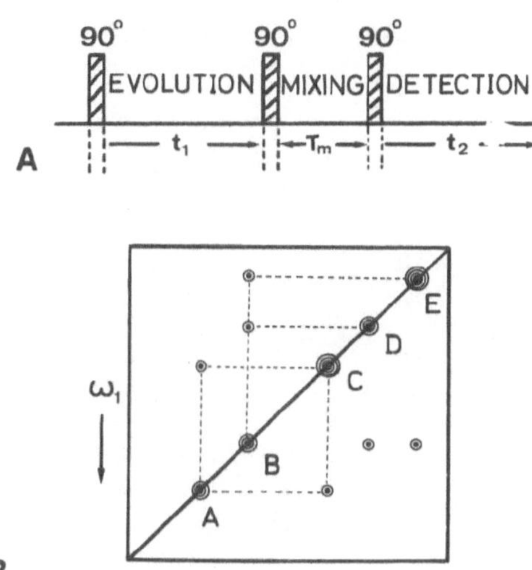

Fig. 10 (A) Experimental scheme for 2D NOE spectroscopy (NOESY).[40,48] The three 90° pulses are separated by the evolution period, t_1, and the mixing period, τ_m. (B) Contour plot of a schematic NOESY spectrum. ω_1 and ω_2 represent the chemical shift (ppm) on the horizontal and vertical axes. Diagonal peaks correspond to the resonance positions in the normal, one-dimensional spectrum. Pairs of symmetrical cross peaks with respect to the diagonal indicate selective NOE's between individual resonance lines, i.e. there are NOE's between A and C, B and D, and B and E. (Reproduced from ref. 40).

the connectivities between protons which are well separated in the covalent structure but closely spaced in the three-dimensional structure. These are illustrated by the cross peaks between the aromatic protons of Tyr-23 and Ala-25, which correspond to the multiplets seen in the TOE difference spectrum of Fig. 6B.

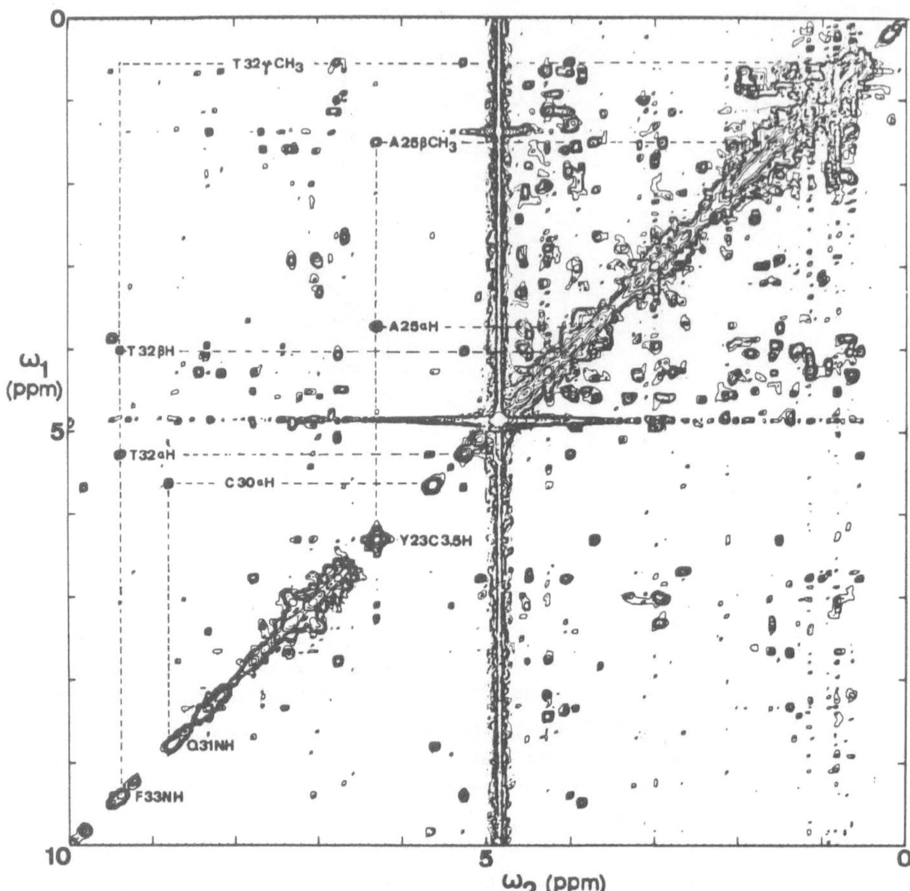

Fig. 11 Contour plot of a 360 MHz ^1H NOESY spectrum of BPTI. The protein concentration was 0.02 M, solvent D_2O, pD = 3.8, T = 18°C, the interior amide protons[45] had not been exchanged with deuterium. The mixing time τ_m was 100msec. The absolute value spectrum is shown. NOE connectivities between selected amino acid residues are indicated by the broken lines (see text). These peaks are identified by the one-letter symbol for amino acids (A = alanine, T = threonine, C = cysteine, Q = glutamine, F = phenylalanine, Y = tyrosine), the position in the amino acid sequence and the type of protons observed. The strong vertical spike and the somewhat less intense horizontal spike at 4.8 ppm are due to the resonance of the residual solvent protons. (Reproduced from ref. 40).

3.5. Two-dimensional J-resolved spectroscopy

In a 2D J-resolved spectrum the J-couplings are manifested on a separate frequency axis perpendicular to the chemical shift axis.[5,49] As a consequence, overlap between the individual multiplets is minimized and a considerable improvement of the spectral resolution can be obtained. From 2DJ spectra accurate measurements of spin-spin coupling constants can be obtained even in the most crowded regions of protein ^1H NMR spectra.[5,49,50] With the use of 2D J-resolved spectroscopy vicinal spin-spin coupling constants, which have so far mainly been exploited for conformational studies of small peptides,[6,25] thus become accessible also for studies of proteins.[50]

4. CONCLUDING REMARKS ON THE POTENTIALITIES OF NMR TO DETERMINE PROTEIN CONFORMATIONS

Section 2.3. presented briefly some notions on various possible avenues for deriving conformational information from NMR parameters. As indicated there, the NOE experiments described in sections 3.1. and 3.4. provide a particularly straight-forward and generally applicable method, since they allow, without perturbation by extrinsic spectroscopic probes, direct measurements of intramolecular, through-space proton-proton distances.[30-35] Obviously, to represent valid contributions for the determination of the protein conformation, the proton-proton distances must be assigned to specific locations in the amino acid sequence or on prosthetic groups, such as heme c (Fig. 5) in cytochromes c. For an illustration we return to the experiment of Fig. 6. The resonance at 6.3 ppm was assigned to Tyr-23 by chemical modification studies[51] an the A_3X spin system of Ala-25 was identified by sequential resonance assignments using two-dimensional NMR[46]. Since the resonance assignments were thus obtained without reference to the crystal structure, the data of Fig. 6 provide a stringent test that the local conformation in crystalline BPTI shown in Fig. 7 is preserved in solution. If on the other hand the resonances had not been assigned to unique locations in the polypeptide chain, a comparison between crystal and solution would have been highly ambiguous, since BPTI contains 4 tyrosines and 6 alanines.

Generally, when approximate distances between numerous amino acid residues in specified sequence positions can be determined from NOE measurements, quite stringent bounds may result for the conformation space available to the polypeptide chain,[52] whereas in the absence of individual resonance assignments detailed structural interpretation of the NOE data would hardly be warranted.

It is further of crucial importance that, unlike most of the work on protein NMR studies published so far,[1,2,6,29] individual resonance assignments are obtained without reference to single crystal X-ray data. Conceptually it has long been quite obvious that sequential, individual assignments of the NMR lines would be a key to the determination of protein conformations by NMR. Very recently, sequential assignments were to a limited extent obtained by studies of BPTI with one-dimensional ^{1}H NMR techniques,[53] and for the peptide apamin with the use of heteronuclear spin decoupling.[54] However, from the presently available experience[46,47] it is quite clear that the use of the 2D NMR experiments described in sections 3.3. - 3.5. provides a much more powerful and efficient way for obtaining individual resonance assignments for polypeptide chains with known amino acid sequence.

Overall we can conclude that high resolution NMR techniques are presently at a stage where sizeable portions of the three-dimensional structures of small and medium size proteins in solution can be elucidated. This opens the possibility to determine protein conformations when no suitable crystals for X-ray analysis are available. When both crystallography and NMR can be applied, meaningful comparisons of the molecular structures in single crystals and in solution may be obtained. In view of further insights into structure-function correlations, studies of surface residues in globular proteins promise to be particularly fruitful, since surface residues appear to have a tendency to undergo changes in both static and dynamic aspects of conformation between crystal and solution[50,55,56] and are often among the groups involved in the specific functions of globular proteins.[28] X-ray crystallography and NMR can further provide complementary information on dynamic aspects of protein conformations.[26,57-59] NMR studies of internal mobility of proteins have recently been extensively discussed.[6,26,27,50,60-63] With the combined use of the fundamental correlations between NMR parameters and internal molecular flexibility[6] and the many-parameter 2D NMR data sets on protein conformations it should eventually be possible to delineate and localize concerted internal fluctuations, which might be unique for proteins and correlated with specific functional properties.

ACKNOWLEDGEMENTS

Development of 2D NMR techniques for biological studies is a joint project with Prof. R.R. Ernst of the ETH Zürich, which is financed by a special grant of the ETH. This report covers results obtained in collaboration with Drs. Anil Kumar, M. Llinás, K. Nagayama and R. Richarz. Financial support was also obtained from the Swiss National Science Foundation (project 3.528.79). I would like to thank Mrs. E. Huber for the careful preparation of the manuscript.

5. REFERENCES

1. R.A. Dwek, I.D. Campell, R.E. Richards and R.J.P. Williams, eds., NMR in Biology, Academic Press, New York (1977).
2. S.J. Opella and P. Lu, eds., NMR and Biochemistry, Dekker, New York (1979).
3. W.P. Aue, E. Bartholdi and R.R. Ernst, J. Chem. Phys. 64, 2229-2246 (1976).
4. R. Freeman and G.A. Morris, Bull. Magn. Reson. 1, 5-26 (1979).
5. K. Wüthrich, K. Nagayama and R.R. Ernst, Trends in Biochem. Sciences 4, N178-N181 (1979).
6. K. Wüthrich, NMR in Biological Research: Peptides and Proteins, North-Holland, Amsterdam (1976).
7. P. Keim, R.A. Vigna, A.M. Nigen, J.S. Morrow and F.R.N. Gurd, J. Biol. Chem. 248, 7811-7818 (1973).
8. R. Richarz and K. Wüthrich, Biopolymers 17, 2133-2141 (1978).
9. R. Richarz and K. Wüthrich, FEBS Lett. 79, 64-68 (1977).
10. A. Allerhand, R.F. Childers and E. Oldfield, Biochemistry 12, 1335-1341 (1973).
11. A. Allerhand, R.S. Norton and R.F. Childers, J. Biol. Chem. 252, 1786-1794 (1977).
12. R. Richarz and K. Wüthrich, Biochemistry 17, 2263-2269 (1978).
13. M. Llinás and K. Wüthrich, Biochim. Biophys. Acata 532, 29-40 (1978).
14. G.E. Hawkes, W.M. Litchman and E.W. Randall, J. Magn. Reson. 19, 255-258 (1975).
15. D. Gust, R.B. Moon and J.D. Roberts, Proc. Nat. Acad. Sci. 72, 4696-4700 (1975).
16. A. Olesker, L. Valente, L. Barata, G. Lukacs, W.E. Hull, K. Tori, K. Tokura, K. Okabe, M. Ebata and H. Otsuka, Chem. Commun. 577-578 (1978).
17. W.W. Bachovchin and J.D. Roberts, J. Amer. Chem. Soc. 100, 8041-8047 (1978).
18. M.W. Baillargeon, M. Laskowski, D.E. Neves, M.A. Porubcan, R.E. Santini and J.L. Markley, Biochemistry 19, 5703-5710 (1980).
19. R. Richarz, H. Tschesche and K. Wüthrich, Biochemistry 19, 5711-5715 (1980).
20. A. Bundi and K. Wüthrich, Biopolymers 18, 285-297 (1979).
21. R.M. Keller and K. Wüthrich, Biochem. Biophys. Res. Commun. 83, 1132-1139 (1978).
22. R.R. Ernst, Advan. Magn. Reson. 2, 1-135 (1966).

23. A. De Marco and K. Wüthrich, J. Magn. Reson. 24, 201-204 (1976).
24. C.C. McDonald and W.D. Phillips, J. Amer. Chem. Soc. 89, 6332-6341 (1967).
25. V.F. Bystrov, Progress in NMR Spectroscopy 10, 41-81 (1976).
26. K. Wüthrich and G. Wagner, Trends Biochem. Sciences 3, 227-230 (1978).
27. G. Wagner and K. Wüthrich, J. Mol. Biol. 134, 75-94 (1979).
28. G.E. Schulz and R.H. Schirmer, Principles of Protein Structure, Springer-Verlag, New York (1979).
29. R.A. Dwek, Nuclear Magnetic Resonance (NMR) in Biochemistry, Clarendon Press, Oxford (1973).
30. J.H. Noggle and R.E. Schirmer, The Nuclear Overhauser Effect, Academic Press, New York (1971).
31. A. Kalk and H.J.C. Berendsen, J. Magn. Reson. 24, 343-366 (1976).
32. W.E. Hull and B.D. Sykes, J. Chem. Phys. 63, 867-880 (1975).
33. S.L. Gordon and K. Wüthrich, J. Amer. Chem. Soc. 100, 7094-7096 (1978).
34. G. Wagner and K. Wüthrich, J. Magn. Reson. 33, 675-680 (1979).
35. A.A. Bothner-By and J.H. Noggle, J. Amer. Chem. Soc. 101, 5152-5155 (1979).
36. K. Wüthrich, G. Wagner, R. Richarz and S.J. Perkins, Biochemistry 17, 2253-2263 (1978).
37. J. Deisenhofer and W. Steigemann, Acta Crystallogr. B 31, 238-250 (1975).
38. R.M. Keller and K. Wüthrich, Biochim. Biophys. Acta 533, 195-208 (1978).
39. H. Senn, R.M. Keller and K. Wüthrich, Biochem. Biophys. Res. Commun. 92, 1362-1369 (1980).
40. Anil Kumar, R.R. Ernst and K. Wüthrich, Biochem. Biophys. Res. Commun. 95, 1-6 (1980).
41. Anil Kumar, G. Wagner, R.R. Ernst and K. Wüthrich, Biochem. Biophys. Res. Commun. 96, 1156-1163 (1980).
42. R.R. Ernst, W.P. Aue, P. Bachmann, A. Höhner, M. Linder, B. Meier, L. Müller, A. Wokaun, K. Nagayama and K. Wüthrich, Proceedings of the XXth Congress Ampère, Tallinn, 15-18 (1978).
43. K. Nagayama, K. Wüthrich and R.R. Ernst, Biochem. Biophys. Res. Commun. 90, 305-311 (1979).
44. K. Nagayama, Anil Kumar, K. Wüthrich and R.R. Ernst, J. Magn. Reson. 40, 321-334 (1980).
45. K. Wüthrich and G. Wagner, J. Mol. Biol. 130, 1-18 (1979).
46. G. Wagner, Anil Kumar and K. Wüthrich, Eur. J. Biochem. 114, 375-384 (1981).
47. K. Nagayama and K. Wüthrich, Eur. J. Biochem. 114, 365-374 (1981).
48. J. Jeener, B.H. Meier, P. Bachmann and R.R. Ernst, J. Chem. Phys. 71, 4546-4553 (1979).

49. K. Nagayama, P. Bachmann, K. Wüthrich and R.R. Ernst, J. Magn. Reson. 31, 133-148 (1978).
50. K. Nagayama and K. Wüthrich, Eur. J. Biochem. (1981) in the press.
51. G.H. Snyder, R. Rowan III, S. Karplus and B.D. Sykes, Biochemistry 14, 3765-3777 (1976).
52. W. Braun, Ch. Bösch, L.R. Brown, N. Gō and K. Wüthrich, Biochim. Biophys. Acta 667, 377-396 (1981).
53. A. Dubs, G. Wagner and K. Wüthrich, Biochim. Biophys. Acta 577, 177-194 (1979).
54. V.V. Okhanov, V.A. Afanas'ev and V.F. Bystrov, J. Magn. Reson. 40, 191-195 (1980).
55. L.R. Brown, A. De Marco, G. Wagner and K. Wüthrich, Eur. J. Biochem. 62, 103-107 (1976).
56. L.R. Brown, A. De Marco, R. Richarz, G. Wagner and K. Wüthrich, Eur. J. Biochem. 88, 87-96 (1978).
57. P.J. Artymiuk, C.C.F. Blake, D.E.P. Grace, S.J. Oatley, C.D. Phillips and M.J.E. Sternberg, Nature 280, 563-568 (1979).
58. H. Frauenfelder, G.A. Petsko and D. Tsernoglou, Nature 280, 558-563 (1979).
59. R. Huber, Trends Biochem. Sciences 4, 271-276 (1979).
60. A. Allerhand, Methods in Enzymology 61, Enzyme Structure, Part H, 458-508 (1979).
61. O. Jardetzky, Biochim. Biophys. Acta 621, 227-232 (1980).
62. R. Richarz, K. Nagayama and K. Wüthrich, Biochemistry 19, 5189-5196 (1980).
63. A.A. Ribeiro, R. King, C. Restivo and O. Jardetzky, J. Amer. Chem. Soc. 102, 4040-4051 (1980).

CLASSICAL AND RESONANCE RAMAN SPECTROSCOPY OF BIOLOGICAL MACROMOLECULES

Warner L. Peticolas

Department of Chemistry
University of Oregon
Eugene, Oregon 92403

INTRODUCTION

Classical and Resonance Raman spectroscopy have been developed over the last 10 years as an important method for studying the conformation and dynamics of biological macromolecules, and membranes containing phospholipids. The method has the advantage that samples may be studied in dilute aqueous solutions, fibers, films or crystals so that correlations between X-ray diffraction determined conformation and other optical spectroscopic methods in dilute solution may be obtained.

The technique of Raman scattering is very simple. Monochromatic light from a laser source is permitted to impinge upon a sample. Light scattered by the sample is collected by means of a lens and focused on the slit of a monochromator. The scattered light is analyzed to obtain its intensity as a function of its frequency. A series of rather sharp bands (10-40 cm^{-1} in width) are found which are shifted to a lower frequency by an amount which usually falls in the range of 3 cm^{-1} to 4000 cm^{-1} (10^{11} sec^{-1} to $10^{14}$$sec^{-1}$). Each of the bands corresponds to a particular molecular vibration. Molecular vibrational frequencies are dependent on the structure (or conformation) of a molecule and the forces between the atoms. The intensities are related to the intensities of the allowed electronic absorption transitions. In the resonance Raman effect, the Raman intensity usually comes entirely from one electronic absorption transition and/or the change of the transition moment with the normal coordinate. The frequency, the intensity and the degree of polarization are useful in characterizing molecular properties.

II. THE CONCEPT OF NORMAL VIBRATIONS

Classically the Hamiltonian of a molecule equals its energy. It is given by :

$$H = \frac{1}{2} \sum_{i=1}^{3N} M_i \dot{X}_i^2 + U(X_1 \ldots X_{3N}) \tag{1}$$

where $X_i = X_i^o - X_i^d$ is the ith cartesian displacement coordinate between X_i^o the equilibrium position and X_i^d the displaced position, i.e. it represents the change in the ith coordinate. The atoms, in the molecule, are numbered consecutively 1...N and the cartesian coordinates $X_1Y_1Z_1X_2Y_2Z_2 \ldots X_NY_NZ_N$ are renumbered to $X_1^o\ X_2^o \ldots X_{3N}^o$.

The mathematics of normal coordinate theory is very simple in mass-reduced coordinates, i.e. $q_1 = \sqrt{M_1}X_1$, $q_2 = \sqrt{M_2}X_2 \ldots q_{3N} = \sqrt{M_{3N}}X_{3N}$. (It is evident that $M_1 = M_2 = M_3$ and $M_4 = M_5 = M_6^{3N}$, etc...) The Hamiltonian may then be written :

$$H = \frac{1}{2}\ \dot{q}^T\dot{q} + \frac{1}{2}\ q^T F^q q \tag{2}$$

where $q = (q_1, q_2 \ldots q_{3N})^T$ is an N-dimensional column vector and U(q) is expanded about the equilibrium position q = 0 to obtain the harmonic approximation (note that $(\delta U/\delta q_i)$ is zero at this position). The quantity F^q is the force constant matrix in q space and is given by :

$$F^q_{ij} = \left(\frac{\partial^2 U(q)}{\partial q_i \partial q_j} \right)_o = F^q_{ji} \tag{3}$$

To obtain the normal coordinates the eigenvalues of F^q and eigenvectors are obtained,

$$F^q A = A\ \Lambda \tag{4}$$

where $\Lambda = \text{diag}\ (\lambda_1 \ldots \lambda_{3N})$.

Since F^q is symmetric (or Hermitean) A is orthogonal (or unitary) so that

$$A^T A = I \text{ and} \tag{5a}$$

$$A^T F A = \Lambda \tag{5b}$$

The transformation to normal coordinates is given by

$$q = AQ \text{ or } q^T = Q^T A^T \tag{6}$$

Putting (6) into (2) we have in view of (5)

$$H = \frac{1}{2} \sum_{i=1}^{N} (\dot{Q}_i^2 + \lambda_i Q_i^2) \tag{7}$$

If this calculation is actually carried out it is found that 6 λ' are zero and correspond to translations and rotations of the molecule in its equilibrium configuration. But 3N-6 λ_i's are non zero. For these λ_i's equation (2) is plainly that of a sum of linear harmonic oscillators if $\lambda_i = \Omega_i^2$ where Ω_i is the circular frequency of the vibration. Consequently a molecule will be undergoing 3N-6 separate oscillations each within a circular frequency Ω_i and each involving the simultaneous displacement of all of the atoms through the relation

$$Q = A^T q \tag{8a}$$

or

$$Q_j = \sum_{k=1}^{3N} A^T_{jk} \; q_k = \sum_{k=1}^{3N} A_{kj} \; q_k \tag{8b}$$

As can be seen from equation (8) any normal mode, Q_j, may involve all N atoms in the molecules and hence all 3N mass reduced cartesian displacements, q_k. However, it often happens that all but one or a few of the A_{kj}'s in equation (8b) are zero. This gives rise to the concept of group frequency. Thus if the molecule contains a carbonyl (C=O) group it will probably contain a Raman active mode at about 1650 cm^{-1} which is primarily the C=O stretching vibration. Thus all of the A'_{kj} are zero for this vibration except those connecting the displacement of the C+O atoms of the C=O bond. Such group frequencies usually remain more or less constant from one molecule to another and are often useful in a qualitative sense to identify specific groups in a large molecule.

For each linear harmonic oscillator in equation (7),

$$H_i = \frac{1}{2} (\dot{Q}_i^2 + \Omega_i^2 \; Q_i^2) \tag{9}$$

the usual transformation to quantum mechanical operators is possible, and for each vibration, i = 1,2 ...3N-6, the quantum mechanical eigenvalue equation

$$H_i |v_i> = (v_i + \frac{1}{2}) \hbar \; \Omega_i |v_i> \tag{10}$$

is obeyed where $|v_i>$ is the harmonic oscillator wave function or state vector, v_i is the vibrational quantum number, v_i=0, 1, 2..., and $(v_i + \frac{1}{2}) \hbar \Omega_i$ is the corresponding quantum mechanical energy.

III. THEORY OF THE RAMAN EFFECT

The absorption of light by a molecule is governed by a single application of the dipole moment operator,

$$\mu = \sum_i e \, r_i$$

where r_i is the position of the ith electron and the sum is over all electrons in the molecule, and e is the electronic charge.

If we start with a molecule in the state of lowest electronic and vibrational energy $|G>$ then the transition probability amplitude to an excited state $|S>$ is proportional to the product of the matrix element, $<S|\mu|G>$, and e_1 the polarization vector of the incident light. However, the molecular wavefunction may be approximated by a product of an electronic wavefunction, $\Psi_g(r,Q)$ and 3N-6 vibrational wavefunctions, $|1>|2>...|3N-6>$.

$$|G> = |\Psi_g(r,Q)> \; |0_1> \; |0_2> \; |0_3> \; ... |0_{3N-6}> \qquad (11a)$$

$$|G> = |\Psi_g> \; | \; 0 \; > \qquad (11b)$$

$$|S> = |\Psi_s \; (r,Q)> \; |v_1> \; |v_2> \; ... | \; v_{3N-6}> \qquad (11c)$$

$$\text{or} \quad |S> = |\Psi_s> \quad |V> \qquad (11d)$$

$|v_i>$ = the harmonic oscillator wavefunction of the ith normal mode in the vth quantum level. If the transition dipole moment $\mu_{sq}(Q)$ is defined by

$$\mu_{sg}(Q) = < \Psi_s(Q,r) \; | \; \mu \; (r) \; | \; \Psi_g(Q,r) > \qquad (12)$$

The energy state $|S>$ over that of state $|G>$ is given by

$$E_{sg} = E^o_{sg} + \sum_j v_j \; \hbar\Omega_j \qquad (13)$$

The absorption transition probability amplitude is given by :

$$P = K \; <V| \; \mu_{sg}(Q) \cdot e_1 | \; 0 \; > \qquad (14)$$

Expanding $\mu_{sg}(Q)$ in a Taylors series,

$$\mu_{sg} = \mu^o_{sg} + \sum_{j=1}^{3N-6} \left(\frac{\partial\mu_{sg}}{\partial Q} \right)_o Q \qquad (15)$$

We see that the transition probability amplitude is proportional to the two terms :

$$P = K \; (\mu^o_{sg} < v|0 > + \left(\frac{\partial\mu_{sg}}{\partial Q} \right)_o < V|Q|0 >) \qquad (16)$$

If μ^o_{sg} is large (a strongly allowed band with large molar absorption coefficient) the first term dominates. However if μ^o_{sg} is small the second term may dominate.

In Raman scattering a photon is annihilated or removed from the incident beam and a photon is created in the scattered beam. This process can occur in either order in time, so we have two terms for α_{12}, the probability amplitude for the scattering process

$$\alpha_{1,2} = \sum_{s,v} \frac{<1|\mu_{qs}\cdot e_2|V><V|\mu_{qs}\cdot e_1|0>}{E_{gs} + \sum_j v_j \hbar\Omega_j - \hbar\omega_1 + i\Gamma_{SV}} + \sum_{s,v} \frac{<1|\mu_{qs}\cdot e_1|V><V|\mu_{qs}\cdot e_2|0>}{E_{gs} + \sum_j v_j \hbar\Omega_j + \hbar\omega_2 + i\Gamma_{SV}} \quad (17)$$

$$<1| = <1_1|<0_2|<0_3| \ldots <0_{3N-6} \quad (18)$$

The subscript 1 is used to label the circular frequency, ω_1, and polarization vector, e_1, of the incident (laser) light while the subscript 2 labels the same quantities for the scattered Raman light. The wavefunction $|1_1>$ corresponds to the final vibrational state of the molecule and equation 18 shows that one of the normal coordinates of the ground state which is arbitrarily labeled the first one has changed quantum numbers $0 \to 1$. There is thus an increase of one quantum of vibrational energy $h\Omega_1$. Thus the energy lost between incident and scattered photon is

$$\hbar\omega_1 = \hbar\Omega_1 + \hbar\omega_2. \quad (19)$$

When the incident light $\hbar\omega_1$ and hence the Raman light $\hbar\omega_2$ lie well below the first excited electronic level E^o_{gs} a sum must be taken over all possible states $|S>$ $|V>$ in the molecule. In this case the first and second terms in equation (17) are equal in magnitude and interchanging the polarization ($1 \rightleftarrows 2$) simply interchanges the two terms so that $\alpha_{12} = \alpha_{21}$. However as the energy of the incident photon, $h\omega_1$, approaches E^o_{gs} of the first excited electronic state, the first term dominates because of the small energy denominator. In this case if there is a large number of overlapping vibrational bands in the electronic states, the sum over the electronic states may be dropped and only the sum over V remains.

$$\alpha_{1,2} = \sum_{V} \frac{<1|\mu_{qs}\cdot e_2|V><V|\mu_{gs}\cdot e_1|0>}{E^o_{gs} + \sum_j v_j h\Omega_j - h\omega_1 + i\Gamma_{SV}} \quad (20)$$

Equation (20) is not very useful and again we make the approximation (15) and keep the first two terms.

$$\alpha^0_{1,2} = (e_1\cdot\mu^o_{qs})(\mu^o_{sq}\cdot e_2) \sum_{V} \frac{<1_1|v_1><v_1|0_1>\prod_{j=2}^{3N-6} <0_j|v_j>^2}{E^o_{qs} + \sum_j v_j h\Omega_j - h\omega_1 + i\Gamma_{SV}} \quad (21)$$

$$\alpha^1_{1,2} = \left((e_1\cdot\mu^o_{qs})\left(\frac{\partial\mu_{gs}}{\partial Q_1}\right)_0\cdot e_2\right) \sum_{V} \frac{<1_1|1_1><1_1|Q_1|0_1>\prod_{j=2}^{3N-6} <V_j|0_j>^2}{E^o_{gs} + \sum_j v_j h\Omega_j - h\omega_1 + \Gamma_{SV}} \quad (22)$$

Equation (21) gives the zeroth order term while Equation (22) gives the first part of the first order term. The second order term involving $(\partial\mu/\partial Q)_0$ twice is mostly used for harmonics and combination bands.

The zeroth order terms are dominated by the Franck-Condon overlap factor, $\langle 1_1 \mid v_1 \rangle \langle v_1 \mid 0_1 \rangle \prod_j \langle 0_j \mid v_j \rangle^2$

These in turn depend upon the shift in the excited electronic state along the normal coordinates, Q_i. If this shift is small then all of the sums over v_1, v_2... condense to two terms involving $v_1 = 0$ and $v_1 = 1$. This is due to the fact that $\langle 1|1\rangle \langle 1|0\rangle$ and $\langle 1|0\rangle \langle 0|0\rangle$ are linear in Δ_{sj}, the shift in excited state geometry while one of the other terms in the summation over v_1 $\langle 1|v_1\rangle \langle v_1|0\rangle$ are higher order in Δ_{sj} and may be neglected if Δ_{sj} is small. Furthermore $\langle 1_1|0_1\rangle \langle 0_1|0_1\rangle = -\langle 1_1|1_1\rangle \langle 1_1|0_1\rangle = \Delta_{sj}$ so that we have the very useful and simple equation :

$$\alpha_{1,2} = K\Delta_{sj}\left(\frac{1}{E_{eo} - h\omega_1 + i\Gamma_{eo}} - \frac{1}{E_{e_1} - h\omega_1 + i\Gamma_e}\right)$$

Since the intensity goes as $|\alpha_{1,2}|^2$ we have :

$$|\alpha_{1,2}^2| = K(\Delta_{sj})^2 \left(\frac{\Omega^2}{((E_{eo} - h\omega_2) + \Gamma^2)((E_{el} - h\omega_2)^2 + \Gamma^2)}\right.$$

Thus for small excited state displacements of totally symmetric modes, the resonant Raman intensity is proportional to the square of the excited state displacement Δ_{sj}, of the jth normal mode in the excited state labeled s [1,2]. For reviews see references 2-4.

IV. RAMAN SPECTROSCOPY OF CHAIN MOLECULES [5,6]

Many of the molecules which are of biological importance are linear sequences of a repeating structure. Thus all-transhydrocarbon chains which are found in model lipids below the Chapman transition, homopolypeptides in an ordered α-helical or β-sheet conformation and homopolynucleotides in a double helix of the A or B conformation are all ordered chain molecules. In each case they may be considered as a small portion of a one dimensional infinite crystal. They are generated by translation and rotation of a chemical repeat unit. If the chemical repeat unit has m atoms it will have 3m normal modes. If M is the number of chemical repeat units in the chain, the number of vibrations of the chain as a whole is 3Mm-6. Each of the 3m interactions is distributed along the chain according to its phase. If all M repeat units are vibrating in phase, the phase angle, θ, is zero. If each unit reaches a minimum amplitude when its neighbor reaches a maximum amplitude θ = Π/2. Between the two points θ = 0 and Π/2 are M-2 other points. The 3mM vibrations are thus distributed as M vibrations on each of 3m curves, called dispersion curves so that one has $\nu_1(\theta)$, $\nu_2(\theta)\ldots\nu_{3m}(\theta)$ where $\nu_1 < \nu_2 < \nu_3\ldots\nu_{3m}$, and

$$\theta = \frac{k\Pi}{2M} \quad k = 0, 1, 2, \ldots M-1$$

For low frequency modes the small forces of interaction between the adjacent monomers are sufficient to change the frequencies considerably so that $\nu_1(\theta)$ etc.. exhibit considerable dispersion (i.e. change in frequency ν, with increasing θ). However, for high frequency modes $\nu(\theta)$ is virtually flat and is almost independent of θ. Hence, the M different vibrations of ν_j will have the same frequency. The important conclusion is that except for the very low frequency modes the Raman spectrum of an ordered chain molecule looks very much like that of the isolated chemical repeat unit and thus only shows those bands of the chemical repeat unit which by virtue of their symmetry are Raman active. Thus the high frequency Raman modes are independent of chain length, beyond a certain chain length.

The low frequency modes (1-100 cm^{-1}) deserve some comment. For chain molecules, the first observable Raman band is always the k = 1 mode of bond-stretching along the chain direction. This is called the longitudinal acoustical mode. Physically it corresponds to an accordion-like mode of the chain. However, for globular molecules like proteins this mode becomes a breathing motion. Low frequency modes have recently been studied in a variety chains and globular proteins[4].

Generally the observable vibrations of biological chain-like macromolecules can be broken down into : (1) low frequency modes (1-100 cm^{-1}) which involve the macromolecule vibrating as a whole, (2) vibrations involving C-C, C-N, S-S, C-S stretching and C-H bending motions in the region 600-1700 cm^{-1} and (3) vibrations involving C-H stretching in the region 2800-3000 cm^{-1}. The S-H stretch occurs at 2570 cm^{-1}, and the OH in water at 3600 cm^{-1}.

V. CLASSICAL VS. RESONANT RAMAN SPECTROSCOPY

We have seen now how a classical Raman spectrum derives its intensity from all the excited states in a molecule while in resonance, one state, the resonant state, is the important one. Practically this means that resonant Raman spectroscopy is 10^3-10^6 times more efficient as a scattering process than is classical Raman scattering because of the small energy denominator in equations 21, 22. Thus resonant Raman scattering allows the use of a wide range of concentrations down to an optical density ($\log I_o/I$) of 1-2, or about 10^{-4} to 10^{-5} M as a lower limit. On the other hand, classical Raman spectroscopy requires concentrations in the range 10^{-1}-10^{-2}M. In biological macromolecules which exhibit the resonant Raman effect, there are specific chromophores which absorb light at a lower frequency than the bulk of the macromolecules. It is the concentration of these chromophores that must be taken into effect.

Classical Raman spectra are useful in obtaining the vibrations of the backbone for chain-molecules (lipids, polynucleotides, polypeptides). In each chain there are classical Raman bands which can be observed, which are sensitive to the backbone conformation. In the forthcoming sections, we will discuss these various vibrations in detail. Because backbone absorption bands do not occur until the far ultraviolet ($\leqslant$ 200 nm), no resonant Raman spectra of backbone vibrations have yet been reported although with lasers of decreasing wavelength becoming available, this is becoming increasingly possible.

Chromophores such as retinal pigments, heme groups and protein-substrate complexes are best studied by resonant Raman spectroscopy which gives only spectra from the chromophore. However, for conformation determination only classical Raman spectroscopy has been useful. Classical Raman spectroscopy may be obtained from crystals, fibers, solutions to a dilution of one percent or about 0.025 M in the chemical repeat group.

VI. CLASSICAL RAMAN SPECTROSCOPY AND THE SECONDARY STRUCTURE OF POLYPEPTIDES AND PROTEINS[7,8]

The Raman spectrum of a protein is a sum of the Raman bands due to the side chain residues of the amino acids and the backbone polypeptide chains. This can be seen in table I where a list of the bands commonly found in proteins is given. As may be seen the aromatic residues -phenylalanine, tryptophan and tyrosine- give prominent bands due to strong preresonance with their $\Pi \rightarrow \Pi^*$ transitions. However for the purpose of discussing the secondary structure, the amide I and amide III bands are most useful. Table 2 gives a list of the amide frequencies before and after deuterium exchange. The amide I vibration which is primarily a C = 0 stretching vibration changes only slightly (about 30 cm^{-1}) upon H $\rightarrow$ D interchange. However, the amide III vibration which is a combination C-N stretch and N-H wag changes remarkably from 1200-1300 cm^{-1} to 900-1000 cm^{-1} upon deuterium exchange.

Because of the overlapping of the frequencies when α-helix, β-sheet and β-turn are considered only semi-quantitative estimates of protein secondary structure is usually possible. This can be helped by deuterium exchange and subtraction of the spectrum of the deuterium exchanged protein from the normal protein spectrum.

Two methods -one due to Lippert et al.[9] and one due to Pezolet et al.[10] are useful for a semi quantitative estimation of %α, %β and % random coil (RC). This last portion includes everything (β-turns etc...) that is not plainly α-helix or β-sheet). The Lippert

Table 1. Raman Frequencies Commonly Observed in Classical Raman Spectra of Proteins [12-16]

Frequency (cm^{-1})		Assignment
20-50		Strong low frequency modes of globular proteins may involve over-all breathing motion but usually over-damped in solutions.
510	C-S-S-C stretching	gauche-gauche-gauche
525		gauche-gauche-trans
540		trans-gauche-gauche
624		ring mode of phenylalanine
630-670		C-S stretching gauche
700-745		C-S stretching trans
760		strong sharp tryptophan ring mode
830-850		Doublet of tyrosine. $\frac{I_{830}}{I_{850}}$ > 1H-bond donor < 1H-bond acceptor
930-940		C-C or C-N strong in α-helical regions
950-960		C-C or C-N strong in disordered regions
1004		strong sharp phenylalanine breathing
1016		tryptophan ring mode strong sharp
1220-1300		Amide III (See Table 2)
1361		tryptophan ring mode
1440		CH_2 bending in side chains
1556		tryptophan
1584		trpt, phen
1605		phenylalanine
1613		trp, tyr, phe
1640-1680		C = 0 Amide I vibration (See Table 2)
2560-2580		SH stretch
2800-3000		C-H stretch

method uses a set of 4 simultaneous equations :

$$C\ I_{1240} = f_\alpha\ I^{\alpha}_{1240} + f_\beta\ I^{\beta}_{1240} + f_{RC}\ I^{RC}_{1240}$$

$$C\ I_{1632} = f_\alpha\ I^{\alpha}_{1632} + f_\beta\ I^{\beta}_{1632} + f_{RC}\ I^{RC}_{1632}$$

$$C\ I_{1660} = f_\alpha\ I^{\alpha}_{1660} + f_\beta\ I^{\beta}_{1660} + f_{RC}\ I^{RC}_{1660}$$

$$f_\alpha + f_\beta + f_{RC} = 1$$

In the above four equations there are four unknown : f_α, f_β, f_{RC} and C. These correspond to the fraction α, fraction β and fraction "random coil" and a constant C which is a scaling factor. Each of the intensities I_{1240}, I_{1632}, I_{1660} is measured from a protein Raman spectra relative to the intensity of the band at 1445 cm^{-1} which is the bending mode of the CH_2 group in the side chains. The values for I^{α}_{1660}, I^{α}_{1632}, etc, are the values determined by Lippert from model compounds to be the intensity one would observe at that frequency if all of the proteins were α-helix, etc... The scaling factor, C, is supposed to take care of the fact that different proteins have different side chains and thus different intensities of the reference band at 1445 cm^{-1}. The method has been applied successfully by Professor Tu to snake toxin proteins (11a-11e). Table 3 gives the standard intensities to use with the 4 simultaneous equations.

Table 2. Summary of Amide I and Amide II Frequencies Based on Model Compounds

Conformation	Amide I* cm^{-1}	Amide III cm^{-1}	Amide III' (D_2O.) cm^{-1}
α-Helix	1650 ± 5	1275-1300	950 ± 10
β-sheet	1670 ± 5	1240 ± 5	985 ± 10
β-turn (type I, two interior peptides)	{1667 1698	{1250 1290	{950 963

* The Amide I' (D_2O) bands are each shifted about 30 cm^{-1} lower in frequency.

Table 3(°). Estimated Standard Raman Intensities for α-helix, β-sheet, and Random Coil at Selected Frequencies

Raman Frequencies (solvent) :	1240 cm^{-1} (H_2O)	1632 cm^{-1} (D_2O)	1660 cm^{-1} (D_2O)
$I^{\alpha}_{frequency}$	0.00	0.80	0.55
$I^{\beta}_{frequency}$	1.20	0.22	0.88
$I^{RC}_{frequency}$	0.60	0.08	0.78

(°) Taken from Lippert, J.L. et al., J. Am. Chem. Soc., 98, 1075 (1976).

VII. THE CLASSICAL AND RESONANCE RAMAN SPECTROSCOPY OF NUCLEIC ACIDS (17-18)

The Raman spectra which are observed from nucleic acids consist of those Raman bands which come from the backbone chain, and those which come from each of the nitrogenous bases. The backbone, a furanose-phosphate polymer, is a fully saturated type molecule which has electronic transitions involving sigma-type bonding electrons. trons. Thus the absorption bands which arise from excitation of the electrons in the backbone chain lie in the vacuum ultraviolet. On the other hand, the bases contain a series of rather broad electronic absorption transitions starting at about 260 nm, and as a result it is possible to obtain both ordinary Raman spectra and resonance Raman spectra from the base molecules. Furthermore, even the ordinary Raman spectra of these bases show many of the characteristics of a strong preresonance effect with the low-lying excited electronic state. Thus it was suggested that there should be decreases in the intensity of the Raman lines from certain nucleic acid base vibrations upon base stacking which would be due to a preresonance Raman effect of these Raman bands with a hypochromic electronic absorption band. It has also been suggested that the term Raman hypochromism be used to describe this phenomenon.[17c,d] In this review we will use the term Raman hypochromism to mean the decrease in Raman intensity which occurs upon the formation of a stacked (i.e. ordered) nucleic acid structure. Actually the effect is usually measured as the increase in Raman intensity upon melting. The vibrations of the backbone are generally discussed in terms of normal coordinate treatments made for the elements of the sugar phosphate chain. Certain strong Raman bands of base molecules are listed in Table 4. In H_2O, uracil has a strong Raman band at about 1690 cm^{-1} which is predominantly due to the carbonyl (C-O) stretching mode. There is also a weaker mode around 1631 cm^{-1}. In D_2O there is a splitting and change of frequency of these bands. In the double-helical form there is a single band at 1680 cm^{-1}, and this apparently splits into two bands at 1660 and 1698 cm^{-1} upon melting of the double-helix. Originally it was assumed that the splitting was simply due to the breakup of the hydrogen bonding between the uracil residue and the adenine residues of double-helix. However, it is now known that it probably is impossible to completely separate the effect of base-stacking from the effect of hydrogen bond interaction on the splitting of this band. Other Raman vibrations of the RNA bases and their hypochromism in simple homopolymers are given in Table 4. Also given in Table 4 are the ultraviolet absorption bands whose wavelength is given in nanometers from which the particular base vibration appears to obtain its intensity through the preresonance phenomenon. One fact of interest is that the 1480 cm^{-1} band which is a longitudinal stretching or wave-like mode involving the entire purine ring along the direction from the C_8 to the N_1 shows little hypochromism in adenine, but substantial hypochromism in self-associated 5'-GMP. In general, the relative intensity of these bands is measured relative to a backbone chain mode at 1100 cm^{-1} which is due to the O-P-O symmetric stretching vibration[14-20]. These hypochromic bands are very useful for measuring independently the amount of base stacking for each type of base when the RNA or DNA undergoes a transition between ordered (helical) and disordered states. Thus it is possible to obtain separate melting curves for each of the 4 bases[17c].

Several of the strongly Raman active bands of the nucleic acid bases derive their intensity from a preresonance effect with low-lying electronic absorption bands. In general, the rule enunciated by Tsuboï and his co-workers is that if the excited electronic state geometry resembles that of the normal mode of the ground state, then the Raman active band will tend to get its intensity from that low-lying electronic state through a preresonance effect[21]. When the bases are placed in juxtaposition, either longitudinally by means of stacking interactions or laterally by means of Watson-Crick hydrogen bonding interactions, the electronic transition moments of the absorption bands may interact to change the intensity both of the electronic absorption bands (UV hypochromism), and consequently the Raman intensities (Raman hypochromism).

There are only two Raman active vibrations belonging to the backbone chain which are clearly identifiable in terms of normal modes. These are the Raman band at 1100 cm^{-1} due to the PO_2^- symmetric stretch and the Raman band at 811 ± 4 cm^{-1}.

This 811 cm^{-1} band is the ester phosphate stretching mode (19c, 20,22). Thus if R_1 is a ribose ring attached to the phosphate at the 5' position and R_2 is either a hydrogen or another ribose ring, this mode involves the chain $[-O-(PO_2^-)-O-R_2-]$. This vibration has been shown to be very conformationally dependent. It only exists in DNA

Table 4. Raman frequencies of base vibrations in double helical ribonucleic acid homopolymers

	Hypochromism			Preresonance
Uracil residues	in Poly(AU)[29]		in Polu A.polyU[18]	(likely excited state)
1680(s)(C=0 str.)	++			
1634(m)(C=0 str.)	no			
1400(m)	++		++	A(265 nm)
1235(s)	++		++	A(265 nm)
785(s)(ring breath.)	++		+	B(210 nm)
Cytosine residue	in Poly C		in GpC	
1657(m)				
1607(m)				
1528(m)	+		+	A(268 nm)
1292(s)	+			A(268 nm)
1240(s)				
782(s)(ring breath.)				B(230 nm)
Adenine residue	in Poly A	in Poly AU	in Poly A. poly U	
1580(s)		no		A(276 nm)
1510(m)	++	no	+	
1484(m)	no	no	-	A(276 nm)
1379(m)	+	++		
1340(m)		no		
1310(s)	++	+	++	
1255(w)	+	no		
729(s)(ring breath.)	+	++	++	C(210 nm)
Guanine residue	in 5'GMP		in GpC	
1582(s)	+			A(276 nm)
1487(s)	++		-	A(276 nm)
1375(m)	+			
1328(m)	++		-	A(276 nm)
670(s)	--		-	Far

or RNA or in the aggregates of 5'-guanosine monophosphate when these chains are in ordered A-type conformation.Although X-ray measurements on fibers of RNA homopolymers always showed these materials to be in the A-type conformation, there was no completely definitive proof that this A-type conformation existed in the double helical structures in solution because until laser Raman spectroscopy there was no technique for structure determination which would work equally well both on fibers or crystals where the X-ray structure is known and also in solution. However by taking X-ray diffraction and Raman spectra of the same fibers, it was possible to correlate the Raman spectrum of the A-type, B-type and C-type forms of DNA[24,28]. Since, in solution the Raman spectrum of ordered RNA always shows the 811 cm^{-1} band of A type DNA (usually shifted to 814 cm^{-1} because of the difference in the sugar substituent at the 2' position) the A-type structure of RNA in solution is definitely established. It has also been shown that the crystals and solution of t-RNA show the same Raman spectrum so that the structure of this material in solution is now well-established.

The 811 cm^{-1} band generally occurs at 814 cm^{-1} in RNA structures and may be used as a measure of the A-type conformation. The 1100 cm^{-1} band is used as an internal standard of reference, and the intensity ratio, I_{811}/I_{1100}, is taken as a measure of the A-type conformation. The value of 1.65 ± 0.05 is generally taken to be 100 % A-type configuration both for DNA and RNA structures[25-27].

In addition to these two main modes, there are also two other modes which are fingerprint bands of unknown origin for the B and C type conformation. In DNA a weak band at 835 cm^{-1} is always obtained when the DNA is in the B-type conformation. When the DNA shifts to the A-type conformation this band disappears with the appearance of the band at 811 cm^{-1}. The C-type structure shows a band at 870 cm^{-1}. Thus we can say rather conclusively the comparison of the X-ray diffraction patterns and the Raman spectrum of the A, B and C forms of DNA allow the establishment of Raman spectra in the region 800 to 1100 cm^{-1} as a determination of the conformation of the nucleic acid. These results are summarized in Table 5.

Table 5. Raman bands of the nucleic acid backbone[23-27]

Frequency (cm^{-1})	Assignment
1100 cm^{-1}	$(R\text{-}O\text{-})_2\text{-}P<^{O}_{O}$ symmetric stretch of - PO_2^-- group
811 ± 4 cm^{-1}	DNA or RNA chain mode for A-type conformation Involves ribose-phosphate diester linkage.
835 cm^{-1}	Weak but prominent Raman mode for DNA in B-form
870-880 cm^{-1}	Similar to mode above but for DNA in C-form

Tables 6 and 7 list the Raman spectral changes observed on the melting of nucleic acids taken from the papers of several authors which represent a summary of the frequencies observed in ribonucleic acids both in H_2O and D_2O. The absolute intensities of these various bands due to the bases will depend on the relative base composition. The increases in intensity which are observed upon melting are the inverse of the decrease of intensity due to helix formation, i.e. Raman hypochromism. However, the 670 cm^{-1} band of guanine actually decreases upon melting. Consequently, it increases upon helix formation and must be regarded as hyperchromic. Other than this band virtually all of the bands show increases in intensity due to the standard Raman hypochromism pre-resonance condition. The changes in the Raman bands due to the changes in the backbone vibrations are also given.

Considerable work on the Raman spectroscopy of viruses has been done by Thomas and his coworkers[31,32]. In general, the viruses show a mixture of bands due to the nucleic acid components and the protein components. Careful study of the intensity and frequency of these bands has shown the relative conformation of both the nucleic acid and protein portions of the spectrum.

The first Raman work done on chromatin (the complex of histone and non-histone space proteins with DNA which is always found in the nucleus of eukaryotic cells was done by Mansy et al.[33] who reported a large fraction of α-helix in the protein part and a predominantly B-type conformation for the DNA in chromatin. In addition, they found that in whole chromatin samples the Raman band at 1490 cm^{-1}; predominantly due to guanine, is measurably weaker than in normal double-helical DNA. As was noted above, and as may be seen in Table 4, this purine vibration shows no hypochromism in the poly A-poly U or poly(A-U) double-helix due to the stacking interactions between the bases. However, a considerable hypochromism is found in the quadruple -helical aggregate, 5'-GMP. Since in helical 5'-GMP there is a hydrogen bond at the N-7 position, the bold assumption was made that H-bonding at the N-7 position of guanine was responsible for the decrease in the intensity of this band rather than base stacking interactions which are normally supposed to be the cause of Raman hypochromism, as discussed above. Recent measurements have confirmed the relatively high α-helical content of the protein portions of chromatin and also confirmed the B-form of the DNA[27,33,34]. In addition it has been shown that the decrease in intensity of the 1490 cm^{-1} band does not occur in purified nucleosomes but only on certain whole chromatin samples. Thus if this decrease in intensity is due in fact to H-bonding at the N-7 position, then it may come from an interaction with guanine and a non-histone component of the chromatin.

Recently, it has been shown that the deoxytetramer, d(pApT)$_2$ contains a very unusual backbone structure in the crystal phase. In particular the furanose rings alternate in structure between C3' endo

Table 6. Raman spectral changes observed on melting of ribonucleic acids

Frequencies (cm^{-1}) H_2O pH7	D_2O pH7	Intensity changes observed upon melting (if any)	Assignment
670	670	Large decrease	G
725	720	Large decrease	A
785	780	Increase due to shift in 814cm^{-1} band	U,C OPO symmetric, if chain is disordered
814	814	Disappears completely, shifts to 785 cm^{-1}	OPO symmetric stretch
867	860	No change	Ribose
915	915	No change	Ribose
975	990	Decrease	Ribose
1003		No change	A,U,C
1047	1045	Small decrease	Ribose-phosphate
1100 (5)	1100 (4)	No change	PO_2^{-}-symmetric stretch
1182 (2)	1185 (1)	Small increase	A,G,C
1240 (6)		Large increase	U
1248 (5)		Small increase	A,C
	1310 (7)	Small increase	C
1320 (7)	1320 (7)	Small increase	A,G
1340 (7)	1340 (7)	Increase	A
1375 (5b)	1370 (3B)	No change	A,G
1420 (2)		No change	G,A
1460 (sh)	1460 (sh)	No change	Ribose
1484 (10)	1480 (8)	Small increase	G,A
1527 (2)	1526 (3)	Small increase	C,G
1575 (8)	1578 (10)	Small increase	G,A
	1658 (4)		C=0 in U,G,C
1692 (4)	1688 (4)		

(A-type) and C2' endo (B-type) i.e. $(C3'-C2')_2$. This had led to speculation that perhaps A-T rich DNA will have similar backbone structures in solution and that such structures could represent a point of identity for regulatory proteins. Recent unpublished experiments in this author's laboratory show that poly(dA-dT) is predominantly C2' endo under usual conditions but that at high salt (0.5 M) and low temperature ($\sim 0°C$) it shows considerable C3' endo band at 816 cm^{-1}. On the other hand poly dA.poly dT shows a strong pair of bands at 840/816 cm^{-1} which seems to indicate that both C3' endo and C2' endo are present -perhaps in roughly equal amounts.

water. Dispersions are larger aggregates (typically 1-4μ diameter) and consist of multiple bilayers. Prolonged ultrasonication followed by chromatographic or centrifugal separation results in small (radius 100-200 Å) unilamellar vesicles. Both unilamellar vesicles and multilamellar dispersions are useful model systems and, as we shall see below, have been shown by Raman spectroscopy to have somewhat different physical properties.

The lipid bilayer constitutes only about half of the membrane. An actual membrane includes about 50% constituent proteins. Intrinsic proteins which are inserted directly into the bilayer may be exposed on either the exterior or interior surface of the bilayer or transverse the entire bilayer with exposure on both surfaces. Extrinsic proteins, on the other hand, are surface-bound and held largely by electrostatic forces.

The Raman spectra of a phospholipid is dominated by the vibrations of the long alkyl hydrocarbon portions of its fatty acyl chains superimposed upon which there are bands from the headgroup. As a consequence assignment of lipid spectra has benefited enormously from the quite extensive literature available on the vibrational spectroscopy of hydrocarbons and polyethylenes. The conformationally sensitive vibrations of hydrocarbons may be divided into three general regions[36,37,38] :

(1) In the low frequency range between 10 cm^{-1} and 300 cm^{-1} a band occurs which is the longitudinal acoustical mode (LAM). This is an accordion-like motion of the entire hydrocarbon chain whose frequency is inversely related to the number of all-*trans* bonds. In DPPC the LAM may be seen as a shoulder occurring at 154 cm^{-1}. This band should, in principle, be very sensitive to conformations of lipids in which this type of conformation occurs. In actual fact because of its low frequency, it is very difficult to observe this band, and many phospholipid dispersions or vesicles simply do not show its existence, a fact that is still not thoroughly understood. Ethanolamines generally show the LAM, for examples rather strongly while lecithins do not [38].

(2) The skeletal-optical (-C-C-stretching) mode between 1000 cm^{-1} and 1500 cm^{-1} are particularly sensitive to the conformational state of hydrocarbons. Of the three bands comprising this region the two bands at 1064 cm^{-1} and 1133 cm^{-1} may be assigned to the B_{1g} and A_g vibrational modes of the all-*trans* chain segments while the third (1090 cm^{-1}) results from structures containing gauche rotations. Basic assignments are summarized in Table 1. This series of three Raman bands is useful for obtaining an estimate of the relative number of *trans* and *gauche* bonds along the hydrocarbon chain 39.

(3) A number of the complex bands in the C-H stretching region

Table 7. Raman spectral changes observed on melting of calf thymus DNA[30]

cm^{-1} in H_2O pH = 7	cm^{-1} in D_2O pH = 7	Changes observed upon melting (if any)	Assignment
672	656	Small increase	T
683	677	Decreases	G
729	716	Large increase	A
750	734	Shifts to lower cm^{-1}	T
	765	Large increase in intensity	
	785	No change	PO_2diester symmetric stretch
786		Small increase in intensity	PO_2 diester symmetric stretch
835	828	Shifts to lower cm^{-1}	PO_2 diester anti-symmetric stretch
879	867	Decreases	Deoxyribose-phosphate
	893	Decreases	Deoxyribose
920		Decreases	Deoxyribose
	966	Decreases	Deoxyribose
	1013	Decreases	C-O stretch
1015		No change	C-O stretch
1051	1047	Decreases and shifts to a higher cm^{-1}	C-O stretch
1094	1091	No change	PO_2^- diestersymmetric stretch
1144		Disappears	Deoxyribose-phosphate
1186		Moderate increase	T
1214		Moderate increase	T
1225		Moderate increase	A
1240		Large increase	T
1259	1260	Small increase	C,A
1303	1300	Moderate increase	A
1340	1343	No change	A
1378	1375	Moderate increase	T,A
1421	1418	Small increase	A,G
1463		Decreases	Deoxyribose-phosphate
	1484	Shifts to lower cm^{-1} deuteration of C-8 proton on A and G	
1491		Moderate increase	G,A
	1501	No change in intensity shifts to lower cm^{-1}	A

Table 7 (continued)

cm^{-1} in H_2O pH = 7	cm^{-1} in D_2O pH = 7	Changes observed upon melting (if any)	Assignment
1521	1520	No change in intensity, shifts to lower cm^{-1}	A
1534		Moderate increase in intensity	C
1579	1575	Large increase in intensity due mainly to G	G,A
	1620	No change	C
1660	1673	Large increase in intensity, shifts to lower cm^{-1}	C=O of T

(Actually the original Raman spectrum of poly d(A-T) showed both the 815 cm^{-1} characteristic of the A-form as well as the 835 cm^{-1} band due to the B-form)[24].

VIII. CLASSICAL RAMAN SPECTROSCOPY OF MODEL MEMBRANE SYSTEMS [35]

The lipid constituents of biomembranes are distinguished by their amphiphilic character as manifested by a strong hydrophilic headgroup and a long hydrophobic tail. Typical phosphoglycerides have the general formula :

```
  H
H-C-O-C(=O)-(CH2)n-CH3
H-C-O-C(=O)-(CH2)n-CH3
H-C-O-P(O2-)-O-X
  H
```

where X may be one of a variety of head-group moieties. For example, if X = $-CH_2CH_2-N^+(CH_3)_3$, the lipid is a diacyl phosphatidyl choline, or lecithin. Other X groups include serine, ethanolamine and glycerol. There are many variations of this type of structure in which the acyl groups are changed. The lipid most frequently studied by Raman spectroscopy is dipalmitoyl phosphatidyl choline (DPPC).

Two varieties of phospholipid preparation have long served as simple model biomembrane systems. Phospholipid dispersions are prepared by mild agitation above the lipid phase transition in excess which extends from 2700 cm^{-1} to 3100 cm^{-1} have been shown to change in shape and intensity distribution upon the disruption of regular chain packing or lateral order. Either melting or dissolution results in a decrease in intensity of the 2890 cm^{-1} band relative to that at 2850 cm^{-1}. The exact assignment of the various peaks of the bands of this complex of bands has been quite difficult. A great deal of work has been carried out to explain exactly the origin of these band intensities . Table 8 summarizes these results[37-44].

Table 8. Observed Raman bands of DPPC and their Temperature Dependence in Dispersions.

Assignment	Marker: Trans	Marker: Gauche	Marker: Packing	Δcm^{-1} at 15° ($x10^{-3}$)	Intensity rel.to 15°	Δcm^{-1} at 37°	Intensity rel.to 15°	Δcm^{-1} at 50°	Intensity rel.to 15°
H_3C-N stretch				718	1.0	715	1.0	717	1.0
SOM	X			1.063	1.0	1.063	0.65	1.063	0.32
SOM *gauche*		X		1.080	1.0	1.070	∿1.1	1.085	> 5
SOM	X			1.098	1.0	1.090	0.95	?	?overlap with 1080 cm^{-1}
SOM	X			1.128	1.0	1.127	0.49	1.122	0.04
CH_2twist		X		1.296	1.0	1.296	0.54	1.301	0.28 *
CH_2scissor			X	1.436	1.0	1.436	0.76	1.439	0.70
CH_2sym.stretch		X	X	2.845	1.0	2.846	0.80	2.850	0.70
CH_2triclinic marker			X	2.860	1.0	2.860	∿0	2.860	∿0
CH_2asym.stretch		X	X	2.881	1.0	2.880	0.73	2.888	0.47
CH_2		X		2.920	1.0	2.920	1.0	2.920	1.25 shoulder
CH_3stretch				2.935	1.0	2.935	1.0	2.935	1.0 shoulder
CH_3stretch				2.961	1.0	2.961	1.0	2.961	1.0
CH_3stretch (choline)				3.039	1.0	3.039	1.0	3.039	1.0

SOM. skeletal optical modes

* Peak broadens

Experimental proof that it is changes in lateral packing order and not changes in the trans-gauche conformation which give rise to these intensity changes in the C-H stretching region is obtained by an experiment in which the hydrocarbon chain is vibrationally decoupled from its neighbors while leaving its chemical and conformational properties unchanged. Thus the preparation of a solution of hexadecane in a crystal matrix of solid perdeutero-hexadecane results in a substantial decrease in the peak height ratio of the asymmetric-symmetric CH_2 stretching bands[38]. Snyder has recently considered in detail the physical origin of this observation which has been used as a semi-quantitative measure of lateral order as will be discussed below[20].

Raman spectra of single component phospholipid systems such as DPPC can be interpreted to give highly detailed information about the structure of the component molecules. Multicomponent systems, however, give spectra in which many of the structurally sensitive modes of different phospholipids or proteins may overlap. Isotopic substitution of one component should, in principle, provide a convenient means by which to separate those common bands arising from distinct species in a mixture and to simultaneously observe their behavior. The CD_2 stretching modes appear in a spectral window free from interference from either perhydro-lipids or proteins. Correspondingly a window is opened in the C-H stretching region permitting direct observation of the lipid headgroup. Basic assignments for the perdeuterated lipids have been given and the sensitivity of the Raman spectrum of the perdeuterated lipid to changes in bilayer conformation is demonstrated by temperature difference spectra[22-23].

Much of the early developmental work in the application of Raman spectroscopy to biomembrane studies adopted the phenomenological approach of monitoring the bilayer liquid crystal phase transitions via changes in peak amplitude and plotting the data in the form of a melting curve. A set of such "Raman melting curves", obtained at a resolution of 1°, show considerable detail. Two transitions may be noted, the well-documented main melting transition (Tm_2) at 41.5°C and the premelting event (Tm_1) at 34.2. A favorable comparison of the Raman data with that obtained by calorimetric and fluorescence procedures has been made [39-41].

Temperature difference spectra created by computer subtraction of Raman data taken at various temperatures show evidence for three distinct bilayer structures characterizing temperatures below Tm_1, between Tm_1 and Tm_2 and above Tm_2. From these different spectra it may be concluded that below the pretransition temperature the hydrocarbon chains assume a nearly all-trans conformation and are well-packed. Between the pretransition and the melting temperature the number of gauche bonds increases slightly to perhaps 1 per chain. The absence of a broad, strong gauche band at 1080 cm^{-1} is evidence for the absence of gauche rotations on adjacent or nearby C-C bonds and suggests that the _gauche_ bonds are highly restricted in this phase and found only at the ends of long all-_trans_ segments. Above the melting transition chain-chain interactions continue to decrease and the number of _gauche_ increases sharply. The appearance of a strong broad band at 1080 cm^{-1} indicates that the restriction on the placement of _gauche_ bonds is no longer present and in this high temperature range the _gauche_ rotamers are free to migrate along the chain.

Another problem to which Raman spectroscopy has been successfully directed is the conjecture that differences should exist between the molecular structure of large multilamellar phospholipid dispersions and small single bilayered vesicles. The proposed structural difference is postulated to arise from disruption of the orderly hydrocarbon chain-packing induced by the small radius of curvature of the vesicles. The problem has been investigated in several laboratories with similar results and basic interpretive agreement. In the skeletal optical region the peak height ratio of 1130 cm^{-1} to 1000 cm^{-1} (_trans_ to _gauche_) is lower for the vesicles than for either the solid or the dispersion. The pattern is confirmed in the C-H stretching region where the asymmetric stretch has lost intensity relative to the symmetric stretch. Thus vesicles are found to be consistently more disordered than multilayer dispersions both in the sense of their lateral or interchain ordering as well as their intrachain ordering.

Yellin and Levin[42c] have employed a somewhat different procedure for estimating the number of chains in the all-_trans_ conformation. Applying the integrated form of the van't Hoff equation to their Raman data in the skeletal optical region at temperatures below the liquid crystal phase transition they estimate an enthalpy difference between the chains in the all-_trans_ conformation and chain conformations containing _gauche_ rotations. Comparing their data to the calorimetrically determined ΔH for the phase transition leads to an estimate of 4-4.5 _gauche_ bonds per chain in dipalmitoylphosphatidyl choline above T_M.

Attempts have been made to treat the Raman data so as to distinguish, insofar as possible, between order due to the intra-chain structure and that due to lateral crystalline interactions. Order parameters have been defined such that S = 1 indicate the highest possible order and S = 0 no order (not necessarily the lowest possible). The _trans_-parameter is defined as

$$S_T = \frac{(I_{1133}/I_{REF})_{observed}}{(I_{1133}/I_{REF})_{DPPC,solid}}$$

and serves to reference the data to a standard of known all-_trans_ chain length. Generation of an order parameter for the lateral interaction presents a more difficult problem since the change in intensity of the 2890 cm^{-1} is not simply a function of the loss of interchain interaction. As a point of departure, however, an order parameter $S_{LATERAL}$ may be defined which compares the peak heights ratios of the sample with those observed for crystalline samples of hexadecane. $S_{LATERAL}$ may then be defined as :

$$S_{LATERAL} = \frac{I_{CH2(observed)} - 0.7}{1.5}$$

where $I_{CH2} = I(2890)/I2850)$. The parameter is semi-quantitative, but does provide insight into the amount of lateral interaction.

Characteristic values are as follows for S_T : DPPC(solid), S = 1.0 ; DPPC (dispersion 30°C), S = 0.18 ; DPPC (vesicles, 30°C), S = 0.52 ; DPPC (dispersion, 41.5°C), S = 0.4 ; DPPC (vesicles, 31°C), S = 0.4 ; DMPC (dispersion 25°C), S = 0.3 ; DMPC (vesicles, 25°C), S = 0.2 ; DPPC (solution in $CHCl_3$), S = 0.2 ; EggPC (dispersion 25°C), S = 0.2.

For $S_{LATERAL}$ solid hexadecane has been taken as 1.0 and liquid hexadecane as 0. On this scale, solid DPPC, S_{LAT} = 0.5, DPPC (dispersion, 30°C), S_{LAT} = 0.5 ; DPPC (vesicles, 30°C), S = 0.2 ; DMPC (dispersion, 25°C), S = 0.2 ; eggPC, (dispersion + vesicles, DMPC, dispersion + vesicles),and DPPC in solution, all have $S_{LAT} < 0.5$. Thus we may conclude that vesicles are less ordered than dispersions almost certainly due to their small radius of curvature. However, vesicles are more highly ordered below their transition temperature than above their transition temperature.

An exhaustive review of Raman spectroscopy of biological membranes has recently been written by Wallach et al.[35] which discusses this field in balanced detail.

IX. RESONANCE RAMAN SPECTROSCOPY OF HEME PROTEINS [45,46]

The Heme group is a planar porphyrin ring containing Fe^{II} or Fe^{III} in the center and usually two axial ligands bound to the iron on each side of the porphyrin plane. Heme containing proteins exhibit two $\Pi \rightarrow \Pi^*$ transitions leading to a strong absorption band at about 400 nm and a weak pair of bands $Q_o(\alpha)$ and $Q_y(\beta)$ at about 500-550 nm laser excitation[47]. The 400 nm (Soret) band is so strong that the resonant Raman spectra obtained from this band are virtually all totally symmetric (Equation 21 is applicable)[48,49] while resonant Raman spectra obtained from the lower bands show a much different set of polarization (Equation 22). Because of the different symmetries (A_{1g} ; B_{1g} or B_{2g} ; A_{2g}) which give the corresponding dipolarization ratios ($q < 0.75$; $q = 0.75$; $q > 0.75$), the various bands can be tracked from model compounds to proteins as their frequencies change due to change in oxydation state and spin state of the iron. There are five resonant Raman porphyrin ring frequencies which are sensitive to either oxydation state or spin state of the iron. Studies have been made on a number of mesoporphyrin IX-iron-ligand complexes in which the oxydation state and nature of the ligand has been varied in a systematic way. The three porphyrin ring bands which are oxydation state markers ($1358\ cm^{-1} \leqslant V_I < 1375\ cm^{-1}$; $1534\ cm^{-1} \leqslant V_{III} \leqslant 1570\ cm^{-1}$; $1617 \leqslant V_V \leqslant 1640\ cm^{-1}$) show a lower frequency for Fe^{II} to highest frequency for Fe^{III}. This is attributed to back-donation of the d_Π electron of Fe^{II} into the Π acceptor orbitals of the porphyrin ring which lowers the bond orders and hence the force constants and frequencies. What is interesting is that axial ligands can withdraw electrons from the Fe^{II} and hence from the porphyrin ring causing the ring frequencies to increase. Thus ligands such as NO which are strong Π acids show Fe^{II} porphyrin frequencies as high as the Fe^{III} porphyrin frequencies. In this regard oxyhemoglobin shows the high oxydation state frequencies (i. e. high) frequencies. This has been used as evidence to support the characterization of O_2Hb as the superoxide adduct of (low spin) Fe^{III} instead of a neutral dioxygen adduct of Fe^{II}. Thus O_2 acts as a Π-acid with drawing electron density from the Fe^{II} prophyrin complex to form Fe^{III}-porphyrin and the O_2^- peroxide ion. There is now quite a lot of vibrational spectroscopic evidence for this model. The electron goes into the Π^* orbital of the O_2 lowering the O-O bond order and stretching frequency to that expected of a superoxide.

Three porphyrin frequencies (V_{II}, V_{IV}, V_V) are spin markers. For example the five coordinate 2-methyl imidazole Fe^{II} mesoporphyrin which is an out-of-plane high spin model heme shows consistantly lower frequencies than the corresponding dimidazole Fe^{II} mesoporphyrin IX. In going from the low spin complex to high spin complex these vibrations change as follows :

$$V_{II} = 1490 \rightarrow 1472\ cm^{-1},\ V_{IV} = 1583 \rightarrow 1558\ cm^{-1}$$

$$\text{and } V_V = 1617 \rightarrow 1606\ cm^{-1}.$$

The change in spin state is believed to be due to an expansion of the porphyrin ring. A similar but smaller change is observed for Fe^{III} low spin $\rightarrow$ high spin complexes.

X. RESONANT RAMAN SPECTROSCOPY OF PIGMENTS RELATED TO VISION

The resonance Raman spectroscopy of visual pigments began with the study by Rimai et al. of different retinal isomers and Schiff bases[51]. Both rhodopsin from the retina of the eye, and bacteriorhodopsin contain a molecule called retinal which is bound to the

C - NH_2 group of a lysine in a protein called opsin by means of a linkage which is called a Schiff base. Retinal is a polyene of 6 conjugated -CH=CH- groups terminating in a carbonyl. The aldehyde retinal reacts easily with aliphatic amines to form the Schiff base linkage :

$$R_1CH = O + H_2N - R_2 \rightarrow RCH = N - R_2.$$

This base can bind proton to give

$$R_1CH = N\text{-}R_2 \rightarrow R_1CH = \overset{+}{N}H - R_2$$

The retinals, Schiff bases and rhodopsins are all very light sensitive. At each double bond there is the possibility of cis $\rightleftarrows$ trans isomerization. Thus starting with all transretinal the number of possible geometric isomers is very large. Furthermore trans → cis isomerization may be induced by the incident laser beam. Thus there is an unfavorable competition between the photochemical event and the Raman event. The quantum yield for resonant Raman spectroscopy is about 10^{-6} Raman photons per photon absorbed[52]. But the quantum yield for the photochemical event in bacteriorhodopsin is about 0,3 molecules isomerised per photon absorbed. With such an unfavorable ratio, it is certain that with a stationary sample one will obtain the RR spectrum of a mixture of photo-induced products. Thus several different methods have been developed to suppress a control the photoinduced isomerization[53-58]. RR=resonant Raman.

Recently several groups have reported Raman studies of bacteriorhodopsin, a retinal-protein complex found in the membrane of halobacteria which is capable of mediating light to energy conversion. BR=bacteriorhodopsin.

It has been found empirically that ν(C=C) in cm^{-1} depends linearly on λmax of the optical absorption band in a way that a red shift in λ max is accompanied by a decrease in the C=C stretching frequency. The very strongest band in BR 570 is the C=C double bond stretch at 1535 cm^{-1}. However, of great importance are the weaker bands, particularly in the 1620-1660 cm^{-1} range. Experiments have shown that the weak band near 1620 cm^{-1} is present when the Schiff base is unprotonated and is the C=N stretch. However, upon protonation the -C=NH+- stretch appears in its place at 1655 cm^{-1} in the retinal Schiff base model compounds but at 1642 cm^{-1} in BR 570. In the BR-570 this 1642 cm^{-1} shifts to 1624 cm^{-1} (-C=ND^+-). Thus this establishes the existence of the Schiff base in the retinal-protein complex in bacterio-rhodopsin. Further experiments have described the nature of the protonation and isomeric states of the various intermediates. Details of these states is very complex and although Raman spectra exist for each isomeric molecule in the pathway, the interpretation of these spectra is still not complete and further theoretical work is needed[58].

XI. RESONANT RAMAN IDENTIFICATION OF CATALYTICALLY IMPORTANT BANDS DURING SYNTHESIS.

Much of the pioneering work in this field has been done by Paul Carey who has reviewed the field recently[64]. The discussion here will be limited to two cases where actual bond or other structural changes have been observed during catalysis. The first system is the hydrolysis of thionohippurate by papain[65]. The reaction may be written schematically as follows :

$\Phi CONHCH_2CS\text{-}O\text{-}CH_3$ + HS-PAPAIN →

$\Phi CONHCH_2CS\text{-}S\text{-}$PAPAIN + CH_3OH $\overset{H_2O}{\rightarrow}$

$\Phi CONHCH_2COSH$ + HS PAPAIN

Carey and coworkers used 337.5 nm excitation to obtain the resonance Raman spectra of the substrate, the papain-substrate complex and the products some of which exhibit intense C=S and C-S stretching vibrations with this frequency exciting source.

For a model of the papain-enzyme complex, the compound $CH_3CS\text{-}S\text{-}CH_2CH_3$ (ethyldithioacetate) was used. The starting material, methyl thionohippurate has features at 623, 707, 1006, 1157, 1207 and 1330 cm^{-1}. This spectrum is not resonance enhanced because the absorption band of the starting material (substrate) lies too far into the ultraviolet (230 nm). On the other hand the model compound ethyl dithioacetate (EDTHA) has an intense absorption band at 307 nm with a tail towards the red so that resonance enhancement at 337.5 seems assured. EDTHA shows two Raman bands, one at 587 cm^{-1} and assigned to the C-S stretching vibration and one at 1192 cm^{-1} assigned to the C=S stretch.

The Raman spectrum of a dithioacyl-enzyme intermediate was obtained in the regions where C-S and C=S vibrations are expected at various times after mixing the substrate and the enzyme. An intense transient peak appears at 1130 cm^{-1} and is assigned to the C=S double bond stretch. General resonance enhanced peaks are seen in the C-S region. A peak at 623 cm^{-1} is a substrate peak, but transient peaks occur at 560 cm^{-1} and 590 cm^{-1} which are assigned to the C-S stretching of the transient intermediate. No resonance Raman peaks can be found from the substrate or product that can be assigned to either C-S or C=S. Since both substrate and product contain only one sulfur, apparently it is the presence of two sulfurs -C(=S)-S- which shifts the absorption band to the region where resonance Raman enhancement can occur.

The other enzyme catalytic process where an important band has been observed is in the resonance enhanced identification of the zinc-oxygen band in horse liver alcohol dehydrogenase-nicotine adenine dinucleotide-aldehyde transient intermediate. The aldehyde chosen was p-dimethylaminobenzaldehyde (DABA) which is quantitatively converted to the corresponding benzyl alcohol by the liver alcohol dehydrogenase (LADH) in the presence of the coenzyme nicotine adenine dinucleotide (NADH).

In dry organic solvents zinc chloride reacts with DABA to give a yellow compound. The change in the Raman spectrum upon complexation is truly striking. The C=O carbonyl stretching band at 1664 cm^{-1} completely disappears in the zinc complex showing that the carbonyl band no longer exists. A new band at 386 cm^{-1} appears which is assigned tentatively to the Zn-O band. Other new very strong bands appear in the complex at 1623 cm^{-1}, 1580 cm^{-1}, and 1547 cm^{-1} which are completely absent in the substrate DABA itself. The complex is dissociated into zinc chloride and DABA, if the organic solvent is not dry.

At pH 9.0 DABA reacts rapidly in the presence of LADH/NADH to form the alcohol, but at pH 9.6 the reaction takes days and a tertiary LADH/NADH/DABA complex is formed which shows exactly the same bands as the anhydrous zinc chloride DABA complex in dry organic solution. Thus it appears that the mechanism of LADH/NADH catalysis involves the formation of a Zn-O band and a positive charge on the rear of the substrate which attracts the hydride ion. These are things which are suggested by the X-ray structure[67], but which Raman spectroscopy can more completely determine.

REFERENCES

1. D.C. Blazej and W.L. Peticolas, Proc. Natl. Acad. Sci. U.S.A. 74:2639 (1977).
2. A. Warshel, Annu, Rev. Biophys. Bioeng. 6:273 (1977).
3. B.B. Johnson and W.L. Peticolas, Annu. Rev. Phys. Chem. 27:465 (1976).
4. T.G. Spiro and P. Stein, Annu. Rev. Phys. Chem. 28; 501 (1977).
5. R. Zbinden, "Infrared Spectra of High Polymers", Academic Press N.Y. 1969.
6. W.L. Peticolas, "Low Frequency Vibrations and the Dynamics of Proteins + Polypeptides Methods in Enzymology. Vol. 61, pp 425-458, C.H.N. Hirs and S.N. Timacheff, Editors, Academic Press NY, NY. (1979).

7. B.G. Frushour and S.L. Koenig, "Advances in Infrared and Raman Spectroscopy." Clark and Hester Editors, Vol. 1 (1975). Heyden London and Philadelphia (35-72).
8. T.G. Spiro and B.P. Gaber, Ann. Rev. Biochem. 46: 553 (1977).
9. J.L. Lippert, D. Tyminski and P.J. Desmeules, J. Am. Chem. Soc. 98:7075 (1976).
10. M. Pezolet, M. Pigeon-Gosselin and L. Coulombe, Biochim. Biophys. Acta 453:502 (1976).
11a. A.T. Tu, B.S. Hong and T.N. Solie, Biochemistry 10:1295 (1971).
11b. A.T. Tu and B.S. Hong, J. Biol. Chem., 246:2772 (1971).
11c. J.B. Bjarnason and A.T. Tu, Biochemistry 17:3395 (1978).
11d. M.L. Raymond and A.T. Tu, Biochim. Biophys. Acta 285:498 (1972).
11e. A.T. Tu, J.B. Bjarnason and V.J. Hruby, Biochim. Biophys. Acta 533:530 (1978).
12. H. Sugeta, A. Go and T. Miyazawa, Bull. Chem. Soc. 46:3407 (1973).
13a. N.T. Yu, Critical Reviews in Biochemistry 4:229 (1977).
13b. N.T. Yu and B.H. Jo, J. Am. Chem. Soc. 95, 5033 (1973).
13c. A.T. Tu, B.H. Jo and N.T. Yu, Int. J. Peptide Prot. Res. 8:337 (1976).
13d. N.T. Yu, B.H. Jo, and D.C. O'Shea, Arch. Biochem. Biophys. 156:71 (1973).
14. T.J. Yu, J.L. Lippert and W.L. Peticolas, Biopolymers, 12:2161 (1973).
15a. J. Jakes and S. Krimm, Spectrochim. Acta, 27A, 19 (1971).
15b. J. Jakes and S. Krimm, Spectrochim. Acta, 27A, 35 (1971).
15c. Y. Abe and S. Krimm, Biopolymers, 11:1817 (1972).
15d. Y. Abe and S. Krimm, Biopolymers, 12:1841 (1972).
15e. S. Krimm and Y. Abe, Proc. Natl. Acad. Sci. USA, 69:2788 (1972).
16a. R.C. Lord, Applied. Spectroscopy, 31:187 (1977).
16b. R.C. Lord and N.T. Yu, J. Biol. Chem. 51: 203 (1970).
16c. R.C. Lord, Pure Appl. Chem. 28, (1971).
16d. R.C. Lord and N.T. Yu, J. Mol. Biol. 51:203 (1970).
16e. R.C. Lord and N.T. Yu, J. Mol. Biol. 50:509 (1970).
17a. W.L. Peticolas, Proc. in Nucleic Acid Research (Cantoni, G.L. and Davies, D.R. Eds) Harper and Row, Vol. 2, pp 94-136 (1971).
17b. W.L. Peticolas and M. Tsuboï, in "Infrared and Raman Spectroscopy of Biological Molecules," Theo. M. Theophanides Ed. D. Reidel Publishing Company, Boston (1978) pp. 153-187.
17c. E.W. Small and W.L. Peticolas, Biopolymers, 10:69 (1971) ;ibid. 10:1377 (1971).
17d. B.L. Tomlinson and W.L. Peticolas, J. Chem. Phys. 52:2154 (1970).
18a. K.A. Hartman, R.C. Lord and G.J. Thomas Jr.,"Physico-chemical Properties of Nucleic Acids"(J. Duchesne, ed.), 2, Chapter 10, pp. 92-143 (1973).
18b. M. Tsuboï, S. Takahashi and I. Harada, in Physico-Chemical Properties of Nucleic Acids, Vol. 2 (J. Duchesne, ed.) Chapter 11, pp. 91-145, Acad. Press, London (1973).
18c. M. Tsuboï et al., "Advances in Infrared and Raman Spectroscopy", Clark and Hester, eds., in press.
19a. M. Tsuboï, J. Am. Chem. Soc. 79, 1351 (1957).

19b. G.B.B.M. Sutherland and M. Tsuboï, Proc. Roy. Soc. (London), A239, 446 (1957).
19c. T. Shimanouchi, M. TsuboI and Y. Kyogoku, in Advances in Chemical Physics, J. Duchesne, Ed, London, Interscience, Vol. VII, pp. 435-498 (1964).
19d. R.C. Lord and G.J. Thomas, Jr., Spectrochim. Acta, 23A, 969 (1967).
20. E.B. Brown, and W.L. Peticolas, Biopolymers 14, 1259-1271 (1975)
21. A.Y. Hirakawa and M. Tsuboi, Science, 188, 359 (1975) ; Tsuboi, M. and A.Y. Hirakawa, J. Raman Spectroscopy, 5, 75 (1976)
22a. J.M. Eyster and E.W. Prokofsky, Biopolymers 13, 2505-2526 (1974); 13, 2527-2543 (1974).
22b. L.L. Van Zandt, K.C. Lu and E.W. Prokofsky, Biopolymers 16, 2481-2490 (1977) ; 16, 2491-2506 (1977).
23. S.C. Erfurth, E.J. Kiser and W.L. Peticolas, Proc. Nat. Acad. Sci. USA, 69:938 (1972).
24. S.C. Erfurth, P.J. Bond, and W.L. Peticolas, Biopolymers, 14, 247, 1259 (1975).
25. K.B. Brown, E.J. Kiser and W.L. Peticolas, Biopolymers, 11, 1855 (1972).
26. G.J. Thomas, Jr. and K.A. Hartman, Biochim. Biophys. Acta, 312, 311 (1973).
27. D.C. Goodwin and J. Brahms, Nucleic Acid Research, 5, 835-850 (1978).
28. M. Tsuboi, S. Takahashi, S. Muraishi, T. Kajiura and S. Nishimura, Science, 174, 1142 (1971).
29. K. Morikawa, M. Tsuboi, S. Takahashi, Y. Kyogoku, Y. Mitsui, Y. Iitaka and G.J. Thomas, Jr., Biopolymers, 12, 790 (1973).
30. S. Erfurth and W.L. Peticolas, Biopolymers, 14, 247 (1975).
31a. G.J. Thomas, Jr., M.C. Chen and K.A. Hartman, Biochim. Biophys. Acta, 324, 37 (1973).
31b. G.J. Thomas, Jr and P. Murphy, Science 188, 1205 (1975).
32. M.C. Chen, R. Geige, R. Lord and A. Rich, Biochemistry, 14, 4385 (1975).
33. S. Mansy, S.K. Engstrom and W.L. Peticolas, Biochem. Biophys. Res. Comm., 68, 1242 (1976).
34. B. Prescott, G.J. Thomas and D.E. Olins, Biophysical Journal, 17, 114a (1977) ; Science, 197, 385-388 (1977).
35. D.F.H. Wallach, S.P. Verme et J. Fookson, Biochem. Biophys. Acta 559, 153-208 (1979).
36. J.L. Lippert and W.L. Peticolas, Biochem. Biophys. Acta 282, 8-17 (1972).
37. K. Larsson and P. Rand, Bioch. Biophys. Acta 326, 245-255 (1973)
38. K.G. Brown, W.L. Peticolas, W.L. and E. Brown, Bioch. Biophys. Res. Comm. 54, 358-364 (1973).
39a. B.P. Gaber and W.L. Peticolas, Bioch. Biophys. Acta 465, 260-274 (1977).
39b. B.P. Gaber, P. Yager and W.L. Peticolas, Biophys. J. 22, 191-207 (1978).

39c. B.P. Gaber, P. Yager and W.L. Peticolas, Biophys. J. 21, 161-176 (1978).
40. M.R. Bunow and I.W. Levin, Bioch. Biophys. Acta 489, 191-206 (1977).
41a. S. Sunder, R. Mendelsohn, H.J. Bernstein, Chem. Phys. Lipids 17, 456-65 (1976).
41b. R. Menselsohn, S. Sunder and H.J. Bernstein, Bioch. Biophys. Acta 419, 563-569 (1976).
41c. R. Mendelsohn, Bioch. Biophys. Acta 290, 15-21 (1972).
42a. B.C. Spiker, and I.W. Levin, Bioch. Biophys. Acta 388, 361-373, (1975).
42b. R.C. Spiker, Jr, and I.W. Levin, Bioch. Biophys. Acta 433, 457-458 (1976).
42c. N. Yellin and I.W. Levin, Biochem. 16, 642-646 (1977).
42d. M.R. Bunow, and I.W. Levin, Bioch. Biophys. Acta 487, 388-394 (1977).
43. P.E. Schoen, S.M. Schnur and J.P. Sheridan, Appl. Specty. 31, 337-339 (1977).
44. R.G. Snyder, R.M. Hsu and S. Krimm, Spectrochim. Acta (in press).
45. T.G. Spiro, in "Infra Red and Raman Spectroscopy of Biological Molecules", T.M. Theophanides, Editor, D. Reidel Publishing Co, Boston (1978) pp 267-274.
46. T.G. Spiro, Biochim. Biophys. Acta 416, 169 (1975).
47. M. Gouterman, J. Chem. Phys. 30, 1139 (1959) ; J. Mol. Spectrosc. 6, 138 (1961).
48. T.C. Strekas, A. Packer and T.G. Spiro, J. Raman Spec. 1, 197 (1973).
49. L.A. Nafie, M. Pezolet and W.L. Peticolas, Chem. Phys. Lett. 20, 563-67 (1973).
50. J.M. Burke and T.G. Spiro, J. Amer. Chem. Soc. 98, 5482 (1976).
51. L. Rimai, D. Gill and J.L. Parsons, J. Am. Chem. Soc. 93:1353 (1971).
52. C. Grundherr and M. Stockburger, Chem. Phys. Lett. 22:253 (1973).
53. B. Aton, A.G. Doukas, R.H. Callender, B. Becher and T.G. Ebrey, Biochemistry, 16:2995 (1977).
54. A. Campion, M.A. El-Sayed and J. Terner, Biophys. J. 20:369 (1977).
55. R.H. Callender and B. Honig, Annu. Rev. Biophys. Bioeng. 6:33 (1977).
56. R. Mathies, T.B. Freedman and L. Stryer, J. Mol. Biol., 109:367 (1977).
57. R. Mathies, A.R. Oseroff and L. Stryer, Proc. Natl. Acad. Sci. USA, 73:1 (1976).
58. M. Stockburger, W. Klusmann, H. Gattermann, G. Massig and R. Peters, "Photochemical Cycle of Bacteriorhodopsin Studied by Resonance Raman Spectroscopy", Biochemistry, 18:4886 (1979).
59. M.A. Marcus and A. Lewis, Biochemistry, 17:4722 (1978).
60. M.A. Marcus, and A. Lewis, Science 195:1330 (1977).
61. R. Cookingham and A. Lewis, J. Mol. Biol. 199:569 (1978).

62. R.E. Cookingham, A. Lewis and A.T. Lemley, Biochemistry, 17:4699 (1978).
63. A. Lewis, J. Spoonhower, A. Bogomolni, R.H. Lozier and W. Stoeckenius, Proc. Natl. Acad. Sci. USA, 71:4462 (1974).
64. P.R. Carey, Quart. Rev. Biophys. 11, 309-370 (1978).
65. A.C. Storer, W.F. Murphy and P. Carey, J. Biol. Chem. 254 3163-3165 (1979).
66. P.W. Jagodzinski and W.L. Peticolas, J. Am. Chem. Soc. 103, 234 (1981).
67. C.I. Brändén, H. Eklund, in "Molecular Interactions and Activity in Proteins", Ciba Found. Symp. Excerpter Med. 60, 63-80 (1978).

CHIROPTICAL METHODS AND THEIR APPLICATIONS TO BIOMOLECULAR SYSTEMS

Ignacio Tinoco, Jr. and Carlos Bustamante

Department of Chemistry and Laboratory of
Chemical Biodynamics
University of California, Berkeley
Berkeley, California 94720

Marcos F. Maestre

Donner Laboratory, Division of Medical Physics
University of California, Berkeley
Berkeley, California 94720

INTRODUCTION

We will describe the various optical methods which have been used to study chiral biomacromolecules and their ordered aggregates such as found in viruses, ribosomes, membranes, etc. It is redundant to use the term chiral biomacromolecule, because all biomacromolecules are chiral. Therefore, we use chiral to characterize the optical method. A chiroptical method is thus one which yields nonzero results only when applied to chiral molecules. Here we will consider:

circular dichroism - the differential absorption of incident left and right circularly polarized light

circular intensity differential scattering - the differential scattering of incident left and right circularly polarized light

optical rotation - the rotation of the plane of polarization of incident linearly polarized light

circularly polarized luminescence - the differential emission of left and right circularly polarized light

fluorescence detected circular dichroism - the measurement of circular dichroism by means of the difference in fluorescence intensity when left and right circularly polarized light is incident

All these phenomena can be measured as a function of wavelength; in principle useful information could be obtained from the X-ray to the microwave region. At present the published results cover the region from the vacuum ultraviolet (120 nm) to the infrared (5 μm). However, synchrotron radiation sources can provide linearly polarized intensity from the X-ray to the visible region. A circular polarizer becomes the limiting factor for the region of interest. A LiF stress-plate modulator can be used down to 100 nm; for shorter wavelengths new types of polarizers will be needed. Circular and linear polarizers are available in the microwave region, but it is not clear what type of information will result from chiral microwave measurements on biological systems.

CIRCULAR DICHROISM

Definition and Description

Circular dichroism is the difference in absorbance for incident left and right circularly polarized light. For circular dichroism to be nonzero the system which absorbs the light must be different from its mirror image. That is, it must be chiral; it must have a handedness.

The circular dichroism is usually measured indirectly. The transmitted light is traditionally measured; it is related to the absorbance by the standard formula

$$I = I_0 10^{-A}$$

The transmitted intensity is I; I_0 is the incident light and A is the absorbance. This formula assumes that light is lost (not transmitted) only by absorbance. This assumption is clearly not correct for samples which scatter a significant fraction of the incident light. We will discuss methods to minimize the errors caused by scattering in succeeding sections. However, a better alternative might be to measure the absorbed energy directly by using photoacoustic detection.[1] If one assumes that only absorption is occurring, then a measurement of the ratio $(I_L-I_R)/(I_L+I_R)$ will lead to the circular dichroism. (A_L-A_R).

$$A_L-A_R = -\frac{2}{2.303}\frac{(I_L-I_R)}{(I_L+I_R)}$$

The relation between absorbance and molar absorptivity or molar extinction coefficient leads to circular dichroism in the usual units of liter/mole cm or $\text{molar}^{-1}\text{cm}^{-1}$.

$$\varepsilon_L-\varepsilon_R = (A_L-A_R)/(\text{path length})(\text{concentration})$$

An equivalent, although now rarely used, method of measuring circular dichroism is to measure the ellipticity of the transmitted light when linearly polarized light is incident. The ellipticity, θ, defined as the arctangent of the ratio of the minor axis of the ellipse to the major axis, is directly proportional to the circular dichroism.

$$\theta \text{ (in radians)} = 2.303(A_L - A_R)/4$$

A useful conversion factor to remember when reading the older literature is that the molar ellipticity in standard units of degrees molar^{-1}cm^{-1} equals $3298(\varepsilon_L - \varepsilon_R)$.

Experimental

The CD apparatus is a spectrophotometer which can measure the difference in absorbance of left circularly polarized light from that of right circularly polarized light. The machines available use the technique of Velluz, Legrand and Grosjean.[2] This technique uses the modulation of linear polarized light by a quarter-wave plate to which an oscillating variable voltage is applied (Pockels' cell or electro-optic modulator), thereby rotating its optical axes by 90°, back and forth, producing a positive or negative 90° phase shift on the x,y components of the incoming light beam. The result is that circularly polarized light is produced, first with one sense of polarization, followed by the opposite sense (i.e., left circularly polarized light followed by right circularly polarized light). The circularly polarized light of intensity I_0 interacts with the material under investigation, and after the interaction, the transmitted intensity, I, is measured. The usual circular dichrograph, following the design of Velluz et al., obtains by electronic means a ratio of intensities

$$\frac{I_L - I_R}{I_L + I_R} = \frac{\text{(ac component)}}{\text{(dc component)}} = \text{signal}$$

If the material in question obeys a Beer-Lambert absorption law for each polarization, we have

$$I_L = I_{OL}\, e^{-c\varepsilon_L d} \qquad I_R = I_{OR}\, e^{-c\varepsilon_R d}$$

and signal = k tanh $(\varepsilon_L - \varepsilon_R)$, with k an instrument constant. For ellipticities of the order of 5° or less,

$$\text{signal} = k(\varepsilon_L - \varepsilon_R) = k\Delta\varepsilon$$

where ε_R = extinction coefficient for right circularly polarized light and ε_L = extinction coefficient for left circularly polarized light. The above equation implies pure absorption phenomena. No scattering (anisotropic or otherwise), fluorescence, or phosphorescence is assumed, and the measuring instrument is perfect (i.e., no stray light, linear birefringence in optics or detectors, or linear dichroism or circular dichroism in the optics or detectors is assumed).

If significant scattering occurs, the above equations must be modified to include these perturbations; two new effects must be considered. (1) There will be scattered light which reaches the detector. The intensity measured by the detector is thus the sum of transmitted light and forward scattered light. (2) Because of scattering in all directions the transmitted light is decreased both by absorption and by the sum of all the scattered intensities.

To take into account the scattered light reaching the detector the signal is written as:

$$\text{signal} = k\,\frac{(I_L - S_L) - (I_R - S_R)}{(I_L - S_L) + (I_R - S_R)}$$

S_L, S_R = intensity of light scattered when left and right circularly polarized light is incident. In general S_L and S_R will depend on the angle between the direction of the incident beam and the direction of the scattered beam. For an oriented sample such as a film, fiber or crystal, it will also depend on the direction of the incident beam. The scattered intensity may depend on the circular polarization of the incident light (as indicated by the subscripts), but it will also depend on any linear polarization present in the incident beam. In actual practice one must remember that the measured signal will also include effects from reflections from the cell windows and the optics, dust in the solutions, etc. In principle the forward scattered light reaching the detector can be eliminated by use of an iris diaphragm in front of the detector, and extrapolation to zero aperture. One can also correct for the dust and reflection artifacts, but it is necessary to be aware of all these problems.

Once the transmitted light is measured correctly for a macromolecular system, we must be careful in its interpretation. The difference in molar extinction coefficients, $\varepsilon_L - \varepsilon_R$, can be a sum of a difference in molar absorptivities, $a_L - a_R$, and a difference in molar scattering coefficients, $s_L - s_R$. The scattering coefficient is proportional to the intensity of scattered light integrated over all directions in space.

$$\varepsilon_L - \varepsilon_R = (a_L - a_R) + (s_L - s_R)$$

For very large particles, with nearly total extinction, other effects such as Duysens flattening[3] can be seen. Certain membranes[4] and red blood cells[5] are examples of systems that have shown this phenomenon. Circular dichroism can routinely be measured to obtain differences of the order of 1 part in 10000 in the extinction of left vs. right circularly polarized light. Consequently this technique is very sensitive to birefringence artifacts in the optical train (lenses, cuvette windows, polarization sensitivity of the detectors), and to misalignment of the optical components. The ability to measure a signal at a given wavelength depends on the optical anisotropy ratio $\Delta\varepsilon/\varepsilon$ (where $\Delta\varepsilon = \varepsilon_L - \varepsilon_R$ and $2\varepsilon = \varepsilon_L + \varepsilon_R$) of the chromophore. The optimal signal-to-noise ratio is obtained for solutions whose optical density is 0.868, if the detection is limited by shot noise. By proper masking of the light beam very small volume cells can be used (~ 4 microliters) for the measurement of scarce materials, although the usual measurement is in 1 cm path length cells of volumes of the order of 3 ml.

The wavelength range is usually limited by the transparency of the solvents. With the most up to date technology, instruments have been constructed which go as low as 120 nm in the vacuum ultraviolet regions and up to 1600 cm^{-1} in the infrared region (vibrational CD) of the spectrum.[6,7]

Theory

It is not necessary to know the theoretical relation between an experimental circular dichroism and a molecular structure to apply circular dichroism. Actually, many useful applications have involved simple, empirical correlations between a measured CD and a molecular state established by some other method. However, maximum information can be obtained from CD experiments, if the theory is understood.

An excellent summary of the quantum mechanical theory and calculations of circular dichroism has appeared recently.[8] It provides an efficient entry into the literature of this field.

The circular dichroism can be written as a sum of contributions from transitions between states.

$$\Delta a = a_L - a_R = \sum_a R_a \, g_a(\lambda)$$

Here R_a is the rotational strength for the transition from the ground state (or other specified state) to state a, and $g_a(\lambda)$ is a function of wavelength which specifies the shape of each CD band. Thus $g_a(\lambda)$ contains the shape of the CD and a number of universal constants relating CD to rotational strength. For each transition the rotational strength can be obtained by integrating over its CD band.

$$R_a = (6909\ hc/32\pi^3 N_0) \int (\Delta a d\lambda)/\lambda$$

The units are cgs with h = Planck's constant, c = the speed of light, and N_0 = Avogadro's number; the rotational strength is obtained with units of esu^2cm^2. The rotational strength is defined as the imaginary part of the dot product of the electric dipole ($\underset{\sim}{\mu}$) and magnetic dipole ($\underset{\sim}{m}$) transition moments.

$$R_a = \mathrm{Im}\langle 0|\underset{\sim}{\mu}|a\rangle \cdot \langle a|\underset{\sim}{m}|0\rangle$$

Nearly all calculations of CD have been based on attempts to find suitable approximations to this expression. As the definition of rotational strength was derived with the assumption that the wavelength of light was large compared to molecular dimensions, we might question the applicability of the expression to macromolecules. The key dimension to compare with the wavelength of light is actually not the length of the molecule, but the distance between two electrons (or chromophores) with significant interaction; that is, the coupling between different parts of the molecule determines whether the usual definition of rotational strength is valid. If it is not valid, we must use the correct transition moment operators to calculate the CD. These involve exponential operators, so it is easier to write the equation for an oriented system. For light incident along unit vector $\underset{\sim}{k}$, the CD is proportional to

$$\underset{\sim}{k} \cdot \langle a|\underset{\sim}{T}|0\rangle^* \times \langle 0|\underset{\sim}{T}|a\rangle$$

with

$$T = \exp(2\pi i \hat{\underset{\sim}{k}} \cdot \underset{\sim}{r}/\lambda)\underset{\sim}{p}$$

p = linear momentum operator

r = position operator

For a randomly oriented system the above expression must be averaged over all possible orientations. The average has been derived by Tobias et al.;[9] it reduces to the usual rotational strength when the exponential is expanded to first order. For most biological systems it is probably appropriate to use the first-order expansions in making calculations; in any case this is what most authors have used. However, if symmetry arguments are used to obtain selection rules for rigid polymers, such as helices, it is imperative that the correct (exponential) transition integrals be considered.

For polymers or macromolecules a practical approximation to the transition moments is obtained by assuming that there is no exchange or transfer of electrons between chromophores. The CD can then be calculated in terms of the properties of the individual chromophores in the polymer, and their interactions. The interactions depend on the geometry of the polymer, thus structural information can be obtained by comparing a calculated CD with experiment. If we accept the assumption about electron exchange as reasonable, the validity of the structural conclusions will depend on knowledge of the chromophore properties and on the accuracy of the coulombic interaction between the chromophore charge densities.

A review of circular dichroism experiments and calculations on polypeptides and polynucleotides by Woody[10] gives full details of recent work. The amide chromophore in polypeptides is fairly well understood at least down to 190 nm. Good agreement is found in calculations of the CD of polypeptides in the α-helix,[11,12] β-pleated sheets[12,13] and β turns.[14] This agreement for different conformations gives us confidence that the general methods used are correct. Calculations of circular dichroism for polynucleotides are more difficult, because there are four chromophores that must be considered: adenine, cytosine, guanine and uracil (or thymine). The assignment and interpretation of the many $\pi\pi^*$ and $n\pi^*$ transitions in these bases is a formidable task.[15,16] However, reasonable agreement with the CD of DNA in A-and B-type conformations is found.[17] The sequence dependence of the CD can also be obtained; better agreement is found for adenine and uracil (or thymine) containing polymers than for guanine and cytosine.[18] The monomer spectra are simpler for the former.

An alternative approach to calculation of polymer optical spectra is a classical oscillator method.[19] Here each chromophore is represented by a polarizability tensor which is complex. The wavelength dependence of the polarizability characterizes the absorption of the isolated chromophore. When the chromophore is part of a polymer, the field which each chromophore senses is determined by all other chromophores in the molecule. This coupling induces a circular dichroism in chiral arrangements of chromophores. For the simplest polymer, a dimer of two identical chromophores, the equation is

$$a_L - a_R = \frac{(6)(6909)}{(3298)} \nu\varepsilon^\circ(\nu) G_{12} R(\nu) \underset{\sim}{R}_{12} \cdot \underset{\sim}{e}_1 \times \underset{\sim}{e}_2 \ / \ D$$

$$D = [1 - G_{12}^2 (R^2(\nu) - I^2(\nu)]^2 + [2G_{12}^2 I(\nu) R(\nu)]^2$$

$\varepsilon^\circ(\nu)$ = chromophore absorption band as a function of frequency, ν, in cm^{-1}

$R(\nu), I(\nu)$ = real and imaginary parts of a principal axis of polarizability

$\tilde{R}_{12}$ = distance between chromophores with principal axes along $\tilde{e}_1$ and $\tilde{e}_2$

G_{12} = $\tilde{e}_1 \cdot \tilde{T}_{12} \cdot \tilde{e}_2$ with $\tilde{T}_{12}$ the dipole interaction tensor.

This method can give good agreement with experiment[20] and is intuitively easy to understand.

Theoretical calculations of $(s_L - s_R)$, the scattering contribution to the extinction, is just beginning. First the preferential scattering of circularly polarized light must be calculated; then this scattering must be integrated over all directions in space. It is too early to know how significant this term may be in general, but it can clearly be dominant in some cases.

The general quantum mechanical expression for CD applies to vibrational transitions as well as electronic transitions. However, the approximations used in evaluating transition moments are very different. The transitions are represented by oscillations of atoms containing partial charges--the fixed partial charge model. So far, agreement with experiment has not been as good as in electronic CD.[21-24] Useful reviews of both theory and experiment have recently appeared.[7,8]

Two-photon CD is the logical extension of two-photon absorption,[25] but at least one photon must be circularly polarized. The theory of two-photon circular dichroism has been presented,[26] but no experimental results have been obtained yet. It will be a difficult measurment to make and it may be many years before applications to biological molecules are available.

Applications

The main application of CD measurements to proteins and nucleic acids is to determine structure and conformation. There are recent reviews[10,27-29] on this subject which provide a wide survey of the field. We will concentrate on a few examples.

The main strategy in the structure analysis of proteins by CD methods is to find a correlation between the CD measurement and a minimum number of structures, or combinations of structures, that describe the protein. Thus, we start with such basic structures as "pure" α-helix, "pure" β-sheet and so-called "random coil" or unordered structure. If the CD spectra of these structures are different enough from each other over a given range of wavelengths, in principle any CD spectrum can be decomposed into a linear combination of spectra as follows:

$$\theta \text{ protein} = \sum_{n=1}^{N} f_n \, \theta_n(\lambda)$$

where $\theta_n(\lambda)$ = CD of nth type of structure

f_n = fraction of type n structure in the protein

N = total number of different types of structures

A criterion for a good fit is then:

$$\sum_{n=1}^{N} f_n = 1$$

A typical application for this type of analysis is protein structure determination by CD in terms of α-helix, β-sheet and random coil fractions by the use of methods developed by Fasman and coworkers[30] and Yang and coworkers.[31,32] These workers have used the CD of synthetic and natural model compounds of known structures to obtain empirical formulae that allow a reasonable fit to the CD of a protein by three contributions: α-helix, β sheet and random coil or unordered structure. Were proteins in nature composed of only different fractions of these three types of basic structures, it would be a simple matter to fit the data (by various least square fitting algorithms) to the experimental curves in the wavelength region of the CD bands of the polypeptides. However, as there are different forms of β-pleated sheet (parallel and anti-parallel), different types of β-turns and various types of random coils, it is difficult to obtain unequivocal fits to such a simple basis vector set. Furthermore, the influence of helix length and the possible influence of side chains are not accounted for by this basis vector set.

Greenfield and Fasman[30] obtained a set of spectra to 190 nm wavelength for poly-L-lysine in three reference conformations (α-helix, β-sheet and "random coil"). With these as a basis vector set, values for the percentage of the three component types were obtained for myoglobin, lysozyme, ribonuclease, ribonuclease S, carboxypeptidase A, chymotrypsin and chymotrypsinogen, and compared to the reported X-ray structures. Good fits were obtained for the α-helical content and less successful fits for the β-sheet and random coil fractions.

A different approach was that of Chen et al.[31,32], which used a set of 5 and 8 proteins, respectively, whose X-ray structure were known, to extract the α-helix, β-sheet and random coil basis vector sets. An attempt was also made to get a measure of the length of the α-helical regions (as an extra parameter) by inclusion of optical rotatory dispersion values of the protein at 233 nm. More recently, Chang et al. (1978)[33] have used the CD spectra of fifteen proteins with known structures to derive a basis vector set composed of α-helix, β-sheet, β-turn and unordered structures.

In all the above attempts the determination of the minimum basis vector sets led to good predictions for the α-helical content and poorer fits for the fractions of β-sheet, β-turn and random coils. The main cause for the poor fit of this type of structure is that in the wavelength region from 230 nm down to 190 nm (which is the available limit for most commercial machines) the CD spectrum of β-sheet, β-turn and random coil contributions is quite weak compared to that of the α-helical components. Indeed, Siegel et al.[34] have shown that for 16 proteins the CD spectrum between 210 and 240 nm is a function mainly of the α-helical content. An improvement on the basis vector set determination occurred when the wavelength region of measurement was extended into the vacuum ultraviolet.[28] Brahms and Brahms[35] and Hennesy and Johnson[36] have now expanded the basis vector set by differing approaches. Brahms and Brahms[35] have used β-turn (type II) reference data determined from the CD spectrum of poly(Ala_2-Gly_2) and the tripeptide N-isobutyl-L-Pro-D-Ala-isopropylamide (L-pro-D-Ala) which show type II beta-turn from X-ray crystallography. Their β-sheet reference spectra came from alternating hydrophylic and hydrophobic containing polymers. The α-helical contribution was extracted from CD data on myoglobin. The CD of the polypentapeptide $(\text{Pro-Lys-Leu-Lys-Leu})_n$ was taken to represent the unordered structure. The CD spectra were computed for proteins containing mainly β-structures, alternating α and β-type structures and proteins which are a mixture of α and β regions. The advantages of the vacuum CD measurements for the study of β-protein structures were obvious when the CD determinations of the β content were compared with the available X-ray data of Levitt and Greer.[37]

A different approach to the determination of reference data is taken by Hennesy and Johnson[36] The CD of 16 proteins was measured by these workers down to 178 nm wavelength and an <u>orthogonal</u> basis vector set was determined by a multicomponent matrix analysis. This process will generate an orthogonal set of CD spectra which can be combined to approximate the CD spectra of any of the 16 original proteins. The hope is that a minimum number of these orthogonal spectra is needed to account for most of the properties of the CD and that by inference this set will also be sufficient to fit the CD of any other protein measured. The relation of these basis vectors with structural characteristics of the proteins is more difficult, since a one-to-one correspondence cannot be established between one basis spectrum and α-helix content, β-turn content, etc. The basis vector set was used by Hennesy and Johnson to estimate the fraction of eight different types of secondary

structure (α-helix, β-parallel, β-antiparallel strands, I, II, III, β-turns, all other β-turns and "other" structure). These authors found that a minimum set of five basis vectors was needed to account for most of the character of the CD spectra and that extending it to a six basis vector set improved the fit little. Each of the vectors determined has a special characteristic; vector 1 is related to the amount of α-helix content; vector 2 has both α-helical and β-strand character and so forth.
Hennesy and Johnson also showed that truncating data to a low wavelength cut-off of 184 nm made little difference in the analysis of the protein structure, but that truncating above 190 nm resulted in striking changes in the determination of β-strand and other conformations. However, the estimates of the α-helical fractions changed very little, indicating that helical content is mainly expressed in the near ultraviolet CD bands and needs little reinforcement from the bands in the vacuum ultraviolet CD region.

The CD of DNA has been used to distinguish between various types of double-strand conformations such as A-, B- or C-conformations.[38] The effect of base sequence on the CD of DNA and double-stranded RNA has been explored,[39-41] and the number of double stranded base pairs in tRNA's has been obtained.[42]
These applications were all based on curve fitting in analogy to the methods used for proteins. If a nucleic acid is not in one of the standard conformations, obviously its CD cannot be interpreted as a combination of known contributions.

Another approach is to use theory to calculate the CD of DNA in any possible conformation. Johnson et al.[43] have calculated the CD of an eleven base-pair helix in all energetically reasonable right-handed conformations.[44] These conformations can be characterized by five parameters: the winding angle (angle between base pairs), the step height between base pairs, the twist angle between the two bases in a base pair, the tilt angle between the base pairs and the helix axis, and the distance from the center of the base pair to the helix axis. Good correlation was found between the calculated CD at 275 nm and two of the parameters: the winding angle and the twist angle. This approach should be useful in interpreting the changes in CD of a DNA when the temperature, salt, solvent, etc. are slightly changed.

Other biopolymers besides proteins and nucleic acids have been studied; they include polysaccharides in the vacuum ultraviolet[45] and the N_3^- group of azidomethemoglobin in the infrared.[46]

OPTICAL ROTATORY DISPERSION

Definition and Description

Optical rotation is the rotation of incident linearly polarized light by a chiral sample. The rotation can be either to the right or the left; it is linear in path length and concentration. It is not measured very often now, but it does have the advantage over CD in that it can be measured outside absorption bands. Inside an absorption band, incident linearly polarized light not only rotates, it becomes elliptically polarized. The wavelength dependence of optical rotation is termed optical rotatory dispersion, ORD. There are exact relations between ORD and CD obtained from Kronig-Kramers transforms.

$$\phi(\lambda) = \frac{2}{\pi} \int_0^\infty \frac{\lambda'\theta(\lambda')d\lambda'}{\lambda^2-\lambda'^2}$$

$$\theta(\lambda) = \frac{-2}{\pi\lambda} \int_0^\infty \frac{\phi(\lambda')\lambda'^2 d\lambda'}{\lambda^2-\lambda'^2}$$

Here $\phi(\lambda)$ is the rotation as a function of angle and $\theta(\lambda)$ is the ellipticity as a function of angle. The main difficulty in applying these equations is the necessity of knowing either the ORD or CD at all wavelengths; however, we will see how these equations can be applied.

Experimental Applications

Historically, ORD was used to characterize compounds whose absorbance bands were outside the wavelength range of the available instruments. Theoretical analysis of the relation betweeen optical activity and structure was concerned with ORD until modern CD machines allowed direct measurement of the CD bands. At present, study of optical activity is limited to CD except for cases in which the CD cannot be measured.

A novel application of ORD to predict the existence of CD bands below the wavelength limit of measurement of commercial CD instruments was attempted by Cassim and Yang.[47] A measurement of both the ORD and CD of an α-helical polymer down to 181 nm allowed these workers to detect a CD band at lower wavelengths. They subtracted the Kronig-Kramers transform of the CD bands from the measured ORD to obtain a residual ORD. This residual ORD was presumably due to the next higher energy bands. The Kronig-Kramers transform of the residual ORD allowed them to predict a positive band at 174 nm for poly-L-glutamic acid at pH 4.5. This band was later confirmed by direct measurement.[48]

FLUORESCENCE DETECTED CIRCULAR DICHROISM

Definition and Description

Fluorescence detected circular dichroism (FDCD) is the use of fluorescence to monitor the preferential absorption of circularly polarized light. The measurement is similar to that of circular dichroism except that instead of measuring the transmitted light when circularly polarized light is incident, the fluorescence is measured. A diagram of the experimental arrangement is shown in Figure 1. The fluorescence signal can be processed by a standard

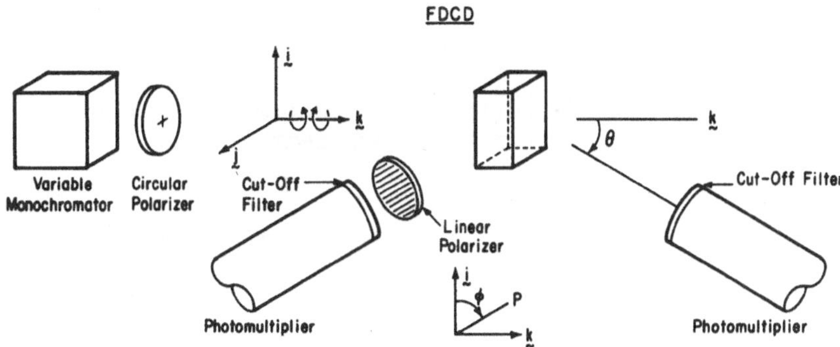

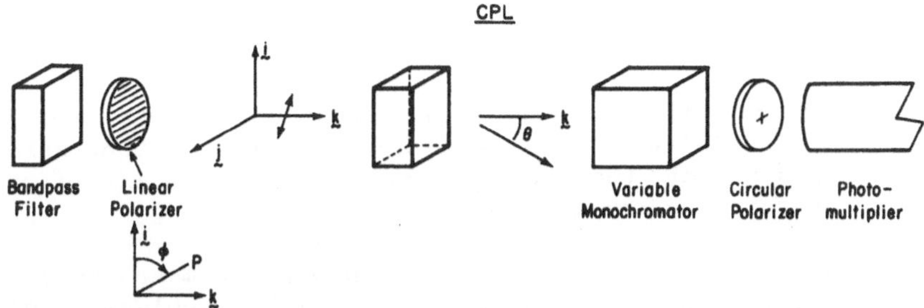

Fig. 1. The measurement of fluorescence detected circular dichroism (FDCD) and circular polarization of luminescence (CPL). In FDCD circularly polarized light excites the sample and the total fluorescence intensity is measured. In CPL unpolarized or linearly polarized light excites the sample and the circular polarization of the fluorescence is measured. Reprinted with permission from reference 102.

CD apparatus so that the apparent ellipticity in degrees, $\theta_F°$, read on a chart can be related to the FDCD. The fluorescence intensity is related to the properties of a fluorophore in a cell of length ℓ by[49]

$$F = I_0 \phi \varepsilon_F c_F \ell (1-10^{-A})/A$$

I_0 = incident intensity

ϕ = fluorescence efficiency of the fluorophore (quantum yield)

ε_F = molar absorption coefficient of fluorophore

c_F = concentration of fluorophore

A = absorbance of the cell; it includes absorption by the fluorophore and other nonfluorescent species

When the ratio $(F_L-F_R)/(F_L+F_R)$ is calculated from the above expression, the apparent ellipticity is:

$$\theta_F^\circ = -14.32\left(\frac{\Delta\varepsilon_F}{\varepsilon_F} - 2R\right)$$

$$R = \frac{\Delta A}{2A} - \frac{2.303\ \Delta A 10^{-A}}{2(1-10^{-A})}$$

$\Delta\varepsilon_F$= the molar CD of the fluorophore

Thus the Kuhn asymmetry factor for the fluorophore, $g_F = \Delta\varepsilon_F/\varepsilon_F$, can be obtained either by extrapolation to zero concentration, or by measurement of the CD, ΔA, and absorbance, A, of the cell in a separate measurement.

There are various applications of FDCD which make the technique particularly helpful with biological molecules. In a protein or nucleic acid which has many chiral groups, but only one fluorophore, the CD of the fluorophore can be separated from the rest of the molecule. If there are more than one fluorescing species in solution, the measured Kuhn asymmetry factor is the quantum yield weighted sum of contributions from each species.

$$\frac{\Delta\varepsilon_F}{\varepsilon_F} = \frac{\Sigma\phi_i c_i \Delta\varepsilon_i}{\Sigma\phi_i c_i \varepsilon_i}$$

These species can be stable molecules, or different conformations of a flexible molecule which are in rapid equilibrium. The measured CD and absorbance of the same system gives:

$$\frac{\Delta\varepsilon}{\varepsilon} = \frac{\Sigma c_i \Delta\varepsilon_i}{\Sigma c_i \varepsilon_i}$$

It is clear then that for a rigid molecule, or a molecule with only one fluorescing conformer, the CD and the FDCD should be the same. Examples are camphor sulfonic acid and tryptophan.[50] However, for molecules which can have several conformers, some of which fluoresce and others do not, the CD and FDCD are very different. An example is dinucleoside phosphates containing fluorescent bases.[51] There are at least two states present in solution. The state with the bases stacked has a characteristic CD of a dimer, but the fluorescence is quenched, therefore the FDCD only measures unstacked conformations.

We have been discussing the CD and FDCD of randomly oriented molecules. It is of course possible to orient macromolecules by electric fields,[52] or hydrodynamic fields,[53] and to measure their CD along different directions. For a molecule whose rotation is slow compared to its fluorescence lifetime, the FDCD depends both on the average CD ($\Delta\varepsilon$) and on the CD along the direction of the fluorescence emission transition vector ($\Delta\varepsilon_{33}$).[54] This effect is photoselection. By using a linear polarizer in front of the detector, the presence of photoselection can be detected, and both $\Delta\varepsilon$ and $\Delta\varepsilon_{33}$ can be obtained.

Experimental Applications

Fluorescence detected CD can be measured using a simple, modified CD machine;[55] however, for the highest accuracy it is important to use square pulse modulation of a Pockels' cell rather than the usual sinusoidal modulation. Sinusoidal modulation produces all states of polarization of the incident light, and linearly polarized components can lead to serious artifacts, particularly when photoselection is present. Lobenstine and Turner[56,57] have designed an FDCD machine patterned after an apparatus to measure Raman optical activity. Two photomultiplier detectors are placed at right angles to each other and to the direction of incidence of the light. This orientation cancels the effects of any residual linear polarization components in the incident circularly polarized light. Lobenstine et al.[58] have recently reported studies of five proteins, each containing one tryptophan fluorophore, which demonstrates that FDCD is a sensitive probe for the environment of a fluorophore in a protein.

A novel application of FDCD is to correct for the scattering components in a CD signal, as reported by Reich et al.[59] and Maestre and Reich.[60] In recent years there has been increasing application of CD techniques to biological structures of more and more complex structural organization and of considerable size with respect to the wavelength of light. Many systems which have been studied by CD, such as bacteriophage, membrane incorporated proteins, DNA in chromosomes, etc., are intense light scatterers.[4, 5, 29, 61-65] Since commercial CD spectrometers are ratio measuring devices, structures which scatter both senses of circularly polarized light with equal efficiency will cause no change in the measured ellipticity. Many light scattering suspensions, however, scatter left and right circularly polarized light to differing extents; i.e., they exhibit differential light scattering. The presence of such scattering can often be evidenced by observed CD signals in regions where the sample does not absorb light. Distorted spectra of this type can, of course, lead to great difficulties in interpretation of the measured signal.

The most common experimental method for correction of differential scattering is to increase the solid angle of detection in order to collect more scattered light. This has been accomplished by increasing the size of the detector, bringing it closer to the sample cell, or both. Configurations of this type can collect light scattered by as much as 90° from the incident direction. Probably the most effective scatter correction method to date is the use of the fluorscat technique.[62,63] Unfortunately, in none of the above experimental methods can the light detector be placed between the sample and the incident beam. Thus, one cannot correct for backscattering by the sample. Maestre et al.[59,60] presented a method, using fluorescence detected circular dichroism (FDCD), in which a nonoptically active fluorescent substance can be used as a CD detector having a solid angle of detection of 4π steradians. Thus, backscattering by the particles can be detected. The CD and FDCD spectra of DNA and poly[d(A-C)•d(G-T)] particles in ethanol solution were measured using the above mentioned techniques for correction of differential scattering.

We now consider a case where the sample to be measured consists of a mixture of a nonoptically active fluorescent substance and an optically active differential scatterer. Although only emission from the fluorophore is being directly measured, the signal is a function of the optical activity of all the components of the sample. This can be understood, if one considers that the relative amount of left and right circularly polarized light reaching a fluorophore at any point in a solution is mediated by the differential absorbance of the medium in front of the fluorescer. In this instance, $\Delta\varepsilon_F/\varepsilon_F$ equals zero, and the FDCD equation can be rearranged to:

$$\Delta A = \frac{2A(10^{A}-1)\theta_{F}}{28.65(10^{A}-2.303A-1)}$$

Thus the CD of the differential scatterer can be measured through the use of a nonoptically active fluorophore. In effect, what is happening is that the scattering particles are surrounded on all sides by fluorescent detectors. These detectors intercept the scattered light in addition to the incident radiation and, thus yield a signal which is a function of all the scattered and transmitted light.

In using the FDCD method to correct for differential scattering, it should be noted that the assumption is being made that the average pathlength of the scattered light is the same as that of the transmitted radiation. This is obviously true for light scattered at an angle of 0° and 180° and, for a square cell, 90°. Other directions lead to either a greater or lesser average optical path than for the aforementioned angles. The exact mean pathlength thus depends on the details of the scattering pattern. In the most extreme cases, the deviation from equality of mean pathlength is about 6%, while for isotropic scattering this reduces to 2%. Thus, the assumption of equal average optical paths is substantially correct. An example of the corrected spectra obtained using FDCD and other methods is shown in Figure 2.

CIRCULARLY POLARIZED LUMINESCENCE

Definition and Description[66-68]

Circularly polarized luminescence, CPL, is the difference in left and right circularly polarized emission of light by a chiral sample (see Figure 1). The sample is excited either by polarized or unpolarized light. The CPL is characterized by a luminescence dissymmetry factor analogous to the Kuhn dissymmetry factor for CD. The theoretical interpretation of the measurements is very similar to that of CD except that the molecular parameters obtained (such as rotational strengths) refer to excited states rather than ground states. In analogy with FDCD, photoselection can affect the CPL. If the rotational relaxation time for a fluorophore is large compared to the fluorescence lifetime, the CPL of an oriented sample is obtained.

Steinberg and his collaborators[67] have made many interesting applications of CPL to lanthanide ions bound to proteins, active sites of enzymes, and antibody conformation.

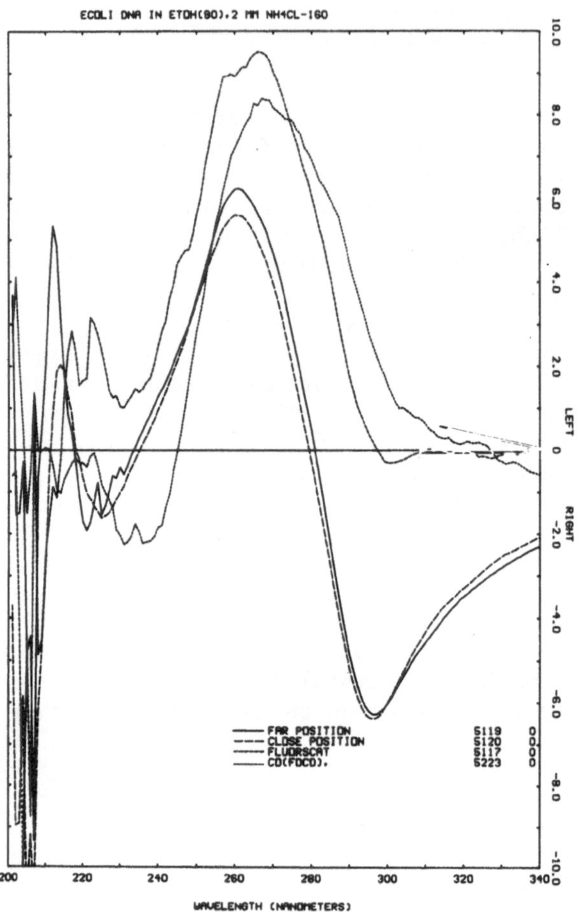

Fig. 2. The apparent CD measured for DNA particles in 80% ETOH. The CD measured by the usual method gives a large apparent CD even at 340 nm where DNA does not absorb. The CD measured using FDCD does not show this scattering contribution and gives a CD as expected for DNA in the A conformation. Reprinted with permission from reference 60.

CIRCULAR INTENSITY DIFFERENTIAL SCATTERING

Definition and Description

Within the last decade, the chiroptical methods discussed in the previous sections have been applied to increasingly complicated systems, such as membranes, DNA precipitates, bacteriophages, nucleosomes, etc. A common feature of all these systsems is the presence of conspicuous anomalies in their CD and ORD spectra. These anomalies range from the presence of long "absorbance" tails extending towards the red, to the existence of unreasonably large ellipticities both inside and outside the absorption bands of the material. Urry and coworkers,[61] studying the aggregation of polyglutamate in solution, suggested that the CD changes observed between the aggregates and the dispersed solution arose from Duysen's flattening effects[3] and what they called the "preferential scattering" of light of opposite circular polarization. Schneider et al.[64] conducted many experiments in an attempt to characterize the CD distortions observed in membrane suspensions, namely the flattening of the spectra, particularly around the 208 nm band. By sonicating ghosts of red blood cells, the authors were able to obtain a general increase in the ellipticity. These authors attributed these distortions to scattering Duysen's flattening artifacts and discussed the possibility of a differential scattering contribution. Maestre et al.,[65] measuring the CD of bacteriophages T_2, T_4, and T_6, found that the CD curves presented long tails extending towards the longer wavelengths of the spectrum up to the visible. These authors attributed correctly the effect to differential scattering. A characterization of the phenomenon was made by Dorman et al.,[62,63] who studied the CD of the T-even bacteriophages using a set of variable acceptance detectors and an integrating light-capturing device (fluorscat-cell) that captures all but the back-scattered light.

Many CD studies of nucleohistones[69,70] and DNA-polylysine complexes[71,72] have been reported in the literature. In all cases a non-negligible differential scattering contribution is present. Paralleling the experimental characterization of the phenomenon, a few theoretical treatments of the differential scattering have appeared in the literature. For a thorough and comprehensive quantum mechanical treatment of the differential scattering of opposite circular polarization by small molecules, the work of Atkins and Barron,[73] and Barron and Buckingham must be consulted.[74] These authors have defined what they call the Circular Intensity Differential Scattering (CIDS) as:

$$\text{CIDS} = (I_L - I_R)/(I_L + I_R)$$

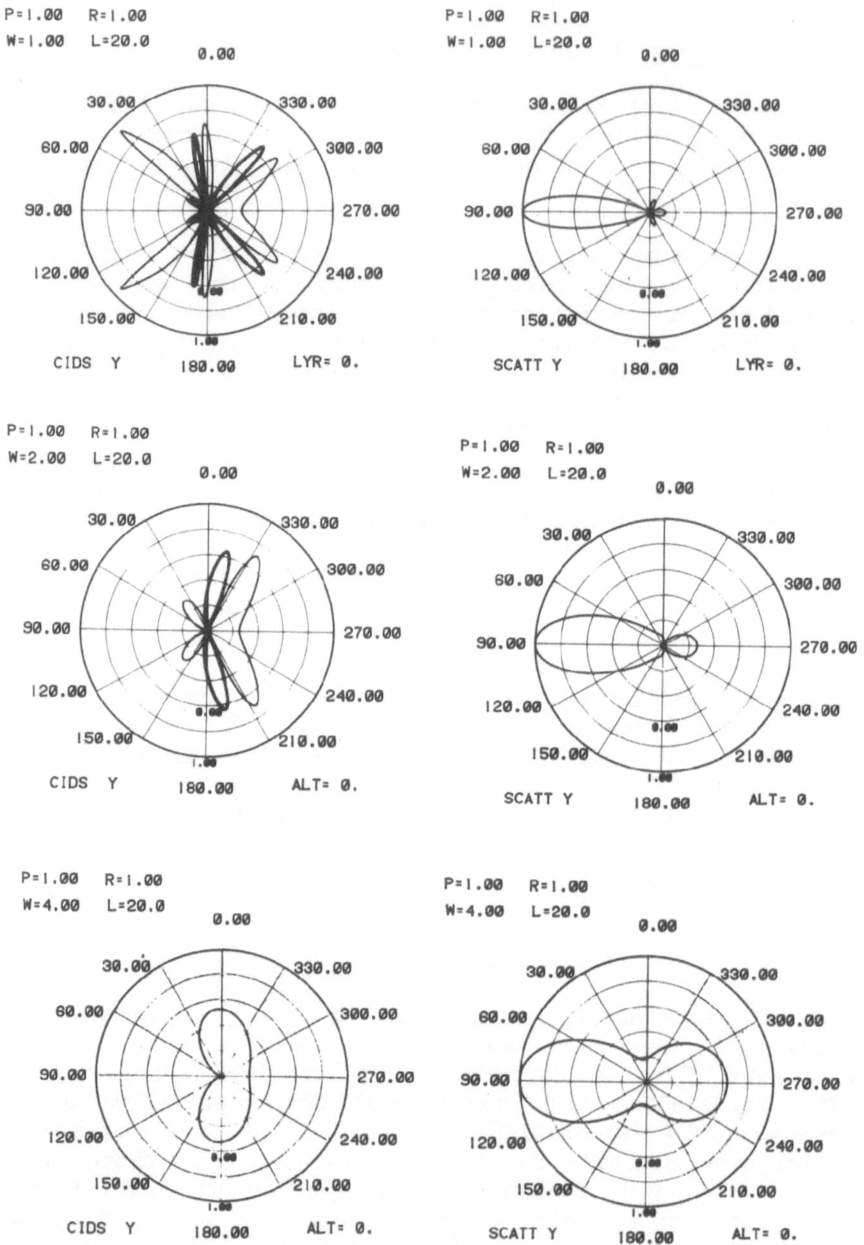

Fig. 3. Polar plots of CIDS and total scattering for increasing values of the wavelength of light while keeping pitch and radius constant. The helix is right handed; a light line means a positive CIDS and a heavy line means a negative CIDS. The light is incident from the right.

where I_L and I_R are the scattered intensities at a given position in space for incident left and right circular polarization. Another quantum mechanical approach has been presented by Harris and McClain,[75] for the case of a polymer of unspecified geometry.

Recently, we have derived a classical approach to CIDS of helical molecules, oriented in space, for wavelengths of incident light within and outside the absorption bands.[29,76-78] The CID scattering of these helical molecules is obtained in terms of the helical parameters and the optical properties of the scatterer. It can be shown that within the dipole-radiation field approximation, the existence of nonvanishing CIDS depends on the symmetry properties of the polarizability tensor of the chiral molecules. Figure 3 shows polar plots of intensities vs. the scattering angle of the total and the CID scattering for a right-handed helix of pitch, P = 1.0, radius, R = 1.0, and with the light of wavelength, W = 1.0, 2.0, 4.0; the light is incident perpendicular to the helix axis. The polarizability is uniaxial and directed along the tangent at each point on the helix. The main feature to recognize in this figure is that the polar plot for the CID scattering has much more detail, and therefore, much more information than the total scattering intensities. For this simple model[77] the general distribution of differentially scattered intensities in space depend only on the geometry of the helix and not on the actual magnitudes of the electronic transitions (induced dipoles) in the scatterer. As a result the details of the scattering patterns, the sign and magnitude of the signal, and its angular dependence give information about the structure of the helix. CIDS is not a differential absorption (dichroic) phenomenon; thus, it is not restricted to wavelengths of light within an absorption band. Indeed, as the wavelength of the incident light changes, different ranges of order and organization in the scatterer can be studied. Characteristic dimensions of the scattering particle which are closest to the wavelength of incident light contribute the most to the CIDS pattern. This makes CIDS complementary to the usual spectroscopic methods of studying structure in solution. CIDS, using visible or near ultraviolet light, is most sensitive to order in the range of 50Å to 5000Å, whereas circular dichroism, absorption and nuclear magnetic resonance are more sensitive to interactions below 50Å. On the other hand, for wavelengths of light within the absorption band of the scatterer, the CIDS and the total scattering present anomalous behavior.[78] In the particular case of helical scatterers, the patterns present well defined symmetry properties in space which can be used to determine the handedness of the structure.

In order to be able to correlate the existing theory of CIDS with the data obtained experimentally, it is desirable to extend the existing theory to include the case of scatterers free to adopt all possible orientations in space, i.e., for a scatterer in solution. To do this it is necessary to carry out the corresponding spatial averages of the total and differential scattered intensities. This task will be accomplished in the next section, where the geometry of the scatterer will be kept completely general throughout the derivation.

Theory for a Randomly Oriented Sample[79]

The scatterer can adopt any random orientation in space; it is made up of an arbitrary number of polarizabilities arranged in an arbitrary geometry. Since the measured signal at the photomultiplier for a tumbling sample is $\langle I_L - I_R\rangle_{av}/\langle I_L + I_R\rangle_{av}$, our goal will be to obtain the corresponding average for the numerator and the denominator. The total scattering intensity for a given sample is the same regardless of the two independent, orthogonal states of polarizations used to describe the incident radiation. Therefore, instead of deriving the averaged total scattered intensity for right and left circularly polarized incident radiation (i.e., $I_L + I_R$), we will derive $(I_\parallel + I_\perp)$, where $I_\parallel$ and $I_\perp$ are the scattered intensities for incident light polarized parallel and perpendicular to the scattering plane, respectively. There are two reasons for this choice: first, the derivations of $I_L + I_R$ are more involved than those of $I_\parallel + I_\perp$, and second, choosing these polarization states, it is easy to calculate also $\langle I_\parallel - I_\perp\rangle_{av}$ from which the quantity $\langle I_\parallel - I_\perp\rangle_{av}/\langle I_\parallel + I_\perp\rangle_{av}$ can be obtained.

The field scattered by a set of polarizable groups in space, for a given incident field E_0, can be written as:

$$\underset{\sim}{E}(r') = Ce^{-ikr'}(1-\underset{\sim}{\hat{k}}\underset{\sim}{\hat{k}})\cdot\sum_j e^{i\Delta \underset{\sim}{k}\cdot \underset{\sim}{r}_j}\, \underset{\sim}{\alpha}_j\cdot \underset{\sim}{E}_0$$

where $\underset{\sim}{r}_j$ is the position vector of group j defined in a suitable molecular frame, $\underset{\sim}{\alpha}_j$ is the polarizability tensor of the jth-group, C is a constant of proportionality containing some inverse distance factors, $\Delta\underset{\sim}{k} = \underset{\sim}{k} - \underset{\sim}{k}_0$, i.e., the momentum transfer vector and $\underset{\sim}{k}$ and $\underset{\sim}{k}_0$ are the wave-vectors of the scattered and incident radiation, respectively. $\hat{\underset{\sim}{k}}$ is a unit vector in the direction of the scattered light.

Now we define a molecule-fixed coordinate system, whose orthogonal unit vectors are i', j', and k'. The polarizability tensor of the j-group can now be written as

$$\underset{\sim}{\alpha}_j = \alpha_j \underset{\sim}{t}_j \underset{\sim}{t}_j$$

where the subscript j labels the group. The $\underset{\sim}{t}$'s are unit vectors in the molecular coordinate system whose three components along this frame determine uniquely the axes of the polarizability of group j. Furthermore, we require that the $\underset{\sim}{k}'$ axis of the molecular frame is oriented along the distance-vector $\underset{\sim}{R}_{ij}$ between groups i^{th} and j^{th}, with $R_{ij} = |\underset{\sim}{r}_i - \underset{\sim}{r}_j|$.

We define also a space-fixed coordinate system whose orthogonal unit vectors are labeled $\underset{\sim}{a}$, $\underset{\sim}{b}$, $\underset{\sim}{c}$. This frame is oriented so that one of its axes (in this case the c-axis) is along the momentum transfer vector of the light, $\Delta\underset{\sim}{k} \equiv \underset{\sim}{k} - \underset{\sim}{k}_0$. Additionally, the frame is rotated around this axis so that $\underset{\sim}{k}$ and $\underset{\sim}{k}_0$ are in the $\underset{\sim}{c}$, $\underset{\sim}{b}$ plane (see Figure 4). From Figure 4 then:

$$\Delta\underset{\sim}{k} = (4\pi/\lambda)\sin\beta\ \underset{\sim}{c}$$

$$\hat{\underset{\sim}{k}} = \cos\beta\ \underset{\sim}{b} + \sin\beta\ \underset{\sim}{c}$$

$$\hat{\underset{\sim}{k}}_0 = \cos\beta\ \underset{\sim}{b} - \sin\beta\ \underset{\sim}{c}$$

where "^" indicates that these vectors are unit vectors along the incident and scattered directions of propagation of the light. From Figure 4, the right and left circular polarization unit vectors of the incident light are:

$$E^0_{L,R} = \underset{\sim}{A} \pm i\underset{\sim}{B} = \frac{1}{\sqrt{2}}[\underset{\sim}{a} \pm i(\sin\beta\ \underset{\sim}{b} + \cos\beta\ \underset{\sim}{c})]$$

Using the last expression, the scattered electric field for the opposite circular polarizations of the incident light can be obtained, from which the differential scattered intensity for left and right circularly polarized light takes the form:

$$I_L - I_R = 2iC^2 \sum_{ij}\sum e^{i\Delta\underset{\sim}{k}\cdot(\underset{\sim}{r}_j - \underset{\sim}{r}_i)} \alpha_j \alpha_i^* [(\underset{\sim}{A} \times \underset{\sim}{B})\cdot(\underset{\sim}{t}_j \times \underset{\sim}{t}_i)][(\underset{\sim}{t}_j\cdot\underset{\sim}{t}_i) - (\underset{\sim}{t}_j\cdot\hat{\underset{\sim}{k}})(\underset{\sim}{t}_i\cdot\hat{\underset{\sim}{k}})]$$

Our task is to perform the spatial averaging of the above equation. The space integrations involved in the averaging process can become very difficult due to the term $\exp[i\Delta\underset{\sim}{k}\cdot(\underset{\sim}{r}_j - \underset{\sim}{r}_i)]$ comprising the product of a space-fixed vector $\Delta\underset{\sim}{k}$ and a molecule-fixed vector $(\underset{\sim}{r}_j - \underset{\sim}{r}_i = \underset{\sim}{R}_{ij})$. Our choice of orienting the space-fixed coordinate system with one of its axes along $\Delta\underset{\sim}{k}$ and the molecule-fixed frame with one of its axes along $\underset{\sim}{R}_{ij}$ can now be understood. Indeed, with this choice, the product $\underset{\sim}{R}_{ij}\cdot\Delta\underset{\sim}{k}$ is a constant for any orthogonal transformation between the two coordinate systems, with the exception of the transformation involving the angle between the polar axes ($\underset{\sim}{k}'$ and $\underset{\sim}{c}$) of

$$\underset{\sim}{c} = \Delta\hat{k} = \frac{\underset{\sim}{k} - \underset{\sim}{k}_o}{|\underset{\sim}{k} - \underset{\sim}{k}_o|}$$

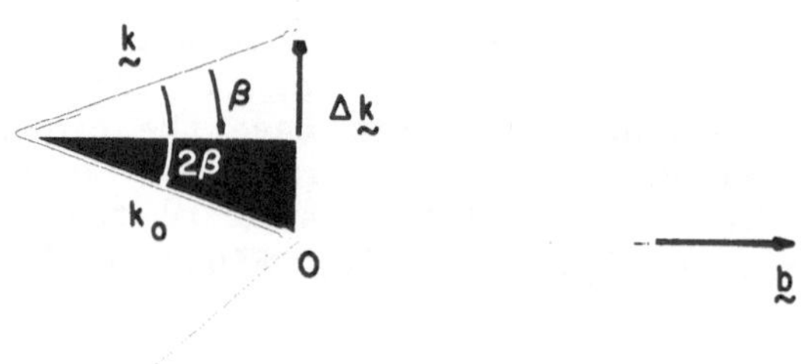

Fig. 4. The space-fixed frame with basis vectors $\underset{\sim}{a}$, $\underset{\sim}{b}$, $\underset{\sim}{c}$ is oriented so that $\Delta k = k - k_0$ is along the direction of the basis vector $\underset{\sim}{c}$. The wavevectors of light (k, k_0) are in the plane ($\underset{\sim}{b}$, $\underset{\sim}{c}$) that coincides with the scattering plane. The polarization vectors of the incident light are defined parallel and perpendicular to this plane.

these cartesian systems. In this way the factor $\exp(i\Delta k \cdot R_{ij})$ can be dealt with in the last step of the spatial integrations evaluated in the averaging process.

The transformation matrix between these frames, often called the Euler-matrix, allows us to express any product between space-fixed and molecule-fixed basis-vectors, in terms of the three Euler angles, θ, ϕ, χ. The spatial averaging involves the integration over the three Euler angles of the quantity to be averaged. Indeed, the averaging of any function $f(\theta,\chi,\phi)$ can be accomplished by:

$$\langle f(\theta,\chi,\phi)\rangle_{av} = \int_0^{\pi}\int_0^{2\pi}\int_0^{2\pi} f(\theta,\chi,\phi)\sin\theta d\theta d\chi d\phi \Big/ \int_0^{\pi}\int_0^{2\pi}\int_0^{2\pi} \sin\theta d\theta d\chi d\phi$$

Using the Euler-matrix the averaging can be done to obtain the results in terms of spherical Bessel functions,[80] $j_0(q)$, $j_1(q)$ and $j_2(q)$ and β, one half the scattering angle.

$$j_1(q) = \frac{\sin q}{q^2} - \frac{\cos q}{q^2}$$

$$j_2(q) = \left(\frac{3}{q^3} - \frac{1}{q}\right)\sin q - 3\,\frac{\cos q}{q}$$

$$q = \frac{4\pi}{\lambda} R_{ij} \sin\beta$$

The result for the averaged $I_L - I_R$ is:

$$\frac{\langle I_L - I_R \rangle_{av}}{C^2} = \sum_{ij}\sum \frac{\alpha_i^* \alpha_j (\underset{\sim}{t}_j \times \underset{\sim}{t}_i)\cdot \hat{R}_{ij}}{2} \{ [(\underset{\sim}{t}_i \cdot \underset{\sim}{t}_j)(j_2/q - j_1)$$

$$-(\underset{\sim}{t}_i \cdot \hat{R}_{ij})(\underset{\sim}{t}_j \cdot \hat{R}_{ij})(5j_2/q - j_1)](\sin\beta + \sin^3\beta)\}$$

Notice that the whole expression is multiplied by a common factor:

$$(\underset{\sim}{t}_j \times \underset{\sim}{t}_i)\cdot \underset{\sim}{R}_{ij} = (\underset{\sim}{t}_i \times \underset{\sim}{R}_{ij})\cdot \underset{\sim}{t}_j$$

This is a form factor that resembles the expression for the rotational strength in optical activity theory:[81] $\underset{\sim}{t}_i \times \underset{\sim}{R}_{ij}$ is related to the magnetic dipole (m_i) associated with the transition dipole $\underset{\sim}{t}_i$ so that:

$$(\underset{\sim}{t}_j \times \underset{\sim}{t}_i)\cdot \hat{\underset{\sim}{R}}_{ij} \sim \underset{\sim}{m}_i \cdot \underset{\sim}{t}_j$$

The last expression is the product of a pseudo-vector with a vector. Pseudo-vectors or axial vectors do not change sign when an inversion of their coordinates is done, whereas pure or polar vectors (such as $\underset{\sim}{t}_j$) do change sign under inversion. The equation will therefore have opposite signs for two molecules that are mirror images of each other. This feature makes CIDS more sensitive to structure than total scattering. The equation is only a function of angle β, while every other directional property of the incident and the scattered radiation has disappeared in the averaging. The differential scattering pattern in space is given by a ring structure of constant intensities, like the powder patterns observed in crystallography.[81] A careful limit analysis can be performed to show that

$$\lim_{\substack{q \to 0 \\ \beta \to 0}} \langle I_L - I_R \rangle_{av} = 0$$

i.e., there is no forward CIDS signal for a tumbling (spatially averaged) sample which can be represented by a sum of polarizabilities. The total scattering can be similarly obtained.

$$\frac{\langle I_\perp + I_\parallel \rangle}{c^2} = \frac{8}{15} \sum_i |\alpha_i|^2 (1 - \sin^2\beta \cos^2\beta) + \sum_{ij} \alpha_i \alpha_j^* \{(\underset{\sim}{t}_i \cdot \hat{\underset{\sim}{R}}_{ij})(\underset{\sim}{t}_j \cdot \hat{\underset{\sim}{R}}_{ij})$$

$$\{[(j_0 - j_1/q) + \sin^2\beta(3j_1/q - j_0)](\underset{\sim}{t}_i \cdot \underset{\sim}{t}_j) - 2(\underset{\sim}{t}_i \cdot \underset{\sim}{R}_{ij})(\underset{\sim}{t}_j \cdot \underset{\sim}{R}_{ij})$$

$$[(j_1/q - 4j_2/q^2) - \sin^2\beta\ j_2/q^2 + \sin^4\beta(j_1/q - 3j_2/q^2)]\} +$$

$$(\underset{\sim}{t}_i \times \hat{\underset{\sim}{R}}_{ij}) \cdot (\underset{\sim}{t}_j \times \hat{\underset{\sim}{R}}_{ij}) \{\tfrac{1}{2}[(j_0 + j_1/q) + \sin^2\beta(j_0 - 2j_1/q)](\underset{\sim}{t}_i \cdot \underset{\sim}{t}_j) -$$

$$\tfrac{1}{2}(\underset{\sim}{t}_i \cdot \hat{\underset{\sim}{R}}_{ij})(\underset{\sim}{t}_j \cdot \hat{\underset{\sim}{R}}_{ij})[(16j_2/q^2 - 3j_1/q + j_0) + \sin^2\beta(8j_2/q^2 -$$

$$3j_1/q + j_0) + 4\sin^4\beta(j_1/q - 3j_2/q^2)]\} - (1 - \tfrac{1}{2}((\hat{\underset{\sim}{R}}_{ij} \times \underset{\sim}{t}_i)^2 +$$

$$(\hat{\underset{\sim}{R}}_{ij} \times \underset{\sim}{t}_j)^2))[(5j_2/q^2 - j_1/q) + \sin^2\beta(j_0 - j_1/q) - \sin^4\beta(j_0 -$$

$$3j_2/q^2)] - ((\underset{\sim}{t}_i \times \underset{\sim}{R}_{ij}) \cdot (\underset{\sim}{t}_j \times \underset{\sim}{R}_{ij}))^2 [(1/8)(-9j_2/q^2 + 2j_1/q + j_0) -$$

$$(1/4)\sin^2\beta(3j_2/q^2 - 4j_1/q - j_0) - (1/8)\sin^4\beta(9j_2/q^2 + 2j_1/q +$$

$$3j_0)] - (\underset{\sim}{t}_j \cdot (\hat{\underset{\sim}{R}}_{ij} \times \underset{\sim}{t}_i))^2 [(1/8)(5j_2/q^2 + 2j_1/q - j_0) + \tfrac{1}{4}\sin^2\beta \cdot$$

$$(3j_2/q^2 + 2j_1/q + j_0) - (1/8)\sin^4\beta(3j_2/q^2 - j_1/q - j_0)]\}$$

This equation gives the total scattering as a function of the scattering angle (2β) for a collection of particles (each described by a set of point polarizable groups) adopting all possible orientations in space. It can be shown that the expressions for both $\langle I_L - I_R \rangle$ and $\langle I_{\parallel} + I_{\perp} \rangle = \langle I_L + I_R \rangle$ are even in β; this means that the scattering patterns have cylindrical symmetry around the direction of incidence of the light--as expected.

Applications

We have done calculations of the average CIDS for a helix made up of point polarizabilities with each polarizability tangent to the helix at each point. The results are seen in Figure 5. Notice that whereas the total scattering intensities change very little when the parameters of the helix are changed with respect to the wavelength of light, the CID scattering patterns show dramatic changes, making CIDS a much more sensitive technique than regular scattering to determine the geometry of the scatterers. Remarkably enough, even for ratios of radius- to - wavelength of 0.1, it is still possible to have at least one zero in the scattering pattern. The value of CIDS in the cases presented ranged from 10^{-5}-10^{-3} and therefore the CIDS of such particles could be, in principle, measured experimentally. It should be noticed that since the average CIDS equations only depend on the absolute value of the angle β, equal to one-half the scattering angle, the CIDS scattering patterns of Figure 5 have a C_∞-axis in the direction of the incident beam (180°-0°). From these calculations, it has become clear that the measurement and interpretation of the CIDS of macromolecular aggregates in solution is both feasible and necessary to fully exploit the power of the technique. A very preliminary experimental measurement of the CIDS of the bacteriophage T4 in aqueous solution is shown in Figure 6. The scattering pattern shows an encouraging amount of structure and a definite zero at about 140° from the incident beam. The magnitudes are of order 10^{-3}. Similar measurements on T5 show a very different pattern with a zero in the forward direction. These results reinforce the idea that CIDS will give useful, new information about structures of large chiral particles in solution.

CONCLUSION

We have quickly reviewed many chiroptical methods, and briefly described the theories and a few applications. To learn more about the theories and the wide range of other applications which we could not discuss, the reader is referred to the recent reviewers cited in the text.[8,10,27-29]

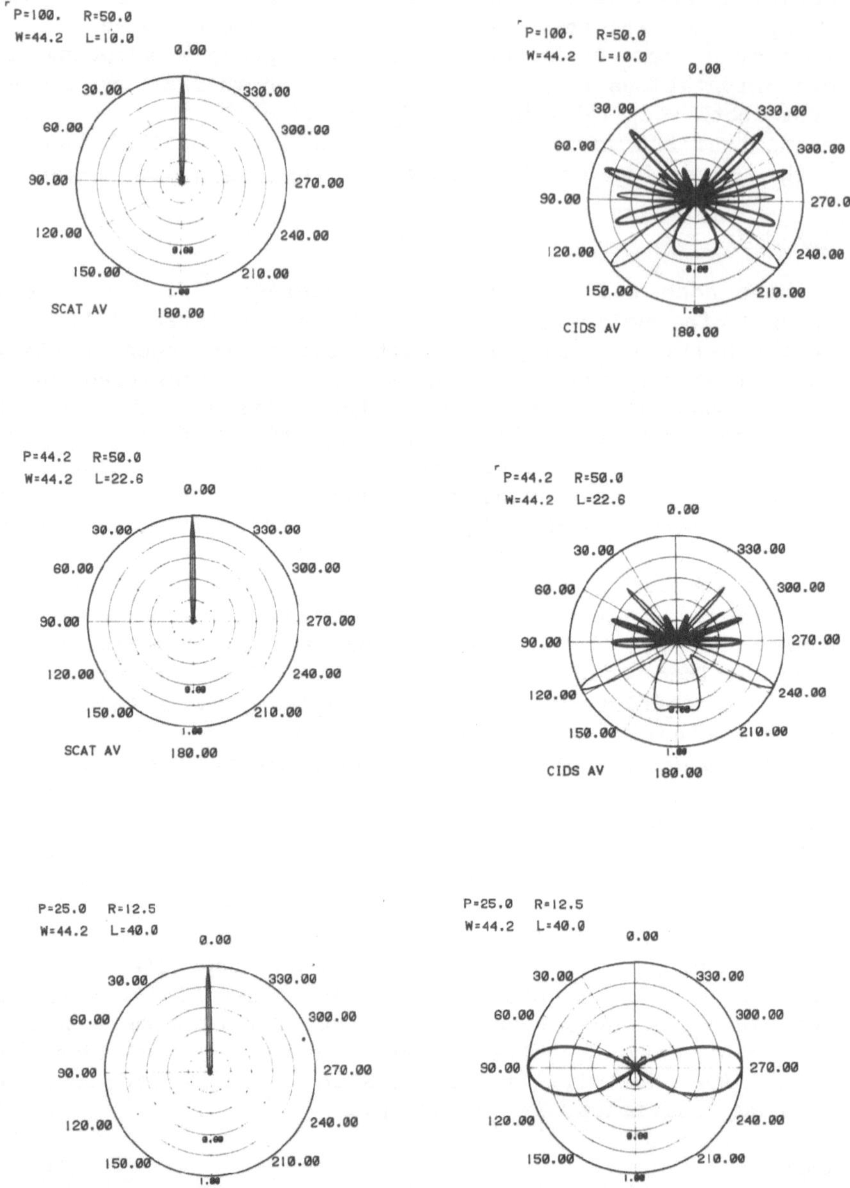

Fig. 5. Calculated CIDS and total scattering for randomly oriented helices for various values of pitch and radius. The pitch is given in units of radius; P = 1.6R or P = 0.5R. The light is incident from the bottom of the figure. The light lines mean negative CIDS; the dark lines mean positive CIDS.

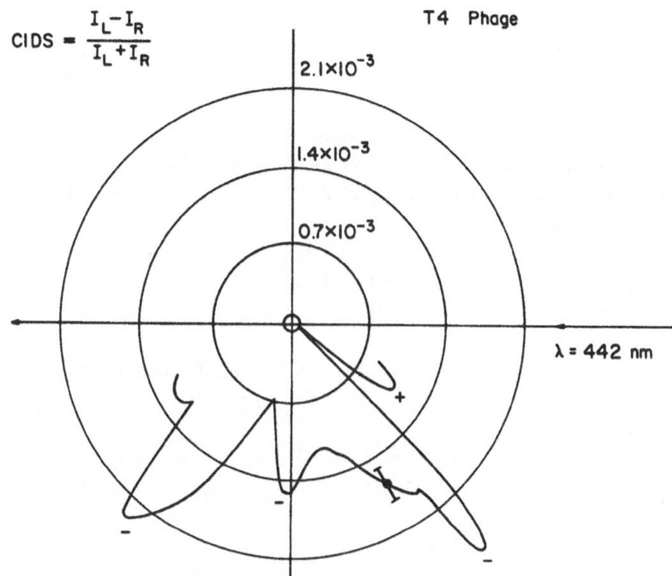

Fig. 6. Approximate measurement of the CIDS of bacteriophage T4 in aqueous solution. The light source was a 40 mwatt He-Cd laser (λ = 442 nm).

REFERENCES

1. A. Rosencwaig, Photoacoustic Spectroscopy, Ann. Rev. Biophys. Bioeng. 9:31 (1980).
2. L. Velluz, M. Legrand, and M. Grosjean,"Optical Circular Dichroism," Academic Press, Inc., New York (1965).
3. L. N. M. Duysens, The Flattening of the Absorption Spectrum of Suspensions, As Compared to That of Solutions, Biochim. Biophys. Acta 19:1 (1956).
4. D. W. Urry, L. Masotti, and J. Krivacic, Improved Ellipticity Data for Several Biological Membranes, Biochem. Biophys. Res. Comm. 41:521 (1970).
5. A. S. Schneider and D. Harmatz, An Experimental Method Correcting for Absorption Flattening and Scattering in Suspensions of Absorbing particles: CD and Absorption Spectra of Hemoglobin in situ in Red Blood Cells, Biochemistry 15:4158 (1976).
6. O. Schnepp, Natural and Magnetic Circular Dichroism Spectroscopy in the Vacuum Ultraviolet, in: "Optical Activity and Chiral Discrimination," S. F. Mason, ed., D. Reidel, London (1979) pp. 87-106.
7. P. J. Stephens and R. Clark, Vibrational Circular Dichroism: The Experimental Viepoint, in: "Optical Activity and Chiral Discrimination," S. F. Mason, ed., D. Reidel, London (1979) pp. 263-287.
8. A. E. Hansen and T. D. Bowman, Natural Chiroptical Spectroscopy: Theory and Computations, Advan. Chem. Phys. 44:545 (1980).
9. I. Tobias, T. R. Brocki, and N. L. Balazs, The Optical Activity of Molecules of Arbitrary Size, J. Chem. Phys. 62:4181 (1975).
10. R. W. Woody, Polypeptides and Proteins; Nucleic Acids, in: "ORD and CD: Theory, Chemical Practices and Biochemical Applications," F. Ciardelli, S. F. Mason, P. Salvadori, and G. Snatzke, eds., John Wiley & Sons, London (1981).
11. R. W. Woody and I. Tinoco, Jr., Optical Rotation of Oriented Helices. III. Calculation of the Rotatory Dispersion and Circular Dichroism of the Alpha and 3_{10} Helix, J. Chem. Phys. 46:4927 (1967).
12. V. Madison and J. Schellman, Optical Properties of Polypeptides and Proteins, Biopolymers 11:1041-1076 (1972).
13. R. W. Woody, Optical Properties of Polypeptides in the β-Conformation, Biopolymers 8:669 (1969).
14. R. W. Woody, Studies of the Theoretical Circular Dichroism of Polypeptides: Contributions of β Turns, in: "Peptides, Polypeptides and Proteins," E. R. Blout, F. A. Bovey, M. Goodman, and N. Lotan, eds., Wiley-Interscience, New York (1974) pp. 338-350.

15. W. Hug and I. Tinoco, Jr., Electronic Spectra of Nucleic Acid Bases. I. Interpretation of the In-Plane Spectra with the Aid of All Valence Electron MO-CI Calculations, J. Am. Chem. Soc. 95:2803 (1973).
16. W. Hug and I. Tinoco, Jr., The Electronic Spectra of Nucleic Acid Bases. II. Out-of-Plane Transitions and the Structure of the Nonbonding Orbitals, J. Am. Chem. Soc. 96:665 (1974).
17. W. C. Johnson, Jr. and I. Tinoco, Jr., Circular Dichroism of Polynucleotides: A General Method Applied to Dimers, Biopolymers 8:715 (1969).
18. C. L. Cech and I. Tinoco, Jr., Circular Dichroism Calculations for Double-Stranded Polynucleotides of Repeating Sequences, Biopolymers 16:43 (1977).
19. H. DeVoe, Optical Properties of Molecular Aggregates. II. Classical Theory of the Refraction, Absorption, and Optical Activity of Solutions and Crystals, J. Chem. Phys. 43:3199 (1965).
20. C. L. Cech, W. Hug, and I. Tinoco, Jr., Polynucleotide Circular Dichroism Calculations: Use of an All-Order Classical Coupled Oscillator Polarizability Theory, Biopolymers 15:131 (1976).
21. G. Holzwarth and I. Chabay, Optical Activity of Vibrational Transitions: A Coupled Oscillator Model, J. Chem. Phys. 57:1632 (1972).
22. J. A. Schellman, Vibrational Optical Activity, J. Chem. Phys. 58:2882 (1973).
23. T. R. Faulkner, C. Marcott, A. Moscowitz, and J. Overend, Anharmonic Effects in Vibrational Circular Dichroism, J. Am. Chem. Soc. 99:8160-8168 (1977).
24. L. A. Nafie and M. Diem, Optical Activity in Vibrational Transitions: Vibrational Circular Dichroism and Raman Optical Activity, Accts. Chem. Res. 12:296 (1979).
25. W. M. McClain, Two-Photon Molecular Spectroscopy, Accts. Chem. Res. 7:129 (1974).
26. I. Tinoco, Jr., Two-Photon Circular Dichroism, J. Chem. Phys. 62:1006 (1975).
27. R. W. Woody, Optical Rotatory Properties of Biopolymers, J. Polymer Sci. Macromol. Rev. 12:181 (1977).
28. W. C. Johnson, Jr., Circular Dichroism Spectroscopy and the Vacuum Ultraviolet Region, Ann. Rev. Phys. Chem. 29:93 (1978).
29. I. Tinoco, Jr., C. Bustamante, and M. F. Maestre, The Optical Activity of Nucleic Acids and Their Aggregates, Ann. Rev. Biophys. Bioeng. 9:107 (1980).
30. N. Greenfield and G. D. Fasman, Computed Circular Dichroism Spectra for the Evaluation of Protein Conformation, Biochemistry 8:4108 (1969).
31. Y.-H. Chen, J. T. Yang, and H. M. Martinez, Determination of the Secondary Structures of Proteins by CD and ORD, Biochemistry 11:4120 (1972).

32. Y.-H. Chen, J. T. Yang, and K. H. Chou, Determination of the Helix and β Form of Proteins in Aqueous Solutions by CD, Biochemistry 13:3350 (1974).
33. C. T. Chang, C. S. C. Wu, and J. T. Yang, Circular Dichroic Analysis of Protein Conformation Inclusion of the β-turns, Anal. Biochem. 91:13 (1978).
34. J. B. Siegel, W. E. Steinmetz, and G. L. Long, A Computer-Assisted Model for Estimating Protein Secondary Structure from Circular Dichroic Spectra: Comparison of Animal Lactate Dehydrogenases, Anal. Biochem. 104:160 (1980).
35. S. Brahms and J. Brahms, Determination of Protein Secondary Structure in Solution by Vacuum Ultraviolet Circular Dichroism, J. Mol. Biol. 138:149 (1980).
36. J. P. Hennessey, Jr. and W. C. Johnson, Jr., Information Content in the Circular Dichroism of Proteins, Biochemistry 20:1085 (1981).
37. M. Levitt and J. Greer, Automatic Identification of Secondary Structure in Globular Proteins, J. Mol. Biol. 114:181 (1977).
38. C. A. Sprecher, W. A. Baase, and W. C. Johnson, Jr., Conformation and Circular Dichroism of DNA, Biopolymers 18:1009 (1979).
39. F. S. Allen, Donald M. Gray, G. P. Roberts, and I. Tinoco, Jr., The Ultraviolet Circular Dichroism of Some Natural DNAs and an Analysis of the Spectra for Sequence Information, Biopolymers 11:853 (1972).
40. D. M. Gray, F. D. Hamilton and M. R. Vaughan, The Analysis of Circular Dichroism Spectra of Natural DNAs Using Synthetic Spectral Components from Synthetic DNAs, Biopolymers 17:85 (1978).
41. D. M. Gray, J.-J. Liu, R. L. Ratliff, and F. S. Allen, Sequence-Dependence of the Circular Dichroism of Synthetic Double-Stranded RNAs, Biopolymers, in press (1981).
42. A. Blum, O. C. Uhlenbeck, and I. Tinoco, Jr., Circular Dichroism Study of Nine Species of Transfer Ribonucleic Acid, Biochemistry 11:3248 (1972).
43. B. B. Johnson, K. S. Dahl, I. Tinoco, Jr., V. I. Ivanov, and V. B. Zhurkin, Correlations Between DNA Structural Parameters and Calculated Circular Dichroism Spectra, Biochemistry 20:73 (1981).
44. V. B. Zhurkin, Y. P. Lysov, and V. I. Ivanov, Different Families of Double-Stranded Conformations of DNA as Revealed by Computer Calculations, Biopolymers 17:377 (1978).
45. E. S. Stephens, Optical Activity in the Vacuum Ultraviolet, Ann. Rev. Biophys. Bioeng. 5:53 (1976).
46. C. Marcott, H. A. Havel, B. Hedlund, J. Overend, and A. Moscowitz, A Vibrational Rotational Strength of Extraordinary Intensity. Azidomethemoglobin A, in: "Optical Activity and Chiral Discrimination," S. F. Mason, ed., D. Reidel, London (1979) pp. 289-292.

47. J. Y. Cassim and J. T. Yang, Critical Comparison of Experimental Optical Activity of Helical Polypeptides and Predictions of the Molecular Exciton Model, Biopolymers 9: 1475 (1970).
48. W. C. Johnson, Jr. and I. Tinoco, Jr., Circular Dichroism of Polypeptide Solutions in the Vacuum Ultraviolet, J. Am. Chem. Soc. 94:4389 (1972).
49. I. Tinoco, Jr. and D. H. Turner, Fluorescence Detected Circular Dichroism. Theory, J. Am. Chem. Soc. 98:6453 (1976).
50. D. H. Turner, I. Tinoco, Jr., and M. Maestre, Fluorescence Detected Circular Dichroism, J. Am. Chem. Soc. 96:4340 (1974).
51. C. Reich and I. Tinoco, Jr., Fluorescence Detected Circular Dichroism of Dinucleoside Phosphates, A Study of Solution Conformations and the Two-State Model, Biopolymers 19:833 (1980).
52. I. Tinoco, Jr., The Optical Rotation of Oriented Helices. I. Electrical Orientation of Poly-γ-Benzyl-L-Glutamate in Ethylene Dichloride, J. Am. Chem. Soc. 81:1540 (1959).
53. S. Y. Chung and G. Holzwarth, Circular Dichroism of Flow-Oriented Nucleic Acids. I. Experimental Results, J. Mol. Biol. 92:449 (1975).
54. I. Tinoco, Jr., B. Ehrenberg, and I. Z. Steinberg, Fluorescence Detected Circular Dichroism and Circular Polarization of Luminescence in Rigid Media: Direction Dependent Optical Activity Obtained by Photoselection, J. Chem. Phys. 66:916 (1977).
55. D. H. Turner, Fluorescence-Detected Circular Dichroism, Methods in Enzymology 49:199 (1978).
56. E. W. Lobenstine and D. H. Turner, Photoselected Fluorescence Detected Circular Dichroism, J. Am. Chem. Soc. 101:2205 (1979).
57. E. W. Lobenstine and D. H. Turner, Further Verification of Fluorescence Detected Circular Dichroism, J. Am. Chem. Soc. 102:7786 (1980).
58. E. W. Lobenstine, W. C. Schaefer, and D. H. Turner, Fluorescence Detected Circular Dichroism of Proteins with Single Tryptophans, J. Am. Chem. Soc., in press (1981).
59. C. Reich, M. F. Maestre, S. Edmondson, and D. M. Gray, Circular Dichroism and Fluorescence-Detected Circular Dichroism of Deoxyribonucleic Acid and Poly[d(A-C)•d(G-T)] in Ethanolic Solutions: A New Method for Estimating Circular Intensity Differential Scattering, Biochemistry 19:5208 (1980).
60. M. F. Maestre and C. Reich, Contribution of Light Scattering to the Circular Dichroism of Deoxyribonucleic Acid Films, DNA-Polylysine Complexes, and DNA Particles in Ethanolic Buffers, Biochemistry 19:5214 (1980).

61. D. W. Urry and J. Krivacic, Differential Scatter of Left and Right Circularly Polarized Light by Optically Active Particulate Systems, Proc. Natl. Acad. Sci. USA 65:845 (1970).
62. B. P. Dorman and M. F. Maestre, Experimental Differential Light-Scattering Correction to the Circular Dichroism of Bacteriophage T2, Proc. Natl. Acad. Sci. USA 70:255 (1973).
63. B. P. Dorman, J. E. Hearst, and M. F. Maestre, UV Absorption and CD Measurements on Light Scattering Biological Specimens: Fluorescent Cell and Related Large-Angle Detection Techniques, Methods in Enzymology, 270:267 (1973).
64. A. S. Schneider, M.-J. T. Schneider, K. Rosenbeck, Optical Activity of Biological Membranes: Scattering Effects and Protein Conformation, Proc. Natl. Acad. Sci. USA 66:793 (1970).
65. M. F. Maestre, D. M. Gray, and R. B. Cook, Magnetic Circular Dichroism Study in Synthetic Polynucleotides, Bacteriophage Structure and DNA's, Biopolymers 10:2537 (1971).
66. I. Z. Steinberg, Circularly Polarized Luminescence, Methods in Enzymology 49:179 (1978).
67. I. Z. Steinberg, Circular Polarization of Luminescence: Biochemical and Biophysical Applications, Ann. Rev. Biophys. Bioeng. 7:113 (1978).
68. F. S. Richardson, Circular Polarization Differentials in the Luminescence of Chiral Systems, in: "Optical Activity and Chiral Discrimination," S. F. Mason, ed., D. Reidel, London (1979) pp. 189-218.
69. G. D. Fasman and M. K. Cowman, The All Nucleus Chromatin, Part B, H. Bush, ed., Academic Press, New York (1978) pp. 55-57.
70. G. D. Fasman, G. Schaffhauser, L. Goldsmith, and A. Adler, Conformational Changes Associated with f-1 Histone-Deoxyribonucleci Acid Complexes. Circular Dichroism Studies, Biochemistry 9:2814 (1970).
71. Y. A. Shin and G. L. Eichborn, Reversible Change in ψ Structure of DNA-Poly(Lys) Complexes Induced by Metal Binding, Biopolymers 16:225 (1977).
72. D. Carrol, Complexes of Polylysine with Polyuridylic Acid and Other Polynucleotides, Biochemistry 11:426 (1972).
73. P. W. Atkins and L. D. Barron, Raleigh Scattering of Polarized Photons by Molecules, Mol. Phys. 16:453 (1969).
74. L. D. Barron and A. D. Buckingham, Raleigh and Raman Scattering from Optically Active Molecules, Mol. Phys. 20:1111 (1971).
75. R. A. Harris and W. M. McClain, Toward a Theory of the Perrin Matrix for Light Scattering from Dilute Polymer Solutions, J. Chem. Phys. 67:265 (1977).
76. C. Bustamante, M. F. Maestre, and I. Tinoco, Jr., Circular Intensity Differential Scattering of Helical Structures. I. Theory, J. Chem. Phys. 73:4273 (1980).

77. C. Bustamante, M. F. Maestre, and I. Tinoco, Jr., Circular Intensity Differential Scattering of Helical Structures. II. Applications, J. Chem. Phys. 73:6046 (1980).
78. C. Bustamante, M. F. Maestre, and I. Tinoco, Jr., Circular Intensity Differential Scattering of Helical Structures. III. General Polarizability and Anomalous Scattering, J. Chem. Phys, in press (May 15, 1981).
79. C. Bustamante, Circular Intensity Differential Scattering of Helical Molecules, Ph.D. Thesis, University of California, Berkeley (1980).
80. G. N. Watson, "Theory of Bessel Functions," Cambridge University Press, London (1958).
81. I. Tinoco, Jr., Theoretical Aspects of Optical Activity, Part Two: Polymers, Advances in Chemical Physics 4:113 (1962).
82. D. Sherwood, "Crystals, X-rays and Proteins," Longman, New York (1976).

77. G. Fournet, M. A. Guinier, and J. [illegible], [illegible] scattering of [illegible] structures, [illegible] (1955).
78. A. Guinier, G. Fournet, C. B. Walker, and K. L. Yudowitch, Small-Angle Scattering of X-Rays, Wiley, New York (1955).
79. [illegible] (1963).
80. [illegible], Theory of [illegible], Cambridge University Press, London (1962).
81. [illegible] Polymers, [illegible] (1967).
82. [illegible] (1976).

CALORIMETRIC METHODS FOR STUDYING BIOMOLECULAR STRUCTURE AND ORGANIZATION

Theodor Ackermann

Institut fuer physikalische Chemie
Universitaet Freiburg
Albertstrasse 23a, D-7800 Freiburg (FRG)

INTRODUCTION

Biopolymers and their synthetic models have a very high degree of intramolecular ordering - a definite arrangement of groups and distribution of secondary bonds. The helical duplex structure[1] (Fig. 1) found for native deoxyribonucleic acid occurs in a whole class of polynucleotides of high helical content. Indeed, this kind of secondary structure, which is stabilized by base stacking and interbase hydrogen bonding, occupies the same central position in the polynucleotide field as does the alpha-helix[2] in the polypeptide field (Fig. 2). In dilute solutions biopolymers have been found to include all gradations of secondary structure, from the completely amorphous to the highly helical. Helical forms other than the duplex have been found to occur for biosynthetic polymers. In studies on polynucleotide systems, ultraviolet hypochroism can be used as a secondary criterion for the helical content of the dissolved polymer molecules; changes in optical density appear to parallel the onset

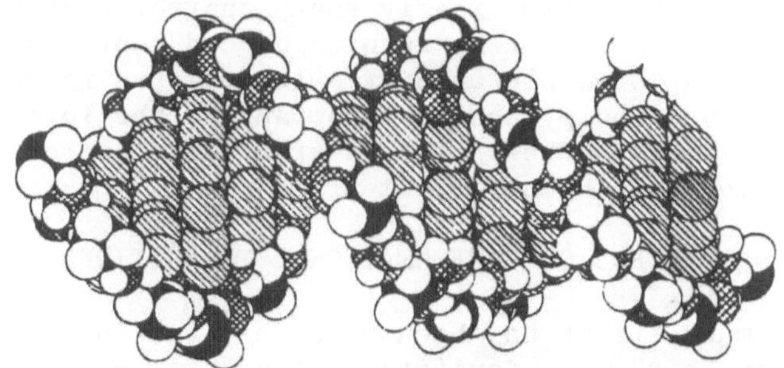

Fig. 1. Model of the helical duplex structure.

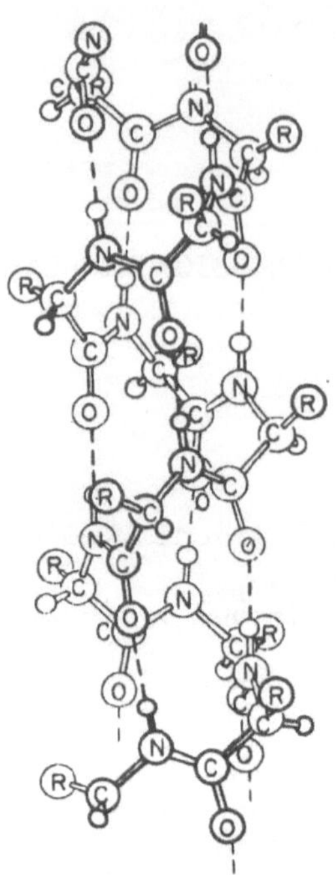

Fig. 2. A drawing showing a possible form of the alpha-helix. The amino acid residues have the L-configuration.

of a disruption of the secondary structure, as judged by other criteria. Most biopolymers and their synthetic models can undergo a transition from the completely ordered helical state to the entirely amorphous situation of a randomly coiled molecule. This helix-random coil transition can be mediated by a change in a parameter that influences the degree of intramolecular hydrogen bonding in the dissolved polymeric molecule. A characteristic feature of the helix-coil transition[3] for polypeptides and polynucleotides of high molecular weight is the dramatically sharp character it often assumes (Fig. 3). In many cases of temperature-induced conformational changes the process is completed over a temperature range of about 5°. In view of the low enthalpy change accompanying the formation of hydrogen bonds, the abruptness of this transition cannot be accounted for by any simple equilibrium treatment.

THERMODYNAMIC PARAMETERS OF CONFORMATIONAL TRANSITIONS

Recently, much effort has been made in determining so-called "stability parameters" for significant structural units of poly-nucleotides and polynucleotide-oligonucleotide complexes in solution. The thermal stability of the secondary structure depends on solute-

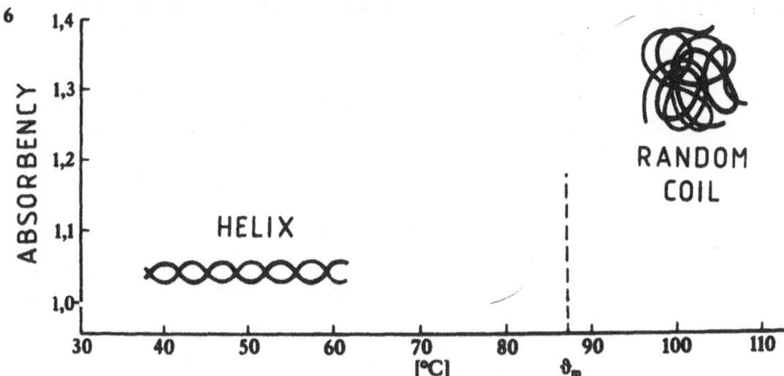

Fig. 3. Variation with temperature of the relative absorbency at 260nm for a helix-random coil transition[3].

solvent interactions as well as on the chain length of the polymer[4] and also on the specific type of a polymer-oligomer interaction (Fig. 4). As demonstrated by spectroscopic measurements, the so-called "melting point" of a dissolved polynucleotide-oligonucleotide complex is shifted to higher temperatures with increasing numbers of monomeric units in the oligomeric ligand molecule[5] (Fig. 5).

Thermodynamic parameters associated with conformational transitions in polypeptides and polynucleotides are of interest for three reasons at least: (1) as a quantitative means of studying

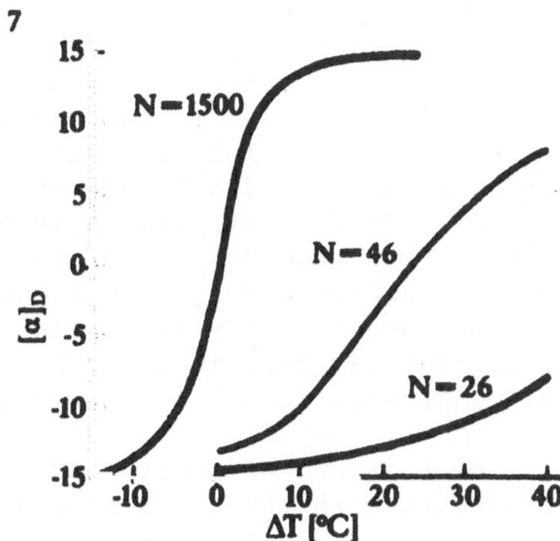

Fig. 4. Curves for fraction of helical residues vs. temperature for samples of poly-γ-benzyl-L-glutamate of various degrees of polymerization[4].

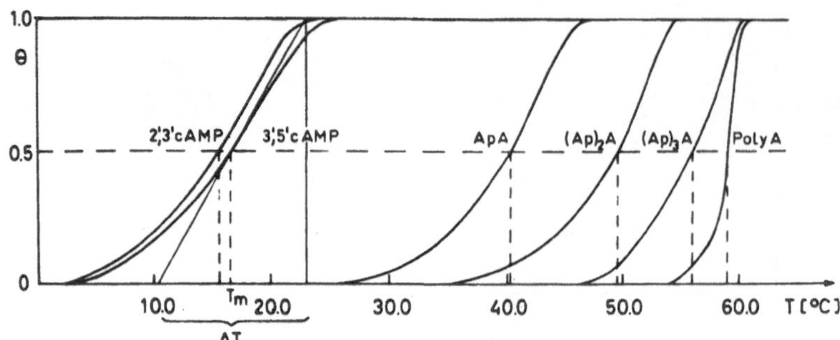

Fig. 5. Transition curves of the (A + 2 Poly U) systems obtained by plotting the molar extinction coefficient at $\tilde{\nu}$ = 1623cm^{-1} against the temperature. Total base concentration 13.4 mmol base/l. Conditions: Concentration of poly U 8.8mmol base/l 0.01 M tris-DCl, pD 7.9, 0.04 M $MgSO_4$ except for poly (A + 2U) where 0.04 M $MgSO_4$ was substituted by 0.1 M NaCl.

changes in molecular interactions that occur during the transitions, (2) as a method of deciding between postulated models and mechanisms, and (3) in providing data with which the validity of the various theoretical treatments of the helix-coil transitions may be tested. As a typical example, one may mention the new class of circular ribonucleic acids having a relatively low molecular weight. These "viroids" which are single-stranded, covalently linked, circular RNA molecules are pathogenic and infectious and have been found to be responsible for various plant diseases such as Cadang-Cadang or the potato spindle tuber disease. The characteristic secondary structure of the potato spindle tuber viroid (PTSV), which has the form of an "extended dumb-bell" was derived simply from an analysis of changes in thermodynamic properties accompanying the temperature-induced denaturation process in solution[6]. The thermal denaturation process proved to be highly cooperative with the half width of the melting curves between 1-3°C. This type of melting curve is characteristic for a "concerted" denaturation process involving many base pairs. It differs strikingly from the "melting" of tRNA which ranges over more than 20°C under identical conditions. In applying a one-helical-section model, it was possible to simulate the shapes of the melting curves fairly well using between 70-100 base pairs for the different viroids, in order to obtain agreement with the measured half widths between 1.4 and 0.9°C.

The presumed existence of one uninterrupted double helix leads to a major discrepancy between the theoretical and experimental results because T_m-values were calculated to be 110 or 125°C (for 50 or 75% GC, respectively), whereas a T_m-value of 76°C was determined experimentally for PTSV in 1M NaCl solution. In order to overcome this discrepancy Riesner and co-workers[6] assumed that the double

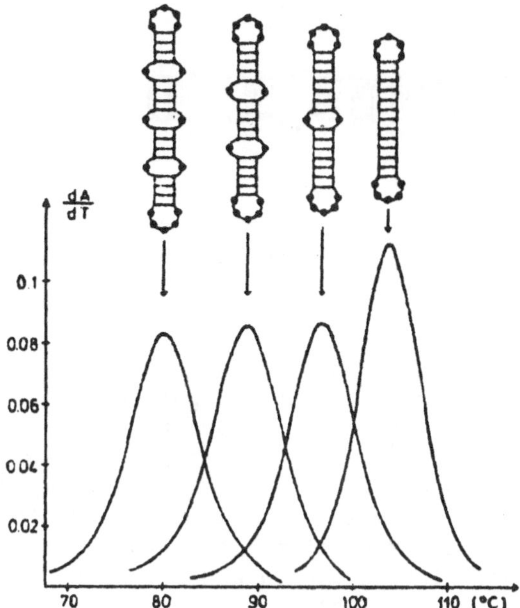

Fig. 6. Simulated melting curves of defective double helices as indicated in the figure. All molecules contain 12 nearest neighbour interactions with 50% G:C.

helix is not completely homogeneous but is interrupted by several internal loops or helix defects. The influence of such helix defects may be seen by calculations of the melting curves of molecules which are much smaller than viroids. The RNA molecules of Fig. 6 have 12 base pairs of the Watson-Crick type either in a homogenous helix or with 1,2, or 3 internal loops. Two properties are evident: The T_m-value is shifted to lower temperatures by each additional internal loop, whereas the cooperativity (see below) is lowered mainly by the first internal loop.

Using such calculations Riesner and co-workers[6] were able to find values of internal loop size and helix length in accordance with the high cooperativity and low T_m-values found on viroids. A schematic drawing of the "defective helix model" is shown in Fig. 7. This model has proved to be in fairly good agreement with the results of a careful determination of the nucleotide sequence of PTSV. This example shows clearly that a precise determination of the molar enthalpy changes associated with helix-random coil transitions and similar processes must be regarded as a requirement for understanding the physico-chemical behaviour of these polyelectrolyte solutions.

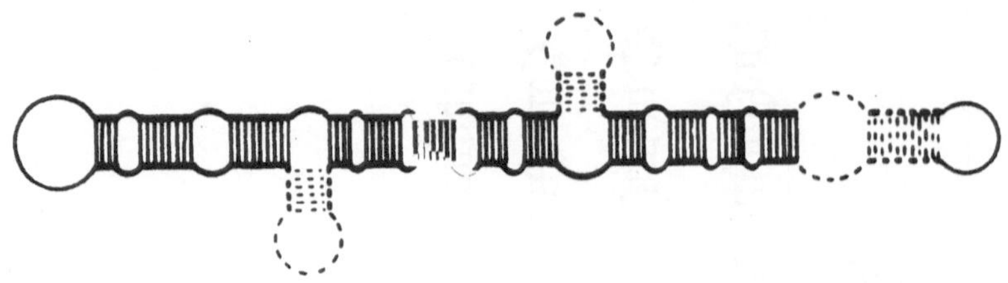

Fig. 7. Secondary structure of the potato spindle tuber viroid (PTSV)

In spite of the great importance of information of this type, calorimetric data on enthalpy changes of biopolymers were lacking for many years. Most of the stability constants were derived from the maximum slope of transition curves obtained in measurements of characteristic physical properties as a function of temperature[7]; the corresponding enthalpy changes derived from van't Hoff plots have a limited significance. Most of the thermal unfolding processes of biopolymers are cooperative phenomena and the slope of the transition curve is strongly influenced by the cooperative "length"[8], which represents the apparent number of interacting adjacent segments of the polymer chain. In order to obtain a value of the molar enthalpy change (i.e. the enthalpy change per monomeric subunit), the apparent enthalpy of transition must be divided by the cooperative length, which is an unknown property in most cases of biochemical interest. The helix-coil transition may be regarded as somewhat similar to a phase transition. Strictly speaking, however, the cooperative change in secondary structure is quite different from a first order phase transition. The imperfect character of the analogy is indicated by the finite breadth of the transition range contrasted with the discontinuous nature of a true phase transition as shown in Figure 8.

CALORIMETRIC METHODS AND THEIR APPLICATIONS

When the first international conference on thermodynamics was held in 1969, only a few papers dealing with reproducible results of exact calorimetric measurements on solutions of biopolymers and related model substances had been published. The pioneering work of Benzinger[9], Kitzinger and Steiner[10], Sturtevant[11], Privalov[12], Karasz and O'Reilly[13], Brandts[14], Mrevlishvili[15] and Rialdi, Biltonen[16] and their co-workers should be mentioned in this context. The general situation at that time may be illustrated by a passage quoted from a book, entitled "Biochemical Microcalorimetry"[17] and published in the same year:

"It can be said without undue optimism that calorimetry is becoming a standard laboratory method in physicochemical biology.

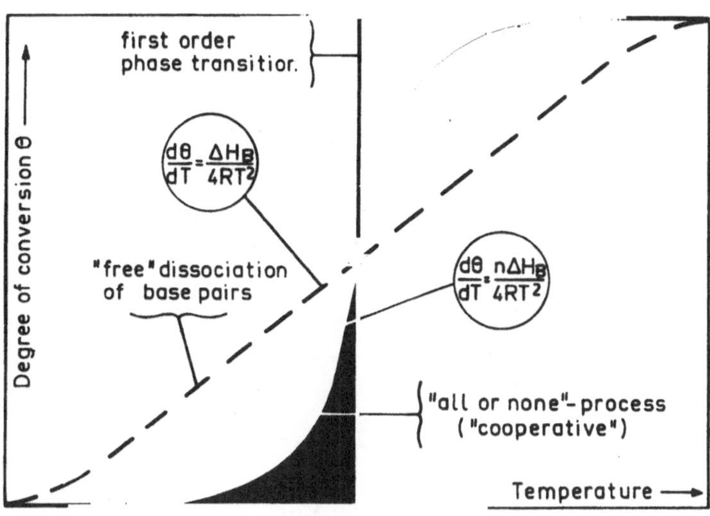

Fig. 8. Influence of cooperativity on the sharpness of transition (schematic).

There is, however, a serious problem that leaves us with the requirement for an improved calorimetric apparatus. In most cases the amount of biochemical samples is too small compared with the amount of sample required for a precise calorimetric measurement. Thus calorimeter designers must focus upon the construction of calorimeters giving a maximum of precision for a reduced amount of sample".

Most of these difficulties have been overcome by using improved experimental techniques with modern electronic equipment. Differential scanning calorimetry (DSC) is the appropriate method for the measurement of enthalpy changes associated with temperature-induced helix-random coil transitions and similar processes. The measurement of the heat capacity results in calorimetric transition curves that are characterized by a maximum value of the additional heat capacity in the temperature range of the thermal transition as shown schematically in Figure 9 for the helix formation of a synthetic polypeptide in a mixture of dichloroacetic acid and ethylene dichloride[18] Calorimetric measurements on model systems of this type have been carried out in our laboratory during the above mentioned period between 1963 and 1969[19]. Optical rotation is a convenient index of conversion, the change in rotation being roughly proportional to the fraction of residues that have undergone the transition. The temperature corresponding to the maximum value of the excess heat capacity is called T_c, originally defined as the midpoint of the thermal conversion in the polarimetric transition curve. The additional heat capacity is proportional to the first temperature

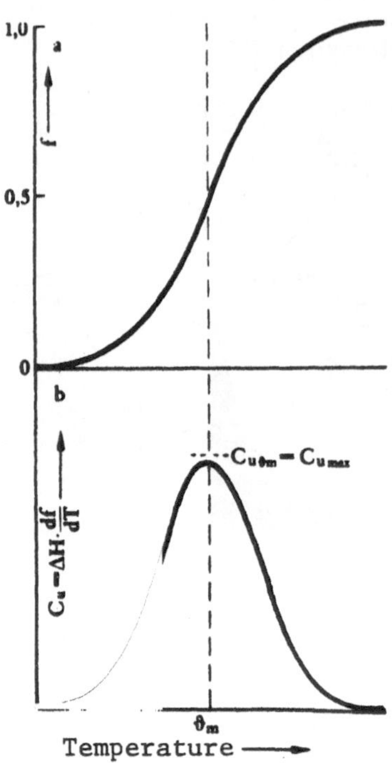

Fig. 9. Degree of helix formation (top) and additional conformational contribution to the heat capacity (bottom) vs. temperature.

derivative of the fractional helical content. The enthalpy of transition is directly proportional to the area under the calorimetric peak and the cooperative length can be calculated from the maximum value of the excess molar heat capacity for a given molar enthalpy change.

The experimental equipment for differential scanning calorimetry is now well established and DSC calorimeters for special purposes are available from various manufacturers. A schematic diagram of a calorimeter which has been developed in our laboratory is shown in Figure 10. The principle of the apparatus is similar to that of the Privalov calorimeter[20] and the figure is self-explanatory. Equivalence of the temperatures in the sample holder and in the reference cell is maintained throughout a run and the additional heating power, which is proportional to the conformational excess heat capacity, is recorded versus temperature. An example shown in

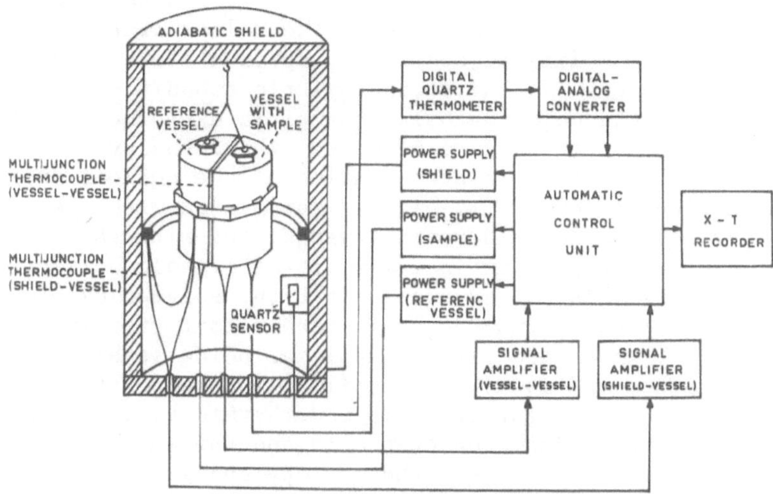

Fig. 10. Schematic drawing of an adiabatic twin-vessel calorimeter assembly.

Figure 11 of the recorder trace obtained from an equimolar mixture of salts of polyriboadenylic acid and polyribouridylic acid in aqueous solution indicates the low thermal lag-time and quick response of the apparatus. Large numbers of calorimetric transition curves for phospholipids, biopolymers and related model compounds have been redetermined using this instrument in the period between 1974 and 1979.

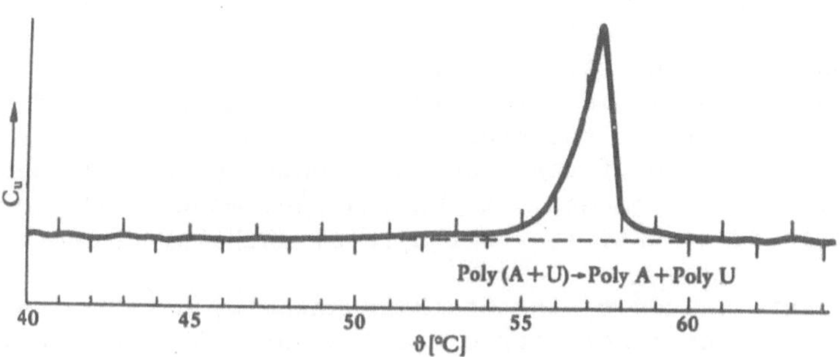

Fig. 11. DSC recorder trace as obtained in a measurement on a solution of an equimolar mixture of polyriboadenylic acid and polyribouridylic acid.

A simplified picture of the field as it presently exists is reflected in the table of characteristic values of transition enthalpies[18] for a limited number of representative processes as summarised in Table 1. The stoichiometric subunit corresponding to the appropriate molar enthalpy change is explained in the third column. It is interesting to note that the enthalpy change per mole base pair accompanying the denaturation of the helical duplex of a polynucleotide is remarkably high in comparison with the enthalpy change per mole amino acid residue associated with the helix-formation of a polypeptide in a mixture of organic solvents. The main reason for this difference is that the solvent dichloroacetic acid (DCA) is involved in the helix-random coil transition of poly-gamma-benzyl-L-glutamate with the polymer being stable as a helix at high temperature and as a random coil at low temperature. The temperature-induced helix-formation is endothermic because hydrogen bonds are formed between dichloroacetic acid and the amide groups in the randomly coiled polypeptide; the entropy change is positive because of the release of DCA molecules into the solvent on forming the helix. The negative enthalpy change resulting from formation of intramolecular hydrogen bonds is opposed by a positive contribution corresponding to disruption of the intermolecular hydrogen bonds between dichloroacetic acid and the amide groups of the polypeptide. An additional negative enthalpy change is due to dimerization of the released DCA molecules.

For aqueous salt solutions of polynucleotides the influence of interactions between solvent molecules and dissolved polymer is less important than for polypeptides. It has been generally assumed that the heat absorbed in the disruption of the hydrogen bonds of the helical duplex is compensated for by the heat produced during the formation of hydrogen bonds between water molecules and polar groups of the single-stranded polynucleotides. So far the calorimetric transition enthalpy has been regarded as a measure of the relative strength of the interactions between stacked bases in the helical duplex. There is a characteristic influence of the relative guanine-cytosine content on the calorimetrically determined enthalpies of denaturation[21] as shown in Figure 12. As there are three hydrogen bonds in the GC pair and two hydrogen bonds in the AT pair (Fig. 13) a significant increase of the transition enthalpies with increasing GC content is to be expected, if the rearrangement of the hydrogen bonds contributes essentially to the total enthalpy change.

Approximate values of the stacking enthalpy have been determined in a series of calorimetric measurements on selected model compounds[22]. Calorimetrists have been able to demonstrate that most of the published van't Hoff enthalpy data on base stacking (in particular, single stranded base-stacking) are in serious error. Actually, the "stacking effect" contributes only 75% to the total enthalpy change per base pair in the unfolding process of DNA in aqueous solution.

Table 1. Order of Magnitude of Characteristic ΔH Values for Different Types of Helix-Random Coil Transitions and Related Processes in Solution

Model Substance	Process	Molar Stoichiometric Subunit	ΔH (kcal/Mol)
Transfer-Ribonucleic Acid	Unfolding of Secondary Structure	tRNA-Molecule	250
Deoxyribonucleic acid	Double Helix-Coil Transition	Base pair	8
Polyriboadenylic acid	Unstacking of Adenine Bases	Nucleotide Residue	2.5
DNA and Ethidium Bromide	"Intercalation"	EB Molecule	5
Poly-γ-Benzyl-L-Glutamate in DCA/EDC Mixture	α-Helix-Coil Transition	Amino Acid Residue	1
(Pro, Pro, Gly)$_{15}$	Polyproline II-Type Helix-Coil Transition	Amino Acid Residue	0.5
Poly N^5-3 Hydroxypropyl-L-Glutamine	α-Helix-Coil Transition	Residue	0.1

DNAs from higher eukaryotes generally melt over a wide temperature range and are "polyphasic". It is interesting to note that some subtransitional detail is resolved in the calorimetric transition curve of calf thymus DNA, as shown in Fig. 14 top. Derivative

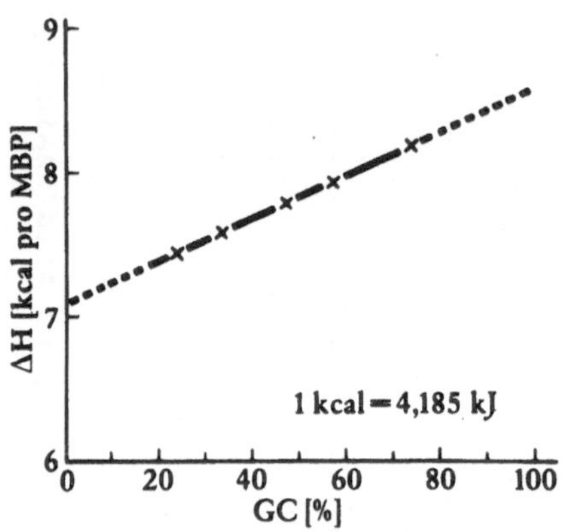

Fig. 12. Influence of GC content on the heat of denaturation of DNA.

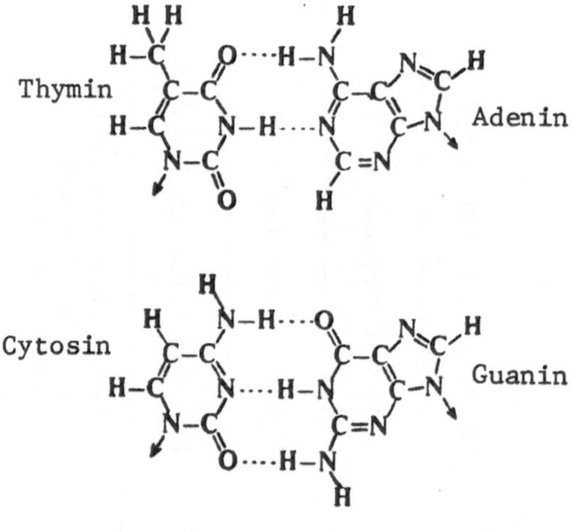

Thymin

Adenin

Cytosin

Guanin

Fig. 13. Hydrogen bonding in base pairs of the Watson-Crick type.

melting profiles of DNA have been obtained directly by recording the difference in absorbance between two identical solutions maintained at a small constant temperature differential[23]. If the DNA is sufficiently short, melting exhibits a fine structure arising from specific subtransitions and, under favourable conditions, may reflect the base composition over regions as small as 300 base pairs. We count at least three distinct peaks and two shoulders in the derivative melting profile of calf thymus DNA recorded at 260nm (Fig. 14 bottom) as well as in the calorimetric transition curve. These "bands" are not subtransitional processes. The bands seen in the

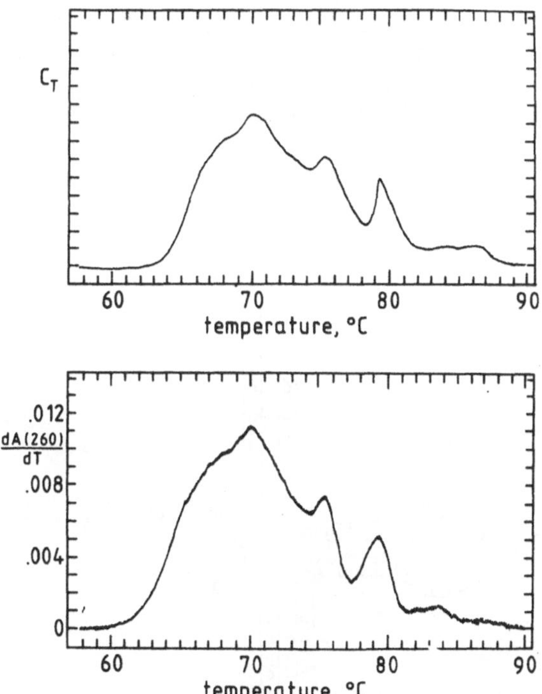

Fig. 14. Comparison of a DSC recorder trace (top) and a high resolution direct differential melting curve[23] (bottom) as obtained in measurements on calf-thymus DNA.

calf thymus DNA profile occur over 3-4°C, approximately ten times the breadth of individual subtransitions. Blake[23] and his co-workers suspect these bands represent the magnified effect of repetitive sequences superimposed on a broad background of unique sequences. One difficulty with this hypothesis is that melting profiles of DNAs from related eukaryotic species are quite different, while the overall base composition remains about the same. This raises interesting questions about the origin and preservation of such bands as putative repetitive sequences, since the range of base composition under any one band is obviously quite narrow. If they are repetitive sequences, they have either evolved quite abruptly and/or are replicated in some special, interdependent fashion so that evolutionary changes in base composition are preserved in all copies of a repeated sequence.

From the comparison of the two melting curves in Fig. 14 it may be concluded that "high resolution direct-derivative melting"[23] and high precision differential scanning calorimetry can be regarded as equivalent methods, at least under favourable conditions.

Modern calorimetric equipment can now be regarded as sufficiently sensitive for a significant discrimination of small differences in transition enthalpies for ribonucleic acids of high structural similarity. Recently, a semi-quantitative correlation between the number of hydrogen bonds in the cloverleaf structure of specific transfer ribonucleic acids and the calorimetrically measured enthalpy

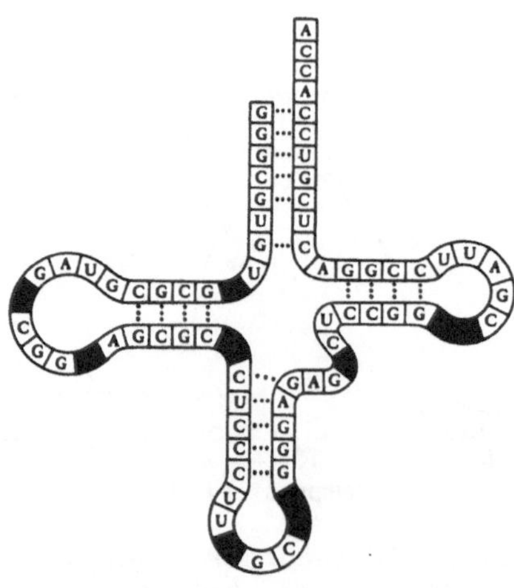

Fig. 15. Cloverleaf model of a specific transfer ribonucleic acid.

changes accompanying the tRNA-unfolding process has been found[24]. A simplified model of the proposed secondary structure of t-RNA described in modern textbooks of biochemistry is shown in Figure 15. Approximate values for the numbers of hydrogen bonds in different specific transfer ribonucleic acids can be derived from a comparison of infrared absorption spectra with simulated spectra generated by a superposition of increment extinction curves for the different nucleotide subunits in the paired or unpaired state. The principle of the method is illustrated in Figure 16 and a comparison of data for a few typical examples is given in Table 2.

A detailed description of the method (Application of heavy water as solvent, hydrogen-deuterium exchange etc.) has been published elsewhere[5]. As an example, a calorimetric transition curve obtained

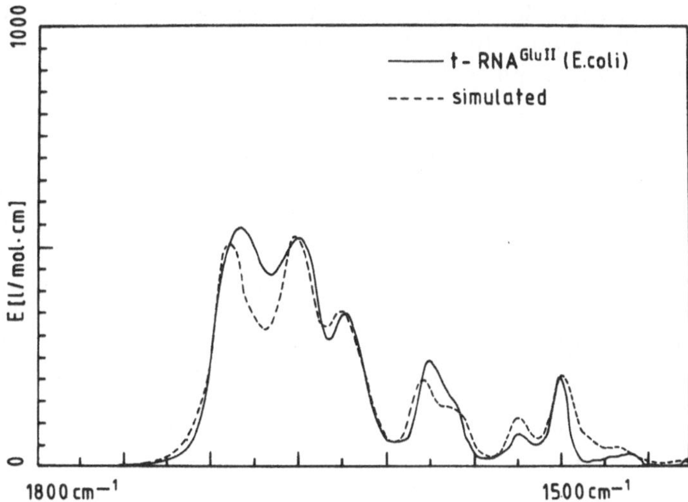

Fig. 16. Verification of the base pairing content by comparing the spectrum of $tRNA^{Glu\ II}$ recorded at 25°C (with the simulated spectrum).

Table 2. Base Pair Content of Specific tRNAs.

Specific tRNA	Number of Base Pairs					
	Cloverleaf Model			This Work		
	AU	GC	Total	AU	GC	Total
$tRNA^{Val}$	5	15	20	6	15	21
$tRNA^{Glu}$	4	16	20	4	14	18
$tRNA^{fMet}$	2	17	19	4	18	22

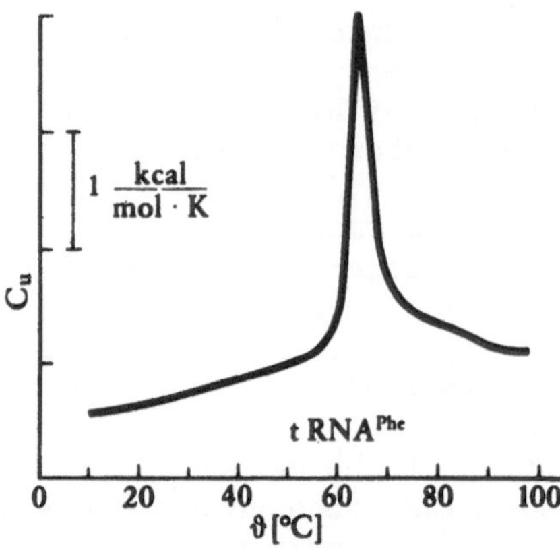

Fig. 17. Temperature dependence of the additional heat capacity of tRNA^{Phe}(yeast) in 5 mM sodium phosphate, 1 mM $MgCl_2$, 150 mM NaCl, pH 7.0[24].

in a DSC measurement of the additional heat capacity of a solution of a specific tRNA[24] is shown in Figure 17. The recorder trace indicates that determination of exact values of the corresponding enthalpy changes is sometimes complicated by baseline problems. Obviously these baseline problems must be regarded as one of the most serious limitations to the general application of the method.

A plot of the calorimetrically determined enthalpy changes vs. the apparent number of hydrogen bonds shown in Figure 18 indicates that the enthalpy of denaturation increases with increasing numbers of hydrogen bonds in the secondary structure. The experimental values are much higher than the estimated values derived from thermodynamic parameters evaluated from calorimetric studies on solutions of model compounds. Although the calorimetric data must be regarded as preliminary results of limited precision, there is no doubt that a large part of the total transition enthalpy is associated with the denaturation of the tertiary structure of tRNA and that, in a thermodynamic sense, this tertiary structure must be very similar for the different transfer ribonucleic acids. The tertiary structure of tRNA has been determined by X-ray crystallography and a diagrammatic representation for tRNA^{Phe} taken from the work of Rich, Kim et al[25] is shown in Figure 19.

"Batch" or "reaction" calorimetry[26] can also serve as an additional tool for the study of samples similar to the systems mentioned above.

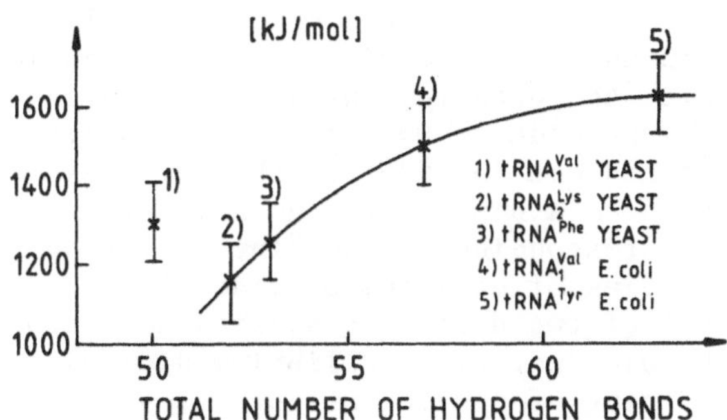

Fig. 18. Enthalpy of denaturation of tRNA samples vs. total number of hydrogen bonds.

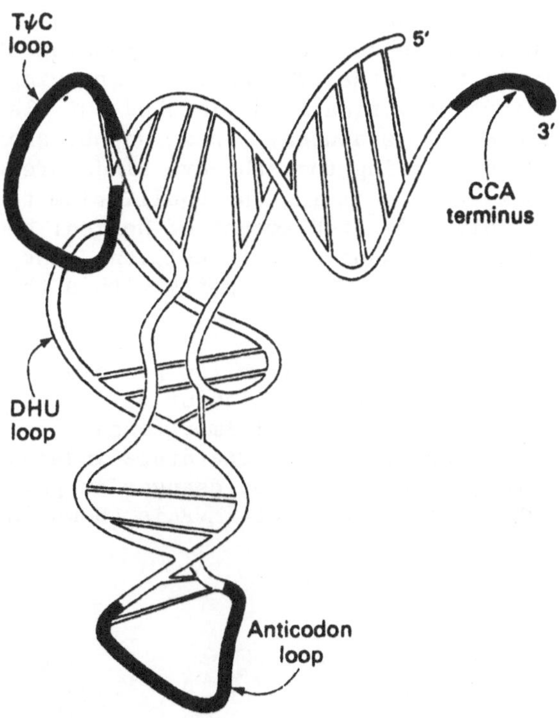

Fig. 19. Tertiary structure of tRNA^{Phe},[25]

The first determination of the enthalpy change associated with the helix formation of polyriboadenylic acid and polyuridylic acid in solution should be mentioned in this context[10]. The batch calorimeter has also been used to determine the so-called "enthalpy of intercalation" (i.e. the enthalpy change accompanying the binding process of ethidium bromide to DNA[27]). An extensive calorimetric study has been carried out in order to obtain a well established set of calorimetric data as a basis for a quantitative discussion of the enthalpy effects corresponding to the "intercalation" phenomena[28]. The enthalpy of binding of ethidium bromide (EB) to double- and single-stranded DNA of two different species has been measured by means of a mixing calorimeter. Equilibrium dialysis served as a tool to determine the extent of binding of EB. At 5, 20, and 35°C the binding enthalpies were -20.3, -23.6, and -33.5 kJ/mol EB, respectively, for the binding to "single-stranded" calf-thymus DNA in SSC-buffer and -21.4, -28.5, and -36.4 kJ/mol EB for the binding to double-stranded calf-thymus DNA.

Recently, a high precision differential scanning calorimeter has been used by Klump and Beaumais[29] to study the influence of ethidium bromide on the melting behaviour of calf-thymus DNA. With these measurements it has been shown that with increasing EB content in the mixtures not only is the enthalpy change accompanying the helix-random coil transition drastically diminished but also the T_m-values are shifted to higher temperatures. These unpublished results may be readily explained by assuming that the dye binds preferentially to single-stranded DNA, thus shifting the equilibrium to the denatured state and leaving only a limited amount of helical section for a thermally induced helix-random coil transition. The higher T_m-values may be due to an increase of GC base pairs within these residual helical sections of the structure.

The self assembly of tobacco mosaic virus protein provides another interesting example for the application of differential scanning calorimetry to systems of biological importance. The TMVP subunit, called the A-Protein, undergoes a temperature induced assembly process, which is strongly influenced by changes in pH. A simplified illustration of the self assembly of TMVP is shown in Figure 20.

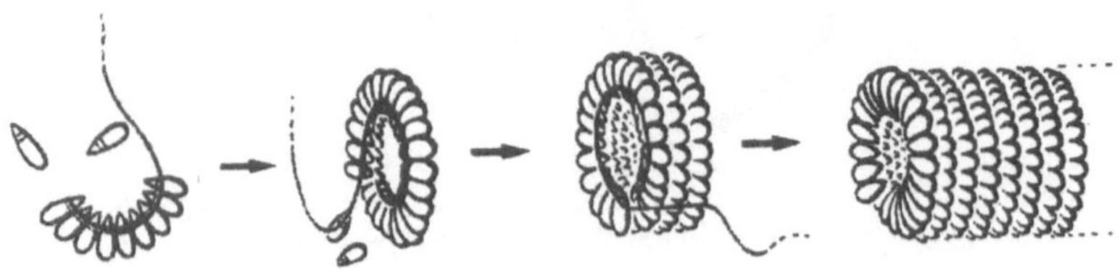

Fig. 20. Self assembly of TMVP (simplified model).

Recently, this system has been re-examined calorimetrically by Jaenicke and Sturtevant[30] in cooperation with a group working in our laboratory; using a calorimeter of the Privalov-type the enthalpies of aggregation have been determined in the pH region 6-7. As expected, the maximum in the calorimetric transition curve is shifted to higher temperatures with increasing pH, and it is interesting to note that the enthalpies of transition per mole of A-protein bound have positive values ranging from 60 to 300 kcal/mole i.e. the temperature-mediated self aggregation is endothermic and the negative entropy change, corresponding to the formation of the ordered helical structure, must be offset by a larger positive contribution associated with the release of water molecules from protein to the solvent.

A few additional remarks should be made concerning the application of differential scanning calorimetry in measurements on aqueous dispersions of phospholipids. Phospholipids are well known as one of the essential components of biomembranes and elementary chemical formulae are shown in Figure 21 for two representative members of this class of compounds.

Information on the structure and function of biomembranes has been helped by studies on model membrane systems or "vesicles" formed by sonication of phospholipids in water. A bilayer is formed with more or less closely packed hydrocarbon chains in the hydrophobic section of the aggregates due to the amphiphilic character of the molecules. A characteristic property of such a membrane is the so-called "phase-transition", the transition from the ordered to the fluid state of the lipid chains. It describes abrupt changes in the fluidity and

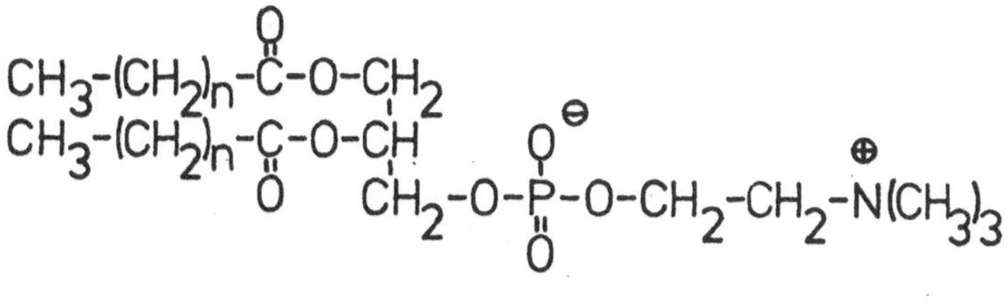

$$\begin{array}{l} CH_3\text{-}(CH_2)_n\text{-}\overset{O}{\overset{\|}{C}}\text{-}O\text{-}CH_2 \\ CH_3\text{-}(CH_2)_n\text{-}\underset{O}{\underset{\|}{C}}\text{-}O\text{-}CH \qquad O^{\ominus} \\ \qquad\qquad\qquad\qquad CH_2\text{-}O\text{-}\underset{O}{\underset{\|}{P}}\text{-}O\text{-}CH_2\text{-}CH_2\text{-}\overset{\oplus}{N}(CH_3)_3 \end{array}$$

PHOSPHATIDYLCHOLIN

$$\begin{array}{l} CH_3\text{-}(CH_2)_n\text{-}\overset{O}{\overset{\|}{C}}\text{-}O\text{-}CH_2 \\ CH_3\text{-}(CH_2)_n\text{-}\underset{O}{\underset{\|}{C}}\text{-}O\text{-}CH \qquad O^{\ominus} \\ \qquad\qquad\qquad\qquad CH_2\text{-}O\text{-}\underset{O}{\underset{\|}{P}}\text{-}O\text{-}CH_2\text{-}CH_2\text{-}\overset{\oplus}{N}H_3 \end{array}$$

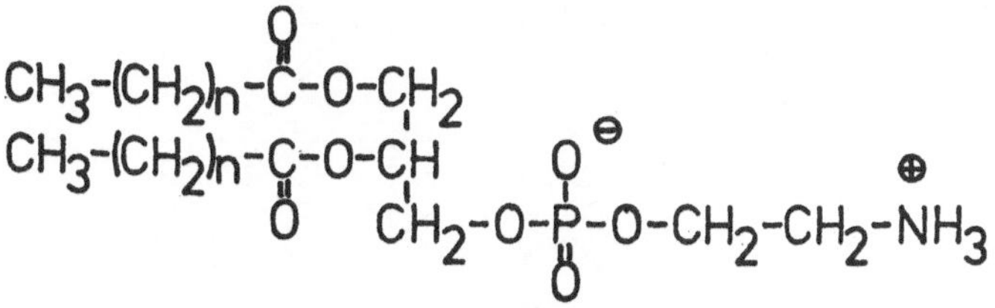

PHOSPHATIDYLÄTHANOLAMIN

Fig. 21. Typical examples for phospholipids.

packing density of the fatty acid chains at a distinct temperature, the transition temperature (T_t). Since lipid phase transitions have been observed by means of various physical methods not only in simple model systems but also in natural membranes, interest has been focused on mechanisms which can lead to changes in lipid chain fluidity and thus control regulatory functions in biomembranes.

A typical DSC-curve of a model membrane system is shown in Figure 22 for an aqueous dispersion of a phospholipid (DPPC). Characteristic changes in the shape of the calorimetric transition curves have been found in many measurements on mixed phospholipid bilayer systems, since Chapman first applied the DSC-method in this field.[32] A few typical examples may be briefly mentioned:

(i) Lysolecithin is a widely distributed surface-active and cytolytic phospholipid which has been shown to induce cell fusion as well as enhanced immune reactions. The calorigram of an equimolar mixture of the two lipids DPPC and DPPE[33] is of the so-called "monotectic" type with almost complete miscibility in the gel phase as shown in Figure 23. Addition of 5 weight % of a synthetic lysolecithin-analog results in a completely different mixing behaviour. The single peak found for the DPPE/DPPC- mixture is split into three peaks (Fig. 23) and separation into three phases with different compositions has taken place.

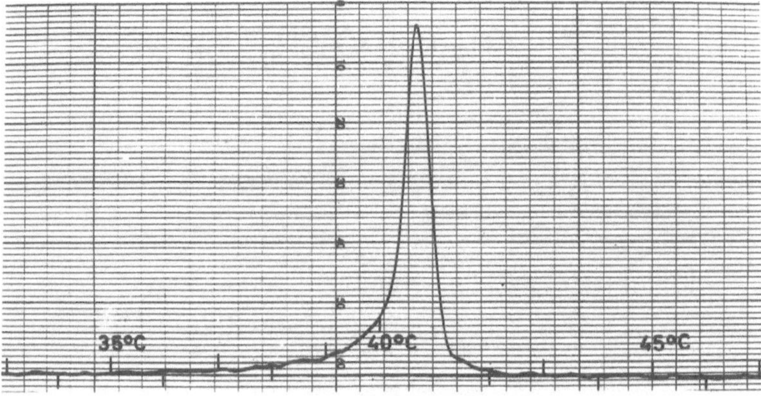

Fig. 22. DSC recorder trace as obtained in a measurement on an aqueous dispersion of a phospholipid (DPPC).

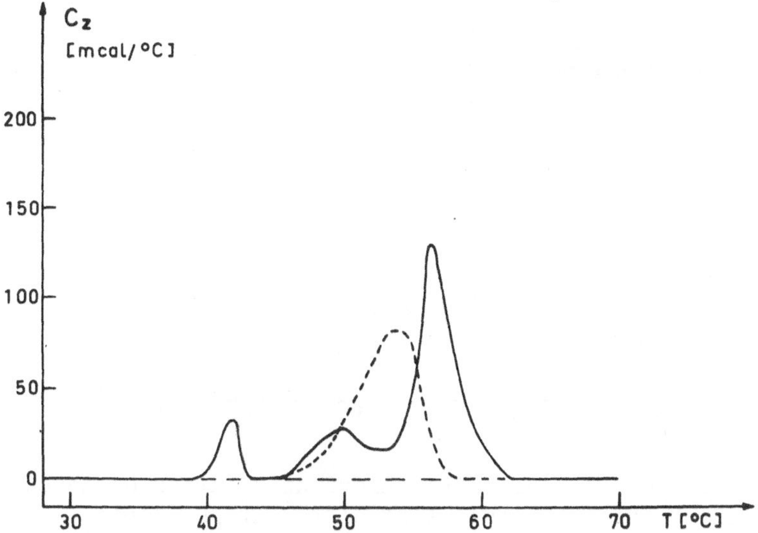

Fig. 23. Calorimetric scans of an aqueous dispersion of 40.3 mg of an equimolar mixture of DPPE and DPPC before (---) and after (——) the addition of 5% (w/w) lysolecithin.

(ii) Calorimetric investigations of the phase transitions of charged phospholipids are rather limited though recently, Blume and Eibl have studied the influence of pH on the transition curves of phosphatidic acid bilayers.[34] As can be seen from the results in Figures 24 and 25 T_t-values of charged phospholipids are strongly influenced by variations of pH. The observed maximum of the transition temperature of DHPA in Figure 25 corresponds to a minimum in the ΔH vs. pH-diagram. At this pH a particular stable bilayer is formed; (full protonation of phosphatidic acids leads to suspensions of microcrystals). The decrease in the transition enthalpy at high pH-values is due to a change in the hydrocarbon chain interactions induced by the doubly charged head groups.

Phospholipids also form two-dimensional monolayers on water. Using a film balance the area of the monolayer versus temperature curves at different surface pressures can be recorded. The characteristic monolayer phase transition temperature is shifted towards higher temperatures when the lateral pressure is increased. Blume has been able to show that the behaviour of the monolayer system is very similar to that of the corresponding bilayer system at a lateral pressure (approximately 30 dyne/cm), where the absolute area and the area change in both systems are nearly the same[35].

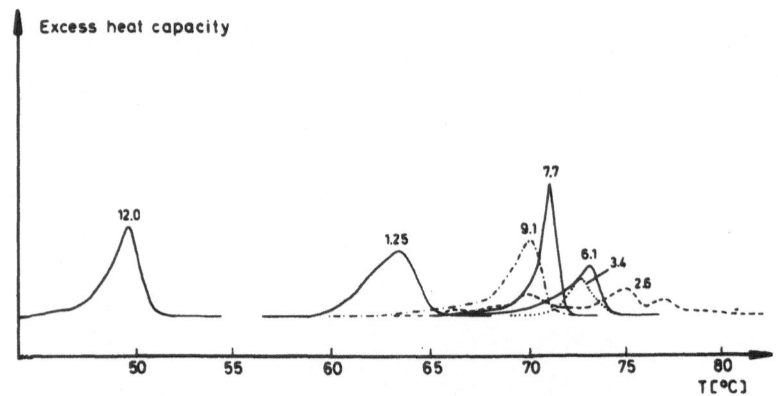

Fig. 24. Differential scanning calorimetry curves of aqueous 2 mM DHPA dispersions at different pH values.

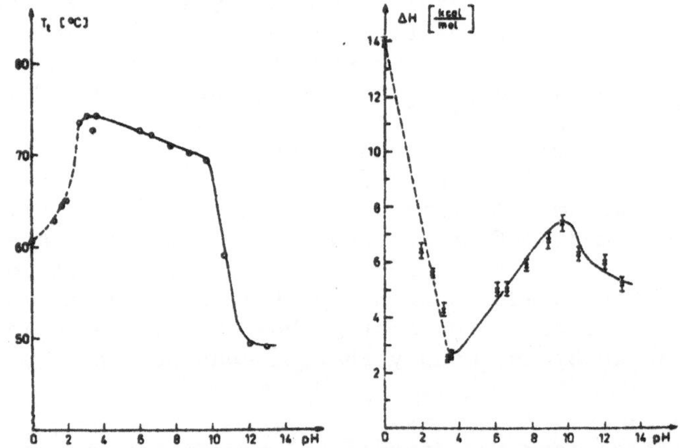

Fig. 25. Dependence of the transition temperature T_t on the pH value of DHPA. In the pH range indicated by the dashed line DHPA does not form stable dispersions. Dependence of the transition enthalpy on the pH value for DHPA dispersions in water. Bars represent standard deviations of at least three different measurements.

Finally, it is pertinent to illustrate the present state of development of various experimental techniques. The number of differential scanning calorimeters which meet the conditions quoted from the book on biochemical microcalorimetry is still rather limited. For this reason, construction and development of appropriate DSC instruments for measurements on small amounts of dilute solutions of biopolymers must still be regarded as a first requirement.

The internal part of a twin calorimeter developed in our laboratory during the last two years is shown in Figure 26. Both the sample holder and the reference cell with its heater wires and filling tubes are mounted in good thermal contact to a multi-junction thermocouple (i.e. to the small thin-walled plate suspended in the center of a mounting ring). Each of the two vessels holds about two ml of solution or solvent respectively. The twin vessel-unit is inserted in an adiabatic shield, as shown in Figure 27. The wires belonging to the shield control multi-junction thermocouple system can be seen clearly in the photograph. Thermal contact is achieved by means of a copper ring fixed to the inner wall of the cylindrical shield. The entire system together with the shield lid is insulated by a polystyrene cover and then inserted in a Dewar vessel which serves as an outer thermal shield (Fig. 28). In an extensive series of measurements, it has been shown that this calorimeter is an excellent tool for precision measurements on solutions of biopolymers. Further details concerning this new instrument, which exhibits nearly the same sensitivity and reproducibility as the Privalov-calorimeter DASM-1 M, will be published elsewhere[36].

Fig. 26. Internal part of the twin-calorimeter (sample holder and reference vessel, see text).

Fig. 27. Twin-vessel inserted in the adiabatic shield.

Fig. 28. Insulated calorimeter inserted in Dewar-vessel.

REFERENCES

1. J.D. Watson and F.H.C. Crick, Genetic implications of the structure of deoxyribonucleic acid, Nature 171:737, 964 (1953).
2. L. Pauling, R. Corey and H. Branson, Atomic coordinates and structure factors for two helical configurations of polypeptide chains, Proc. Natl. Acad. Sci. U.S. 37:205 (1951).
3. J. Marmur and P. Doty, Determination of the base composition of deoxyribonucleic acid from its thermal denaturation temperature, J. Molec. Biol. 5:109 (1962).
cf. also Bresch, C. and R. Hausmann, Klassische und molekulare Genetik, 3rd Ed., Springer, Berlin, Heidelberg, New York (1972).
4. B. Zimm, P. Doty and K. Iso, Determination of the parameters for helix formation in poly-benzyl-glutamate, Proc. Natl. Acad. Sci., 45:1601 (1959).
5. U. Schernau, S. Marcinowski and Th. Ackermann, Infrared spectroscopic studies of the interaction between polyuridylic acid and adenosine mono- and oligonucleotides, Zeitschrift für Physikalische Chemie Neue Folge 117:11-18 (1979).
6. J. Langowski, K. Henco, D. Riesner and H.L. Sänger, Common structural features of different viroids: Serial Arrangement of double helical sections and internal loops, Nucleic Acids Research 5:1589 (1978).
cf. also H. Klump, D. Riesner and H.L. Sänger, Calorimetric studies on viroids, Nucleic Acids Research 5:1581 (1978).
7. D.M. Crothers, in Physical Chemistry of Nucleic Acids, Harper & Row, Publishers, New York, Evanston, San Francisco, London 1974, p.350.
8. J. Applequist, On the helix-coil equilibrium in polypeptides, J. Chem. Physics 38:934 (1963).
9. T.H. Benzinger, in Analytical methods of protein chemistry, Volume 5, Pergamon Press, Oxford 1969, p.93.
10. R. Steiner and C. Kitzinger, Heat of reaction of polyriboadenylic acid and polyribouridylic acid, Nature 194:1172 (1962).
11. M.A. Rawitscher, P. Ross and J.M. Sturtevant, The heat of reaction between polyriboadenylic acid and polyribouridylic acid, J. Am. Chem. Soc. 85:1915 (1963).
12. P.L. Privalov, K.A. Kafiani and D.R. Monaselidze, Study of thermal denaturation of DNA with the aid of an adiabatic microcalorimater, Dokl. Akad. Nauk SSSR 156:951 (1964).
13. F.E. Karasz and J.M. O'Reilly, Enthalpy changes in the helix-coil transition of poly-γ-benzyl-L-glutamate, Biopolymers 5:27 (1967).
14. J.F. Brandts, W.M. Jackson and T. Yao-Chung Ting, A calorimetric study of the thermal transitions of three specific transfer ribonucleic acids, Biochem. 13:3595 (1974).
15. P.L. Privalov, G.M. Mrevlishvili, Biofizica 12:22 (1966).
16. G. Rialdi, J. Levy and R. Biltonen, Thermodynamic studies of transfer ribonucleic acids. I. Magnesium binding to yeast phenylalanine transfer ribonucleic acid, Biochemistry 11:2472 (1972).

17. Biochemical Microcalorimetry, H.D. Brown, Ed., Academic Press, New York 1969.
18. Th. Ackermann, Kalorimetrie von Biopolymeren, Chemie in unserer Zeit 11:97 (1977).
19. Th. Ackermann and H. Rüterjans, Kalorimetrische Messungen zur Helix-Knäuel-Umwandlung von Nucleinsäuren und synthetischen Polypeptiden in Lösung, Ber. d. Bunsenges. 68:850 (1964).
20. M. Grubert and Th. Ackermann, Ein empfindliches adiabatisches Differenzkalorimeter zur Untersuchung kooperativer Strukturumwandlungen in Lösungen, Zeitschrift für physikalische Chemie Neue Folge 93:255 (1974).
21. H. Klump and Th. Ackermann, Experimental thermodynamics of the helix-random coil transition. IV. Influence of the base composition of DNA on the transition enthalpy, Biopolymers 10:513 (1971).
22. H. Klump, A calorimetric study of polyguanylic acid at neutral pH, Biophysical Chemistry 5:359 (1976).
23. R.D. Blake and St. G. Lefoley, Spectral analysis of high resolution direct-derivative melting curves of DNA for instantaneous and total base composition, Biochimica and Biophysica Acta 518:233 (1978).
24. F.J. Schott, M. Grubert, W. Wangler and Th. Ackermann, A comparative calorimetric study on tRNA unfolding, Biophysical Chemistry, in press (1980);H.-J. Hinz, V. Filimonov and P.L. Privalov, Calorimetric studies on melting of $tRNA^{Phe}$ (yeast), Eur. J. Biochem. 72:79 (1977).
25. S.H. Kim, G.J. Quigley, F.L. Suddath, A. McPherson, D. Sneden, J.J. Kim, J. Weinzierl and A. Rich, Three dimensional structure of yeast phenylalanine transfer RNA:Folding of the polynucleotide chain, Science 179:285 (1973).
26. I. Wadsö, in Biochemical Microcalorimetry, H.D. Brown, Ed., Academic Press, New York 1969, p.83.
27. F. Quadrifoglio, V. Creszenzi and V. Giancotti, Calorimetry of DNA-dye interactions in aqueous solution. I. Proflavine and ethidium bromide, Biophys. Chem. 1 319 (1974).
28. W. Burkart. Ph.D Thesis, Freiburg (1980) cf. also W. Burkart und H. Klump, Thermodynamische Messungen der Bindungsenthalpien kleiner Liganden mit Nukleinsäuren. 1. Anlagerung von Ethidiumbromid an DNA und Apurinsäure, Ber. d. Bunsenges, in press (1981).
29. H. Klump and J. Beaumais, private communication (1981).
30. R. Jaenicke and J.M. Sturtevant, personal communication cf. also W. Schulz, Masters Thesis, Freiburg (1973).
31. A. Blume and Th. Ackermann, A calorimetric study of the lipid phase transitions in aqueous dispersions of phosphorylethanolamine mixtures, FEBS Letters 43:71 (1974).
32. D. Chapman, R.M. Williams and B.D. Ladbrooke, Chem. Phys. Lipids 1:445 (1967).

33. B. Arnold and H.U. Weltzien, Effects of a synthetic lysolecithin analog in the phase transition of mixtures of phosphatidylcholine FEBS Letters 61:199 (1976).
34. H.J. Eibl and A. Blume, The influence of charge on phosphatidic acid bilayer membranes, Biochimica and Biophysica Acta 553:476 (1979).
35. A. Blume, A comparative study of the phase transitions of phospholipid bilayers and monolayers, Biochimica and Biophysica Acta 557:32 (1979).
36. M. Grubert and Th. Ackermann, Angew. Chemie, to be published (1981).

METHODS FOR STUDYING FAST KINETICS IN BIOLOGICAL SYSTEMS

Dietmar Pörschke

Max-Planck-Institut für
biophysikalische Chemie
34 Göttingen, W.-Germany

1. INTRODUCTION

Biological systems have been selected during evolution for high response rates to their environment. Systems with a high response rate obviously have a clear selective advantage compared to systems, which cannot adapt as quickly to their environment. As a consequence many reactions encountered in biological systems are very fast and can only be analysed with the aid of special kinetic techniques.

A large number of these techniques for the investigation of rates and mechanisms of fast reactions have been developed during the last 20 years. The development of techniques was almost parallel to - and partly driven by - the great progress in molecular biology and biophysics. It will not be possible to cover all these techniques in the present short review. The main emphasis will be on relaxation techniques and their present state of the art. The aim of this contribution will be to provide some basic information about the principles, possibilities and also the limits of the various techniques. It will be shown that a wide spectrum of techniques is available, which should be sufficient to analyse and characterise almost any reaction encountered in biological systems.

2. EXPERIMENTAL PROCEDURES

2.1 Mixing Methods

The classical procedure for the analysis of reaction rates and mechanisms is mixing of the components and subsequent recording of some parameter indicating the reaction of the system. Since complete mixing in a usual cuvette, for example, takes about one second, the time resolution is limited and the analysis restricted to relatively slow reactions. However, the mixing time can be reduced considerably by the construction of special mixing devices (Hartridge and Roughton[1]). The most useful device is the so called "stopped-flow", which allows fast mixing experiments at a relatively low demand of material[2-4]. The solutions are pushed from syringes through a mixing chamber with a set of jets for maximal turbulence into an observation cell and finally into a stop-syringe. When the stop-syringe is filled, the flow cannot continue and the recording of the reaction by some spectroscopic signal is initiated (Fig. 1).

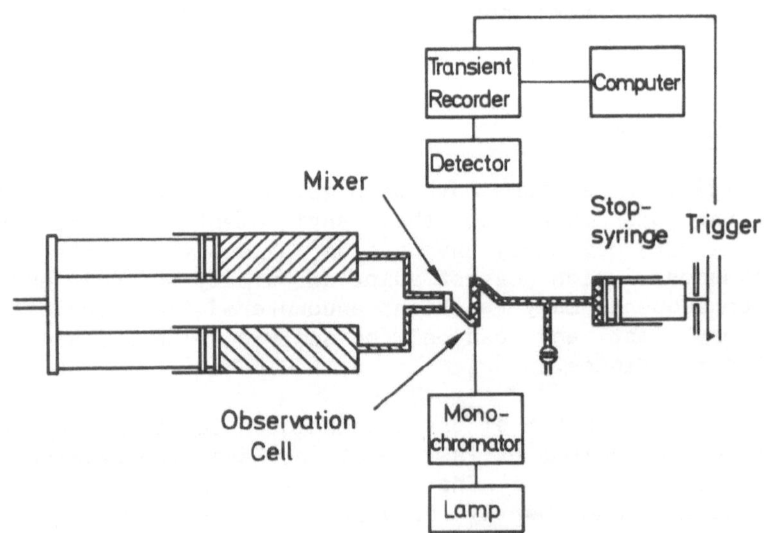

Fig. 1 Scheme of stopped flow

With this procedure it is possible to reduce the mixing time to a hundred microseconds. The time resolution of conventional machines, which are available commercially, usually does not exceed the millisecond range.

Stopped-flow reactions are mainly followed by absorbance or fluorescence detection. Fluorescence detectors can also be used for straylight measurements. Some special signals like the circular dichroism[5,6] or the reaction enthalpy[7] (calorimetry) have also been used to follow stopped-flow reactions. However, in these cases it is more difficult to attain reasonable signal to noise ratios.

Recently a special "jet flow" technique[8] has been described with mixing times down to about 5 µs. However, it will be difficult to use this technique for systems with relatively small changes in the optical signals as usually found in biological systems.

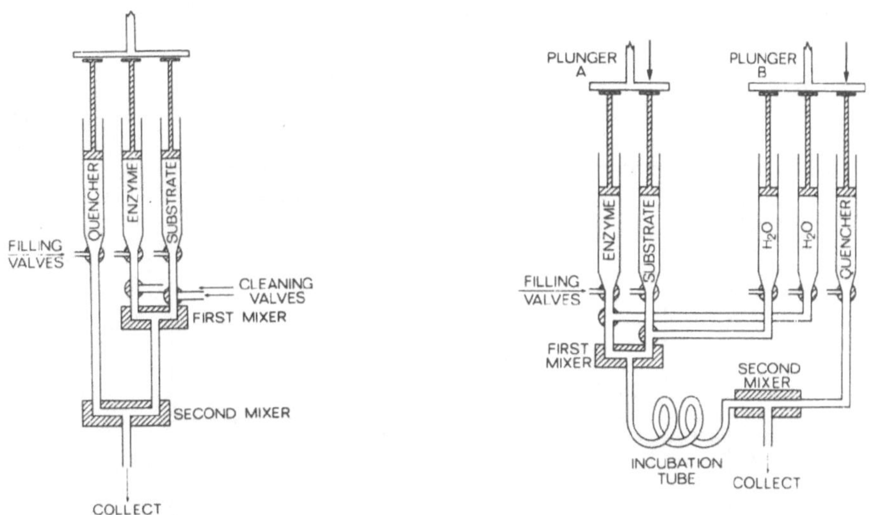

Fig. 2 Scheme of quenched flow apparatus, a) conventional, b) pulsed (reproduced with permission from ref. 9).

In the case of enzyme reactions it may be useful to stop the digestion of a substrate after a certain time by addition of a "quencher". This may be done with a time resolution close to a ms by a quenched flow apparatus. Enzyme and substrate are mixed together in a first mixing chamber and after flowing through a tube of variable length the reaction mixture is quenched by addition of the quenching solution in a second mixing chamber.

This procedure[9] permits a time resolution up to 5 ms. When the reaction times are not as short, another procedure, the "pulsed quenched flow", may be more useful. Here the enzyme substrate mixture can be incubated for a given time in an incubation tube. At the end of the incubation the solution is washed out of the tube and mixed with the quenching solution (second mixer). The samples may then be analysed by any method, including chromatographic or electrophoretic techniques for example.

2.2 Transient Relaxation Methods

Owing to the development of fast mixing techniques the time range for kinetic analysis was extended by at least three orders of magnitude and yet many reactions remained "too fast". These reactions can only be analysed by techniques which provide the information on the dynamics from measurements on homogeneously mixed solutions. Among these, the relaxation techniques have proven to be particularly useful[10]. Their basis is relatively simple: the equilibrium constants K of chemical reactions are dependent upon "external" parameters like the temperature T, pressure p or electric field strength E according to

$$\delta \ln K = \left(\frac{\partial \ln K}{\partial T}\right)_{P,E} \delta T + \left(\frac{\partial \ln K}{\partial p}\right)_{T,E} \delta p + \left(\frac{\partial \ln K}{\partial E}\right)_{T,p} \delta E \qquad (1)$$

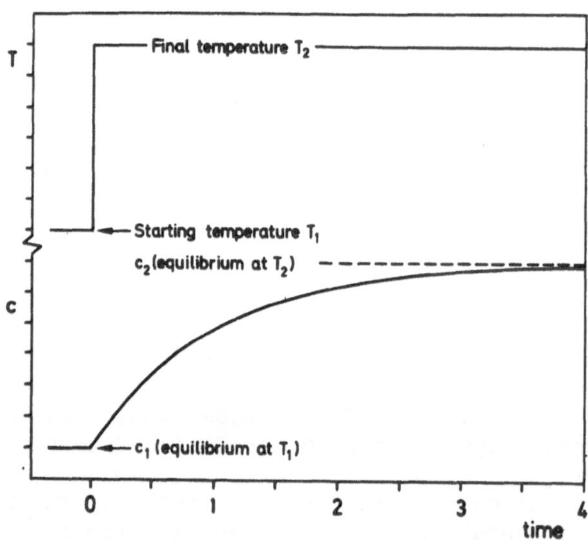

Fig. 3 Scheme of a relaxation experiment. A sudden change of the external parameter T induces chemical relaxation to a new equilibrium at T_2. The time scale is given in units of the relaxation time constant (cf. equation 6 in section 3).

Thus any change of one of these parameters will result in a shift of the equilibrium and will induce a reaction towards the new equilibrium. If the external parameter is changed within a very short time t, it is possible to observe reactions with time constants very close to t. The time resolution is determined by the rate of change in the external parameter and, of course, also by the response time of the detector used for the measurements. In the following sections the relaxation methods will be dicussed according to the external parameter used for perturbation.

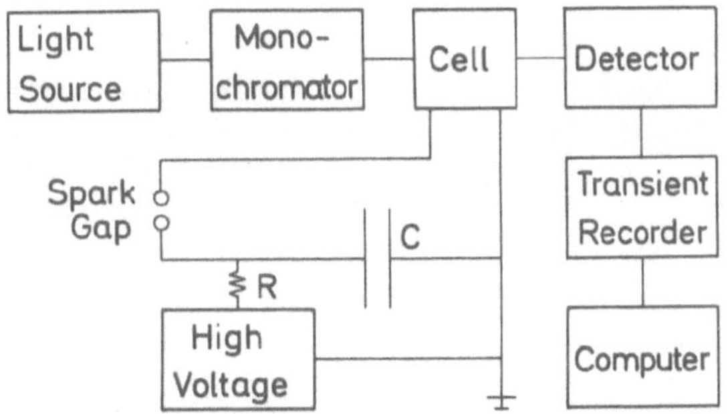

Fig. 4a General scheme of temperature jump apparatus.

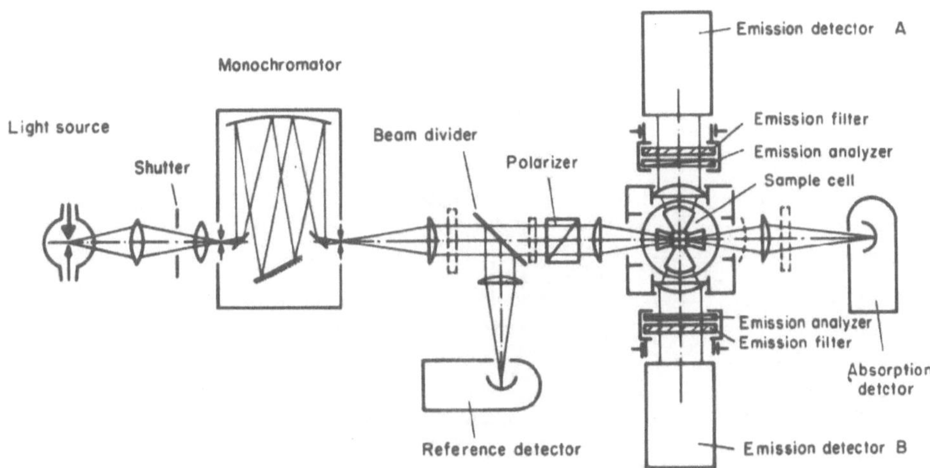

Fig. 4b Optical scheme of a fluorescence temperature jump apparatus[14] (courtesy of C.R. Rabl).

2.2.1 Temperature jump

The most convenient method to produce an increase of temperature in a solution in a short time is Joule heating. Usually the required electric energy is accumulated in a capacitor, which is then discharged through the sample (Fig. 4)[4,10-13]. For a fast discharge the ohmic resistance R of the sample should be low, which can be adjusted by addition of neutral salt. In most cases the ionic strength used for the analysis of biological systems is around 0.1 M, which is sufficiently high for fast Joule heating. The heating time constant t_h is given by

$$t_h = \frac{1}{2} \cdot R \cdot C \tag{2}$$

where C is the capacity. For fast heating the capacity has to be small ($\sim 10^{-8}$ Farad); as a consequence high voltages ($\sim$30 KV) are required, in order to get sufficiently high temperature jumps. The cells containing the sample have to be designed for homogeneous heating. Temperature jump cells are available with sample volumes between 5 ml and 50 µl (cf. Fig. 5).

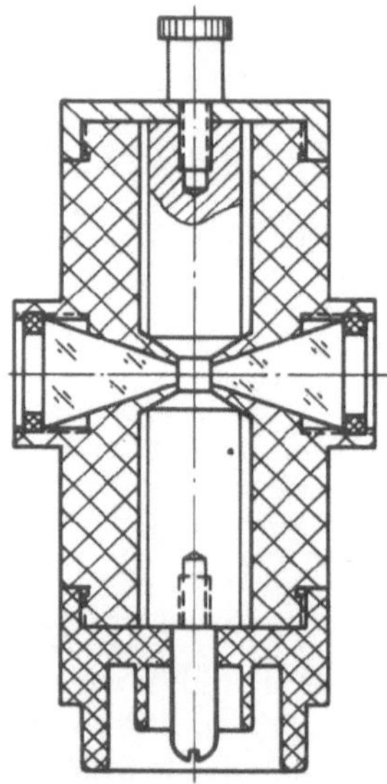

Fig. 5 Construction of temperature jump cell with a small sample volume. This cell is used both for conventional and cable temperature jump measurements.

After the temperature jump the reaction towards the new equilibrium can be observed by changes in the absorbance, fluorescence or straylight intensity. Other modes of detection have not been as successful. In particular polarimetric or CD measurements[15] are difficult due to optical artifacts induced by shock waves following the temperature jump. Such problems may be avoided by measuring at low temperatures around the density maximum of aqueous solutions.

"Conventional" temperature jump machines can be used in the time range from some seconds to 10^{-6} seconds. The upper limit is given by the rate of heat exchange of the sample with its surrounding and the lower limit by the rate of heating. The particularly wide time range over 6 orders of magnitude, which became accessible for kinetic analysis by the relatively simple temperature jump instrument, is the main reason for its success. Another advantage of the temperature jump technique is the low demand of material, which is particularly important for biochemical investigations. Moreover the same sample can be used for many temperature jump experiments.

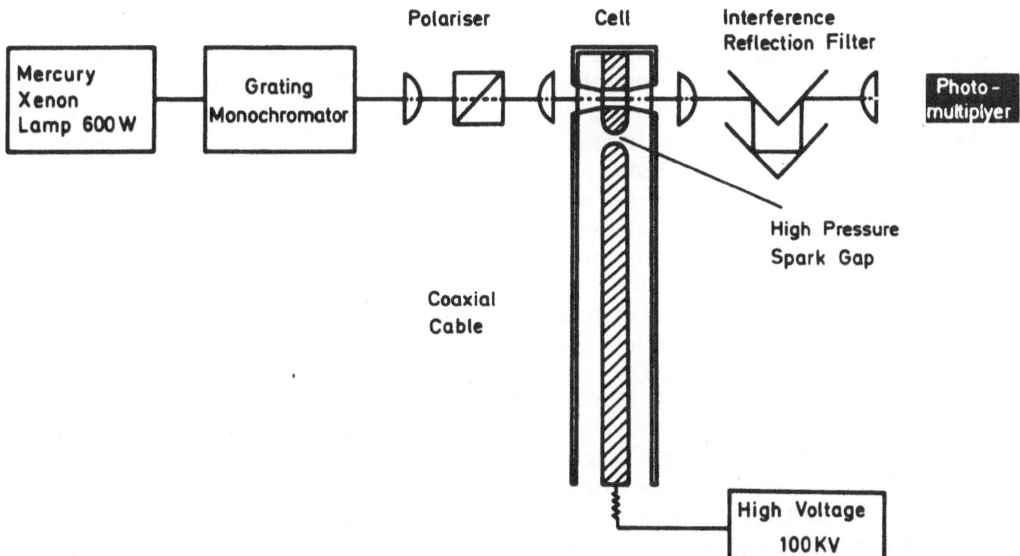

Fig. 6 Cable temperature jump apparatus. The jump is started by release of pressure from the spark gap. Absorbance changes due to orientation are supressed by polarised light (magic angle), electroluminence effects by an interference reflection filter.

The time resolution of joule temperature jump instruments can be extended beyond the μ-sec range by using a special type of discharge[16,17]. When a coaxial cable is used instead of a capacitor and the impedance of the sample is matched with the cable impedance, the discharge is completed in the time required for the traveling of the electric signal twice along the cable. Thus short cables allow extremely short heating times. This principle has been used in the cable temperature jump instrument (Fig. 6), which can resolve relaxation processes down to 10 ns.

Other procedures for heating like microwave-[18] or laser-pulses[19] are not as widely used, probably because the instruments are more expensive and more difficult to handle. Nevertheless these procedures have some advantages for special applications. Both microwave- and laser heating do not require high salt concentrations for fast temperature jumps and thus are very useful for high time resolution at low ionic strengths.

2.2.2 Pressure jump

As described in the previous section temperature jump instruments always produce an increase in the temperature because it is difficult or practically impossible to decrease the temperature in a very short time interval. In the case of pressure jumps it is possible to produce fast changes in both directions. Both ways have been used and some machines also produce pressure pulses in a repetitive manner[20]. The release of pressure by a bursting metal membrane (Fig. 7) or fast opening of a mechanical valve can be achieved in times of 50 to 100 μs, whereas the characteristic time constants for the pressure rise by the valve method are between 2 and 40 ms. A special instrument for small periodic pressure changes[21] induced by piezoelectric crystals can be used for relaxation measurements down to 20 μs. One of the advantages of the pressure jump technique is the fact that the measurements are not limited to "short" observation times, but can in principle be extended to "infinite" times. Of course, the method can only be applied for reactions with a significant volume change (although the pressure jump usually produces a simultaneous change of the temperature). However, this is not much of a restriction, since many reactions are associated with volume changes. This may be illustrated by some examples of biophysical applications of the pressure jump technique, which are of very different nature[20]: association behaviour of ribosomal subunits, assembly of tubulin, self association of glutamate dehydrogenase and the phase transition of lipid bilayers.

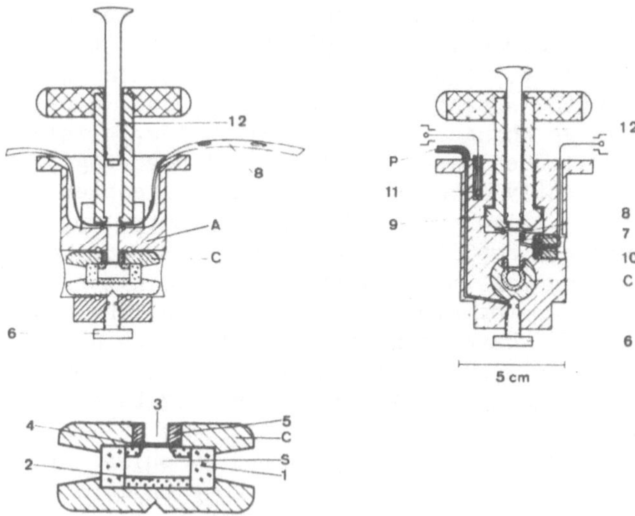

Fig. 7 Pressure jump cell with optical absorption detection. Hydrostatic pressure is applied to the sample S, which is separated by an elastic membrane (3), until a metal membrane (8) is bursting. The limit pressure is defined by the thickness of the membrane. The relaxation is observed through window (1) (reproduced with permission from ref. 20).

The detection of pressure induced relaxation processes is possible by as many techniques as applied in the case of temperature jump instruments. Pressure jump instruments[20] have been constructed with detection by absorbance, fluorescence (straylight), circular dichroism (optical rotation), conductometric and also calorimetric measurements.

2.2.3 Field jump

The field jump technique is not as widely used as the other techniques described above. This is partly due to the fact that the application is limited to samples with a low conductivity, i.e. low ionic strength. Furthermore the construction of a decent field jump instrument is more difficult than the construction of the other instruments discussed above. Finally field jump experiments with samples containing macromolecular components as usually found in biochemical systems may show a superposition of chemical relaxation and orientation effects. However, it will be shown below that the last problem can be avoided by using polarised light for the measurements. With this special trick the field jump technique has proven to be very useful for the analysis of reactions involving macromolecules.

As in the case of pressure jumps, field jumps can be easily applied in both directions. A particular advantage of the field jump technique is its high time resolution, since field pulses can be produced with extremely short rise and decay times. For sufficiently large changes in the equilibrium, field pulses of very high amplitudes are required. High voltage pulses may be produced conveniently with the aid of coaxial cables[22]. With the cable technique (Fig. 8) it is possible to apply high voltage pulses of constant amplitude even in the case of conducting samples, as long as the pulse time does not exceed twice the travelling time for the electric signal along the cable. Field jump instruments[22] have been constructed, which allow a variation of pulse lengths between microseconds and milliseconds. The time resolution goes down to 10 ns[23] and may be extended to 1 ns. Since electric fields increase reaction rates in many cases, it is possible to induce equilibrium shifts by pulses of about 100 µs also in the case of reactions, which take much more time at zero field strength. The reaction rate, which is not perturbed by the electric field, can then be studied after the pulse is turned

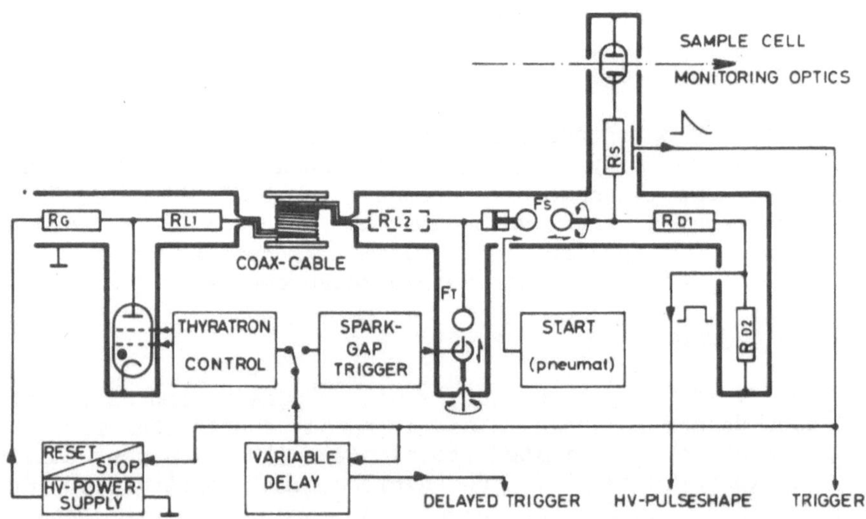

Fig. 8 Diagram of coaxial high field pulse generator. The coax-cable is charged up; then the pulse is applied to the sample by pneumatic closing the spark gap F_s. Termination of the pulse is either by a triggered spark gap F_T or by a thyratron (reproduced with permission from ref. 22).

off. Thus the time range available for field jump experiments extends from nanoseconds to at least milliseconds and in special cases even to seconds.

Two different types of reactions are known to be sensitive to electric fields and thus may show field induced relaxation effects[24,25]. The first type of reaction is associated with a change in the dipole moment. It can be calculated that rather high changes in the dipole moment are required for any appreciable changes in the equilibrium at the field strengths which are accessible in the usual instruments. Reactions of this type with sufficiently high changes in the dipole moment are quite rare. The second type of reaction, which may be perturbed by field pulses, is associated with some change in the state of ionisation. As calculated by Onsager[26] and found experimentally, this type of reaction is quite sensitive to electric fields. Already a simple $A^+ + B^- \rightleftharpoons AB$ reaction may be perturbed by electric fields of moderate strength to a considerable degree, such that the relaxation can be easily characterised. Much larger amplitudes are found for reactions with higher changes in the number of charges. In the case of polyelectrolytes, which may show particularly high changes in the charge density, large amplitudes may be induced already at relatively low field strengths[27]. Since most biopolymers are also polyelectrolytes, their reactions can be analysed with the aid of field jump instruments.

In spite of these possibilities the field jump technique has not been used so much for the analysis of reactions involving biopolymers. One of the main reasons seems to be the complication due to orientation effects. Biopolymers are usually aligned by electric fields and as a consequence of their optical anisotropy the orientation is reflected by changes in the absorbance or also fluorescence. However, these changes may be suppressed completely by using polarised light. As calculated by Labhart[28] the contribution of orientation effects to absorbance changes disappear at the "magic angle", i.e. at an orientation of 54.8° of plane polarised light with respect to the vector of the electric field. Corresponding relations can be derived for the case of fluorescence. It has also been shown by various experiments that changes observed at the magic angle are exclusively due to "chemical" contributions, provided that certain conditions concerning the magnitude of the signals are fulfilled. With this magic angle technique it has been possible to study many reactions involving biopolymers[29] by field jump experiments, which could not be studied by any of the other methods. The magic angle technique has also been used to suppress orientation effects in temperature jump experiments[17,57].

2.3 Stationary Relaxation Methods

2.3.1 Sound absorption

The first relaxation measurements have been performed with the ultrasound technique: an unusual effect in the propagation of sound waves in seawater could be explained by an excess sound absorption due to an ion complex reaction[30] of $MgSO_4$. This was the starting point for the development of the ultrasound techniques and of other relaxation techniques. The coupling between sound waves and chemical reactions results from the periodic change of pressure in the sound waves. Sound waves may be described as a series of small pressure jumps. Thus the equilibrium of pressure dependent reactions may be shifted by sound waves (Fig. 9) as long as the reaction is fast enough to follow the period of the sound wave. The coupling between sound and reaction then leads to an excess absorption. The excess absorption is not observed when the sound period is short compared to the time constant of the chemical reaction. As soon as the chemical reaction can "follow" the periods of the sound waves, an excess absorption can be determined. Thus the information on the rate of the chemical reaction can be obtained from the frequency dependence ("dispersion") of the sound absorption. The absorption per wavelength $\alpha\lambda$ as a function of the angular frequency $\omega = 2\pi f$ for a single relaxation process with a time constant τ is described by

$$\alpha\lambda = A \cdot \frac{\omega\tau}{1+(\omega\tau)^2} + B\omega \qquad (3)$$

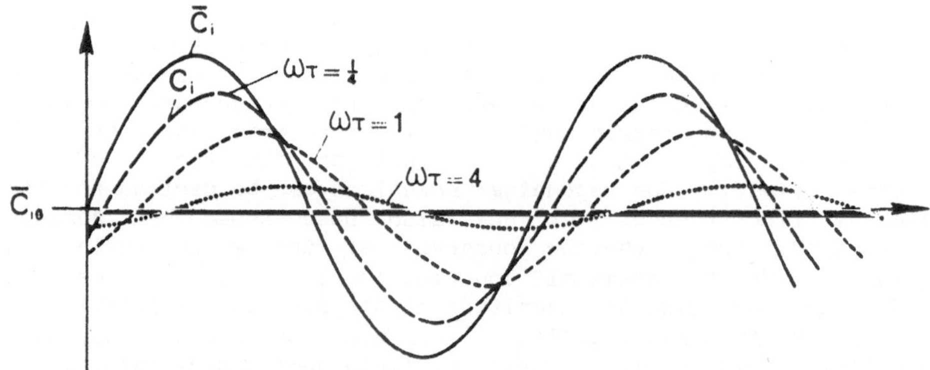

Fig. 9 Chemical relaxation by a periodic forcing function. The amplitude and the phase shift is dependent upon the reciprocal relaxation time $1/\tau$ compared to the frequency of the forcing function.

The amplitude factor A is given by

$$A = \Gamma \frac{\pi}{RT\kappa} [\Delta V - \frac{\alpha_p}{\rho \cdot c_p} \Delta H]^2 \qquad (4)$$

with ΔV molar volume change, ΔH reaction enthalpy, κ adiabatic compressibility, α_p thermal expansion coefficient and c_p specific heat at constant pressure. Γ is an amplitude factor, which is characteristic for the chemical reaction (cf. section 4). In aqueous solutions the amplitude contribution from the ΔH term is usually below 10 per cent.

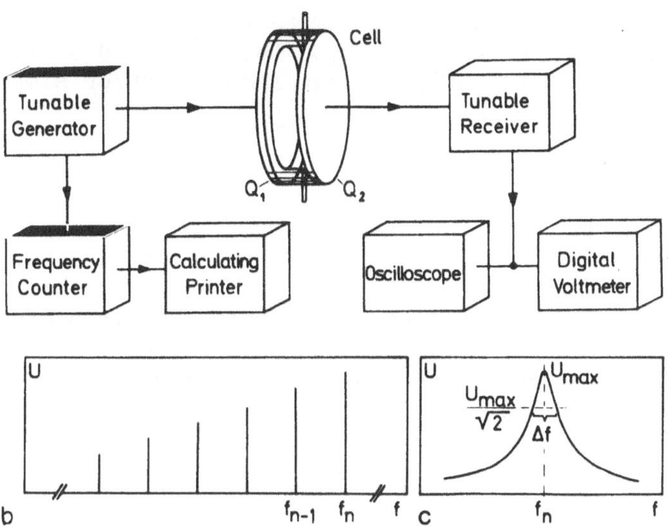

Fig. 10 a) Diagram of the resonator method. b) Output voltage U from receiver quartz QZ as a function of frequency. c) Single resonance peak of quartz-liquid resonator, cf. text (reproduced with permission from ref. 31).

The quantitative characterisation of chemical reactions by the sound absorption technique require measurements over a broad range of frequencies[31]. Unfortunately these measurements often cannot be performed with a single experimental setup, since sound absorption cells can usually be applied only in a restricted frequency range. Two major procedures[31] have been developed for ultrasound measurements. The resonator method operates in the frequency range from 0.1 to 25 MHz, corresponding to relaxation times between 2 µs and 6 ns. The resonator consists of 1) a tunable sine wave generator, which drives a sender quartz Q_1 and excites standing sound waves at particular frequencies and 2) a receiver quartz Q_2 which is in resonance at these frequencies. The bandwidth of the resonances is a function of the absorbance and is recorded at as many resonance frequencies as possible (Fig. 10).

The second procedure is called the pulse method and is used in the frequency range from 15 to 150 MHz corrresponding to relaxation times between 10 ns and 1 ns. In this procedure single pulses of ultrasonic waves are send through the liquid to a receiver and the decay of amplitude is measured as a function of the frequency. Both the resonator and the pulse method require reference measuments with the solvent. Thus ultrasound measurements are quite laborious and time consuming.

Nevertheless sound absorption measurements are very useful in many cases, when the other relaxation techniques described above do not provide sufficiently large signals and/or simply do not have a sufficient time resolution.

2.4 Processing of Relaxation Data

The main practical problem in relaxation experiments is the correct assignment of relaxation times and amplitudes. This is not a trivial problem, especially in the case of multiexponential relaxation curves. A correct assignment requires experimental data of a high signal to noise ratio S/N. Since this ratio S/N decreases with decreasing response time of the detector according to

$$S/N \sim \sqrt{I \cdot \tau_D} \qquad (5)$$

the noise problem is particularly serious for measurements which require a high time resolution and thus a small τ_D. As shown by equation (5) S/N may be increased by increasing the light intensity I (in photoelectric detection). However, many biological systems may suffer photodamage by too high light intensities. In these cases S/N may be increased by signal averaging. Another possibility is an increase of τ_D close to the chemical relaxation time constant. By this procedure S/N is increased according to equation (5). Then the "convolution" of the chemical relaxation effect with the detector time constant has to be considered in the evaluation. The "deconvolution" is relatively easy in the case of simple exponential time constants[10,11]. Deconvolution procedures cannot be avoided, when relaxation time constants are determined, which are close to the limit of time resolution of the apparatus used for the measurements.

Before computers were available, the evaluation of exponentials was quite tedious and time consuming. With the aid of computers the evaluation can be rather comfortable. For example, programs are available which examine automatically experimental relaxation curves for the number of processes and determine both time constants and amplitudes[32].

3. THEORETICAL

In most relaxation experiments the chemical equilibrium is perturbed only to a relatively small degree. Owing to the small perturbation it is possible to linearise the rate equations, which simplifies the calculations considerably, especially for systems with several coupled reactions[10,11]. A single reaction step is reflected by one relaxation process (cf. Fig. 3)

$$c(t) = c_1 + (c_2-c_1)*(1-\exp(-t/\tau)) \qquad (6)$$

with a time constant τ dependent upon the rate constants and the concentrations. The type of reaction can be characterised by measurements of the concentration dependence, as can be seen from the compilation of theoretical expressions for the reciprocal relaxation time constants for several types of reactions (Table I). From the concentration dependence of the relaxation times it is possible to determine the rate constants, i.e. also the equilibrium constants.

When several reactions are coupled to each other, one relaxation effect is expected for each independent reaction step. The coupling between the different reactions is then analoguous to the coupling of vibrations in a system of several oscillations. The resulting resonance frequencies correspond to the "normal modes" of vibration rather than to the single modes of isolated oscillators. In a similar way the relaxation times of coupled reaction systems correspond to the normal modes of reactions. The calculation of these normal modes is simple in principle. Difficulties due to relatively complex arithmetic expressions may usually be avoided by restriction to limit conditions like high concentration of one of the reactants compared to the others. It is also possible, of course, to calculate complete relaxation spectra without approximations by using computers; easy-to-use programs are available[33].

In addition to the time constants the relaxation curves contain another very useful information: the amplitudes directly depend upon thermodynamic parameters and thus can be used for the determination of e.g. equilibrium constants. In the case of a temperature jump experiment the change of the concentration induced by a temperature change is given by

Table I

Reaction	Reciprocal relaxation time
$A \underset{k_{-1}}{\overset{k_1}{\rightleftharpoons}} B$	$1/\tau = k_1 + k_{-1}$
$A+B \underset{k_{-1}}{\overset{k_1}{\rightleftharpoons}} C$	$1/\tau = k_1(c_A+c_B) + k_{-1}$
$2A \underset{k_{-1}}{\overset{k_1}{\rightleftharpoons}} B$	$1/\tau = 4\ k_1\ \bar{c}_A + k_{-1}$
$nA \underset{k_{-1}}{\overset{k_1}{\rightleftharpoons}} B$	$1/\tau = n^2 k_1 (c_A)^{n-1} + k_{-1}$
$A+B \underset{k_{-1}}{\overset{k_1}{\rightleftharpoons}} C+D$	$1/\tau = k_1(\bar{c}_A+\bar{c}_B)+k_{-1}(\bar{c}_C+\bar{c}_D)$

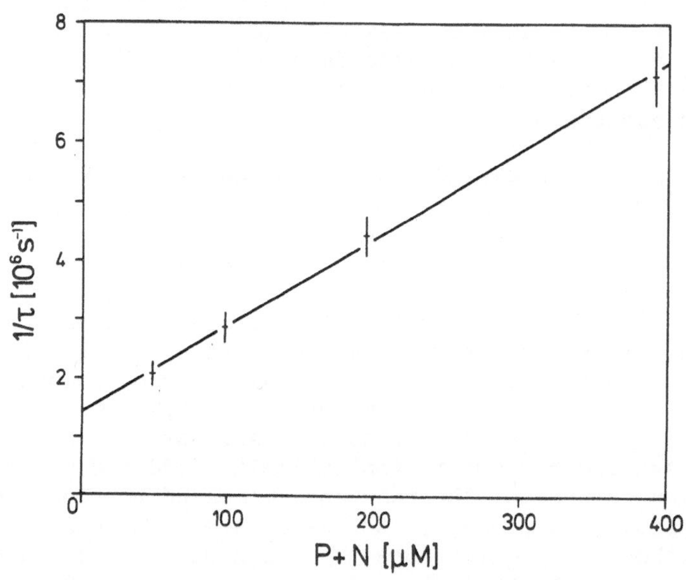

Fig. 11 Reciprocal relaxation time $1/\tau$ as a function of the sum of free concentrations for the binding of the oligopeptide $(Lys)_3$ to the oligonucleotide $I(pI)_5$. For this simple $A+B \rightleftharpoons C$ reaction the rate constant of recombination $k_R = 1.47 \cdot 10^{10}\ M^{-1}s^{-1}$ corresponds to the slope and the rate constant of dissociation $k_D = 1.4 \cdot 10^6\ s^{-1}$ to the intercept of the straight line defined by the experimental data[60]

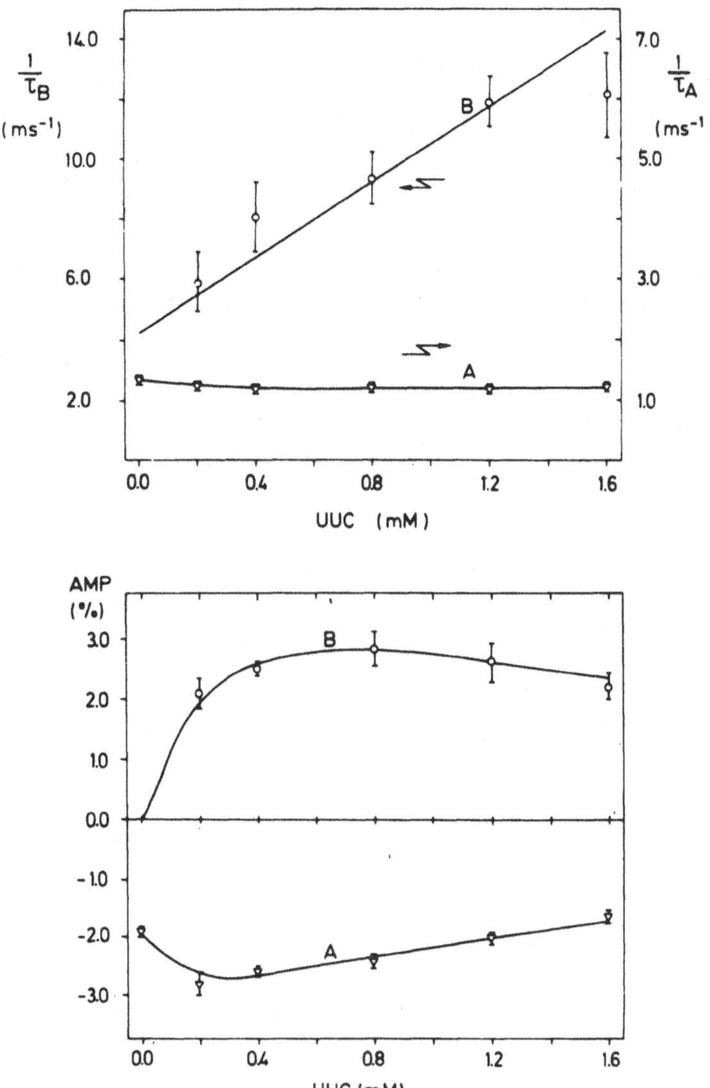

Fig. 12 Binding of UUC to $tRNA^{Phe}$: example of relaxation time constants and amplitudes for a two step reaction. The anticodon loop of the tRNA exists in two conformations F and H; one of these (H) has a preferential affinity for the codon. One relaxation process (A) is already observed for tRNA alone; addition of UUC leads to some change of process (A) and the appearance of a second process (B), which essentially represents the bimolecular step. The solid lines show the representation of data according to the following mechanism (from ref. 50):

$$tRNA_F \underset{93\ s^{-1}}{\overset{740\ s^{-1}}{\rightleftharpoons}} tRNA_H \qquad tRNA_H + UUC \underset{2600\ s^{-1}}{\overset{5*10^6 M^{-1} s^{-1}}{\rightleftharpoons}} tRNA_H*UUC$$

Table II

		Detection*	Time range	Volume
Stopped Flow	Mixing	A, F, S, R, CD	1ms - "∞" (5µs)	200 µl/ experiment
Temperature jump	capacitor discharge	A, F, S, R, CD	1 µs - 5 s	50 µl - 10 ml
	cable discharge	A	10ns - 10ms	50 to 100 µl
	laser pulse	A, C	10ns - 10ms	100 µl
	microwave	A, C	1 µs - 10ms	100 µl
Pressure jump	rupture of metal membrane	A, F, S, C, T	100µs - "∞"	0.3 to 5 ml
Field jump	cable technique	A, F, S	10ns - 10ms	100µl - 5ml
Sound absorption	ultrasonic wave	dissipation of energy	1 ns - 2 µs (30ps-30µs)	1ml (- 100ml)
Dielectric absorption	high electric field with high frequency superimposed	dielectric loss	5 ns - 1 µs	1 ml

*Abbreviations: A absorbance, F fluorescence, S straylight, R optical

Type of Reaction	Advantages	Disadvantage	References
all types of complexation reactions	simple, large signals	large volumes, linearisation of rate equations usually not possible	2 - 7
all temperature-dependent reactions, (e.g. proton transfer, metal complexes, enzyme-ligand, antibody-hapten, nucleic acid helix coil)	widely applicable	high salt required	10 - 15
	high time resolution	high salt required	16, 17
	high time resolution also for low salt	relatively complex apparatus	19
	"	"	18
reactions with volume change (e.g. ion reactions, protein association, membrane phase transition)	simple, does not require addition of salt, measurements up to time "∞"	not very fast	12, 20, 21, 35
electrolyte reactions (e.g. proton transfer, metal complexes, field induced reactions in biopolymers	high time resolution, selective perturbation of electrostatic interactions	relatively complex apparatus only for low salt	22 - 25
reactions with volume change (e.g. ion reactions, hydrophobic complex formation	high time resolution	laborious and time consuming, more than a single apparatus required	30, 35
dipole reactions in unpolar solvents (e.g. hydrogen bonding)	high time resolution	difficult, only non-conducting solvents	24, 25

rotation, CD circular dichroism, C conductometry, T thermometry

$$\Delta C = \Gamma \cdot \frac{\Delta H}{RT^2} \cdot \Delta T$$

with ΔH reaction enthalpy, T absolute temperature, R gas constant. Γ is a factor, which depends upon the stoichiometry of the reaction. In the case of a simple reaction

$$A + B \rightleftharpoons C$$

$$\Gamma = \frac{K \cdot \bar{c}_A \cdot \bar{c}_B}{1 + K(\bar{c}_A + \bar{c}_B)}$$

where K is the association constant and $\bar{c}_A$ ($\bar{c}_B$) the equilibrium concentration of species A(B). From the concentration dependence of the amplitude it is possible to determine the equilibrium constant K and also the enthalpy change ΔH. In many cases the amplitude procedure provides more accurate information than other procedures (for a recent detailed discussion of amplitude analysis cf. ref. 34).

4. COMPARISON OF METHODS FOR THE ANALYSIS OF FAST REACTIONS

Due to the limited space available the present contribution can give only a very broad survey over the various techniques available for the analysis of fast reactions. Nevertheless the general background given in the present survey should be useful for a first orientation about possible applications of fast reaction techniques. Essential information about the various procedures, their advantages and limitations, can be obtained with the aid of Table II. The readers, who want to know more details, should consult the literature given in the last column of this Table. Some other techniques, which have not been discussed in the present survey, can also be used to investigate reaction kinetics. For example nuclear magnetic resonance, electromagnetic resonance, fluorescence decay, electrochemical methods and pulse or flash radiolysis may be useful in some cases[35].

5. APPLICATIONS

Fast reaction techniques have been used for the investigation of many different reactions involving all kinds of molecules found in biological system. It will not be possible to discuss these applications in detail. Instead some reactions, which are typical for various classes of biological molecules, are listed below together with the methods used for their analysis and references for further information.

Proteins:
α-helix coil transition * ultrasound[36], field-jump[37]
de- and re-naturation * temp.-jump[38]
ligand binding * temp.-jump[39]
allosteric transition * temp.-jump[40]
antibody-hapten interactions * temp.-jump.[41]

Nucleic Acids
syn-anti transition in mononucleotides * ultrasound[42]
base-stacking of monomers * ultrasound[43]
Base stacking in single stranded nucleic acids * cable temp.-jump[44], laser temp.-jump[45],
fraying of helix ends * cable temp.-jump[46]
double helix coil transition * temp.-jump[47,48]
anticodon-anticodon interaction of tRNA * temp.-jump[49]
codon recognition by tRNA * temp.-jump[50]
ion binding * field-jump[51]
intercalation * temp.-jump[52]

Nucleic Acid - Protein Interaction:
peptide binding to polynucleotides * field-jump[28,60]
ribosome association dissociation * press.-jump[53]
tRNA-synthetase * quenched flow[9] temp.-jump[54]
DNA-polymerase * temp.-jump[55]

Membranes:
carrier transport * temp.-jump, ultrasound[56]
phase transition * stopped flow[58], press.-jump[59]

ACKNOWLEDGEMENT

The author had the opportunity to "relax" many years in the department, where most of the techniques described above have been developed by M. Eigen and his coworkers. Most of the author's knowledge on chemical relaxation (and on many other fields of science) comes from M. Eigen.

REFERENCES

1. H.Hartridge and F.J.W.Roughton, A Method of Measuring the Velocity of Very Rapid Chemical Reactions, Proc. Roy. Soc. (London) A 104:376 (1923)

2. B.Chance, Rapid Flow Methods, in: "Techniques of Chemistry", Vol. VI, Part II, G.G.Hammes, ed., Wiley Interscience, New York (1974)

3. Q.H.Gibson , Rapid Mixing: Stopped Flow, in: "Methods in Enzymology", Vol. XVI, K.Kustin, ed., Academic , New York (1969)

4. K.Hiromi, "Kinetics of Fast Enzyme Reactions", Wiley, New York (1979)

5. P.M.Bayley and M.Anson, Stopped-Flow Circular Dichroism: A New Fast Kinetic System, Biopolymers 13:401 (1974)

6. I.Luchins and S.Beychok, Far-Ultraviolet Stopped-Flow Circular Dichroism, Science 199:425 (1978)

7. B.Balko, P.Brown, R.L.Berger and K.Anderson, Fast Stopped-Flow Microcalorimeter, J. Biochem. Biophys. Meth. 4:1 (1981)

8. P.Davidovits and S.C.Chas, Kinetics with Microsecond Mixing of Liquid Reactants, Anal. Chem. 52:2435 (1980)

9. A.R.Fersht and R.Jakes, Demonstration of Two Reaction Pathways of the Aminoacylation of tRNA. Application of the Pulsed Quenched Flow Technique, Biochemistry 14:3350 (1975)

10. M.Eigen and L.DeMaeyer, Relaxation Methods, in: "Technique of Organic Chemistry", Vol. VIII, Part II, A.Weissberger, ed., Wiley Interscience, New York (1963)

11. C.F.Bernasconi, "Relaxation Kinetics", Academic Press, New York (1976)

12. H.Strehlow and W.Knoche, "Fundamentals of Chemical Relaxation", Verlag Chemie, Weinheim (1977)

13. G.G.Hammes, Temperature-Jump Methods, in: "Techniques of Chemistry", Vol. VI, Part II, G.G.Hammes, ed., Wiley Interscience, New York (1974)

14. R.Rigler, C.R.Rabl and T.M.Jovin, A Temperature Jump Apparatus for Fluorescence Measurements, Rev. Sci. Instrum. 45:580 (1974)

15. P.Bayley, S.Martin and M.Anson, Temperature-Jump Circular Dichroism: Observation of Chiroptical Relaxation Processes at Millisecond Time Resolution, Bio. Bio. Res. Comm. 66:303 (1975)

16. G.W. Hoffman, A Nanosecond Temperature-Jump Apparatus, Rev. Sci. Instrum. 42:1643 (1971)

17. D.Pörschke, Cable Temperature Jump Apparatus with Improved Sensitivity and Time Resolution, Rev. Sci. Instrum. 47:1363 (1976)

18. E.F.Caldin and I.E.Crooks, A Microwave Temperature-Jump Apparatus for the Study of Fast Reactions in Solution, J. Sci. Instrum. 44:449 (1967)

19. T.D.Dewey and D.H.Turner, Raman Laser Temperature-Jump Kinetics, Adv. Mol. Relax. Int. Proc. 13:331 (1978)

20. B.Gruenewald and W.Knoche, Recent Developments and Applications of Pressure Jump Methods, in: "Techniques and Applications of Fast Reactions in Solution", W.J.Gettins and E.Wyn-Jones, eds., Reidel (1979)

21. R.M.Clegg and B.W.Maxfield, Chemical Kinetic Studies by a New Small Pressure Perturbation Method, Rev. Sci. Instrum. 47:1383 (1976)

22. H.H.Grunhagen, A High Power Square Wave Pulse Generator for the Investigation of Fast Electric Field Effects in Solution, Messtechnik 19:23 (1974)

23. H.H.Grunhagen, Fast Spectrophotometric Detection System for Coupled Physical and Chemical Electric Field Effects in Solution, Biophysik 10:347 (1973)

24. L.C.M.DeMaeyer, Electric Field Methods, in: "Methods in Enzymology", Vol. XVI, K.Kustin, ed., Academic, New York (1969)

25. L.DeMaeyer, Electric Field Methods, in: "Techniques of Chemistry" Vol. VI, Part II, G.G.Hammes, ed., Wiley Interscience, New York (1974)

26. L.Onsager, Deviations from Ohm s Law in Weak Electrolytes, J. Chem. Phys. 2:599 (1934)

27. D.Pörschke, Threshold Effects Observed in Conformation Changes Induced by Electric fields, Biopolymers 15:1917 (1976)

28. H.Labhart, Methoden der Zuordnung von Absorptionsbanden von Farbstoffen zu berechneten Uebergaengen, Chimia 15:20 (1961)

29. D.Pörschke, The Binding of Arg- and Lys-peptides to Single Stranded Polyribonucleotides and its Effect on the Polymer Conformation, Biophys. Chem. 10:1 (1979)

30. M.Eigen, G.Kurtze and K.Tamm, Zum Reaktionsmechanismus der Ultraschallabsorption in waessrigen Elektrolytloesungen, Z. Elektrochem. 57:103 (1953)

31. F.Eggers and T.Funck, Ultrasonic Relaxation Spectroscopy in Liquids, Naturwissenschaften 63:280 (1976)

32. S.W.Provencher, A Fourier Methods for the Analysis of Exponential Decay Curves, Biophys. J. 16:27 (1976)

33. G.Ilgenfritz, Theory and Simulaton of Chemical Relaxation Spectra, in: "Chemical Relaxation in Molecular Biology", I.Pecht and R.Rigler, eds., Springer, Berlin (1977)

34. R.Winkler-Oswatitsch and M.Eigen, The Art of Titration. From Classical End Points to Modern Differential and Dynamic Analysis, Angew. Chem. Intern. Ed. 18:20 (1979)

35. G.G.Hammes, ed., Investigation of Rates and Mechaniams of Reactions, Part II, in: "Techniques of Chemistry", Vol. VI, Wiley Interscience, New York (1974)

36. G.Schwarz and J.Engel, Kinetik kooperativer Konformationsumwandlungen von linearen Biopolymeren, Angew. Chem. 84:615 (1972)

37. T.Sano and T.Yasunaga, Kinetics of the Helix-Coil Transition of Polypeptides in Solution by the Relaxation Methods, Biophys. Chem. 11:377 (1980)

38. T.Y.Tsong, R.L.Baldwin and E.L.Elson, The Sequential Unfolding of Ribonuclease A: Detection of a Fast Initial Phase in the Kinetics of Unfolding, Proc. Nat. Acad. Sci. US 68:2712 (1971)

39. K.Kirschner, E.Gallego, I.Schuster and D.Goodall, Co-operative Binding of Nicotinamide-Adenine-Dinucleotide to Yeast Glyceraldehyde- 3-Phosphate Dehydrogenase, J. Mol. Biol. 58:29 (1971)

40. K.Kirschner, M.Eigen, R.Bittman and B.Voigt, The Binding of Nicotinamide-Adenine Dinucleotide to Yeast D-Glyceraldehyde-3-Phosphate Dehydrogenase: Temperature Jump Relaxation Studies on the Mechanism of an Allosteric Enzyme, Proc. Nat. Acad. Sci. US 56:1661 (1966)

41. I.Pecht and D.Lancet, Kinetics of Antibody-Hapten Interactions, in: "Chemical Relaxation in Molecular Biology", I.Pecht and R.Rigler, eds., Springer, Berlin (1977)

42. L.M.Rhodes and R.R.Schimmel, Nanosecond Relaxation Processes in Aqueous Mononucleoside Solutions, Biochemistry 10:4426 (1971)

43. D.Pörschke and F.Eggers, Thermodymanics and Kinetics of Base-Stacking Interactions, Eur. J. Biochem. 26:490 (1972)

44. D.Pörschke, The Nature of Stacking Interactions in Polynucleotides. Molecular States in Oligo- and Polyribocytidylic Acids by Relaxation Analysis, Biochemistry 15:1495 (1976)

45. T.G.Dewey and D.H.Turner, Laser Temperature Jump Study of Solvent Effects on Poly(Adenylic Acid) Stacking, Biochemistry 19:1681 (1980)

46. D.Pörschke, A Direct Measurement of the Unzippering Rate of Nucleic Acid Double Helix, Biophys. Chem. 2:97 (1974)

47. D.Pörschke, Elementary Steps of Base Recognition and Helix-Coil Transitions in Nucleic Acids, in: "Molecular Biology, Biochemistry and Biophysics" Vol. 24, I.Pecht and R.Rigler, eds., Springer, West Berlin (1977)

48. M.E.Craig, D.M.Crothers and P.Doty, Relaxation Kinetics of Dimer Formation by Self Complementary Oligonucleotides, J. Mol. Biol. 62:383 (1971)

49. H.J.Grosjean, S.DeHenau and D.M.Crothers, On the Physical Basis for Ambiguity in Genetic Coding Interactions, Proc. Nat. Acad. Sci. US 75:610 (1978)

50. D.Labuda and D.Pörschke, Multistep Mechanism of Codon Recognition by Transfer Ribonucleic Acid, Biochemistry 19:3799 (1980)

51. D.Pörschke, The mode of Mg^{++} binding to Oligonucleotides. Inner Sphere Complexes a Markers of Recognition?, Nucl. Ac. Res. 6:883 (1979)

52. J.Ramstein, M.Ehrenberg and R.Rigler, Fluorescence Relaxation of Proflavin-Desoxyribonucleic Acid Interaction. Kinetic Properties of a Base-Specific Reaction, Biochemistry 19:3938 (1980)

53. E.Schulz, R.Jaenicke and W.Knoche, Pressure Jump Relaxation Studies of the Association-Dissociaton Reaction of E.coli Ribosomes, Biophys. Chem. 11:253 (1976)

54. G.Krauss, D.Riesner and G.Maass, Mechanism of Discrimination between Cognate and Non-Cognate tRNAs by Phenylalanyl-tRNA Synthetase from Yeast, Eur. J. Biochem. 68:81 (1976)

55. T.M.Jovin and G.Striker, Chemical Relaxation Kinetic Studies of E.coli RNA Polymerase Binding to Poly d(AT) using Ethidium Bromide as a Fluorescence Probe, _in:_ "Chemical Relaxation in Molecular Biology", I.Pecht and R.Rigler, eds., Springer, Berlin (1977)

56. E.Grell and I.Oberbaumer, Dynamic Aspects of Carrier-Mediated Cation Transport through Membranes, _in:_ "Chemical Relaxation in Molecular Biology", I.Pecht and R.Rigler, eds., Springer, Berlin (1977).

57. M.Dourlent, J.F.Hogrel and C.Helene, Anisotropy Effects in Temperature Jump Relaxation Studies on Solutions Containing Linear Polymers, _J. Amer. Chem. Soc._ 96:3398 (1974)

58. U.Strehlow and F.Jaehnig, Electrostatic Interactions at Charged Lipid Membranes. Kinetics of the Electrostatically Triggered Phase Transition, _Bio. Bio. Acta_ 641:301 (1981)

59. B.Grunewald, A.Blume and F.Watanabe, Kinetic Investigations of the Phase Transition of Phospholipid Bilayers, _Bio. Bio. Acta_ 597:41 (1980)

60. D.Pörschke, Thermodynamic and Kinetic Parameters of Oligonucleotide- Oligopeptide Interactions, _Eur. J. Biochem._ 86:291 (1978)

POTENTIAL ENERGY FUNCTIONS FOR STRUCTURAL MOLECULAR BIOLOGY

Shneior Lifson

Department of Chemical Physics

Weizmann Institute of Science, 76100 Rehovot, Israel

INTRODUCTION

Structural molecular biology is first and foremost an experimental science. This is quite understandable. As long as we did not know *what* is the structure of enzymes how could we ask questions on *how* do they obtain their structure or how is their structure determining their catalytic function? However, as our knowledge of facts about biological structures grows now at a tremendous rate, there is an ever-growing need to apply theory, and calculations based on theory, to supplement the experimental study of structural molecular biology.

The potential energy of intra- and inter-molecular interactions is one of the central concepts in a theoretical approach to problems of structure and function in biology. What do we know about the molecular potential energy in biomolecules and related organic compounds? Where does this knowledge come from? How reliable or approximate is it? This is our first subject. We then will ask what are the potentialities and limitations of its uses? How do we relate the potential energy to experimental data on structural, dynamic and thermodynamic properties of molecules? What are the theoretical considerations and which are the computational algorithms available? Finally, we shall point out the advantages of an objective search for energy functions and energy parameters, and of establishing a "bench-mark" for comparing alternative force fields.

INTRA-MOLECULAR POTENTIAL ENERGY FUNCTIONS

We start our discussion of the use of potential energy functions in structural biology by a concise review of the quantum mechanical and the empirical basis of our qualitative understanding and quantitative determination of the potential functions of molecular interactions. This is necessary because an intelligent and useful application of potential functions in biology requires a feeling for the power and limitations of both the theoretical and the empirical approach, and for the complex and subtle nature of their interrelations.

The nature of molecular forces is in principle well understood, thanks to quantum mechanics, which "made theoretical chemistry a branch of applied mathematics" according to the famous exaggeration by Dirac. Atoms and molecules are made of nuclei and electrons, and therefore the potential energy of interatomic and intermolecular interactions originates from the Coulomb (electrostatic) interactions

$$V_{Coulomb} = \frac{1}{2} \sum_{i,j} e_i e_j / r_{ij} \tag{1}$$

between the nuclei and electrons which form the atomic and molecular assemblies. The dynamic behavior of such assemblies is, however, controlled by the laws of quantum mechanics, i.e. by the Schrödinger equation, in which the kinetic energy of the Hamiltonian of the system becomes a differential operator while the potential energy is given by Eq. (1), and the eigenvalues represent the energy states of the system. An exact solution of the Schrödinger equation is available only for very few simple models, and various approximations have been introduced for treating more complicated cases. Among these the Born-Oppenheimer (BO) approximation is of great importance for the understanding of the nature of potential functions and their application. It is based on the fact that electrons are lighter than nuclei by several orders of magnitude (the mass of a proton is ~2000 times the mass of an electron). The BO approximation treats the Schrödinger equation on two levels: electronic and nuclear.

At the first level one assumes that the nuclei are at rest and considers the quantum mechanical behavior of the electrons in the

*This is indeed a very good approximation, as can be seen from the spectra of molecules. The frequencies of electronic transitions are orders of magnitude higher than those of molecular vibrations. Consequently, the nuclei may be considered stationary on the time scale of electronic vibrations.

"external" field of the nuclear charges.* Thus the electronic eigenvalues are considered as continuous functions of the molecular structure, where the molecular structure is represented by the Cartesian coordinates of the atoms, or alternatively by the internal coordinates (bond lengths, bond angles and torsional angles) of the molecular systems. In other words, for electronic states of a molecule, the energy of the molecule is represented in the BO approximation by the so-called "energy surface", or "Born-Oppenheimer surface" of the molecule. The lowest point on this surface corresponds to the stable structure (configuration or conformation) of the molecule, while other, local minima correspond to metastable structures.

At the second level of the Born-Oppenheimer approximation, the Hamiltonian of the Schrödinger equation for the dynamic behavior of the nuclei is comprized of the kinetic energy for the atoms of the molecule as differential operator, and of the BO surface as the potential energy of interatomic interaction. Thus the BO potential energy replaces the Coulomb energy as the potential of atomic interactions. The empirical and semi-empirical energy functions used in energy calculations in chemistry and molecular biology are, from the quantum mechanical point of view, approximations to the BO surface. There is, therefore, no sound theoretical basis for the common belief that empirical potentials are "classical", as against *ab initio* and other molecular orbital calculations which are "quantum mechanical". The true distinction between these concepts or approaches is that molecular orbital calculations produce approximate values of points on the BO surface, i.e. aim at the first level of the BO approximation, while empirical functions are approximations to the BO potential energy surface, i.e. aim at the second level of the BO approximation.

Quantum chemistry of semi-empirical and *ab initio* methods have advanced during the years, improving the solutions of the Schrödinger equation in the BO approximation for small and medium size molecules of up to about 10-20 atoms, with the aid of modern computers. By spending sufficient computer time, it is possible now to obtain a numerical mapping of BO surfaces. However, quantum chemistry is incapable, at least in the foreseeable future, to yield mathematical analytical description of BO surfaces of polyatomic molecules, and this is, as we shall see below, the reason for the need for and importance of empirical potential energy functions. Let us now see how the various energy potentials used in molecular calculation have been derived, either from theoretical quantum mechanical or from empirical considerations. We shall start with the bond potential.

The Bond Potential

A notably successful guess of the analytic form of a BO surface of diatomic molecules is due to Morse[1], who proposed the function now called the Morse potential, for the electronic ground state of a diatomic molecule as a function of its bond length b:

$$V_M(b) = D[\exp(-\alpha(b-b_o))-1]^2-D \tag{2}$$

where D, α and b_o are adaptable constant parameters. $V_M(b)$ has a minimum, -D, at b_o and vanishes at $b=\infty$, thus identifying D and b_o with the equilibrium bond energy and bond length, respectively. Morse solved the Schrödinger equation for the atomic nuclei (the second level of the Born-Oppenheimer approximation). He obtained, using only one more adaptable parameter, α, a good approximation for the whole molecular vibration spectrum of diatomic molecules, including its anharmonic overtones and combinations, a unique and conspicuous success of the merger of empirical and quantum mechanical considerations.

Although the Morse potential was derived for diatomic molecules only, it probably describes fairly well the potential of the chemical bond in polyatomic molecules as well. However, the variability of the chemical bonds has a very narrow range, of the order of 0.1Å. Therefore, for applications to structural molecular biology, chemical bonds may be assumed as rigid, or else a quadratic (harmonic) approximation to the bond potential is amply sufficient

$$V_{bond} = \frac{1}{2} K_b(b-b_o)^2 \tag{3}$$

where K_b is related to the parameters of the Morse potential by $K_b=2D\alpha^2$. This is not the case when vibrational spectra are concerned. Here the Morse potential is significantly superior over the harmonic potential.

BO Surfaces of Polyatomic Molecules: Bond Angles and Torsional Angles

Bond angles variability is much more important to structural analysis of polyatomic molecules than the variability of bond lengths, but no theoretical basis for an analytic presentation of bond angle energy potentials has been offered as yet. However, quantum mechanics has given us a comprehensive conceptual insight into the nature of the forces which determine bond angles and their variability. Thanks to quantum mechanics we know that atoms tend to have closed valence electron shells; that bonds are formed by bond orbitals of valence electron pairs, one electron pair for each valence; that electrons in the closed valence shells which do not participate in chemical bonds appear always in pairs, and that these "lone pair electrons" form "lone pair orbitals" which resemble to some extent bond orbitals. We know that the regions of lowest energy on the BO surface, which belong to the observed equilibrium structure of molecules, possess certain symmetry properties with respect to the bond orbitals, whose origin is the mutual repulsion between adjacent orbitals. We shall now see how these trends affect the bond angles and the torsional angles.

The Bending Potential

The angles between bonds connected to a given atom are essentially determined by the tendency of the bonds to be as far apart as possible. Thus, if only two bond orbitals are connected to a single atom, as in BeH_2, they will be colinear. Similarly, three orbitals connected to an atom are always coplanar, with bond angles of around 120°. Each bond angle separately would tend to increase, but since it is impossible for one of the three bond angles in the plane to increase without decreasing the others, the minimum of the total bending potential energy is reached when the three bond angles are about equal. (The angles would be exactly equal for idential bonds, as one would expect to observe in CH_3^+). Similarly, four bonds connected to an atom form a tetrahedral structure*, with all six angles between the four bonds equal to the tetrahedral angle 109,47° when the four bonds are identical, as is the case in CH_4. Deviations from the tetrahedral angle occur when the four bonds are not equal, as is the case in most tetrahedral carbon atoms. (The deviations are usually small, 2 to 3°, unless the molecule is strained by other forces, as is the case in a number of ring molecules, like cyclobutane, cyclodecane and other cycloalkanes, where the closure of the ring imposes significant deviations from the tetrahedral angles, or in overcrowded molecules like tri-tertiary butyl-methane). In the same way, when five bonds are connected to an atom, as in pentavalent phosphorus compounds, three bonds are coplanar (equatorial) and two are perpendicular to the plane of the others (axial). This is the highest symmetry available for such a structure, and the fact that it is an energy minimum point can be derived from symmetry considerations.

Deviations from these rules occur when lone pairs of electrons form bond-like orbitals. For example, the two OH bonds in the water molecule form an angle of 105°, but there are two lone pairs of electrons which are connected to the oxygen atom and form lobes which behave like chemical bonds in the sense that they repel each other and the adjacent OH bonds. Thus, the two OH bonds and the two lone pairs form a tetrahedral structure around the oxygen atom. Similarly, the three NH bonds and the one lone pair of NH_3 form a tetrahedral structure, so that NH_3 is pyramidal, not planar.

*Tetrahedral structure means that the central atom is considered to be at the centre of a tetrahedron while the 4 atoms bonded to it are located at the vertices.

Thus, the qualitative nature of the intramolecular bending energy may be considered to be well understood. However, we are still unable to describe this energy quantitatively. Therefore, the common empirical representation of the bending potential of a polyatomic molecule is given in terms of the deviation of the bond angles θ from the "equilibrium value" or "reference value" θ_o, and is usually assumed to be quadratic, i.e. of the form

$$V_{angle}(\theta) = \frac{1}{2} K_\theta (\theta-\theta_o)^2 \qquad (4)$$

where K_θ is the "bending-force constant", a term taken from the theory of vibrational spectra, and θ_o is the reference angle. The reference angle is assumed to be 120° for coplanar trivalent structures and 109.47° for tetrahedral structures; alternatively, θ_o may be used as an adaptable parameter, like K_θ, where the values are adapted to give the best fit with the experimental data.

By definition, this empirical potential is good for small deviations from the equilibrium angle. For larger deviations, like those mentioned above for some cyclodecane molecules, it is not satisfactory. Unfortunately, various efforts to replace it by some better functions have not so far been particularly successful. It is, however, worth bearing these limitations in mind.

The Torsional Potential

When a bond connects two polyatomic groups in a molecule, as for example the C-C bond in CH_3-CH_3, it serves as an axis for the rotation of the two groups relative to each other. Such a rotation is called internal rotation, and the angle of internal rotation is called torsional angle (or also dihedral angle). The torsional potential is due to repulsive interactions between the orbitals of the two groups. When they are nearest, e.g. in the "eclipsed" state of ethane, the potential is highest. When they are farthest, as in the "staggered" state, the potential is lowest. Much is known about the torsional potential in hydrocarbons, mainly due to Pitzer[2] in honor of whom it is often called "the Pitzer potential". The simple empirical form of the torsional potential is

$$V_{torsion}(\phi) = \frac{1}{2} K_\phi (1+\cos n(\phi-\phi_o)) \qquad (5)$$

where ϕ is the torsional angle, K_ϕ is the height of the potential, n is its periodicity, and ϕ_o is the reference angle where V_ϕ is maximum. For tetrahedral carbons, as for example in the alkane chains $\cdots$-CH_2-CH_2-$\cdots$ occuring in many R groups of amino acids, K_ϕ is about $\sim$3kcal and the periodicity is 3. The potential is very high, K_ϕ=$\sim$40 kcal/mole and the periodicity is n=2 for rotation around the double bond -C=C- as in ethylene and its derivatives. The peptide bond C' ⩦ N in proteins is known to have a partial double bond character and its torsional potential height is about 20 kcal/mole, with n=2. On the other hand, the torsional potentials of the two other bonds of the backbone of proteins, N-C^{α} and C^{α}-C' are very low, because only the orbitals around C^{α} form a tetrahedral structure, while the orbitals around C' and N are planar, the first because of the double bond C'=O, and second because of the partial double bond character of the C' ⩦ N bond. In fact, little is known about these torsional potentials, and even the periodicity and location of the minima are disputable.

Out-of-Plane Torsion

When three bond orbitals are confined to be planar, as are the bonds around the ethylenic double-bonded carbons, or the bonds around the C' and N which form the planar peptide unit, there is a force which resists the distortion of the planar structure. The out-of-plane torsion can be envisaged as the angle χ between the planes through the points (1,2,3) and (1,2,4) in the structure $1{=}2{<}^{3}_{4}$, which intersect along the axis 1=2. The potential of such out-of-plane distortion may be represented, for small angles, by

$$V_{\chi} = \frac{1}{2} K_{\chi} \chi^{2} \tag{6}$$

Spectroscopic analysis of ethylenes tells us that the stiffness of V_{χ} is about the same as that of V_{ϕ} for rotational torsion of the C=C double bond. It has been noted that the peptide bond in proteins is often distorted, i.e. values of the torsional angle ω of the peptide bond up to around 10° may occur as a result of internal strains in the molecule. Out-of-plane distortions of the peptide bond are as likely and as abundant as those of the torsional distortions.

The Non-Bonded Intramolecular Potential

Interactions between groups of atoms in a large molecule which are close to each other but not linked by chemical bonds constitute the last contribution in our list of intramolecular interactions. These are, however, not different by their nature from the intermolecular interactions. Indeed the study of intermolecular interactions in gases and crystals has been a rich and reliable source of information about these interactions, since in the study of intermolecular interactions the intramolecular energy may be neglected in many instances, when the molecular structure may be considered to be rigid so that the intramolecular energy remains constant.

INTERMOLECULAR POTENTIAL ENERGY FUNCTIONS

Intermolecular interactions, like the intramolecular interactions, are of electrostatic origin. We distinguish between electrostatic interactions between charged or polar groups, which obey the Coulomb law, and other interactions which exist always, even in completely non-polar molecules. The latter are often called Van der Waals interactions because Van der Waals was the first to estimate them quantitatively in his famous study of the deviations of real gases from the Boyle-Mariot law of ideal gases. Van der Waals interactions are repulsive at the short range of atomic contacts, and attractive at long range.

The Attractive Dispersion Force

The quantum mechanical theory of the long range attraction was derived by London[3], and although its mathematical derivation is rather elaborate, its physical basis and qualitative explanation are rather simple and clear. Consider two electro-neutral atoms at a distance long enough so that their electron clouds do not overlap. At first approximation the electrostatic force between them is zero. However, the electrons of both atoms are in constant motion and therefore possess instantaneous, fluctuating, dipole moments whose magnitude depends on the atomic polarizabilities. The fluctuating dipoles of the two atoms interact with each other, thus producing a net attractive force. London showed that the energy function of this attraction at a distance r has the form of the power series

$$V_{London} = -C/r^6 - C'/r^8 - C''/r^{10} \ldots \quad (7)$$

where the first term is the leading one at large enough distances. An extensive literature exists on the so-called dispersive or London attraction potential. Various theoretical derivations for the coefficient C have been proposed, and the one best fitting to experiments on noble gasses (within 15-20%) is that due to Slater and Kirkwood[4]

$$C_{a,b} = (3eh/2m_e^{1/2})\alpha_a\alpha_b[(\alpha_a/N_a)^{1/2} + (\alpha_b/N_b)^{1/2}]^{-1} \quad (8)$$

where e, h and m_e are universal constants, a and b denote the kinds of atoms, α their polarizabilities, and N their number of electrons (preferably only p electrons) in the outer shell. London's theory is strictly speaking applicable to pairs of atoms with spherically symmetric distribution of electrons. However its importance goes far beyond this limitation. Since dispersion forces are pair-wise (approximately) additive, they are the basis of our understanding of the intermolecular forces which hold molecules of solids and liquids together, as well as the intramolecular forces between non-bonded atoms in polyatomic molecules. It has been consequently generally assumed, although never rigorously proved, that dispersion interactions in polyatomic molecules are given by the sum of such interactions between non-bonded atoms, both between the molecules and inside each molecule. In the molecule atoms are defined as non-bonded, for both the attractive and the repulsive interactions, if they are separated by at least 3 or 4 consecutive bonds. The choice between 3 and 4 is in fact arbitrary, however if it is 3 then its contribution to the torsional barrier (Equation (5)) has to be recognized.

From Equation (8) it follows that the coefficients $C_{a,b}$ obey a combination rule which relates *a-b* interactions between atoms *a* and *b* to *a-a* and *b-b* interactions, i.e.

$$2\alpha_a\alpha_b/C_{a,b} = \alpha_a^2/C_{a,a} + \alpha_b^2/C_{b,b} \quad (9)$$

The application of combination rules is essential when the coefficients are determined by empirical methods. Consider, for example, only the most frequent atoms in bio-molecules, H,C,N,O,S,P. Without a combination rule we would need 21 instead of 6 empirical parameters, and since these are not independent by their nature, they cannot be uniquely defined by empirical methods. The common practice in potential energy calculations has been to use a geometric mean combination rule

$$C_{a,b}^2 = C_{a,a}\, C_{b,b} \, . \quad (10)$$

Its practical advantage is that it avoids the use of the polarizabilities α; its disadvantage is that it has no theoretical justification and that it is inaccurate. Kramer and Herschbach[5] have compared the two combination rules with a set of accurate published calculations of 153 unlike coefficients. The root mean square deviation from experiment was 3.25% when Equation (9) was used, but 73.5% when Equation (10) was used! Whether the use of Equation (9) is justified depends on estimates of the overall accuracy of the various contributions to the total set of potential energy functions, a subject to which we shall return later.

The Repulsive Force

Short range repulsion forces arise whenever atoms or molecules come so near each other that their electron clouds interpenetrate. Quantum mechanics offers a satisfactory explanation for such repulsions in terms of Pauli's exlusion principle and perturbation theory, but no formula of general applicability is available. It is commonly accepted that the leading factor implied by theory is exponential, so that a simple form for the repulsive energy function is

$$V_{repulsive} = A \exp(-br) \quad (11)$$

The constants A and b are, however, not derivable in terms of atomic properties, and are therefore indeterminate.

Moreover, since the repulsive potential is very steep, and it is very difficult to isolate it from other interactions experimentally, the determination of A and b are not a simple matter. Lennard-Jones[6] avoided this difficulty in his classical study of the Van der Waals interactions, by proposing the empirical formula for such interactions, now known as the Lennard-Jones potential

$$V_{LJ}(r) = A/r^n - C/r^6 \tag{12}$$

where A and n are empirical constants, and C is the dispersion coefficient . Lennard-Jones examined various values of n by fitting A and C to experimental data, and found no distinction between different values of n, or between A/r^n and the exponential form of Equation (11), provided the calculated repulsion is sufficiently steep, i.e. n>>6. He therefore chose n=12 for computational convenience, and this form of the potential became the most common in energy calculation. The common combination rule for A is similar to that given in Equation (10) for C, namely the geometric average

$$A^2_{a,b} = A_{a,a} A_{b,b} . \tag{13}$$

The heuristic nature of this potential is evident in another way. It is not rigorously correct just to simply take C/r^6 for the attractive term, since this is the leading term in Equation (7) for dispersion interactions only for large distances where A/r^n is negligible, while at contact where the interaction energy is highest, other dispersion terms might be significant. Nevertheless, Equation (12) has been proven to be extremely useful in a wide range of applications, particularly when A and C are considered empirically adaptable parameters, since it represents faithfully the essential features of non-bonded interactions, namely steep repulsion at short distances, minimum energy at an intermediate distance, and r^{-6} behaviour at long distances. If we denote by r* the distance for which V_{LJ} is at minimum, and by $-\varepsilon$ the value of this minimum, the Lennard-Jones potential takes the form

$$V_{LJ}(r) = \varepsilon(n-6)^{-1}[6(r/r^*)^{-n} - n(r/r^*)^{-6}] \tag{14}$$

which is less handy for computations but has the advantage that its parameters r* and ε are more directly related to the observable properties of molecular systems and therefore their numerical values are more meaningful. Also the combination rule $r^*_{a,b}=(r^*_{a,a}+r^*_{b,b})/2$, in conjunction with Equation (9), is perhaps more physically justified than the geometric average for A (Equation (13)), although this suggestion is still awaiting an empirical test.

The Three-Body Force

Intermolecular forces are commonly believed to be pair-wise additive. That is, the energy of an assembly of atoms or molecules is equal to the sum of the interaction energies of all pairs of interacting particles. However, this assumption has been challenged. It is certainly not strictly valid for condensed systems, namely also for systems of interest in molecular biology. The reason is that when an atom is polarized simultaneously by two other neighbouring atoms, its polarization is proportional to the vectorial sum of the electrostatic forces, thus the dispersion energy of three neighbouring atoms must depend not only on the three interatomic distances but also on the three angles between them. According to Axilrod and Teller[7], this three body interaction potential, namely the correction to the pairwise additive potential energy of the dispersion interaction, is

$$V_3 = C_{abc}(1+3\cos\theta_a \cos\theta_b \cos\theta_c)(r_{ab}r_{ac}r_{bc})^{-3} \qquad (15)$$

where C_{abc} is a constant and the r's and θ's are the sides and angles of the triangle formed by the atoms a,b and c. While it is common practice to neglect this term in calculations of non-bonded interactions, we should realize that they are by no means always negligibly small. For example, Kestner and Sinanoglu[8] concluded that three body forces reduced the dispersion interaction between base pairs in the DNA double helix by 28%.

Electrostatic Interactions

Most chemical bonds are not purely covalent, but possess a partial electrostatic character. Whenever the electronegativities of of two bonded atoms are different, the charge distribution of the bonding electrons is shifted partly towards the more electronegative atom, thus making the total charge (nucleus and electrons) in the vicinity of this atom equivalent to a partial negative charge, while the less electronegative atom obtains a partial positive charge. The detailed charge distribution may be obtained from the wave functions of the Schrödinger equation. However, the calculation of electrostatic interactions from continuous charge distributions is extremely cumbersome, involving multidimensional integrals. Since the available solutions to the Schrödinger equation are approximate and so are the charge distributions, there is no reason to attempt such calculations. A major simplification of the problem is obtained if the continuous charge is represented approximately by point charges located on the atoms. This is done, for example, by the so called Mulliken population analysis[9], where the charges in a given molecular orbital are assigned to the individual atoms, according to the extent to which the atomic orbitals of each atom contribute to the molecular orbital. There is a vast literature on the Mulliken analysis of various molecules, and also on other similar methods. Unfortunately, the partial charges thus obtained depend heavily on the method or the type of approximation employed. For example, Mulliken population analysis of *ab initio* calculations yields widely different values of the partial charges when different basis sets are used in amides[10] and carboxylic acids[10,11]. They all indicate, however, that electrostatic interactions between partial charges are an important part of the non-bonded interactions, in particular in molecules such as are the subject of structural molecular biology, in which highly polar bonds and groups are abundant.

The electrostatic potential energy of interaction between the partial charges q_i is given by the Coulomb law

$$V_{electrostatic} = \frac{1}{2} \sum_{i,j} \frac{q_i q_j}{r_{ij}} \tag{16}$$

where r_{ij} is the distance between atoms i and j. For intermolecular interactions the summation is obviously over the atoms i of one molecule and j of the other. For intramolecular interaction one is confronted with a difficulty: If the summation would extend over all pairs i,j ($i \neq j$), the electrostatic forces would be considered as if they acted even between bonded atoms, over and above the bond potential, which would make little sense. On the other hand, if the definition of non-bondedness for Van der Waals interactions is adopted also for electrostatic interactions, namely, the summation extends over atoms separated by, say at least 3 consecutive bonds, then the interacting atoms do not obey electroneutrality (for example this rule would retain only H···H interactions in ethane). There is not, as yet, an agreed solution for this difficulty.

Equation (16) represents electrostatic interactions in vacuum, i.e. the dielectric constant has been given the value $\varepsilon=1$. Some authors have used $\varepsilon>1$, arguing that interactions in a polarizable medium are weaker than in vacuum. This argument seems, however, to be based on a misconception. When the charge-carrying atoms constitute the medium, then their interaction must be considered as interaction in vacuum. Thus intramolecular interactions or intermolecular interactions in crystals must be represented by Equation (16) with $\varepsilon=1$. On the other hand intermolecular electrostatic interactions between molecules dissolved in a solvent of dielectric constant ε must be reduced by a factor ε if (and only if) the solvent is considered as a continuous medium. The problem of the meaning of a dielectric medium at microscopic distances is very complex and could not be discussed here.

The following *Gedanken Experiment* will support this view: Consider a molecular crystal whose molecules carry atomic partial charges q_i; assume that the molecules have no permanent dipole moments or that their dipole moments are fixed in the crystal. The dielectric constant ε of the crystal is then determined only by the polarizability of the molecules.

Imagine now that it would be possible to freeze the polarizability and then release it at will. When it is frozen then $\varepsilon=1$ and when it is released then $\varepsilon>1$. In the first case Equation (16) is certainly valid. If in the second case the electrostatic energy would be devided by $\varepsilon>1$ then it would be *smaller* if the total sum would be positive, and *larger* (smaller in absolute value) if the sum would be negative. However, when the frozen polarizability is released, the crystal adapts spontaneously to the gained freedom, and may respond only by decreasing the total electrostatic energy in both cases!

The Hydrogen Bond

The hydrogen bond is a special case of electrostatic interactions. It is special in several ways. First, it is ubiquitous in biological structures, on the molecular level as well as on higher levels of structural biology, and plays a very important role in controlling biological specificity and life processes. Second, it is one of the strongest among non-bonded interactions, if strength of interaction is to be evaluated by the ratio of attractive energy to number of atoms involved. It is comparable in strength with the salt bridge which is, however, much less abundant. It is also special in that although it has been recognized about 80 years ago, the question whether it is merely a strong non-bonded interaction or whether it is more like a weak chemical bond has been the subject of some controversy or soul searching until recently.

Molecular orbital calculations attributed for a long time to the hydrogen bond a significant quantum mechanical effect of "charge transfer" indicating a chemical bond character[12]. Such calculations have been recently revised, using extended basis sets. The result has been the reduction of the estimates of the covalent character of the hydrogen bond in water, for example, from about 10 kcal/mole to about 1-2 kcal/mole[13,14].

From the point of view of potential energy calculations, the question has been whether a special potential function is required to represent the hydrogen bond, or whether it could be adequately represented by the functions of Van der Waals and electrostatic interactions. In the past, various functions were introduced to represent the hydrogen bond in terms of the "bond" distances and/or "bond" angles, which we shall not review here except to state that these were arbitrary guesses not based on theoretical considerations. A closer analysis of the experimental behaviour of the hydrogen bond, however, showed that electrostatic and Van der Waals non-bonded interaction potentials are both necessary and sufficient to represent the hydrogen bond[15,11]. (See below).

The Total Energy of Molecular Systems: Aditivity and Transferability

Summing up the above discussion of the various intra- and inter-molecular energy potentials we shall state now the important though seemingly self-evident assumption, that the total potential energy of a molecular system is the sum of all the various potential energy functions reviewed above, namely that

$$V_{total} = \Sigma V_{bond} + \Sigma V_{angle} + \Sigma V_{torsion} + \Sigma V_{LJ} + \Sigma V_{electrostatic} + \ldots \tag{17}$$

where the summations are extended over all bond lengths, bond angles, torsional and out-of plane torsion angles, and over all inter-atomic distances of non-bonded atoms, and the dots on the right indicate that we omit some less important contributions (or rather, that we believe that these contributions are less important), such as the inter-molecular three-body forces mentioned above; or potential functions of more than one internal coordinate, which are crucial for vibrational normal mode analysis to be discussed later.

In fact the additivity of these single variable potential functions is only approximate, otherwise we would not insert the three dots to cover-up the shortcomings of the assumption of additivity. Furthermore, the assumption of additivity was not deduced or supported by quantum mechanical considerations. It was mainly deduced from a very large body of experimental data from many branches of physical chemistry, some of which will be discussed in the next section, when the application of energy calculations to the analysis of experimental data will be introduced.

The concept of transferability of potential functions is closely linked with that of additivity. Transferability means that the energy functions which give a satisfactory description of the physico-chemical properties of one molecule, may be used with confidence to describe the properties of any other molecule of *similar structure.* Thanks to the transferability of potential functions it is possible to determine them by studying simple, small-size molecules, and then use them in complex, large molecules such as proteins and nucleic acids. However, in doing so we should always remember that transferability, like additivity, is an empirical and approximate rule, valid only if the concept of *similar structure* is used properly.

THE CALCULATION OF OBSERVABLE PROPERTIES

The first major application of inter-molecular energy functions was related to the statistical mechanics of real gases. It was one of the great break-throughs of statistical mechanics that the "second virial coefficient" was derived in terms of inter-molecular interaction functions, such as the Lennard-Jones potential. A large literature has been devoted to this subject which was focussed, however, only on the simplest molecules, mainly noble gases. The reason was that intermolecular distances in fluids (gases and fluids) are in constant dynamic change, and therefore the relation between the inter-molecular forces and properties of fluids require a statistical averaging process which erases the details of the functional form of the intermolecular or interatomic interactions. Thus potential energy functions are suitable for use in statistical mechanical calculations, however their validity must be judged by other observable properties, more directly related to the functional form of the potentials. Such observables are best introduced and discussed by considering the main features of the total potential energy function V_{total} (Equation (17)) as a whole, ignoring at this level the details of its components.

V_{total} of any molecular system is naturally a function of structure. The structure may be represented in various ways. One way is to specify the Cartesian coordinates of all the atoms which comprise the system. If the system is a single molecule, whether small or large, the positions of its n atoms are specified by 3n Cartesian coordinates, which we shall denote symbolically by a single vector $\underset{\sim}{r}$. If the system is, say, a crystal, then $\underset{\sim}{r}$ may be considered as representing the atomic coordinates of the atoms in one unit cell together with the translation vectors of the unit cell. Another way to represent the molecular system is by specifying the internal coordinates (bond lengths, bond angles, torsional angles), from which all interatomic distances may be calculated by standard mathematical devices. Whatever representation we choose, $\underset{\sim}{r}$ represents the detailed structure of the molecular system. Now consider the molecular system to be at equilibrium. Then its vector $\underset{\sim}{r}$ has an equilibrium value, $\underset{\sim}{r}_o$, for which the total energy function $V_{total}(\underset{\sim}{r}_o)$ is at its minimum. Let us now develop Vtotal in a Taylor series around $\underset{\sim}{r}_o$.

$$V_{total}(\underset{\sim}{r}) = V_{total}(\underset{\sim}{r}_o)+\sum_i(\partial V_{total}/\partial r_i)_o\ \delta r_i + \\ + \frac{1}{2}\sum_i\sum_j(\partial^2 V_{total}/\partial r_i\partial r_j)_o\ \delta r_i\delta r_j + \ldots \qquad (18)$$

where r_i and δr_i are the components of r and of small deviations from $\underset{\sim}{r}_o$, respectively. A close examination of this expression will show a whole world of applications to energy calculations of molecular properties.

Molecular and Crystal Energies and the Corresponding Thermodyanic Functions

$V_{total}(\underset{\sim}{r}_o)$, the first term in the Taylor expansion (Equation (18)), is the molecular energy at equilibrium. The corresponding experimental, measurable property is the heat of atomization (or some related functions such as the heat of formation, or the heat of combustion) of the system. In comparing the two, we have to take into consideration their differences. The heat (or enthalpy) of atomization is a thermodynamic function related to thermal agitation and temperature, while $V_{total}(\underset{\sim}{r}_o)$ represents the energy of a single microscopically defined state. The heat of atomization's major component is the quantum mechanical energy of formation (or dissociation) of the chemical bonds, which is represented by Morse potentials for the various types of bonds (Equation (2)) as well as by other terms of intramolecular interactions included in V_{total}. It includes, however, other contributions which must be taken into account. First, there is the zero-point energy, which is the vibrational energy of the lowest vibrational quantum state corresponding to zero absolute temperature. Then there are the energies of molecular translations, rotations, and vibrations in the gas phase, or the packing energy of the crystal lattice and the molecular and lattice vibrations in the solid phase. These quantities are functions of thermodynamic variables such as temperature. They can be calculated from V_{total}, as we shall see below.

In many instances we are interested in the changes in energy accompanying conformational or structural changes in the molecular system. Examples are the cis-trans isomerism of the peptide bond, the chair-half-chair-boat conformational transitions of sugar rings, or the trans-gauche transition in alkanes. In such cases, the zero-point energy and the other thermodynamic contributions may cancel out to a good approximation, and the correspondence to experiment is then simpler.

When intermolecular interactions are included in V_{total}, it is possible to consider $V_{total}(\underset{\sim}{r})$ as representing macroscopic condensed phases, i.e. solids, liquids and solutions. A very useful and simple example is the lattice energy of crystals, or the sublimation energy, i.e. the energy of crystal-to-gas transition. For rigid molecules, which maintain the same structure in gas and solid phases, the sublimation process involves work against inter-molecular forces only. The intramolecular energy and the enthalpy of molecular vibrations remain invariant. Therefore, the heat of sublimation is an important source of information on intermolecular energy potentials. Note, however, that here again, when energy calculations of crystal packing are compared with experimental heat of sublimation, enthalpy corrections, for translations, rotations and the pV term in the gas and for lattice vibrations in the solid, are required. Such corrections may be estimated satisfactorily by classical thermodynamics[15].

Equilibrium Structure of Molecules and Crystals

The equilibrium structure of molecules, as well as that of molecular crystals is calculated from the second term of the Taylor expansion of $V_{total}(\tilde{r})$, Equation (18). Since the equilibrium structure is the structure for which the energy function V_{total} obtains a minimum value, all derivatives of V_{total} with respect to the components of $\tilde{r}$, namely the gradient of V_{total} (denoted grad V_{total}) must vanish at $\tilde{r}_o$. The set of equations

$$\partial V_{total}(\tilde{r})/\partial r_i = 0 \quad i=1,\dots,3n \tag{19}$$

may therefore be used to solve for $\tilde{r}_o$. Computer programs are available to obtain numerical solutions of Equation (19), and various algorithms for such solutions have been studied extensively. The general idea of all such algorithms is the same: One starts at any non-equilibrium value $\tilde{r}$ and calculates grad V_{total}. If the gradient does not vanish, $\tilde{r}$ is changed to $\tilde{r}+\delta\tilde{r}$, where $\delta\tilde{r}$ is calculated in such a way as to move $\tilde{r}$ in the direction of $\tilde{r}_o$. The process is iterated until finally the minimum is reached to the desired precision.

The choice of the appropriate minimization method is very important in energy calculations. The simplest method of solving Equation (19) is the "steepest descent" method. At each iteration the vector grad V_{total} is calculated, and a "line search" is performed along the direction of the gradient. That is to say that the value of V_{total} is calculated at several points along the direction of the gradient, and a minimum point is obtained by interpolation between these points. The method is "stable", in the sense that each iteration leads to a new point of lower value of V_{total}. However it usually gets stuck by progressively slow convergence.

Fast convergence is obtained if grad V_{total} is expanded in a Taylor series around the (yet unknown) point of minimum $\tilde{r}_o$. In matrix notation such expansion is written as

$$\text{grad } V_{total}(\tilde{r}_o) = \text{grad } V_{total}(\tilde{r}) + F(\tilde{r})\delta\tilde{r} + \dots \tag{20}$$

where $\delta \underset{\sim}{r} = \underset{\sim}{r}_o - \underset{\sim}{r}$ and $\underset{\sim}{F}$ is a matrix whose components are $\partial^2 V_{total}(\underset{\sim}{r})/\partial r_i \partial r_j$. Since grad $V_{total}(\underset{\sim}{r}_o)$ is zero by definition, one obtains an equation for $\delta \underset{\sim}{r}$

$$\delta \underset{\sim}{r} = -\underset{\sim}{F}^{-1}(\underset{\sim}{r})\text{grad } V(\underset{\sim}{r}) \tag{21}$$

This method, called the Newton-Raphson method, yields the minimum in one step, provided the higher terms in the Taylor series are negligible, which is *approximately* true for points near the minimum of any well behaved function. Therefore, starting near the minimum, a few iterations of Equation (21) converge progressively to the minimum. Such a convergence is called "quadratic convergence" since for a quadratic function the higher terms are precisely zero and the exact minimum is reached by Equation (21) in one step. There are however difficulties with this algorithm. First, when $\underset{\sim}{r}$ is far away from the minimum the method may be unstable, namely $\delta \underset{\sim}{r}$ derived by Equation (21) may move $\underset{\sim}{r}$ away from minimum. Second, the inverse of $\underset{\sim}{F}$ does not exist if $\underset{\sim}{F}$ is singular, so that $\underset{\sim}{F}^{-1}$ in Equation (21) must be replaced by $\underset{\sim}{F}^{+}$, the "generalized inverse" matrix[16]. Third, when V_{total} is a function of a great number of variables, as is the case for V_{total} of biological macromolecules, the calculation, storage and manipulation of the matrix $\underset{\sim}{F}$ becomes technically unmanageable.

To overcome these difficulties, new methods have been developed which combine the advantages of the steepest descent and the Newton-Raphson methods but avoid their pitfalls. Such methods are both stable and fast converging. Among these the most suitable for energy calculations in molecular biology is the so called "conjugate gradient" method[17]. Its great advantage is that like the steepest descent method its requirements for computer memory space is proportional to the number of variables (3n for a molecules of n atoms), while other methods, like the Newton-Raphson, require memory space proportional to the square of the number of variables. We shall explain here the basic ideas behind the "conjugate gradient" method, while skipping the mathematical details.

The method starts at the first iteration like the steepest descent, by a "line search" along the direction of the gradient. At the second, third, or in general at the i'th iteration, the new direction of the line search is determined by a linear combination of the gradient at the present point (i) and the direction of line search at the previous point (i-1). The coefficients of the linear combination are chosen, very cleverly, such that if the function to be minimized is quadratic to a good approximation then two major goals are achieved: 1) Each new direction of search is linearly independent of all previous directions of search. 2) The minimum along the direction of the line search at each iteration is also the minimum of the function in the whole subspace spanned by all previous directions. That is to say, after i iterations we obtain the minimum of the function in the subspace spanned by the first i directions. Therefore, if V is a quadratic function of N variables, the total minimum must be reached in exactly N steps. The fast convergence and modest requirements for memory space make the conjugate gradient method a best choice for minimizing the energy of proteins and other biopolymers.

Vibrational Modes in Molecules and Crystals

We come to the third term in the expansion of the energy in a Taylor series around equilibrium, Equation (18). It represents the energy of the system due to small deviations of the (atomic or internal) coordinates from their equilibrium values. Such deviations must exist, because the uncertainty principle of quantum mechanics implies that molecules possess a vibrational energy even at absolute zero temperature (zero point energy).

The derivation of vibrational energy of molecules and crystals is linked to the concept of normal modes of vibrations. Consider a single polyatomic molecule in a gas, composed of n atoms. Each atom has three degrees of freedom, therefore the molecule has 3n degrees of freedom. Molecular translations and rotations use up 3 degrees of freedom each. The remaining 3n-6 are vibrational degrees of freedom, in which the atoms participate collectively, since an atom cannot move within the molecule without exerting forces on its neighbours. Thus to each vibrational degree of freedom there corresponds a collective, "normal", or "fundamental" mode of vibration with a characteristic frequency.

The calculation of these normal modes can be performed in the "harmonic approximation" by using the third (last) term of Equation (18). The higher, neglected terms would then supply the "anharnomic corrections" to the harmonic normal modes.

There is an extensive literature on the normal mode analysis of polyatomic molecules. The classical treaties by Herzberg[18] and by Wilson[19] make use of symmetry properties of polyatomic molecules and the theory of group representations. However, the application of potential functions to normal mode analysis[20] is made much simpler and straight-forward, mostly thanks to the availability of fast computers with large memory-space. Symmetry properties of the system are neither needed nor assumed, rather, they are derived for both the equilibrium structure and the normal vibrations around equilibrium.

We give here a brief summary of the derivation of normal modes from potential energy functions in its simplest form. The reader is referred to (20) for further details. The potential energy of vibrations around the equilibrium conformation $\underset{\sim}{r}_o$ is taken from Equation (18), omitting the second term since it vanishes at equilibrium. It is presented in matrix notation by

$$V_{total}(\underset{\sim}{r}_o+\delta\underset{\sim}{r}) - V_{total}(\underset{\sim}{r}_o) = \frac{1}{2}\,\delta\underset{\sim}{r}'\underset{\sim}{F}(\underset{\sim}{r}_o)\,\delta\underset{\sim}{r} + \ldots \qquad (22)$$

The vector $\underset{\sim}{r}_o$ of the equilibrium structure is represented here in Cartesian coordinates. Its 3n components are the x,y and z coordinates of the n atoms of the molecule, chosen in any convenient order; since $\delta\underset{\sim}{r}$ represents the deviations from equilibrium during the vibrational motion, it is a function of time. The row vector $\delta\underset{\sim}{r}'$ is the transpose of the column vector $\delta\underset{\sim}{r}$; $\underset{\sim}{F}(\underset{\sim}{r}_o)$ is the matrix ot the second derivatives of V_{total} at $\underset{\sim}{r}_o$. Cartesian coordinates are used because only in these coordinates is the kinetic energy given in the simple form

$$T(\delta\dot{\underset{\sim}{r}}) = \frac{1}{2}\,\delta\dot{\underset{\sim}{r}}'\;\underset{\sim}{M}\;\delta\dot{\underset{\sim}{r}} \qquad (23)$$

where $\delta\dot{\underset{\sim}{r}}=d\delta\underset{\sim}{r}/dt$ is the vector of atomic velocities. $\underset{\sim}{M}$ is a diagonal matrix of the atomic masses, each atomic mass appearing thrice, once for each Cartesian coordinate of that atom.

From the potential and kinetic energies, the simultaneous equations of motion for all atoms in the molecule are readily obtained:

$$\underset{\sim}{M}\;\delta\ddot{\underset{\sim}{r}} + \underset{\sim}{F}(\underset{\sim}{r}_o)\,\delta\underset{\sim}{r} = 0 \qquad (24)$$

where $\delta\ddot{\underset{\sim}{r}}$ is the vector of atomic accelerations. This is in fact a very simple form of Newton's equations of motion. It says that for each atom i the acceleration $\delta\ddot{r}_i$ times the mass M_i is equal to the force $-\Sigma_j F_{ij}\delta r_j$ exerted by all other atoms. A normal mode of vibration with a frequency ν and a vector amplitude $\delta\underset{\sim}{r}_o$, whose components are the amplitudes of the atoms participating in the normal mode, is expressed in vector notation by $\delta\underset{\sim}{r}=\delta\underset{\sim}{r}_o\exp(2\pi i\nu t)$. The corresponding acceleration vector is $\delta\ddot{\underset{\sim}{r}}=-(2\pi\nu)^2\delta\underset{\sim}{r}_o\exp(2\pi i\nu t)$. When these expressions are inserted in Equation (24), a set of algebraic equations

$$[\underset{\sim}{F}(\underset{\sim}{r}_o) - (2\pi\nu)^2\underset{\sim}{M}]\,\delta\underset{\sim}{r}_o = 0 \qquad (25)$$

is obtained for the amplitudes of the normal modes and their corresponding frequencies. In the standard mathematical language this is a typical "eigen value problem". Since the order of the matrices is 3n, there must be 3n solutions. Each solution yields a characteristic vector, or "eigenvector", which is the vector-amplitude of a normal mode. To each "eigenvector" there belongs an "eigenvalue" representing the characteristic frequency of this normal mode. Six of the 3n frequencies are zero, corresponding to the 3 translations and 3 rotations of the molecule, and the remaining 3n-6 are the molecular harmonic frequencies. Once the eigenvectors are derived in Cartesian coordinates, they can be transformed to internal coordinates where the symmetry properties of the normal modes are easily recognized.

This formalism for calculating vibrational normal modes is particularly suitable for large computers. It requires one and the same program for any molecule, provided sufficient computer space and time are available. It is, however, much more than a convenient formalism. Its main innovation is that it links together the calculation of the static properties - the equilibrium energy and the equilibrium structure, and the dynamic properties - the vibrational modes, and brings out their intrinsic interdependence.

Molecular vibrations contain much information about the molecular energy surface in the whole vicinity of the equilibrium conformation, since they depend on the curvature of the energy surface around equilibrium. Therefore, the representation of V_{total} as a sum of functions of single variables, as given by Equation (17), must be supplemented for the calculation of vibrational spectra by functions of adjacent internal coordinates. Such functions could not be deduced by theory, therefore simple guess-functions have been introduced, such as the bi-linear function $K_{b\theta}(b-b_o)(\theta-\theta_o)$ which couples bond stretching and angle bending. These are called "cross terms" or "interaction terms", and are reduced to the bare minimum required to fit experimental data, but they still increase the number of functions and adaptable parameters more than necessary for calculations of molecular structure alone. However, it should be noted that if cross terms are omitted from V_{total}, then the coefficients of the single-variable functions cannot be transfered from spectroscopy.

We discussed hitherto the variables of single molecules. By extending the potential energy V_{total} to include the lattice energy of molecular crystals, the vibrations of molecules in a lattice are obtained[21]. The 6 degrees of freedom of molecular translations and rotations in the gas phase are replaced in condensed phases by lattice vibrations and molecular librations. Further, some molecular vibrations are modified by interactions with neighbour molecules, and frequencies of degenerate modes are split by the asymmetric environment of the molecules in the crystal.

Thermodynamic Functions

The energy of a system in the gas phase contains, as noted already, V_{total} as well as the energies of molecular translations, rotations and vibrations. In condensed systems, from solids to living cells, the vibrational energy is the main component variable with temperature, and the main contributor to the thermodynamic functions such as enthalpy, free energy, specific heat and thermal expansion.

According to the *equipartition law* of classical statistical thermodynamics, the kinetic and potential energies of translation, rotation and vibration are equally distributed over all degrees of freedom of a system, each obtaining $\frac{1}{2}$kT of energy at a temperature T. Purely kinetic degrees of freedom like molecular translations, molecular rotations and free (unhindered) internal rotations obtain $\frac{1}{2}$kT, while molecular vibrations, including hindered rotations, obtain 1kT, (half for kinetic and half for potential energy). For example, alanine has 13 atoms, i.e. 39 degrees of freedom. Assuming free rotation around ϕ and ψ and hindered rotation around the $C^{\alpha}-C^{\beta}$ bond, 8 degrees of freedom are purely kinetic (3 translations, 3 rotations and 2 internal rotations) and 31 degrees of freedom are vibrational. Therefore alanine has, by classical theory, an energy of (4+31)kT per molecule in the gas phase due to translations rotations and vibrations, namely over and above its equilibrium potential energy $V_{total}(\mathbf{r}_o)$. According to classical statistical thermodynamics, the energy of a system depends on its number of degrees of freedom, and is independent of its vibrational frequency distribution. However, the law of equipartition is not supported by experiment. The recognition of this fact was strongly linked to the origin of the quantum theory. It led Planck to suggest his theory of black body radiation, and it led Einstein and Debye to produce the theory of specific heats of solids, all of which came to replace the equipartition law.

According to quantum statistical thermodynamics[22], the contribution of each normal mode of frequency ν to the thermodynamic functions depends on the ratio of $h\nu$ to kT, and the contributions are additive. For example the vibrational energy is given by

$$\frac{1}{2} h\nu + h\nu/(\exp(h\nu/kT)-1)$$

which tends to the classical limit kT either at high temperatures or at low frequencies ($h\nu \ll kT$), and to the zero-point energy limit $\frac{1}{2}h\nu$ for $h\nu \gg kT$. The vibrational free energy is given by:

$$\frac{1}{2} h\nu + kT \ln(1-\exp(-h\nu/kT)$$

while the vibrational entropy is obtained from the difference between energy and free energy.

CHOICE OF ENERGY PARAMETERS

Theoretical, quantum-mechnical considerations guided us generally in finding the functional form for many of the potential energy functions, but not the numerical values of their coefficients, or energy parameters. These have to be determined, as a rule, by fitting calculated results to experimental ones. The number of parameters to be fitted this way for all atoms and groups of atoms involved in biological molecules is rather large. The experimental data on which the parameter fitting rests is varied. In some cases there are not enough data, in other cases they are not sufficiently accurate. Sometimes different authors derive different parameters from the same data, which is understandable because the research objectives as well as personal biases may affect the method by which the parameters are determined. Consequently, there is not yet a generally agreed "force field" for molecular biology. (The term "force field" denotes a set of energy functions and their parameters, fitted for a given family of molecules). It is to be expected, however, that force fields will improve as their application in chemistry and molecular biology will advance, and that Darwinian natural selection and survival of the fittest will lead to better force fields which will be gradually accepted according to agreed-upon standards.

One useful way to further such a purpose is to establish a "bench-mark" for an objective comparison of force-fields, namely an extensive data-base of high-quality experimental results, with efficient computer programs by which the corresponding theoretical results from different force-fields can be calculated and compared. Such a bench-mark for amides and carboxylic acids has been published recently[23], and may serve as a first step in this direction. It was a natural outcome of the efforts of our group in obtaining a "consistent force field" for organic molecules and biopolymers.

The Consistent Force Field (CFF) Method

This method consists of a systematic selection of energy functions, and an objective empirical determination of their parameters. Computer programs are prepared for calculating various observable properties from V_{total} and a large and varied body of reliable experimental data, pertaining to whole families of similar molecules is accumulated for comparison with the calculated results. The energy parameters are optimised by a non-linear least squares algorithm[24] to obtain the best fit between theory and experiment.

We started with a force-field for conformations, vibrations and enthalpies of cyclic and normal alkanes[20], and extended it to crystal structures, sublimation energies, lattice vibrations and thermal expansions of alkanes crystals[21]. It was then further extended to include conformations, vibrations and heats of hydrogenations of non-conjugated olefins[25]; conformations and vibrations of amide-alkane rings (lactams)[26]; heats of sublimation and crystal structures of hydrogen bonded crystals of amides[15] and carboxylic acids[11], as well as some of their dipole moments[15,11] and enthalpies of dimerization in the gas phase[27].

The families of alkanes, amides and carboxylic acids contain the constituents which comprise most of the amino acid residues of protein chains, and indeed the major motivation for deriving a consistent force field for these families was its potential application for biopolymers. However, a comprehensive consistent force field for biological molecules should include more groups e.g. aromatic and heterocyclic rings, carbohydrate, phosphate and sulfur groups; it should also extend considerably the consistent analysis of intra-molecular energy parameters pertaining to polypeptide and polynucleotide chains. This could be done fully only if more experimental data could be accumulated to supply the specific needs of the consistent force field analysis.

Among the past accomplishments of the consistent force field method, perhaps the most interesting are those related to the non-bonded interactions, namely the Van der Waals repulsions and attractions, as well as the Coulomb forces between partial charges and the hydrogen bond.

The Lennard-Jones (LJ) potential is commonly considered to be a "12-6" potential, that is, the exponent n in the repulsive term Ar^{-n} is taken to be n=12 (see above). Examining the LJ potential for alkanes, we found that the LJ parameters optimized for intramolecular interactions were too low for intermolecular interactions, while the LJ parameters optimized for intermolecular interactions were *too high* for intramolecular interactions. These trends indicated that the r^{-12} dependence is too steep. Trying various values of n we found that the "9-6" potential (i.e. n=9) was the best for alkanes[21] The same trend was observed in carboxylic acids, and to a lesser extent in amides.

Extending the consistent force field to hydrogen bonded amide crystals[15], we put all hydrogen-bond potentials available in the literature to the least squares test. The unequivocal, at that time surprising, result was that there is no need for a special potential for the hydrogen bond. The objective parameter-fitting algorithm of the least-squares led to reasonable Lennard-Jones and partial charge parameters, while the special functions, representing the expected covalent character of the hydrogen bond, received neglibibly small or ill-determined parameters (i.e. having large standard-deviations). Whether such functions were included or omitted did not affect the ultimate fit of calculated and measured crystal properties.

The same conclusion was obtained again when the consistent force field was extended to carboxylic acids[11]. We could then conclude with confidence that, to a reasonable approximation, the hydrogen bond is a non-bonded interaction between two polar bonds, where the attractive forces are mainly electrostatic. The Van der Waals radius of the hydrogen atom in the OH or NH bond was found to be so small as to be practically engulfed by the large Van der Waals radius of the electronegative atoms N or O. This allowed the contact distance to become small enough, and the electrostatic attraction large enough, to make this non-bonded interaction look like a weak bond. As noted above, recent extended *ab initio* calculations support this new concept of the hydrogen bond[13,14].

The essence of the CFF method, may be illustrated by Table 1 from the "bench mark" paper[23]. The table represents the root mean

Table 1. Root Mean Square Deviations of Properties Calculated for Carboxylic Acids and Amides by Three Force Fields

Property	Units	No. of terms	rms dev 12-6-1	rms dev 9-6-1	rms dev MCMS
		Acids			
energy	kcal/mol	12	2.486	2.053	2.118
UCV length	Å	42	0.489	0.307	0.604
UCV angle	deg	17	3.456	2.856	4.465
volume	$Å^3$	14	15.911	16.772	18.876
d<4	Å	14	0.247	0.190	0.322
H•••O dist	Å	16	0.062	0.072	0.058
O•••O dist	Å	16	0.047	0.071	0.041
C-O•••O angle	deg	16	11.071	9.881	14.048
O•••O=C angle	deg	16	7.843	7.760	11.786
H•••O=C angle	deg	16	12.362	12.144	17.985
180°-O-H•••O	deg	16	8.491	7.732	11.710
		Amides			
energy	kcal/mol	6	1.574	1.930	8.446
UCV length	A	36	0.208	0.235	0.261
UCV angle	deg	14	1.824	1.261	2.385
volume	$Å^3$	12	7.057	17.797	13.951
d<4	Å	12	0.145	0.145	0.164
H•••O dist	Å	30	0.049	0.059	0.056
N•••O dist	A	30	0.055	0.055	0.076
C-N•••O angle	deg	22	3.337	3.575	4.071
N•••O=C angle	deg	22	5.931	5.502	9.257
H•••O=C angle	deg	30	5.830	5.609	7.329
180°-NH•••O	deg	30	4.396	3.894	4.093

square deviation (rms dev) between calculated and measured observables for three force fields: CFF "12-6-1" (i.e. "12-6" for LJ and "1" for Coulomb), CFF "9-6-1" and MCMS[28,29], semi-empirical force field which uses a special function for the hydrogen bond, and theoretical values for dispersion parameters and partial charges. The data base comprises 14 carboxylic acids and 12 amides. The 11 observables represented are: Lattice energies (heats of sublimation), UCV (unit-cell-vector) lengths, UCV angles, cell volumes, close inter-atomic contact distances (d<4Å), and 6 hydrogen-bond ovservables - 2 distances and 4 angles.

A close examination of this table and those which follow in the bench mark paper[23] can tell us not only which force field is better and which observables are better represented. One can look for systematic trends of rms deviations, search for their cause or origin, and indicate the way to replace poorer potentials by better ones. Thus, for example, it is noted that rms deviations are smaller in amides than in carboxylic acids in both 12-6-1 and 9-6-1. The reason could be traced to two facts: Amide crystals contain twice as much hydrogen bonds than carboxylic acids (there are 2 NH bonds per amide vs one OH bond per carboxyl), and hydrogen bond distances are better fit to experiment than alkane contact distances (most terms in "d<4"). The indication is that the alkane non-bonded interactions need a reevaluation. Indeed we are trying out some ideas to improve the Van der Waals potential for alkanes. Such an improvement may be important for a better representation of hydrophobic, non polar groups. Hydrophobic regions, typical of the interior of globular proteins, are most suitable for potential energy calculations related to protein structure, since they do not involve solvent (water) interaction with proteins. Therefore, reducing the errors in the Van der Waals potential to the level of that of the polar interactions is a desirable goal from the point of view of potential energy calculations in structural molecular biology.

REFERENCES

1. P. M. Morse, Phys.Rev. 34:57 (1929).
2. K. S. Pitzer, Disc.Faraday Soc. 10:66 (1951).
3. F. London, Z.Physik.Chem.B 11:222 (1930).
4. J. C. Slater and J. G. Kirkwood, Phys.Rev. 37:682 (1931).
5. H. L. Kramer and D. R. Herschbach, J.Chem.Phys. 53:2792 (1970).
6. J. E. Lennard-Jones, Proc.Roy.Soc. London A 106:463 (1924).
7. B. M. Axilrod and E. Teller, J.Chem.Phys. 11:299 (1943).
8. N. R. Kestner and O. Sinanoglu, J.Chem.Phys. 38:1730 (1963).
9. R. S. Mulliken, J.Chem.Phys. 23:1833 (1955).
10. A. T. Hagler and A. Lapiccirella, Biopolymers 15:1167 (1976).
11. S. Lifson, A. T. Hagler and P. Dauber, J.Am.Chem.Soc. 101:5111 (1979).

12. P. Schuster, in "The Hydrogen Bond", P. Schuster, G. Zundel and C. Sandorfy, editors, Vol. I. Ch. 2, North Holland Publishing Co., 1976.
13. See note added in proof, p. 154 in Ref. 12. Also, last section in Ref. 11.
14. H. Umeyama and K. Morokuma, J.Am.Chem.Soc. 99:1316 (1977).
15. A. T. Hagler, E. Huler and S. Lifson, J.Am.Chem.Soc. 96:5319 (1974).
16. See R. Fletcher, Computer J. 10:392 (1968); also Ref. (21) page 585.
17. R. Fletcher and C. M. Reeves, Computer J. 7:149 (1964).
18. G. Herzberg, "Molecular Spectra and Molecular Structure", Vol. II, "Infrared and Raman Spectra of Polyatomic Molecules", Van Nostrand Co., New York (1945).
19. E. B. Wilson Jr., J. C. Decius and P. C. Cross, "Molecular Vibrations" McGraw-Hill, New York (1955).
20. S. Lifson and A. Warshel, J.Chem.Phys. 49:5116 (1968).
21. A. Warshel and S. Lifson, J.Chem.Phys. 53:582 (1970).
22. See, for example, T. L. Hill, "An Introduction to Statistical Thermodynamics", Addison-Wesley Inc. (1960).
23. A. T. Hagler, S. Lifson and P. Dauber, J.Am.Chem.Soc. 101:5122 (1979).
24. The non-linear least squares method is originally due to D.W. Marquardt, J.SIAM, 11:431 (1963); for its various applications to CFF see Ref. (20); Ref. (21); A. T. Hagler and S. Lifson, Acta Crystallogr.B 30:619 (1974); S. Lifson and M. Levitt, Computers and Chem. 3:49 (1979).
25. O. Ermer and S. Lifson, J.Am.Chem.Soc. 95:4121 (1973).
26. A. Warshel, M. Levitt and S. Lifson, J.Mol.Spect. 33:84 (1970).
27. A. T. Hagler, P. Dauber and S. Lifson, J.Am.Chem.Soc. 101:5131 (1979).
28. F. A. Momany, L. M. Carruthers, R. F. McGuire and H. A. Scheraga, J.Phys.Chem. 78:1595 (1974).
29. L. A. Dunfield, A. W. Burgess and H. A. Scheraga, J.Phys.Chem. 82:2609 (1978).

THE STRUCTURE AND DYNAMICS OF WATER IN GLOBULAR PROTEINS

J.L. Finney, J.M. Goodfellow & P.L. Poole

Department of Crystallography, Birkbeck College,
Malet Street,
London. WC1E 7HX

INTRODUCTION

Water plays more than a passive role in biomolecular processes that occur in aqueous media. Much of its involvement is often assigned to the so-called "hydrophobic interaction", a process which, although well-characterised thermodynamically[1], is poorly understood at the molecular level. In addition, there are other potentially large contributions to free energy changes accompanying such processes as protein-substrate binding or protein folding. These relate to hydrogen-bonding and entropic changes involving the solvent region[2] and are similarly poorly understood.

In any given biomolecular process A → B (eg protein folding), we are presently unable to link unambiguously the measured thermodynamic changes with the changes that occur at the molecular level. Although this inability to connect structure and thermodynamics involves more than solvent effects, it is these aspects that we understand least. Except in one or two very favourable cases, we do not have an adequate characterisation of the solvent "organisation" - a term which we use to cover both structure and dynamics - for both initial (A) and final (B) states. It is with the first preliminary step of obtaining this information for one very limited class of biomacromolecular systems - aqueous globular proteins - that we are concerned with here.

The second, and biologically much more interesting problem, is to try to understand the *changes* in solvent organisation during the A → B process, and ultimately to connect these changes - together with the changes in the non-solvent part of the system - with thermodynamics. Attempts to lay the groundwork for this using computational techniques will be discussed briefly in the final section.

COMPLEXITY OF PROTEIN-WATER SYSTEMS

A major problem we face when trying to extract experimental information on solvent organisation - the instantaneous "structure" ' and the dynamics of the molecules - is that of the *complexity* of the heterogeneous, multicomponent assemblies that are our experimental samples. In comparison, a protein molecule of itself is much simpler: we know its sequence, and a considerable amount of stereochemistry provides a strong constraint which assists our understanding of the molecule's structure and dynamics. For example, in solving a protein structure by X-ray crystal structure analysis, we can compensate for the usual insufficient numbers of data by applying restraints which are consistent with this stereochemical information. The stereochemical restraints we can apply to the solvent are very much weaker: with water, we are dealing with a large number of small and mobile molecules which interact only weakly through a hydrogen bond whose distance and orientational dependencies are imperfectly known.

This problem of complexity has two immediate consequences. First it forces us to use as many different experimental techniques as possible (see Table 1) to obtain complementary information; a

Table 1. Sample Techniques Used to Investigate Water-Protein Interactions

Experimental
X-ray diffraction
Neutron scattering - high and low angle (contrast variation)
Inelastic neutron scattering
Infra-red and Raman spectroscopy
NMR
Isopiestic techniques
Volume measurements
Calorimetry (especially heat capacity)
Theoretical
Model building
Statistical mechanics
Computer simulation

satisfactory model must be shown to be consistent with *all* the experimental information from those techniques that are relevant. To paraphrase Felix Franks, the problem cannot be hit on the head, it must rather be slowly and carefully backed into a corner. Secondly the interpretation of much experimental data is strongly model-dependent: so much so that a change in the interpretational model may lead to a very different picture of the process being examined. This is particularly the case for dynamic measurements using, eg NMR, dielectric relaxation, and neutron inelastic scattering, as will be discussed later.

Our brief here is to illustrate how the application of many of the techniques used in structural molecular biology have been used to throw light on the structure and dynamics of water close to globular proteins. In particular, we want to underline many of the problems of interpreting the data obtained on such complex heterogeneous systems and to suggest how in some cases changes in the interpretational model may lead to fundamental changes in our concept of the system. However, we need first to fix a working framework of the unperturbed, pure solvent - liquid water.

A WORKING MODEL OF WATER

Water is a liquid. At the molecular level, this means we must consider two aspects: *molecular mobility* (related to fluidity) and *instantaneous structure*. We require a description of both these aspects in the bulk liquid, before we can discuss possible deviations from the bulk characteristics when water is in contact with a protein or indeed any other surface.

In inert gas liquids such as the theoretician's favourite argon, the spherically-symmetrical Van der Waals intermolecular interaction is relatively simple. To a first approximation, we can simplify the interaction further and consider the atoms as hard spheres. A crystal of these hard-sphere atoms has the familiar characteristic long-range order illustrated in fig 1(a). In contrast, the instantaneous structure of a liquid of hard-sphere atoms can be represented as in fig 1(b) - *an irregular, densely-packed structure of spheres which contains no crystalline regions*[3]. The structural differences between the regular *pile* of atoms representing the crystal, and the dense, inregular *heap* that is our model of the liquid, are illustrated in fig 1(c). An essential point to note is that the liquid is not in any sense a disordered crystal: its instantaneous structure is one which is almost as dense as the crystal, and which contains no crystalline regions: *yet it is a structure which is consistent with the potential function which describes the intermolecular interactions* (in this case a hard-sphere)[4,5].

For water, we can take a similar approach, and ask the question

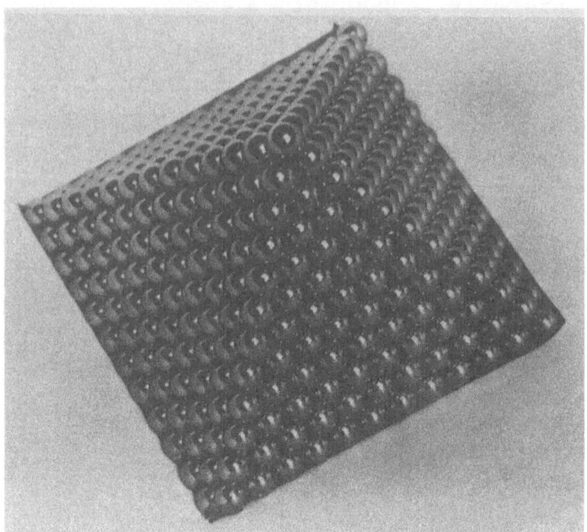

Fig. 1a. An idealised hard sphere crystal.

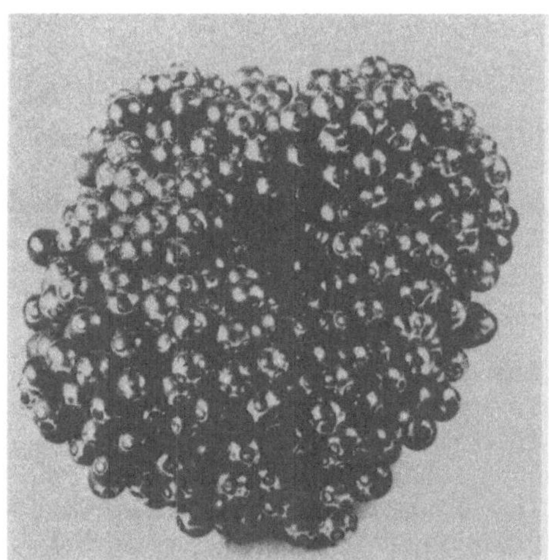

Fig. 1b. An idealised hard sphere liquid.

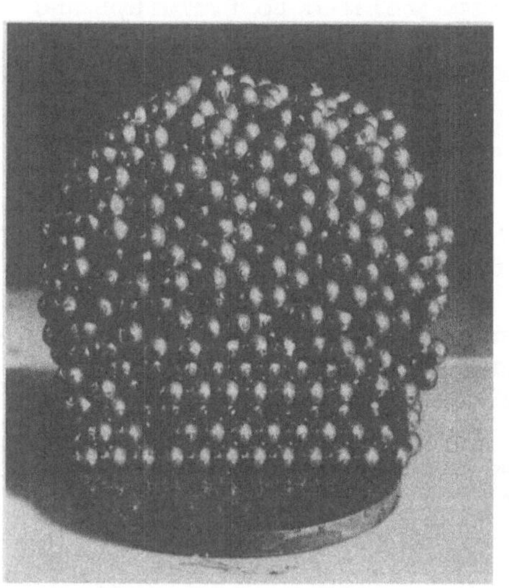

Fig. 1c. An idealised hard sphere liquid (upper) contrasted with an idealised hard sphere crystal (lower).

what structural organisation is consistent with the potential function describing the water-water hydrogen bond, yet is non-crystalline (ie it is *not* ice-like) and is of adequate density?

The first problem encountered in answering this question arises from our imperfect knowledge of the water-water hydrogen bond. A first approximation can be obtained by taking a simple picture of the water molecule in which the two protons and two lone pairs adopt an approximately tetrahedral arrangement; we might then argue that each water molecule is capable of forming four hydrogen bonds to neighbouring molecules in an approximately tetrahedral geometry as is found in hexagonal ice (see fig 2(a)).

Using this first order tetrahedral model of the local intermolecular geometry, we can build hypothetical models which fulfil the desired constraints of density and non-crystallinity. Such an idealised model - termed a *random tetrahedral network* - was first built by Bernal in the early 1960s[3]; the theory of the concept has been developed more recently by Sceats and Rice[6]. Figure 2(b) illustrates a laboratory realisation of the structural model obtained: it is clearly non-crystalline and has none of the long-range order of the ice structures. Figure 2(c) shows a two-dimensional three-coordinated analogue which shows the random network/crystalline network contrast more clearly.

Although this conceptual random network model is adequate for our present purposes, it should be underlined that it is only a first order model, and that the quantitative picture of the hydrogen-bond which has been used to develop the model is itself inadequate in at least two major ways. First, high level *ab initio* quantum mechanical calculations suggest that the two lone-pair regions are not as well-separated from each other as has been assumed above[7]; this is consistent with observations in many crystal structures[8] which show the directional control on the hydrogen bond by the lone-pairs to be much weaker than that exerted by the proton. We would expect this effect to reduce the degree of tetrahedrality in water, although the presence of two proton donors per water molecule will still be a strong driving force to average four-coordination. Secondly, there is strong evidence for *cooperativity* in the water-water hydrogen bond[9-11] and this may have significant effects on both our model of bulk water itself, and more particularly on the organisation of water close to a protein or other surface. The structural consequences of these two effects are the subject of active research on pure water[12-14].

The *molecular mobility* in bulk water presents fewer difficulties and the various time constants can be obtained from NMR and dielectric relaxation measurements. Three correlation times can usefully be defined[15]:

- τ_D : a correlation time for translational diffusion, taken as

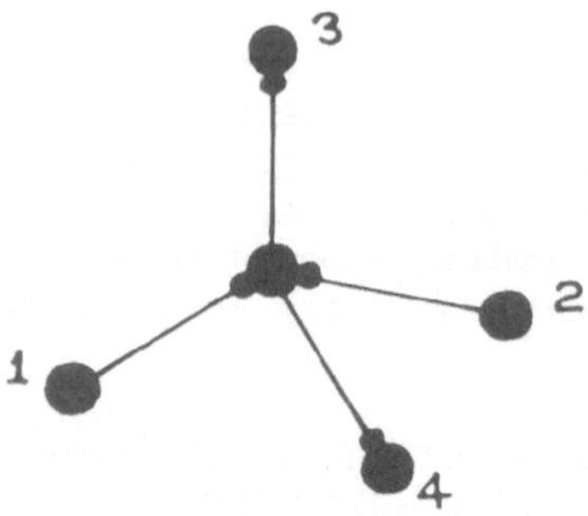

Fig. 2a. Ideal local tetrahedral water coordination in ice Ih.

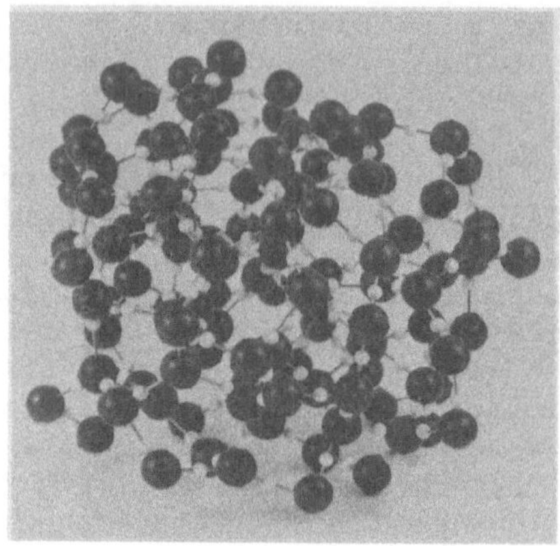

Fig. 2b. A four-coordinated random network model of water after Bernal[3]

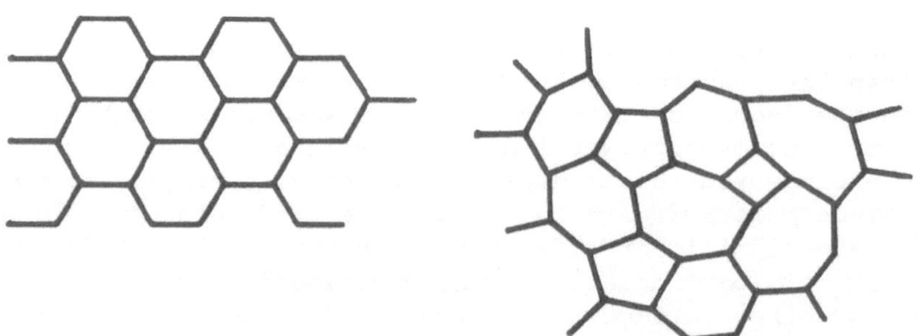

Fig. 2c. Schematic representation of a three-coordinated two-dimensional crystal (left) and liquid (right) network.

$\sim 3 \times 10^{-12}$s.

- τ_R : a correlation time for water molecule reorientation, $\sim 10^{-11}$ to 10^{-12}s. The precise value depends on the model used for the motion, but it appears that $\tau_R \sim \tau_D$.

- τ_E : the lifetime of a proton on a water molecule between intermolecular exchange events. $\tau_E \sim 0.9$ms at pH 7.

In summary, liquid water can be regarded to a first approximation as a *non-crystalline* "random network" consistent with approximately tetrahedral coordination (fig 2(b)). It is a liquid, with each instantaneous network continuously transforming into a statistically equivalent network on a time scale of 10^{-11} - 10^{-12}s. *There are no crystalline regions resembling ice*, as has often been assumed through a misinterpretation of the term "iceberg" coined by Henry Frank[16]. Although this conceptual picture is a useful reference point, we should bear in mind that the *tetrahedral*, *pair-additive* hydrogen bond picture used is oversimplified and subject to modification. In particular, when the balance between the two proton donors and two proton acceptors (lone pairs) that we find in bulk water is removed - eg at a protein-water interface - the driving force towards tetrahedrality will be weakened. This further underlines the weakness of the stereochemical constraints that we may be tempted to use to assist our understanding of the geometry of instantaneous water networks at an interface. *The water molecule is small, interacts weakly, and has a relatively weak angular dependence in its intermolecular interactions.* Its versatility at the molecular level implies that an uncritical application of over-simplified ideas of tetrahedrality may be misleading especially in heterogeneous systems.

"BOUND" WATER?

Characterisation of Solvent Perturbation at Interfaces

The existence of surface tension illustrates that a bulk liquid is perturbed in some way even at a free surface. Our protein-water interface is much more complex; and we would like to characterise quantitatively how the bulk water organisation is perturbed by the presence of the protein. Such a perturbation has both dynamic (mobility) and static (structural) aspects.

Changes in molecular mobility close to an interface can in principle be characterised by changes in characteristic correlation times such as τ_D and τ_R. In addition, we can introduce quantities such as lifetimes of association of a given water molecule with the protein or its constituent exposed groups, and consider the possibility of directional anisotropy in τ_D and τ_R[16]. Thus we can develop models of mobility changes which will have significant consequences for biomolecular processes. For example, diffusion of substrate close to the protein will be affected, and there will be non-negligible entropic changes on the release of water from the interfacial contact region to the bulk liquid when two macromolecules associate.

Quantifying perturbations of the bulk liquid structure is more difficult. For a molecular assembly which itself is characterised by restricted structural disorder, it is not easy to define an order parameter which gives a useful realistic index of the structural perturbations caused by the interface; this compares with a relative ease of describing distortions of or imperfections in crystalline structures, where the reference structure is well-defined and can be described simply. Much relevant literature attempts to describe *changes* in liquid structure in terms of increased or decreased "ordering", yet without clearly defining an order parameter. The general result of such imprecision is confusion, which is often increased by the fact that "ordering" can be considered also in a dynamical context. Unless precise static and dynamic order parameters can be defined for a given system - and this is extremely difficult for the complex heterogeneous systems we are considering - statements concerning relative degrees of order are best avoided.

Characterisation of so-called "Bound Water"

Historically, the perturbation of the solvent close to a bio-macromolecule interface has often been described in terms of a concept termed "bound water". Although the meaning of this term at the molecular level never seems to have been precisely defined, it implies the specific binding of water to groups that are exposed on the protein surface. There is also an implication that bound water residence times on the protein are long compared to the rotational and translational correlation times of bulk water.

Attempts to measure the amount of "bound water" associated with a given protein have therefore - not surprisingly - given rise to problems[18]. Table 2(a) shows the amounts of so-called "bound" water associated with lysozyme and haemoglobin obtained using a variety of experimental techniques. From the variations observed between values *for the same protein*, we conclude that the results depend upon the experimental technique used: in effect, we can say that *"bound" water is defined operationally by the technique used to measure it.* In addition, the fraction of water that is perturbed by the protein surface may depend upon the form of the sample: the fraction as measured in dilute solution may differ from that measured in a protein film of much lower water content.

Consequently, we would argue strongly for the term "bound water" to be abandoned. Instead of using such a poorly-defined term, we should aim to understand more clearly what perturbation of the bulk solvent is being measured by each technique, and to link the results to possible changes at the molecular level.

Table 2(a). "Bound Water" Fractions According to Different Experimental Techniques

Fraction of Water "Bound" in g water/g protein*								
	Viscosity	Diffusion	Sedimentation	NMR	DSC	Dielectric	(Isopiestic +)	Range
Lysozyme	0.34	0.52	0.52	0.34	0.30	0.30	(0.25)	0.25–0.52
Haemoglobin	0.62	0.52	0.75	0.42	0.32		(0.37)	0.32–0.75

* Data from refs 17 and 19
\+ Assumed water activity of 0.92

Table 2(b). Non-Freezing Water Fraction From Different Experimental Techniques[20]

Fraction of Non-Freezing Water Fraction in g Water/g Protein			
	DSC	IR	NMR
Lysozyme	0.32±0.02	0.31±0.02	0.34±0.02

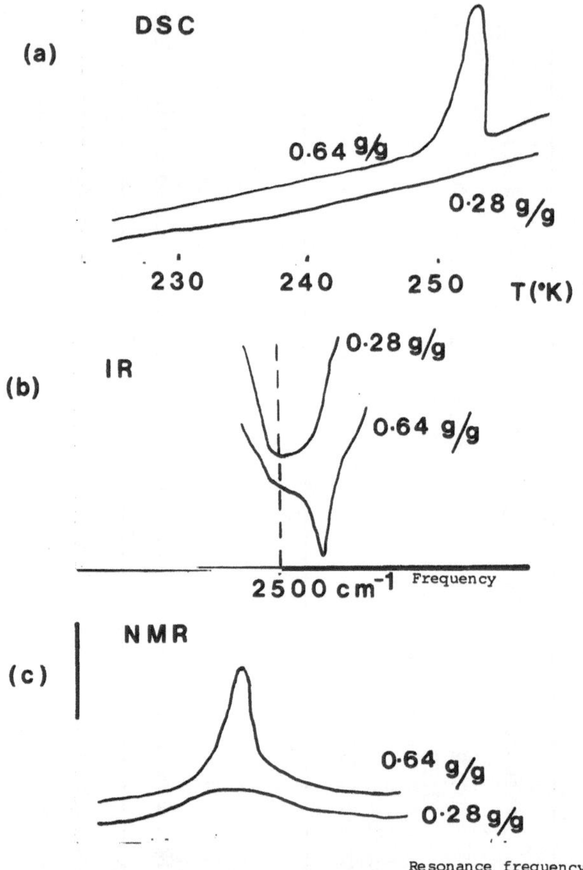

Fig. 3. Schematic illustration of the detection of non-freezing water using three different experimental techniques.

The Non-Freezing Fraction

One example of this latter approach is a series of measurements of that fraction of water which does not freeze in lysozyme films taken down to liquid nitrogen temperature[20]. Considerable care was taken to produce protein film samples of controlled hydration in as similar a way as possible for three measuring techniques: infra-red, differential scanning calorimetry and NMR. The principles of the applications of these experimental techniques are illustrated in fig 3.

Differential Scanning Calorimetry (fig 3(a)). A sample of known water content is taken down to liquid nitrogen temperature and then warmed up at a given rate in the scanning calorimeter. For water contents greater than a critical value, a melting peak is observed as in the upper curve. As the water content of the sample is decreased, a point is reached at which no melting peak is observed; the water content of this sample gives an estimate of that fraction of water which does not freeze. In practice, the area of the melting peak is plotted against water content (fig 4(a)), and a least-squares line used to obtain a value for the critical hydration.

Infra-red (fig 3(b)). This makes use of the shift from 2500 cm-1 to 2420 cm-1 observed in the uncoupled O-D stretch peak on freezing liquid water (containing 5% D_2O) to ice Ih, a technique first applied to DNA by Falk[21]. Spectra of samples of known water content are taken at liquid nitrogen temperature, and the critical hydration for which no ice band is present is obtained. Again, plots of the area of the ice band can be plotted against hydration (fig 4(b)).

NMR (fig 3(c)). This technique, first used by Kuntz[22], relies on the fact that the proton resonance in ice is very broad, compared to the sharp signal observed for liquid water. Thus, observation of the proton resonance at sub-ambient temperature (here 233K) as a function of hydration leads to an estimate of the non-freezing fraction.

The results obtained for hen egg-white lysozyme are shown in table 2(b). In contrast to the "bound" water data of table 2(a), the results are consistent between the different techniques within the estimated error limits, and indicate for lysozyme a non-freezing fraction of about 0.32 ± 0.02 g water/g protein. There is also an indication of a change in slope of the IR plot (fig 4(b)) suggesting some change in behaviour of the water at about 0.4-0.45g/g. A similar slope change is found also in the NMR longitudinal relaxation time (T_1) plotted as a function of hydration at room temperature[20].

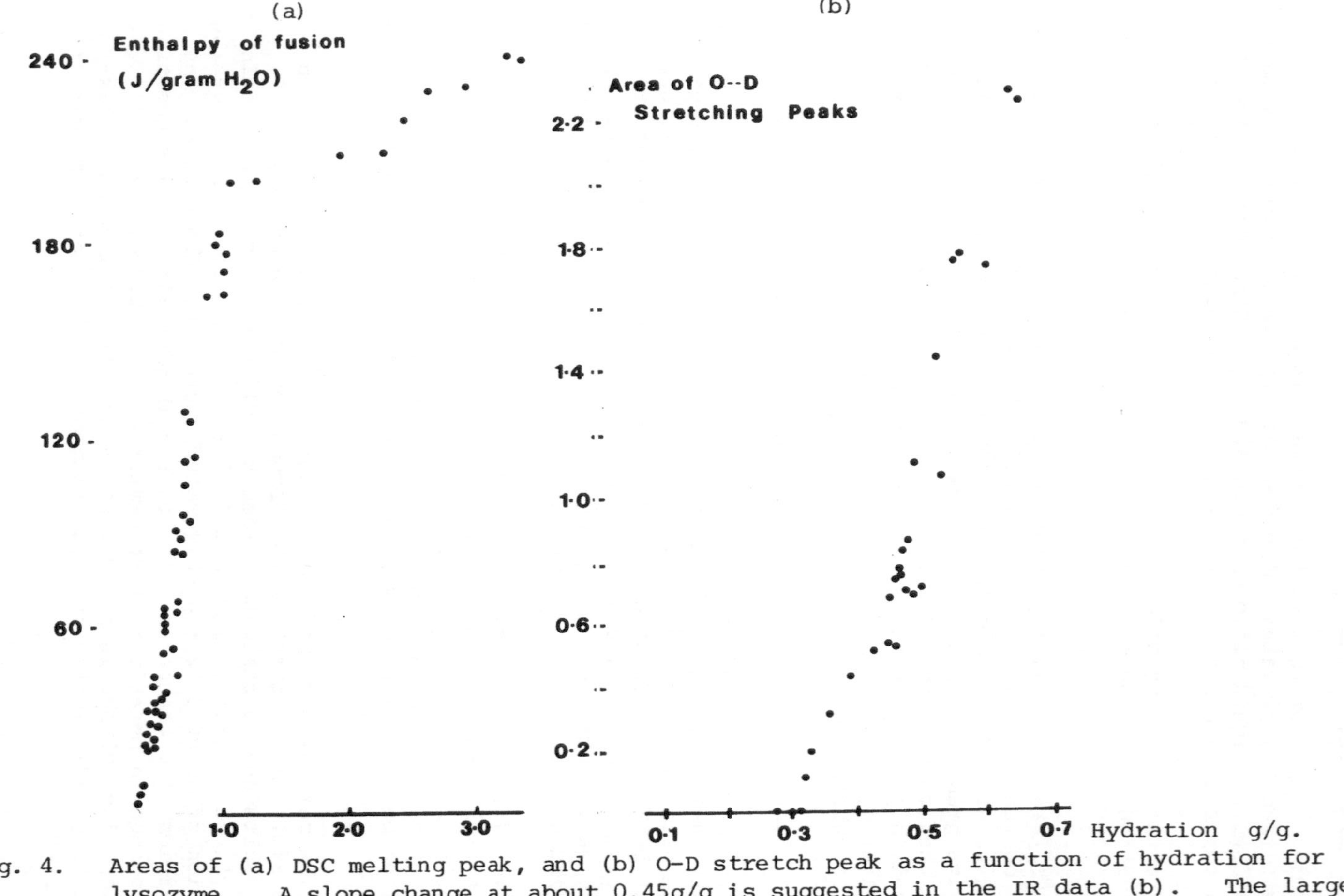

Fig. 4. Areas of (a) DSC melting peak, and (b) O–D stretch peak as a function of hydration for lysozyme. A slope change at about 0.45g/g is suggested in the IR data (b). The large change in slope of the DSC plot at about 1g/g is at the glass-solution transition[20].

A Possible Model for the Non-Freezing Fraction: Plug-Compatibility

What does this non-freezing fraction correspond to at the molecular level? In attempting to answer this question, we should remember that the information from the experiments described is *very* limited - it consists merely of a figure for the total amount. Therefore, any model used to explain this value must be a very tentative one, and must be subject to modification or even rejection in the light of further evidence.

The first step in developing our model is to examine the detailed molecular nature of the protein surface, using atomic coordinates obtained from X-ray crystallography. We may then assess the possible interactions between water and the protein surface groups. To do this, we use a modification of the method of Lee and Richards[23] to calculate the surface exposures of various groups in the molecule[21, 24,25], and by comparison with other aqueous systems[24] identify those surface sites that are theoretically capable of hydrogen bonding to a water molecule. We then assess the consequences of "*plugging in*" a water molecule into each of these sites.

For lysozyme, we find 245 possible water sites. This is equivalent to a hydration level of 0.31 g/g, a value remarkably close to the measured non-freezing fraction of 0.32 ± 0.02 g/g. Thus, although the model is crude, it leads to an apparently reasonable hypothesis that those water molecules directly hydrogen-bonded to the protein surface may make up the non-freezing fraction; in addition, an estimate of the available remaining exposed protein surface (apolar groups) suggests an additional 130-150 water molecules are required for complete coverage; when this is added to the 0.31 g/g, the total estimated monolayer coverage corresponds to about 0.46 g/g, interestingly close to the (rather indeterminate) break points of the IR plot (fig 4(b)) and the NMR T_1 data[20].

There are several problems which can be raised at this stage which suggests this simple model should be treated circumspectly. First, we have ignored geometrical situations in which two surface polar sites may be close enough to be bridged by one single water: an estimate of this effect reduces the polar-group-associated water to 0.25 g/g. Secondly no account has yet been taken of the ionisation of charged groups, which would be expected to interact with a slightly larger number of water molecules than in their uncharged state. Attempting to correct for this brings us back to 0.31 g/g! Thirdly, we have assumed the protein is static, with atomic coordinates as found in the crystal, yet we know from H-D exchange studies that many of the apparently inaccessible polar group hydrogens do exchange and therefore are at some time accessible to solvent. Which if any of these groups are hydrogen bonded to water so as to contribute to the non-freezing fraction is impossible to say: it may even be argued that at these low temperatures in dense film preparations, such internal exchangeable sites are inaccessible to the surface, and hence may not contribute at all to the non-freezing water fraction.

These uncertainties illustrate the problems of using this simple plug-compatibility model as any more than an interesting, though possibly useful, working hypothesis. For it to have any more validity than numerology, we require additional evidence. Possibly relevant experimental data from difference infra-red and other techniques is discussed subsequently below.

"STATICS" OF PROTEIN HYDRATION

We can in principle extract information on solvent locations close to a protein from careful analysis of a single-crystal X-ray study. All protein crystals contain upwards of 25% solvent (in most cases water plus a variety of ions). Where the refinement can be taken to very high resolution (preferably 1.5Å or better), a significant number of water molecules can be located in the vicinity of polar and charged groups on the surface, and sometimes internally. However, great care must be taken in interpreting this region: unlike the protein, whose stereochemistry is reasonably well understood, we cannot safely apply much in the way of geometrical restraints on water molecule positions (see above).

The problems met with in locating water molecules by diffraction methods together with the data available up to 1979, have been reviewed at length elsewhere[26,27]. In addition, more high-quality solvent data is becoming available, eg on *Streptomyces griseus* protease A[28], human lysozyme (C. C. F. Blake, personal communication), vitamin B_{12} coenzyme (H. J. Savage, unpublished neutron and X-ray work at 0.9Å), and avian pancreatic polypeptide (J. E. Pitts *et al*, unpublished). Here we summarise the main points concerning both the technical problems and the results obtained.

Problems of Diffraction Location

(1) Only data from high resolution refinements (preferably 1.5Å or better) seem to give reasonably reliable information on water positions. Changes in apparent solvent organisation have been seen in going from 1.9Å to 1.5Å in insulin (G. Dodson, personal communication) and even from 1.5Å to 1.2Å in rubredoxin[29].

(2) Heavy atom phases are usually inadequately reliable at resolutions better than 2Å and thus should not be used for serious solvent work.

(3) Ideally, any refinement aimed at locating solvent should not - at least at the stage when solvent positions are being concentrated upon - make any assumptions about either the protein

or the solvent stereochemistry. This argues for the use of unrestrained refinement procedures, which require a favourable ratio of intensities measured to parameters being refined. These conditions can be fulfilled only at high resolution.

(4) Disorder in the solvent region - for example the occupation of two mutually exclusive sites at different times or in different unit cells - may be difficult to distinguish from noise in the electron density map. Refinement of partial occupancies may improve the agreement between the model and the data (ie reduce the R factor), but this may reflect no more than an absorption of errors.

In summary, the reliability of solvent data can be variable. For well-ordered water molecules, they may in favourable cases be located as clearly as well-ordered protein surface groups. As we move further away into the solvent region, the data becomes much less reliable, and must be treated with very great care indeed.

Summary of Water Organisation in Protein Crystals

In all high resolution studies, water molecules are found which make hydrogen bonds of apparently reasonable geometries to surface polar and charged groups, and often to other well-ordered water molecules. In only a few cases does this ordered water region extend significantly further away from the protein surface; whether or not this is a real effect, or a consequence of the problem of extracting signal from noise in this region, is presently not clear.

Fig 5 shows the distribution of first neighbour polar group-water distances in three different proteins at different resolutions. The highest resolution data of rubredoxin *C. pasteurianum* (1.2Å) shows a very reasonable distribution of hydrogen bond distances peaking at about 2.9Å; the wider spreads of the other two histograms seem to reflect the greater reliability of the 1.2Å data. However, even at this very high resolution, there are several apparently very close approaches: these must reflect problems with the model, eg difficulties in sorting out partially occupied mutually-exclusive water sites. Fig 6 shows water-protein contacts for insulin (G. Dodson, personal communciation). Of particular interest is the apolar group - water distances which peak quite reasonably between 3.5 and 4.0Å. Again some of the shorter distances $\sim$ 2.5Å probably reflect inadequacies in the model.

No detailed analysis of hydrogen-bond angles has yet been published. However, from cases we have examined, considerable deviations from tetrahedrality are found, with extensive three-coordination as would be expected from the previous discussion on the angular dependencies of the water-water hydrogen bond. Further work is in progress on this point.

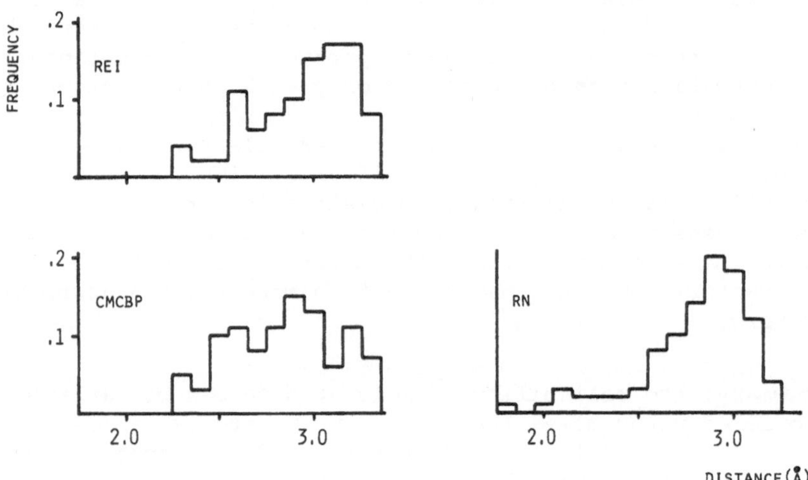

Fig. 5. Water-polar group hydrogen bond distances in three protein crystals at different resolution.[27] REI : Bence-Jones REI protein (2.0Å; CMCBP : carp muscle calcium-binding parvalbumin (1.85Å); RN : rubredoxin *C. pasteurianum* (1.2Å).

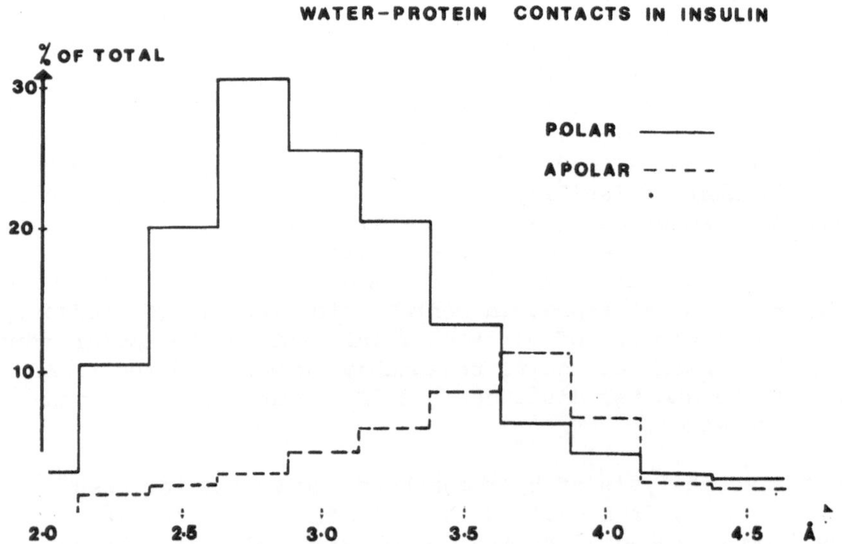

Fig. 6. First neighbour water-protein contacts in insulin.

The crystallographic data, therefore, is generally consistent with what we expect from our (weak) knowledge of hydrogen bonding in water and to the relevant polar and charged groups on the protein surface. Water is found to bind specifically to metal ions internally, to charged and polar groups both internally and on the surface, both singly and in networks. Waters have been found which mediate (otherwise impossible) hydrogen bonds within a molecule, between molecules, between sub-units and between neighbouring protein molecules in crystals , and between enzyme and substrate. Water molecules on the surface connect to single polar groups, may form single or multiple bridges between different polar groups on the same molecule, and are often found in active site clefts. Within the limits of the data, hydrogen bond distances found are reasonable and geometries are variable, with angles deviating significantly from the tetrahedral.

The possible functional role of these waters is discussed in detail elsewhere[2,26,27].

DYNAMICS OF THE HYDRATION SHELL

In a protein-protein or protein-substrate interaction, water will be expelled from the intermolecular contact region to the bulk liquid. If the water close to the protein is distinguishable thermodynamically from the bulk, there will be a contribution to the free energy of the process from the expulsion of the solvent. In order to assess the magnitude and therefore relative significance of this contribution, we need information concerning (i) how the *motions* of the water molecules at the protein surface are affected and (ii) how far from the surface the influence of the surface is felt.

Several techniques have been used to try to obtain such information. *In all cases, however, the interpretation of the data is strongly model-dependent, and different models may result in apparent residence times of water molecules close to the proton which differ by up to six orders of magnitude.* Thus, great care is needed in drawing conclusions from the experimental data. We will consider several of these techniques in turn: dielectric relaxation, NMR, inelastic neutron scattering, and Raman spectroscopy.

Dielectric Relaxation

This technique probes the relaxations of dipole moments in the protein-water system as a function of frequency. Fig 7 shows a typical plot of ε' (the real part of the dielectric constant) against frequency for a protein solution.

Three relaxation processes are evident, of which the lowest (β) and highest (γ) frequency relaxations can be interpreted fairly straightforwardly. The β-process at about 1MHz depends upon the size and shape of the protein, and thus can be assigned to the rotation of the protein molecule. The γ-relaxation at about 20GHz is essentially indistinguishable from the relaxation observed for bulk water, although there have been interpretations of this process in terms of a spread of relaxations centred on about twice the free water relaxation.

The δ-process at about 200MHz is not easily assigned to a particular relaxation. The two main candidates that have been discussed are (a) the relaxation of water molecules restricted at the protein surface[32,33], and (b) the relaxation of polar side chains of the protein[33-35]. The evidence does not seem to favour conclusively either assignment: however, some of the consequences of identifying the δ-process solely with restricted water relaxations leads to several awkward problems as follows[35].

(a) The low frequency dielectric permittivity ε_o *increases* with temperature for lysozyme crystals[32] and also for collagen[35]. However, water and ice both show a slight *decrease* of dielectric constant with temperature, implying the δ-process is unlikely to be purely a water relaxation.

(b) The dielectric relaxation time τ *increases* with temperature[32]. For this to be consistent with a water relaxation we need to invoke a *negative* activation energy for water rotation at the protein surface implying there is less hydrogen bonding at the surface than in bulk water[32]. Although this view can be defended, the observed temperature dependencies of ε_o and τ lead to uncomfortable consequences if the water assignment is to be accepted.

The alternative assignment of the observed relaxation to mobile polar side chains gives rise to no such problematical consequences, and is consistent with proton NMR studies on collagen in both H_2O and D_2O. Moreover, the observed temperature dependencies of ε_o and τ appear qualitatively consistent with the side chain assignment[35]. For example, the side chain contribution to the dielectric permittivity will be proportional to the mean square fluctuation of the dipole moment; as temperature increases, the motional freedom of the side chain is expected to increase and a larger fraction of the side chain will contribute to the permittivity with a larger dipole moment fluctuation. Moreover, the increased mass and length of the mobile part of the side chain would lead us to expect an increase in the correlation time of the dipole fluctuation and hence τ is expected to increase as is observed. The side chain assignment appears consistent also with both the observed decrease in τ with increased water content and the apparent net negative enthalpy of relaxation[35].

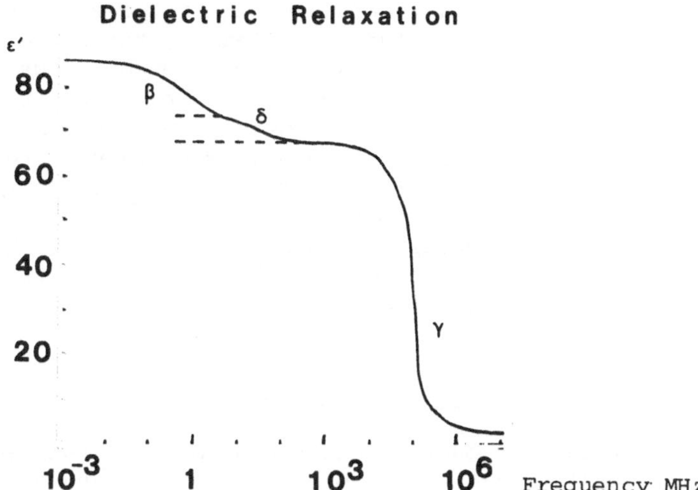

Fig. 7. Dispersion of the real part of the dielectric constant for a typical protein solution.

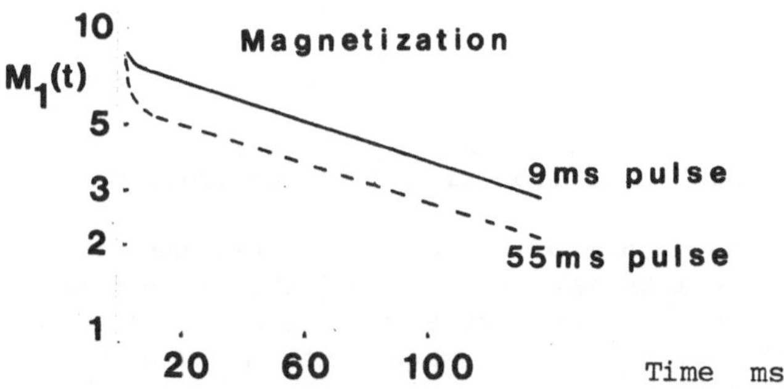

Fig. 8. Decay of water proton longitudinal magnetisation for a hydrated protein, showing pulse width dependence[44].

Although the side-chain assignment of the δ-process may be the most consistent interpretation, the issue cannot be regarded as settled: Pennock and Schwan[34], for example, argue that the process involves *both* polar side-chain and restricted water relaxations. However, in the light of this uncertainty, the dielectric data on proteins cannot be regarded as providing unambiguous positive evidence for restriction of the mobility of the water close to the protein. More work is needed on simpler model systems to assist our interpretation of the data on large molecules. For example, recent work of Hallenga *et al*[36] has examined dielectric relaxation in alcohol solutions. Their conclusion that there is an increase of the dielectric correlation time of water molecules by a factor of about two to three close to apolar groups has clear significance for the water close to such groups on a protein surface.

Nuclear Magnetic Resonance

A very large amount of published NMR work has tackled the problem of water mobility close to a protein surface; the conclusions are even more model-dependent than those drawn from the dielectric relaxation measurements. Most of the work until the last five years (see refs 15, 19, 37 for reviews) was interpreted in terms of an almost standard two (or more) fast exchange model, in which the rotational correlation time of that water fraction close to the protein (τ_P) was thought to be considerably larger than and in fast exchange with, the surrounding bulk water (τ_B). However two relatively recent investigations have led to a serious questioning of the standard chemical exchange model.

In the first of these, Bryant and co-workers measured the longitudinal proton relaxation time T_1 in lysozyme crystals and powders[38-43]. A typical plot of the decay of the water proton longitudinal magnetisation is shown in fig 8. This clearly non-exponential relaxation behaviour can be interpreteted in terms of a two phase *slow* chemical exchange model. This model, however, leads us into considerable difficulties[39,43], as follows.

(1) A consequence of the model is that the proton lifetimes in the immediate vicinity of the protein must be very long, of the order of several milliseconds at room temperature. This is in conflict with other measurements of solvent lifetime close to the protein.

(2) The slow exchange model should be pulse-width independent: experimentally, a significant pulse-width dependence is observed (see fig 8).

(3) We would expect the relaxation behaviour to be independent of the nucleus observed. However, the measured deuteron

relaxation data are clearly exponential[43,44], indicating a chemical exchange model to be inappropriate.

These difficulties are overcome by abandoning a chemical exchange model in favour of one in which the coupling between solvent protons and protein protons is via a *magnetic* interaction with no material exchange (cross-relaxation)[39,42,44]. Within this new framework, the data on lysozyme crystals and powders can be interpreted to give a clearer view than previously of the water dynamics at the protein surface: the perturbed solvent region is *not* extensive, and the rotational correlation time of the water at the protein surface is shorter than nanoseconds[44]. Thus the presence of the protein surface slows down the water relaxation time by less than two orders of magnitude.

The second set of measurements concerns the frequency dependence of the water proton longitudinal relaxation time in protein solutions by Koenig and coworkers[45-46], a typical curve being represented in fig 9. By changing the size and shape of the protein, the large dispersion (the A component) can be demonstrated to be related to the protein reorientation; hence we can conclude immediately that the solvent protons must feel the protein rotation. The mode of coupling between water and protein again appears at first sight consistent with a two (or more) component fast exchange model with water reorientation times close to the protein slower than in the bulk by up to six orders of magnitude[45].

As for the NMR work on lysozyme crystals and powders, however, several consequences of this exchange model lead to problems. First, the model predicts that the number of waters which are affected by the nearby presence of the protein is very small - of the order of 3-10 for a protein the size of lysozyme, depending upon how rigidly these molecules are assumed to be bound. Secondly, the analysis of the exchange conditions assumed by the model leads to a strong inconsistency when the results of proton, deuteron, and ^{17}O relaxation measurements are considered together[46].

This inconsistency arises when attempts are made to narrow down the possible range of values of τ_M, a model parameter indicating the lifetime of a protein-associated water molecule. We can set lower and upper bounds on τ_M as follows[46]:

(a) $\tau_M \gtrsim \tau_R$, the rotational relaxation time of the protein. This condition is necessary if the protein-resident waters are to sense the protein motion.

(b) $\tau_M \lesssim \tau_{1P}$, the nuclear relaxation time of the bound water molecule. If this condition were not valid, the protein-associated molecule could not communicate its information on protein rotational motion to the bulk solvent.

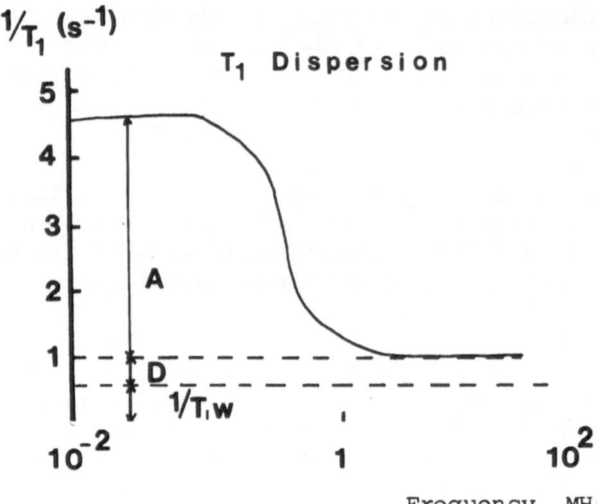

Fig. 9. A typical NMR dispersion curve for a protein solution[46].

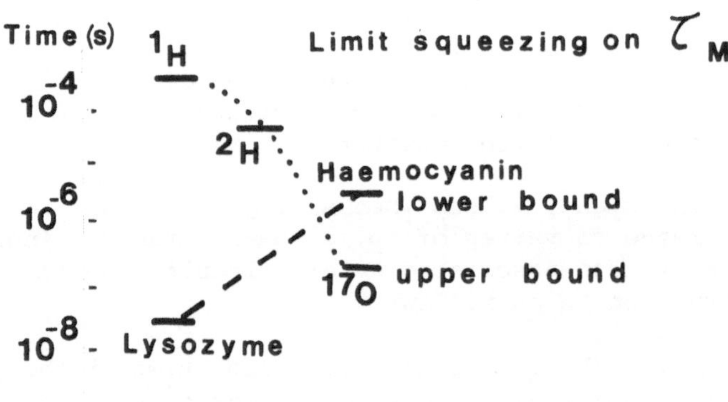

Fig. 10. Squeezing the upper and lower bounds on τ_M[46].

By performing relaxation experiments using a larger protein, Koenig *et al*[46] were able to raise the lower bound on τ_M. Moreover, by measuring ^2H and ^{17}O relaxations (these nuclei in water relax about 10 and 1000 times respectively faster than protons), the upper bound could be lowered. The results of squeezing the limits on τ_M are given in fig 10: *the upper and lower bounds cross!* In the light of this inability to define τ_M consistently, it is difficult to continue to accept the two-phase fast exchange model as a model for interpreting this data. Additional problems - such as the insensitivity to pH of the number of waters associated with the protein, and an inability to associate an activation energy with τ_M[45,47] - further encourage an abandonment of the model.

As an alternative to one based on chemical exchange, Koenig[47] proposes a model involving a hydrodynamic interaction mechanism between solute protein and solvent water molecules which is consistent with the non-specific nature of the proton T_1 dispersion data. This mechanism envisages the solvent molecules (regarded as a continuum fluid) surrounding the protein molecules which are undergoing rotational Brownian fluctuations as "sloshing" about in an anti-correlated manner to conserve angular momentum. Through solvent-solvent interactions, some of this angular momentum will be converted from that of the continuum fluid to individual solvent molecules. As a result, the solvent molecules will have a small thermal motion component that is related to the Brownian motion of the protein molecules. In addition to this proposed mechanism, the variation of the A process as the protons are diluted with deuterons argues for a contribution through exchange of magnetisation between protein protons and hydration shell protons (cross-relaxation again)[41].

Reference to fig 9 shows a residual high frequency contribution (the D term) whose relaxation is not shown, but which disperses in the range of 100-300MHz[47]. This is about the same frequency as the δ-process observed in dielectric relaxation measurements (see previous section). Hallenga and Koenig[47] favour an interpretation in terms of the water in the hydration layer: on this basis they estimate correlation times of this water $\sim 10^{-9}$s, consistent with the conclusions reached earlier on the basis of T_1 measurements on protein powders[44]. However, in the light of the uncertainty of the interpretation of the dielectric δ-process discussed above, we should bear in mind the possible contribution from protein side-chains, which have been observed to relax in this region[35].

It is clear from a consideration of the T_1 measurements for protein powders and crystals and the T_1 dispersion data for protein solutions that the conclusions we can draw concerning water molecule dynamics close to the protein are extremely model-dependent. The fact that both sets of measurements appeared to be consistent with two (or more) phase chemical exchange models which later had to be abandoned underlines the necessity of examining and testing the consequences of a particular model.

The resultant picture we can draw of water dynamics close to a protein interface is consequently rather sketchy. The water molecule motion at the protein surface is fast, being slowed down by no more than a factor of 100 with respect to the bulk, although this conclusion might be affected if the possible contribution of relaxing protein side-chains is considered as in the dielectric case. If this slowing down is due to steric hindrance to water diffusion from the presence of charges on the protein surface itself, the extent of the restricted water region can be estimated to be about one to two layers of water[48]. Finally, if the strong evidence for a cross-relaxation mechanism is accepted, much published proton relaxation data in many biological systems will require re-examination*.

Other Methods

For completeness, the following brief discussion of other work related to water dynamics in proteins is given. The state of development of technique, data, and/or interpretational models in these cases is insufficiently advanced to materially change the conclusions on water dynamics set out immediately above.

Preliminary results have been published on neutron inelastic scattering measurements on deuterated C-phycocyanin at different H_2O hydration levels. As with NMR and dielectric relaxation measurements the interpretation of the neutron data is strongly model-dependent; in addition there are problems of instrumental resolution, in that the quasi-elastic broadenings measured (up to 0.003 cm^{-1}) are less than the instrumental resolution width (quoted as 0.003 to 0.01 cm^{-1} [45]).

* The NMR interpretation offered here is subject to further modifications in the light of very recent work. First, the hydrodynamic coupling argument (the "slosh effect") now seems difficult to sustain in the light of a more detailed theoretical examination (R. P. Bryant personal communication). Secondly, recent ^{17}O measurements[69], which avoid the cross-relaxation complication, have been analysed successfully in terms of an *anisotropic* two-state fast exchange model in which approximately two "layers" of water molecules reorient less than one order of magnitude slower than in bulk water. Further experiments are in progress to resolve the remaining problems. The current (July 1981) state of the NMR interpretation is given in the proceedings of the Cambridge Working Conference on the Biophysics of Water (ed. F. Franks).

In making an interpretation of the momentum transfer dependence of the broadening, the assumption is made that the effects observed are due to the water protons; as in the NMR and dielectric discussions previously, however, one would expect a contribution from the movements of relevant parts of the protein itself - in this case the motions of those hydrogens which exchange with deuterons originally on the protein. In addition, there is evidence[50,51] that the motions of the protein are a function of hydration especially at low hydration levels: any such changes are not allowed for in the current first order interpretation of the neutron data.

Middendorf and Randall use a jump-diffusion model to assist their interpretation of the data[49]. On this basis, their major conclusions are that the characteristic hydration site residence time of a water molecule is about 5-30 ns, with a characteristic jump length of 6-9Å. In the light of the NMR conclusions which seem to argue strongly against such a long definite residence time, it would seem that the results of examining the neutron data in the light of alternative models would be of interest. Clearly neutron inelastic studies of water in proteins is at a very early stage: further work is necessary, preferably on systems whose structures are known from crystallography and with improved instrumentation.

A small amount of work has been done using Raman spectroscopy to probe protein-water interactions. Two studies have examined the intensity changes of the OH stretching band in lysozyme solutions as a function of concentration. Unfortunately, the *experimental results* of the two studies disagree. Cavatorta et al[52] find a fall in the OH intensity with increased protein concentration, which they interpret in terms of a reduced freedom of water molecules in the "overlapping hydration shells" of two neighbouring protein molecules. In contrast, Samanta and Walrafen[53] find an increase in the OH stretching band intensity with increased protein concentration, which they explain in terms of increased hydrogen bonding at the protein surface as compared to in the bulk. In addition to the discrepant experimental results, it seems that more work requires to be done in developing stricter models for the interpretation of the OH stretching band intensity in these systems.

Both Raman and far IR studies can be used to probe the low frequency restricted translation region in an attempt to identify eg the intermolecular O-H...O stretch ($\sim$ 170 cm^{-1}) and the hydrogen bond bend ($\sim$ 60 cm^{-1}) bands. Any shifts in these bands could tell us something about the restriction of water motions close to the protein. The two Raman studies discussed above[52,53] both observe a feature at about 170 cm^{-1}; this frequency is indistinguishable from bulk water, and therefore tells us nothing about how the surface water H-bond stretching motions might be affected by the protein. Genzel *et al*[54] found indications of peaks at both $\sim$ 160 cm^{-1} and 75 cm^{-1}, but again the evidence for any significant shift in these frequencies from those in bulk water is not strong. Attempts to find a $60cm^{-1}$ feature in the far infra-red by Golton[20] were unsuccessful.

SEQUENTIAL HYDRATION OF DRY PROTEINS

Although the aqueous phase seems to be necessary to some degree before an enzyme can be active, exactly how much water is necessary, and why, is a subject of much discussion. It is clear that *all* the aqueous phase is not essential: for example, provided the ionic strength and dielectric constant of the solvent is maintained approximately constant, changing the solvent medium and lowering the temperature does not appear to disrupt the activity of an enzyme and would therefore appear not to cause a significant denaturation of the active tertiary structure[55-59].

A variety of attempts have been made to examine in detail the nature of the protein hydration process and how much conformation and activity are thereby affected. Questions to which answers would be useful include:

- what molecular events occur on rehydrating a dry globular protein?
- which sites hydrate preferentially?
- what is the thermodynamics of the various absorption processes?
- what is the effect of water addition on: enzyme activity?
protein conformation?
protein stability?

Several different experimental techniques have been used in trying to answer these and related questions. Some of the most useful ones are listed briefly below.

Calorimetry. Heat capacity values and changes with hydration and temperature give unambiguous thermodynamic information[19,37,50,51, 60-62]. Interpretation in molecular terms is strongly model dependent. Recent measurements of heats of solution as a function of hydration have been interpreted in terms of the relative strengths of binding of different protein sites[63].

Difference Spectroscopy (mainly IR, but also UV and Raman are relevant). Only a limited amount of work has been performed on protein film samples[51,60,61,64]. Shifts in intensities and frequencies of various bands (eg N-H and C=O stretching or bending modes) upon hydrogen bonding give some information on the hydration of the group concerned. Interpretations may be only weakly model-dependent.

Absorption isotherms. An extensive literature exists on such studies, the work pre 1973 being reviewed critically and thoroughly in ref 19. A major problem with much of this work concerns interpretation of the data in terms of a standard theoretical isotherm such as BET. Parameters meant to have physical significance which are derived from such data fitting should be treated with very great care: for example, most BET-fitted data suggests only 0.05-0.10 g water/g protein are required for monolayer coverage. Such figures are clearly in strong disagreement with information from other sources discussed above.

In conjunction with selective chemical modification, absorption studies have proved very powerful, as in the work of Westerman and Rochester[65-67], who concluded that the relative abilities of charged or polar groups to sorb water was in the sequence $COO^- > NH_3^+ > NH_2 > COOH$. Contrary to earlier work, they demonstrated that absorption at peptide groups must be assumed in order to explain the data at moderate to high vapour pressures. They also suggest that at these higher relative humidities, their data is best fitted by the clustering of water molecules around the initial absorption sites and that this continued clustering depended little on the nature of the site.

Low angle neutron scattering measurements using contrast variation with the help of cosolvents (eg alcohols, DMSO) have recently been made[68]. The technique can in principle give information on any preferential partitioning of water between the bulk (mixed) solvent and the protein surface region, and with further development promises to be a powerful method.

Other useful techniques include enzyme activity measurements, compressibility, diamagnetic susceptibility and EPR.

Fig 11 attempts to summarise what we have learned about the rehydration of lysozyme using these and other techniques: included also are some of the results presented in the discussion of the non-freezing fraction (above). The following points are emphasised for discussion.

1. The first major event is the ionisation of charged groups; this is detected through changes in the COO^-[51] or COOH[61] IR bands. Although this proton redistribution process appears to be approximately completed by about 0.11 water/g protein, the greatest changes occur at much lower hydrations[61]. The heat capacity changes around .05g/g may reflect this proton redistribution, though film structure changes or even conformational readjustments may be responsible.

2. As judged from changes in the amide I and amide II bands, hydration of the C=O and N-H groups seem to occur approximately in parallel; the amide hydration however seems to go almost to completion earlier (∿ 0.15 g/g cf ∿ 0.29 g/g for

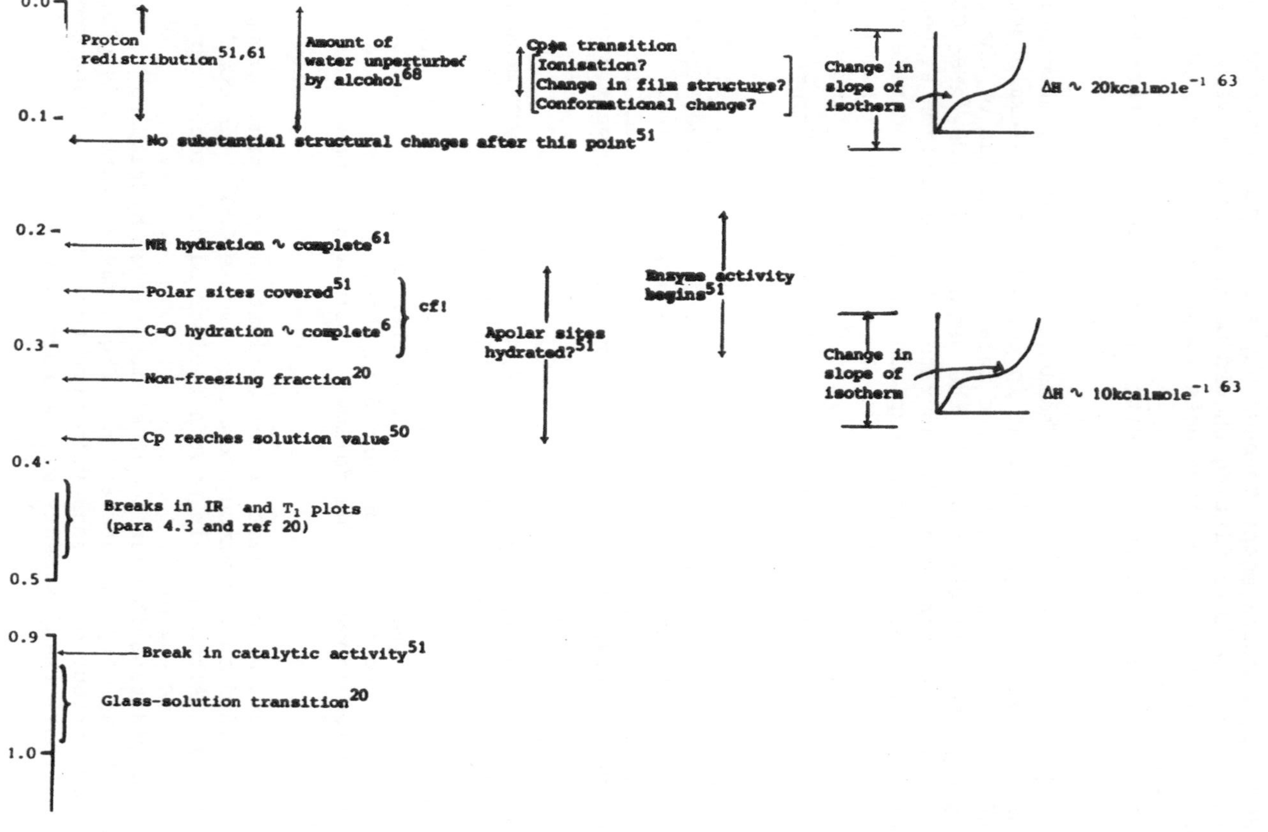

Fig. 11. Lysozyme hydration events.

C=O)[61]. There is a discrepancy between Poole[61] and Careri et al[51] in that the latter assert polar group coverage occurs earlier (∿ 0.25 g/g).

3. Onset of catalytic activity corresponds almost - but not quite - to completion of hydration of polar sites[51]. It might be postulated that this amount of water is necessary to allow the protein the flexibility it requires for its activity.

4. The heat capacity reaches the solution value by 0.38 g/g, implying that the thermodynamics of the protein are effectively fully established at this hydration level. Rupley interprets this point as corresponding to the completion of the coverage of the apolar sites on the protein surface[51]. Similarly the IR, DSC and T_1 plots of Golton[20] showed a change in slope at 0.40-0.45 g/g, which was similarly interpreted in terms of apolar group coverage.

5. There are no signs of large conformational changes. This suggests the tertiary structure is in effect "frozen in" as water is removed on drying; without adequate water, the protein is inadequately mobile to denature. Moreover, in the complete absence of water, a denatured conformation with unsatisfied hydrogen bonds would be energetically unstable, though the unknown entropic contribution would act in the opposite direction.

6. The break in catalytic activity observed at about 0.9 g/g is interestingly close to the glass-solution transition which occurs at about 1.0 g/g[20]. The closeness of these two figures is consistent with an interpretation in terms of increased mobility of the product.

COMPUTER SIMULATION TECHNIQUES

In the introductory section, our interest in water structure and dynamics near biomacromolecules was related to the contribution of changes in water "organisation" to the thermodynamics and kinetics of a given biomolecular process A→B, such as hormone-receptor binding, enzyme-substrate interaction, or protein denaturation. The first stage in improving our knowledge of solvent involvment is to try to understand how the solvent is organised (structurally and dynamically) in both the initial (A) and final (B) states. In particular, we would like to characterise the deviations from the bulk liquid structure and dynamics that are induced by the presence of the protein or other macromolecule.

We have discussed in the preceding sections our current knowledge of the state of the solvent close to macromolecular surfaces. As a result we have a semi-quantitative idea of how the water is perturbed. However, we have not been able to devise a valid, thermodynamically useful order parameter capable of quantifying the deviation of an assembly from the ideal bulk water reference state. Consequently, it is difficult to relate the molecular level changes in our model A→B process to the thermodynamics.

The obvious route *through* this problem is via statistical mechanics. The complexity of our model system is however just too great for us to hope at present to tackle the problem theoretically. Alternatively, we may be able to find a path *around* the problem using computer simulation techniques: in effect, we can do statistical mechanics numerically in the computer. Using this approach, we aim initially to specify states A and B (inclusive of solvent). We would then direct attention to the changes occurring in the A→B process, in particular to the calculation of associated free energy changes. Work is in progress on this latter step. Here, we restrict ourselves to the problems of simulating the initial state A: the structural and dynamic organisation of the water around biomolecules.

Methods

Two computer simulation methods are commonly used in the study of liquid assemblies[71]. In *Monte Carlo* calculations, the energy difference between successive randomly-generated configurations of atoms is calculated. The new configuration is accepted or rejected according to a Boltzmann weighting. This procedure results in the sampling of a series of accessible phase states from which average structural properties and thermodynamic quantities can be calculated. In *molecular dynamics*, atoms are allowed to move with time according to Newton's laws of motion and the intermolecular forces operating between the various atoms. Again information on structural and thermodynamic properties of the model system can be obtained; in addition, molecular dynamics gives access to dynamic quantities such as the diffusion coefficient.

We discuss in what follows some of the problems of and results obtained using Monte Carlo simulations of water organisation in biomolecule hydrate crystals. We address in particular, the central problem of devising an adequate set of intermolecular potential functions and underline the sensitivity of predictions to the potentials used. Crystals are chosen as test systems as the known positions and thermal parameters of the waters allow us to test the predicted results against experiment before we can justifiably extend the calculations to larger molecules under solution conditions.

Interatomic Potential Functions

In addition to a knowledge of the three-dimensional coordinates of the biomolecule, we require a detailed understanding of the potential functions which describe the interactions between the several types of atoms in the model system. Thus we require adequate models of both water-water interactions, and of those between water and each atom type found in amino acids (eg carbonyl oxygens, amide nitrogens, tetrahedral carbons).

For water itself - especially at interfaces - we need a model which can describe not only the main properties of liquid water, but of the vapour phase and crystal also. The calculations discussed here therefore use the polarisable electropole (PE) model[12, 72], in which the molecular charge distribution is represented by the molecular dipole and quadrupole moments. This representation removes the problem of over-tetrahedrality discussed in section 3 previously, and also attempts to include known cooperative effects by allowing the dipole moment to change under the influence of a surrounding polarising field from the neighbouring molecules.

The next stage is to extend the water model to describe water-biomolecule interactions[73,74]. However, the energy parameters we need to describe the interactions of each (eg) protein atom are not well defined. Even for the relatively 'well-known' 6-12 form of the Lennard-Jones non-bonded interaction, a wide range of values is found in the literature[75-77]. Moreover, transferability of these values from the small molecule crystal structures for which they are calculated to the atoms of a protein molecule may be limited. We also have to combine a set of such Lennard-Jones coefficients for protein atoms with those from our water model. This may again lead to uncertainties.

We also require electrostatic parameters for the biomolecule. Partial atomic charges can be obtained from quantum mechanical calculations on small molecules. However, at the CNDO/2 level (which is routinely available) the partial charges are known to be inaccurate[77] compared with large basis set *ab initio* calculations, relatively few of which are computationally feasible at present. Dipole moments and polarisabilities[78] are available for atoms in amino acid residues but again transferability of these parameters may be limited. The problems of transferability and combination rules together with the more general problem of the form of the hydrogen-bond potential are reviewed by Professor S. Lifson in his lecture at this Summer School.

Sensitivity of Predicted Results

Before drawing conclusions from simulation it is advisable to examine the sensitivity of the predicted results to the values used for the input parameters of the interatomic potentials. This is especially true for solvent networks around proteins in which there are many weakly interacting components whose intermolecular interactions are poorly known. For this reason we have undertaken a series of full Monte Carlo simulations on test systems in which the water molecules are well ordered[74]. Using amino acid hydrate crystals[79-81] whose three-dimensional structures have been found at high resolution by X-ray and neutron diffraction, predicted solvent organisation can be compared with the experimentally-derived positions and thermal factors.

In each simulation, the input parameters have been varied in a specific way with respect to:

(a) The set of Lennard-Jones coefficients for protein atom-protein atom interactions.

(b) The combination and scaling of these protein-protein Lennard-Jones coefficients to those for water-water interactions.

(c) The fractional partial atomic charges used to describe the amino acid molecule.

In comparing the predicted with the experimental solvent organisation, several criteria were used:

(i) The root mean-square deviation between predicted and experimental positions.

(ii) The root mean-square deviation around the final predicted position: this is directly related to the experimentally-derived thermal factors.

(iii) The coordination of the water molecules in predicted and experimental networks.

(iv) The hydrogen bond lengths and angles in the solvent network, given that the coordination is the same in the predicted as in the experimental structure.

For each of three kinds of input parameter variations (Lennard-Jones coefficients, scaling to water, and the partial charges) the predicted structure was found sensitive to the input parameters. An example is given in table 3, where the change in root mean-square deviation between experimental and predicted water molecule positions caused by changes in the Lennard-Jones σ parameter is shown. Moreover in only four of the twenty simulations performed was criterion (iii) satisfied - ie the water *coordination* predicted successfully.

Table 3. Effect of Variation in Lennard-Jones Coefficients on the RMSD of Water Molecules from Mean Positions in Homoproline Tetrahydrate

	$\sigma_{polar-water}$ Å	rmsd Å
1	2.93	0.78
2	2.97	0.80
3	3.02	0.58
4	3.17	1.15

Using one set of partial charges from CNDO/2 calculations, the input Lennard-Jones parameters and their scaling to the water model (variations (a) and (b) discussed previously) were optimised in a series of Monte Carlo calculations. The resulting optimised potential gave the best agreement with experimental hydrogen bond distances. The comparisons for homoproline tetrahydrate are given in table 4.

Cooperative Effects in Water Biomolecule Systems

Monte Carlo simulations on a series of small to medium-sized biomolecules have been carried out using the optimised potentials described in the last section. In this study we have looked at the non-pair-additive (cooperative) effects in these systems by monitoring the enhancement of the water molecule dipole moments. (This important feature of the PE model has been described above). We have also studied the effect on the predicted results of certain simplifications to the polarisable electropole potential in order to make simulations computationally feasible for much larger water-protein assemblies.

The predicted water molecule dipole moments are summarised in table 5. In each system studied the mean dipole moment has increased above the monomer value of 1.855D. The size of this mean enhancement varies between systems and can be as high as 50% of the monomer value. For serine monohydrate[79] a high value of 2.77D is found whereas for azidopurine monohydrate[82] the dipole increases only to 2.1D. These numbers should be compared with the mean dipole moment for simulated bulk water which is 2.5D. From table 5 we conclude that *the mean dipole moment of the water molecules is a property of*

Table 4. Comparison of Predicted with Experimental Hydrogen Bond Distances in Homoproline Tetrahydrate

		X-Ray	Simulation
OW1 -	N7	2.82	2.89
	O26	2.91	2.82
	O36	2.80	2.82
	OW5	3.24	2.88
OW2 -	OW3	2.81	3.11
	OW11	2.80	3.18
	OW12	2.77	2.87
	OW16	2.79	3.10
OW3 -	OW2	2.81	3.11
	OW4	2.71	3.18
	OW10	2.79	3.17
OW4 -	O36	2.73	2.84
	OW3	2.71	3.18
	OW10	2.78	3.24
	OW14	2.79	3.10

each system and not an intrinsic property of the water molecule alone.

A further important feature of the results presented in table 5 is the spread or range of dipole moments within each system. For small systems (eg L-arginine dihydrate) the range is not significant. With increasing water content from 6 waters (α-cyclodextrin)[83], 17 waters (vitamin B_{12})[84], to 25 waters (dCpG)[85], we find that the range of values increases significantly until we reach bulk water (216 molecules) with a range from 1.9 to 3.1D.

In order to ascertain the importance of the spread in dipole moments, Monte Carlo simulations were run with the moments fixed at the mean value for the given system. As the number of water molecules in the system increased, so did the discrepency in the final energies between these 'fixed dipole and the original 'unfixed' simulations (table 6). From this result, it appears that even if we know the mean dipole moment for the water molecules in a given system, it would lead to inaccurate results if this fixed value was used in the simulation.

A further possibility is neither to use 'fixed' or totally

Table 5. Predicted Water Molecule Dipole Moments

	μ/Debye
Water - monomer	1.855
Water - liquid	2.50 (range 1.9-3.1)
Serine monohydrate	2.77 ± 0.13
Azidopurine monohydrate	2.11 ± 0.04
L-arginine dihydrate	2.43 ± 0.09
	2.49 ± 0.09
Homo-proline tetrahydrate	2.58 ± 0.03
	2.45 ± 0.23
	2.48 ± 0.23
	2.46 ± 0.20
α-cyclodextrin hexahydrate	2.07 ± 0.11
	2.09 ± 0.10
	2.15 ± 0.09
	2.15 ± 0.09
	2.18 ± 0.08
	2.32 ± 0.11
Vitamin B_{12} coenzyme ≈ 17 waters/au	Mean 2.30 (range 1.9-2.6)
dCpG 25 waters/au	Mean 2.41 (range 1.9-2.75)

'unfixed' dipole moments but to allow them to increase every n Monte Carlo moves ($n > 1$: for normal Monte Carlo, $n = 1$). The optimum value of n would presumably depend upon the number of water molecules in the assembly. The results for n = 50, 200, 500 configurations are tabulated in table 7 for three crystal systems. It appears that the larger the number of water molecules the more often we have to update the dipole moments in order to obtain results indistinguishable from the original 'unfixed' simulations. This method leads to a major decrease in computational speed and makes it feasible to study large water-globular protein crystals using a non-pair-additive water potential.

Table 6. Effect of Fixing the Water Molecule Dipole Moments at the Average Value for Each System (energies in kcal mole^{-1})

System	Unfixed	Fixed	% Difference
Monohydrate	- 3.47 ± 0.05	- 3.42	1
Dihydrate	- 2.23 ± 0.02	- 2.18	2
Hexahydrate	- 0.405 ± 0.016	- 0.366	10
25 H_2O (dCpG)	- 2.47 ± 0.01	- 2.36	4
216 H_2O (Bulk)	- 7.90 ± 0.03	- 6.87	13

Summary and Prospects

Preliminary studies using computer simulation techniques to study water organisation around biomolecules have demonstrated the feasibility of the method, but that our knowledge of intermolecular potential functions was inadequate for predictions of positions to better than 1Å[86,87,88,89]. In our studies summarised here, we have concentrated on the problems of the potentials, using a water model which incorporates cooperative effects and is not overtetrahedral, together with an extension of the model, to describe water-protein potential energy functions. From a detailed study of the sensitivity of the predicted results to the value of the input energy parameters, a set of optimised potentials have been derived which have then been used to study properties of several small to medium-sized crystal systems.

Although a great deal of detailed work remains to be done to provide totally adequate (probably non-pair-additive) potential functions, the optimised set described above is now being extended to examine the possible involvement of water in several biomolecular systems and processes. Of particular interest is the comparison of water organisation around several proteins of known (high resolution) structure and between a protein in the crystal and in solution. Molecular dynamics calculations can probe the dynamics of the water neighbouring various exposed groups, as well as the influence of solvent on the dynamics of the protein. The way is also open to study the phase space restriction of water in enzyme active sites and at protein association surfaces; extension of the methods of Valleau and Torrie[90], Quirke[91] and others to calculate free energies will lead us towards a relating of molecular level changes and thermodynamic

Table 7. Effect of Updating Dipoles every n Monte Carlo Moves

	ENERGY kcal $mole^{-1}$	RMSD [(a)] Å	$\overline{U^2}^{\frac{1}{2}}$ [(b)] Å	DIPOLE Debye
Serine				
n= 1	- 3.47 ± 0.05	0.33	0.21	2.77 ± 0.13
50	- 3.47 ± 0.06	0.28	0.17	2.73 ± 0.14
500	- 3.50 ± 0.04	0.33	0.16	2.74 ± 0.16
L-arginine				
n= 1	- 2.23 ± 0.02	0.57	0.21	2.45
50	- 2.18 ± 0.05	0.51	0.21	2.52 ± 0.12
200	- 2.13 ± 0.03	0.62	0.21	2.51 ± 0.14
α-cyclodextrin				
n= 1	- 0.405 ± 0.016	0.44	0.22	2.11 ± 0.15
200	- 0.424 ± 0.008	0.63	0.36	2.20 ± 0.14
500	- 0.451 ± 0.013	0.75	0.29	2.20 ± 0.13

(a) RMSD is the root mean square deviation between predicted and experimental structure.

(b) $\overline{U^2}^{\frac{1}{2}}$ is the root mean square deviation around the final predicted structure.

driving forces. Provided our potential functions, computational methods and computing power are adequate.

REFERENCES

1. F. Franks, in: "Water: A Comprehensive Treatise", F. Franks, ed., Plenum, New York 4: 1 (1975).
2. J. L. Finney, B. J. Gellatly, I. C. Golton and J. M. Goodfellow, Biophys J., 32: 17 (1980).
3. J. D. Bernal, Proc. Roy. Soc., A280: 299 (1964).
4. J. L. Finney, Proc. Roy. Soc., A319: 479 (1970).
5. J. L. Finney, Proc. Roy. Soc., A319: 495 (1970).
6. M. G. Sceats and S. A. Rice, J. Chem. Phys., 72: 3260 (1980).
7. G. H. F. Diercksen, Theor. Chim. Acta, 21: 335 (1971).
8. I. Olovsson and P.-G. Jönsson, in: "The Hydrogen Bond", P. Schuster, G. Zundel and C. Sandorfy, eds., North-Holland, Amsterdam, 2: 393 (1976).

9. D. Hankins, J. W. Moskowitz and F. H. Stillinger, J. Chem. Phys., 53: 4544 (1970).
10. E. Clementi, W. Kolos, G. C. Lie and G. Ranghino, Inter J. Quant. Chem., XVII: 377 (1980).
11. J. L. Finney, Farad. Disc. Chem. Soc., 66: 80, 86 (1978).
12. P. Barnes, J. L. Finney, J. D. Nicholas and J. E. Quinn, Nature, 282: 459 (1979).
13. P. Barnes, D. V. Bliss, J. L. Finney and J. E. Quinn, Farad. Disc. Roy. Soc. Chem., 69: 210 (1980).
14. B. J. Gellatly and J. E. Quinn; in preparation.
15. K. J. Packer, Phil. Trans. Roy. Soc., B278: 29 (1977).
16. H. S. Frank and M. W. Evans, J. Chem. Phys., 13: 507 (1945).
17. F. Franks, Phil. Trans. Roy. Soc., B278: 89 (1977).
18. H. J. C. Berendsen, in: "Water: A Comprehensive Treatise", F. Franks, ed., Plenum, New York, 5: 293 (1975).
19. I. D. Kuntz and W. Kauzmann, Adv. Prot. Chem., 28: 239 (1973).
20. I. C. Golton, PhD thesis, University of London (1980).
21. M. Falk, A. G. Poole and C. G. Goymour, Can. J. Chem., 48: 1536 (1970).
22. I. D. Kuntz, T. S. Brassfield, G. D. Low and G. V. Purcell, Science, 163: 1329 (1969).
22. B. Lee and F. M. Richards, J. Mol. Biol., 55: 379 (1971).
24. J. L. Finney, J. Mol. Biol., 119: 415 (1978).
25. B. J. Gellatly and J. L. Finney; in preparation.
26. J. L. Finney, Phil. Trans. Roy. Soc., B278: 3 (1977).
27. J. L. Finney, in: "Water: A comprehensive Treatise", F. Franks, ed., Plenum, New York, 6: 47 (1979).
28. M. N. G. James, A. R. Sielecki, G. D. Brayer, L. T. J. Delbaere and C.-A. Bauer, J. Mol. Biol., 144: 43 (1980).
29. K. D. Watenpaugh, T. N. Margulis, L. C. Sieker and L. H. Jensen, J. Mol. Biol., 122: 175 (1978).
30. O. Epp, E. E. Lattman, M. Schiffer, R. Huber and W. Palm, Biochemistry, 14: 4043 (1975).
31. P. C. Moews and R. H. Kretsinger, J. Mol. Biol., 91: 201 (1975).
32. S. C. Harvey and P. Hoekstra, J. Phys. Chem., 76: 2967 (1972).
33. H. P. Schwan, Ann. N. Y. Acad. Sci., 125: 344 (1965).
34. B. E. Pennock and H. P. Schwan, J. Phys. Chem., 73: 2600 (1969).
35. J. R. Grigera, F. Vericat, K. Hallenga and H. J. C. Berendsen, Biopolymers, 18: 35 (1979).
36. K. Hallenga, J. R. Grigera and H. J. C. Berendsen, J. Phys. Chem. 84: 2381 (1980).
37. R. Cooke and I. D. Kuntz, Ann. Rev. Biophys. Bioeng., 3: 95 (1974).
38. E. Hsi, J. E. Jentoft and R. G. Bryant, J. Phys. Chem., 80: 412 (1976).
39. E. Hsi and R. G. Bryant, Arch. Bioch. Biophys., 183: 588 (1977).
40. B. D. Hilton, E. Hsi and R. G. Bryant, J. Amer. Chem. Soc., 99 : 8483 (1977).
41. S. H. Koenig, R. G. Bryant, K. Hallenga and G. S. Jacob, Biochemistry, 17: 4348 (1978).
42. R. G. Bryant, Ann. Rev. Phys. Chem., 29: 167 (1978).

43. R. G. Bryant and W. M. Shirley, in: "Water in Polymers", ed., F. Rowland, Amer. Chem. Soc., Washington DC, p 147 (1980).
44. R. G. Bryant, Biophys. J., 32: 80 (1980).
45. S. H. Koenig and W. S. Schillinger, J. Biol. Chem., 244: 3282 (1969).
46. S. H. Koenig, K. Hallenga and M. Shporer, Proc. Nat. Acad. Sci., 72: 2667 (1975).
47. K. Hallenga and S. H. Koenig, Biochemistry, 15: 4255 (1976).
48. S. H. Koenig, in: "Water in Polymers", F. Rowland, ed., Amer. Chem. Soc., Washington DC, p 157 (1980).
49. H. D. Middendorf and Sir J. Randall, Phil. Trans. Roy. Soc. B290: 639 (1980),
50. P.-H. Yang and J. A. Rupley, Biochemistry, 18: 2654 (1979).
51. G. Careri, E. Gratton, P.-H. Yang and J. A. Rupley, Nature, 284: 572 (1980).
52. F. Cavatorta, M. P. Fontana and A. Vecli, J. Chem. Phys., 65: 3635 (1976).
53. S. R. Samanta and G. E. Walrafen, J. Chem. Phys., 68: 3313 (1978).
54. L. Genzel, F. Keilmann, T. P. Martin, G. Winterling, Y. Yacoby, H. Frölich and M. W. Makinen, Biopolymers, 15: 219 (1976).
55. B. Bielski and S. Freed, Bioch. Biophys. Acta, 89: 314 (1964).
56. H. P. Kasserra and K. J. Laidler, Can. J. Chem., 48: 1793 (1970).
57. S. J. Singer, Adv. Prot. Chem., 17: 1 (1962).
58. A. L. Fink, Biochemistry, 13: 277 (1974).
59. A. L. Fink and I. A. Ahmed, Nature, 263: 294 (1976).
60. J. A. Rupley, P.-H. Yang and G. Tollin, in: "Water in Polymers", F. Rowland, ed., Amer. Chem. Soc., Washington DC, p 111 (1980).
61. P. L. Poole; unpublished results.
62. J. Suurkuusk, Acta Chem. Scand., B28: 409 (1974).
63. R. Almog and E. E. Schrier, J. Phys. Chem., 82: 1701 (1978).
64. G. Careri, A. Giansanti and E. Gratton, Biopolymers, 18: 1187 (1979).
65. C. H. Rochester and A. V. Westerman, J. Chem. Soc. Farad. Trans. I, 72: 2498 (1976).
66. C. H. Rochester and A. V. Westerman, J. Chem. Soc. Farad. Trans. I, 72: 2753 (1976).
67. C. H. Rochester and A. V. Westerman, J. Chem. Soc. Farad. Trans. I, 73: 33 (1977).
68. M. S. Lehmann, unpublished results.
69. B. Halle, T. Anderson, S. Forsén and B. Lindman, J. Amer. Chem. Soc., 103: 500 (1981).
70. J. L. Finney and J. M. Goodfellow, unpublished.
71. D. W. Wood, in: "Water: A Comprehensive Treatise", F. Franks, ed., Plenum, New York, 6: 279 (1979).
72. P. Barnes, in: "Progress in Liquid Physics", C. A. Croxton, ed., Wiley, Chichester, p 391 (1978).
73. J. M. Goodfellow, J. L. Finney and P. Barnes, Proc. Roy. Soc. B., in press.
74. J. M. Goodfellow; in preparation.

75. S. Lifson, A. T. Hagler and P. Dauber, J. Amer. Chem. Soc., 101: 5111 (1979).
76. M. Levitt, J. Mol. Biol., 82: 393 (1972).
77. F. A. Momany, L. M. Carruthers, R. F. McGuire and H. A. Scheraga, J. Phys. Chem., 78: 1595 (1974).
78. R. Pethig, "Dielectric and Electronic Properties of Biological Materials", Wiley, Chichester (1979).
79. M. N. Frey, M. S. Lehmann, T. Koetzle and W. C. Hamilton, Acta Cryst., B29: 876 (1973).
80. M. S. Lehmann, J. J. Verbist, W. C. Hamilton and T. F. Koetzle, J. Chem. Soc. Perkin II, 1973: 133 (1973).
81. S. K. Bhattacharjee and K. K. Chacho, Acta Cryst., B35: 396 (1979).
82. J. P. Glusker, D. van der Helm, W. E. Love, J. A. Minkin and A. L. Patterson, Acta Cryst., B24: 359 (1968).
83. B. Klar, B. Hingerty and W. Saenger, Acta Cryst., B36: 1154 (1980).
84. H. F. J. Savage; unpublished results.
85. H.-S. Shieh, H. M. Berman, M. Dalvan and S. Neidle, Nucleic Acids Res., 8: 85 (1980).
86. A. T. Hagler and J. M. Moult, Nature, 272: 222 (1978).
87. J. Hermans and M. Vacatello, in: "Water in Polymers", F. Rowland, ed., Amer. Chem. Soc., Washington DC, p 199 (1980).
88. E. Clementi, G. Corongui, B. Jönsson and S. Romano, FEBS Lett., 100: 313 (1979).
89. J. D. Nicholas, P. Barnes and J. L. Finney, in: "Abstracts of 4th European Crystallographic Meeting", Oxford, p 375 (1977).
90. G. M. Torrie and J. P. Valleau, Chem. Phys. Lett., 28: 578 (1974).
91. N. Quirke, in: "The Proceedings of NATO Summer School on Superionic Conductors", Odense, Denmark (1980).

DYNAMICS OF PEPTIDES AND PROTEINS*

Martin Karplus**

Department of Chemistry
Harvard University
Cambridge, Massachusetts 02138

An essential question in any attempt to understand the biological function of peptides and proteins concerns the dynamics of their structural fluctuations and conformational changes. The motions that occur range from small local atom displacements, sidechain reorientations and main chain fluctuations, through almost rigid body movements of domains and subunits to the folding and unfolding transition involving the entire polypeptide chain. It is important to know how these motions take place and what the time scales of the motions are.

For most globular proteins, the biological function includes an interaction between one or more small molecules (ligand, hormone, substrate, coenzyme chromophore, etc.) and the macromolecule. Whether reactive or nonreactive systems are being considered, there can be important conformational alterations in the small molecule that is bound and concomitant changes in the structure of the macromolecule to which the binding occurs. These conformational changes are the essential element in some cases; in others, they play a less significant role. In hormone-receptor binding, for example, the

---

*Supported in part by grants from the National Science Foundation and the National Institutes of Health.

**This manuscript was prepared while the author was Visiting Professor at the Laboratoire D'Enzymologie Physicochimique et Moléculaire, Universite de Paris-Sud, Orsay, France and at the Laboratoire de Chimie des Interactions Moléculaires, Collège de France, Paris, France.

structural changes induced in the receptor are fundamental to the transmission of information. Correspondingly, the conformational transition induced by ligand binding in hemoglobin is an integral part of the cooperative mechanism. In many other systems, smaller structural changes have been observed (e.g., the differences between the liganded and unliganded structure of enzymes) that are likely to be involved in the function.

The most direct approach to the internal motions of peptides and proteins is provided by the molecular dynamics simulation method. It can be used to determine the details of the motions on a subnanosecond time scale. For certain systems an understanding of such short time phenomena is sufficient, but there are many biologically significant events (e.g., enzyme reactions with high barriers, folding and unfolding transition) that take much longer and require special techniques for their investigation. In what follows, I shall present some illustrative applications of a number of different theoretical methods for studying the dynamics. They include reaction path calculations, molecular dynamics, harmonic dynamics, stochastic dynamics, and activated dynamics. For more details, the reader is referred to a recently published general review of experimental and theoretical studies of protein dynamics and to the references cited therein.[1]

DYNAMICS OF A DIPEPTIDE IN WATER

As a preliminary to the study of the internal dynamics of proteins, the behavior of the alanine dipeptide ($CH_3C'ONHCHCH_3C'ONHCH_3$) in a box of 195 water molecules (ST2 model) was simulated; i.e., the equation of motion of the dipeptide solute and the solvent water molecules were solved simultaneously to obtain a phase space trajectory for a system with periodic boundary conditions equilibrated at 300°K.[2] This relatively simple solute is a neutral molecule that includes the polar (peptide) groups and nonpolar (methyl) groups characteristic of proteins. During the entire simulation the dipeptide was found to remain in the C_7^{eq} conformation (one internal hydrogen bond) that is the global minimum in vacuum. Comparison of the average geometry and of the magnitude of the fluctuations in the internal coordinates obtained from the solution simulation and from a corresponding simulation in vacuum shows that the solvent perturbation is small; a sample of the internal coordinate results is listed in Table I. As to the time dependence of the fluctuations, examination of the correlation functions indicates that solvent damping is negligible, except for the methyl group rotations. They are strongly perturbed by the solvent and the rotational motion in solution can be described by a Langevin equation for a damped oscillator,[3] which has the form

$$I \frac{d^2\theta}{dt^2} + f \frac{d\theta}{dt} + I\omega_o^2\theta = N_r(t) \tag{1}$$

where θ is the angular displacement from the equilibrium position, I is the effective moment of inertia of the methyl group, f is the friction coefficient due to the solvent, ω_o is the harmonic vacuum frequency (35 ps^{-1}), and $N_r(t)$ represents the randomly fluctuating torque characteristic of Brownian motion. The harmonic oscillator in vacuum involves only the first and third term; the second and fourth term arise from the presence of solvent.

Table I. Average Solute Structure

A^a	$\langle A \rangle$ vacuum	$\langle A \rangle$ solution	$\langle \Delta A^2 \rangle^{1/2}$ vacuum	$\langle \Delta A^2 \rangle^{1/2}$ solution
Bonds[b]				
C'_L-O_L	1.235	1.237	0.023	0.028
N_L-H_L	0.994	0.997	0.018	0.012
$C_\alpha-C_\beta$	1.544	1.542	0.041	0.035
Bond Angles[b]				
$C_L-C'_L-O_L$	122.31	121.89	3.30	3.23
Dihedral Angles[b,c]				
ϕ	-67.21	-63.96	9.67	7.83
ψ	63.45	59.33	11.53	22.57
ω_L	-179,25	-179.67	8.74	9.67
ω_R	-179.68	178.08	14.72	12.39
χ	-59.10	-62.88	9.96	32.25

[a]Structural parameters, A, refer to the dipeptide with atom labels $C_LH_3-C'_LO_L-N_LH_L-C_\alpha HC_\beta H_3-C'_RO'_R-N_RH_R-C_RH_3$ and the angles ϕ, ψ and ω follow the standard definitions with ω_L (ω_R) associated with the left (right) peptide group; the C_7 ring involves a hydrogen bond between O_L and H_R. $\langle A \rangle$ = mean value, $\langle A^2 \rangle^{1/2} = \langle (A-\langle A \rangle)^2 \rangle^{1/2}$.

[b]Bonds in Ångstroms; bond angles and dihedral angles in degrees.

[c]The vacuum minimum occurs at $\phi = 66.2°$, $\psi = -65.3°$, $\omega_L = 179.2°$, $\omega_R = 179.9°$.

The results obtained for the dipeptides in water suggest that the short time (subnanosecond) dynamics of this system, and by extension a protein, can yield results of interest even when the solvent is not treated explicitly. In the protein case, the external sidechains that are the analog of the methyl groups of the dipeptide clearly would have their motion damped by the solvent but the internal dynamics is expected to be relatively insensitive to solvent perturbations (see below).

In contrast to the dynamics, the thermodynamics of the dipeptide is strongly dependent on the presence of the solvent. Table II shows a decomposition of the energy terms that contribute to the enthalpy of solution of the dipeptide as calculated from the molecular dynamics simulation. In the table, the water molecules included in the simulation have been separated into a first solvation shell and the rest of the molecules, referred to as "bulk" water (161 molecules). The water molecules in the first solvation shell are divided into a class that interacts mainly with the two peptide groups ("polar" water; 14 molecules) and a class that interacts mainly with the three methyl groups ("nonpolar" water; 20 molecules). It can be seen that the polar water molecules have a very strong interaction with the peptide, due primarily to hydrogen bonding (there are two strong hydrogen bonds with one C=O and two weak ones with the other, plus one hydrogen bond with the NH not involved in the internal hydrogen bond). The nonpolar water molecules have a relatively weak interaction with the dipeptide, as do the bulk water molecules; however, their large number leads to a significant contribution to the solvation energy. Much of the solvent-dipeptide interaction energy is balanced by the change in

Table II. Energy of Solution of Dipeptide (kcal per mol of dipeptide)

Water-Dipeptide	ΔE	
Nonpolar	- 3.60	
Polar	-15.82	
Bulk	- 3.22	
		-22.64
Water-Water		
Nonpolar	- 3.4	
Polar	+19.32	
Bulk	~0	
		+15.92
	TOTAL	- 6.72 kcal/mol

the water-water energy of the polar waters; that is, in forming stabilizing water-dipeptide hydrogen bonds, the polar water forms fewer water-water hydrogen bonds. The simulation results show, in fact, that the polar waters form essentially the same total number of hydrogen bonds (3.28 with a cut-off energy of -3 kcal/mol) as do the bulk water molecules (3.45). Thus, the net solvation energy is much smaller (-6.72 kcal/mol) than the water-dipeptide interaction energy (-22.64 kcal/mol). Although the quantitative values obtained from the simulation may not be accurate due to statistical and potential function errors, the qualitative point concerning the cancellation between water-dipeptide and water-water energy terms is valid and must be considered in interpreting any theoretical or experimental solvation energy studies.

THE α-HELIX: HARMONIC DYNAMICS AND UNFOLDING

Early evidence for motion in the interior of proteins or their fragments come from analysing vibrational spectroscopic studies. It is generally assumed in interpreting such data that a harmonic potential and the resulting normal mode description of the motions is adequate. This approximation is most likely to be correct for the tightly bonded secondary structural elements, like α-helices and β-sheets. We have studied the fluctuations of a finite α-helix (hexadecaglycine) from the normal modes of the system.[4] At 300°K, the root mean square fluctuations of ϕ and ψ about their equilibrium values were equal to ~12° in the middle of the helix and somewhat larger near the ends; the dihedral angle fluctuations were significantly correlated over two neighboring residues; these correlations tend to localize the fluctuations. Fluctuations in the lengths between adjacent residues (defined as the projection onto the helix axis of the vector connecting the centers of mass of adjacent residues) ranged from about 0.15 Å in the middle of the helix to about 0.25 Å at the ends. These length fluctuations were negatively correlated for residue pairs (i-1,i) and (i,i+1) so as to preserve the overall length of the helix; positive correlations were observed for the pairs (4,5), (8,9) and (8,9), (12,13), suggesting that the motion of residue 8 is coupled to the motions of residues 4 and 12 to retain optimal hydrogen bonding. When the above calculations were repeated with a model in which the peptide dihedral angles (ω) were fixed, the rms fluctuations of ϕ were reduced by about 2° and the pattern (but not the range) of dihedral angle fluctuation correlations was changed somewhat.

More recently,[5] a full molecular dynamics calculation has been performed for a decaglycine helix as a function of temperature between 5° and 300°K and the results compared with those obtained in the harmonic approximation. For the mean square positional fluctuations, $\langle \Delta R^2 \rangle$, of the atoms, it is found that the harmonic approximation is valid in the classical limit below 100°K, but that there are significant deviations above that temperature; e.g., at 300°K, the average value of $\langle \Delta R^2 \rangle$ obtained for the α-carbons from the full dynamics is more than twice that found in the harmonic model.

As an approach to the helix-coil transition in α-helices, a simplified model for the polypeptide chain has been introduced to permit a dynamic simulation on the submicrosecond time scale appropriate for this phenomenon.[6] In the model,[7] each residue is represented by a single interaction center ("atom") located at the centroid of the corresponding sidechain and the residues are linked by virtual bonds.[8] The potential function used includes harmonic bond length and bond angle terms, torsional angle terms, which account for the averaged atomic nonbonded interactions between nearest neighbor residues in the primary sequence, and central-force pair potential terms, which account for the averaged van der Waals and solvent-mediated interactions between residues separated by more than two virtual bonds. The diffusional motion of the chain "atoms" expected in water was simulated by using a stochastic dynamics algorithm based on the Langevin equation with a generalized force term replacing the harmonic interaction in Eq. (1). Starting from an all-helical conformation, the dynamics of several residues at the end of a fifteen-residue chain were monitored in several independent 12.5 ns simulations at 298°K. It was found that the mobility of the terminal residue is quite large, with a rate constant $\approx 10^9$ s^{-1} for the transitions between coil and helix states. This mobility decreases for residues further into the chain; unwinding of an interior residue requires simultaneous displacements of residues in the coil, so that larger solvent frictional forces are involved. The coil region does not move as a rigid body, however; the torsional motions of the chain are correlated so as to minimize dissipative effects. A typical 12 ns trajectory is illustrated in Figure 1; the terminal dihedral angles Φ_{10}, Φ_{11}, and Φ_{12} are shown as a function of time. In the figure, the region around 40° corresponds to the α-helical geometry and that around 200° to the random coil. It can be seen that angles Φ_{11} and Φ_{12} make an almost simultaneous transition from the helix to the coil region at 3.5 to 4 ps and remain there for most of the simulation. With Φ_{11} and Φ_{12} in the coil conformation and Φ_{10} in the neighborhood of the helical geometry, there are clear, positive correlations between Φ_{10} and Φ_{11} and negative correlations for Φ_{10} and Φ_{12}.

X-RAY TEMPERATURE FACTORS

The first molecular dynamics simulations of a protein determined the mean square fluctuations in the atomic positions of the bovine

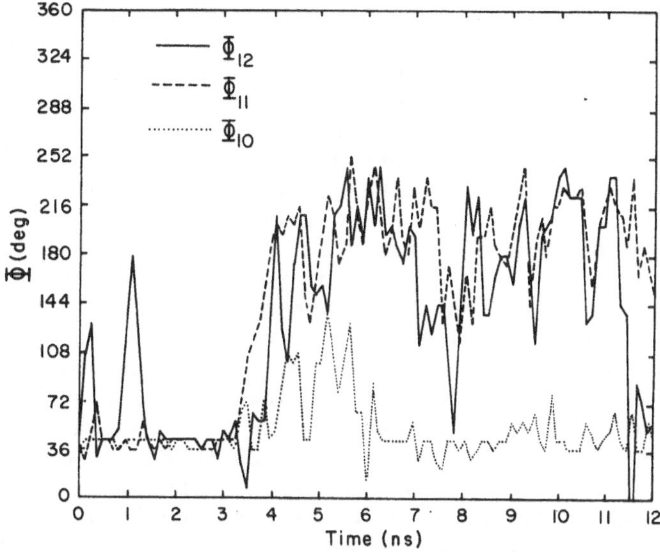

Fig. 1. Helix-coil transition simulation; dihedral angle var_ation as a function of time with Φ_{12} (—), Φ_{11} (---), and Φ_{10} (···).

pancreatic trypsin inhibitor (PTI) and pointed out their relation to the temperature factors available from X-ray diffraction analyses of protein structures.[9] Since that work, similar theoretical studies have been or are being made on ferrocytochrome c (cyt c),[10] myoglobin,[11] and lysozyme.[12] Concomitantly, protein crystallographers, who had generally regarded the experimental temperature factors as being of little interest because they were presumed to be dominated by other than motional contributions (crystal disorder, errors in refinement, etc.), have begun to publish their values and to attempt to analyse them;[10,13,14] in fact, a plot of mean square fluctuations or temperature factors versus residue number, analogous to that presented in reference 9, has now become an integral part of papers on the high-resolution structures of proteins. There are two primary reasons for focusing on the temperature factors. The first, experimental in nature, is that their dynamic part provides a way of measuring the lowest-order term in a description of the equilibrium atomic fluctuations of a protein. In a general sense, the distribution function for the atomic positions of a protein (or any other molecule) can be expressed in terms of its moments. All of the past work in protein crystallography has concentrated on the first moment, $\langle \vec{r} \rangle$, which gives the average positions of the atoms in the crystal. The importance of such positional data, which serve as the basis of much of our present understanding of protein structure and function, does not have to be repeated here. The mean square atomic fluctuations provide the second moment of the distribution function, $<|\vec{r} - <\vec{r}>|^2>$, and as already mentioned correspond to the lowest order motional contribution. In principle, the temperature factors include the effects of the higher moments as well, but the analysis of the experimental values has had to be simplified by assuming that the second moment makes the dominant contribution and, until now, that the motion is isotropic and harmonic. Clearly, a knowledge of the temperature factors should aid in our understanding of the motional contribution to protein function. The second reason for the interest in temperature factors, which is of importance from the theoretical viewpoint, is that they can serve as a check on the molecular dynamics simulations. Calculations of the mean square fluctuations can be compared on an atom-by-atom basis with the experimental data. Any differences between the two may be the result of crystal effects (the simulations have mainly been done in a vacuum environment, but see below), crystal disorder and other contributions to the experimental temperature factor, as well as to errors in the calculations. In fact, one of the aspects of the comparisons between experimental and theoretical temperature factors is an evaluation of the purely motional contribution to the former. Further, since the X-ray data give only the magnitudes of the fluctuations, it is necessary for the calculations to supply information concerning their time dependence.

The atomic scattering factor, B, determined by X-ray diffraction, can be related to the mean square positional fluctuation, $<\Delta r^2>_{dyn}$, with the assumption of isotropic and harmonic motion by the relation

$$<\Delta r^2>_{dyn} = \frac{3B}{8\pi^2} - <\Delta r^2>_{dis} \tag{2}$$

where $<\Delta r^2>_{dis}$ is the contribution to B from lattice disorder and other effects that are difficult to evaluate experimentally. For a number of proteins,[10,13,14] the measured value of $(3B/8\pi^2)$ averaged over all of the non-surface atoms of the protein is in the range 0.48 - 0.58 Å^2. Comparison of this result with the mean value of $<\Delta r^2>_{dyn}$ from the simulations (0.28 - 0.36 Å^2) suggests that the non-motional contribution to the B factor, $<\Delta r^2>_{dis}$, is in the range 0.20 - 0.25 Å^2. The only experimental estimate of $<\Delta r^2>_{dis}$ is from Mössbauer data for the heme iron in myoglobin; for that one atom a somewhat smaller value (0.14 Å^2) was obtained.[13] Thus, in the cases examined approximately half of the experimental B factor is associated with thermal fluctuations in the atomic positions and half with other sources. However, some protein and many nucleic acid crystals, particularly those with a high percentage of water, have a larger disorder contribution (e.g., tortoise lysozyme).

For most proteins studied, there is an increase in the magnitude of the experimental and theoretical fluctuations with distance from the center of the molecule. This might be misinterpreted as due to an overall rotational motion of the molecule, but cannot be that in the calculations which are done for a molecule with no net linear or angular momentum. As to more detailed comparisons, there is a good, though not perfect, correlation between the dynamic part of the experimental B values and the calculated results on a residue-by-residue basis (i.e., the individual atom values averaged over residues to damp out unphysical fluctuations in the data).

The magnitudes of the rms fluctuations range from ~0.4 Å for backbone atoms to ~1.5 Å for the ends of long sidechains. As to their time scale, the calculations show that they have relaxation times between 0.25 and 1.5 ps.[15] This means that the fluctuations are associated with motions at "frequencies" in the range 25 to 150 cm^{-1}, which correspond to the lowest frequency torsional modes of proteins. The short relaxation times of the rms fluctuations suggests that a 100 ps simulation can provide an adequate sampling of their average values.

Some examples are given of the results obtained for cyt c and for PTI. Figure 2 shows a comparison of the calculated and experimental rms fluctuations for each residue in cyt c.[10] The experimental values have been corrected for an estimated disorder contribution by subtracting from all of them $\langle \Delta r^2 \rangle_{dis} = 0.25$ Å^2, obtained from the average calculated results for the protein interior. As is evident, there is generally a very good correlation between the experimental and theoretical results. The most prominent differences involve the residues that are calculated to have very large fluctuations; these are all charged sidechains (particularly lysines) that protrude from the protein and so are not correctly treated in the present vacuum simulation. A study of PTI in a Lennard-Jones solvent and in a crystal environment[15] shows that the motion of such outside residues is significantly perturbed by the surrounding medium; in particular, the interaction between charged sidechains of a given protein and its crystal neighbors can produce a reduction in the rms values. The results for external residues contrast sharply with those for the protein interior where the medium effects on the amplitude of fluctuations are found to be very small.

Of interest also are the results from the dynamic simulation concerning deviations of the atomic motions from the isotropic, harmonic behavior assumed in the X-ray analysis. The motions of many of the atoms in the proteins examined are found to be highly anisotropic and somewhat anharmonic. Some specific information for PTI obtained in a 25 ps, room temperature simulation is given in Table III.[15] The average anisotropy for all atoms and for the C_α atoms alone is indicated by the ratio of the largest to the smallest

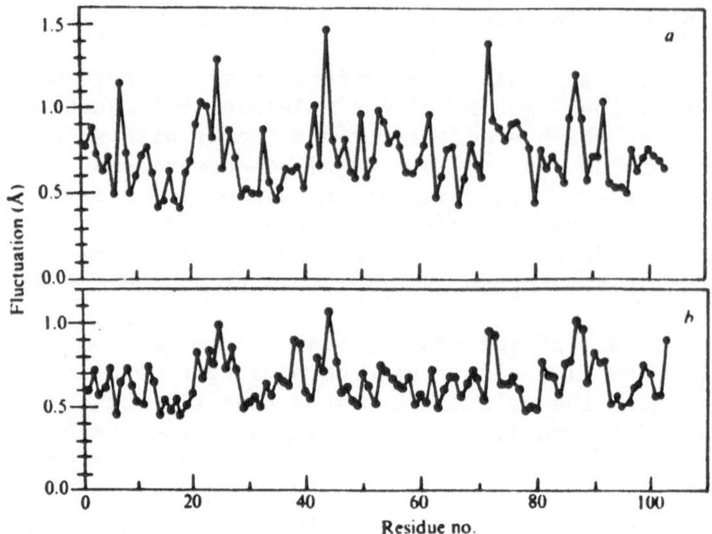

Fig. 2. Calculated and experimental rms fluctuations of ferrocytochrome c; residue averages are shown as a function of residue number: (a) molecular dynamics simulation; (b) X-ray temperature factor estimation corrected for mean disorder contribution (see ref. 10 for details).

eigenvalue of the mean square fluctuation matrix for the Cartesian coordinates. As to deviations from harmonic behavior, the ratio in Table III would be unity if the distribution for all the Cartesian components were Gaussian. It is evident for PTI that the all-atom and the C_α atom averages have a significant deviation from harmonic behavior and that the deviation is in opposite directions for the two cases.

Table III. Anisotropic and Anharmonic Effects in the Pancreatic Trypsin Inhibitor[a]

	$\langle \Delta x^2_{max} \rangle^{1/2} / \langle \Delta x^2_{min} \rangle^{1/2}$	$\dfrac{\langle \Delta x^4 \rangle + \langle \Delta y^4 \rangle + \langle \Delta z^4 \rangle}{3[\langle \Delta x^2 \rangle^2 + \langle \Delta y^2 \rangle^2 + \langle \Delta z^2 \rangle^2]}$
All atoms	2.51	2.24
C_α atoms	2.34	0.554

[a]See reference 15.

Figure 3 gives for PTI a qualitative picture of the fluctuations observed in the molecular dynamics simulation; only the α-carbon skeleton plus the three disulfide bonds are shown. The left-hand drawing represents the X-ray structure and the right-hand drawing an instantaneous picture of the equilibrated structure after 3 ps.[9] It is evident that the two structures are very similar but that there are small differences throughout. The largest displacements appear in the C-terminal end, which interacts with a neighboring molecule in the crystal, and in the loop in the lower left, which has rather weak interactions with the rest of the molecule. Corresponding behavior and deviations from the X-ray structure would be observed in "snap-shots" taken at any other time during the simulation.

Although many of the individual atom fluctuations observed in the simulations or obtained from temperature factors are probably not in themselves important for protein function, they contain information that may be of considerable significance. The calculated fluctuations are such that the conformational space available to a protein at room temperature includes the range of local structural changes observed on substrate (inhibitor) binding for certain enzymes; hinge bending and other such more global transformations have to be treated by special methods (see below). The calculated and experimental fluctuations show further that there are local correlations and that some regions of the protein are more flexible

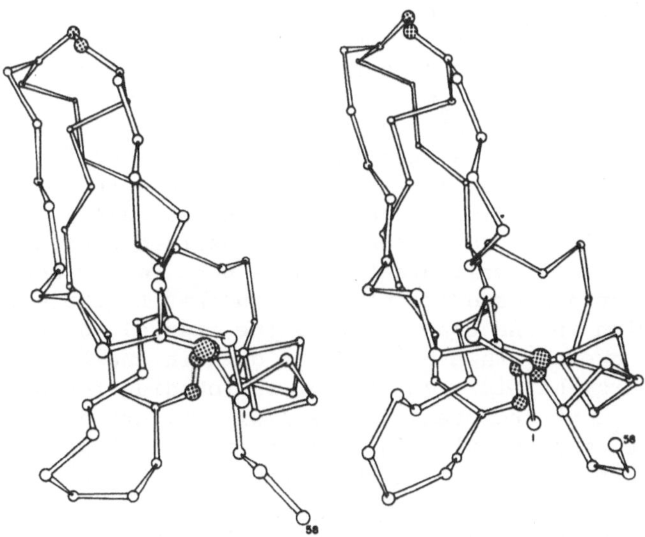

Fig. 3. Drawing of α-carbon skeleton plus S-S bonds of PTI; left-hand drawing is the X-ray structure and right-hand drawing is a typical "snap-shot" during the simulation.

than others. This is found to be true not only in comparing the inside and outside of a protein but one interior region with another. In addition, it may be possible to extrapolate from the calculated rms fluctuations to larger protein motions. Finally, changes in the fluctuations induced by perturbations (e.g., ligand binding) may be important in some cases.

LIGAND-PROTEIN INTERACTION IN MYOGLOBIN

A biological problem where protein fluctuations are important concerns the manner in which ligands like carbon monoxide and oxygen are able to get from the solution through the protein matrix to the heme group in myoglobin and hemoglobin and then out again. These transport proteins are systems in which the ligand is unchanged before and after the interaction with the macromolecule, though a detailed description of the heme iron-ligand bond would clearly require a quantum-mechanical treatment. What makes this problem interesting is that examination of the high-resolution X-ray structure of myoglobin[16] does not reveal any path by which ligands such as O_2 or CO can move between the heme binding site and the outside of the protein. Since this holds true both for the unliganded and liganded protein (i.e., myoglobin and metmyoglobin), structural fluctuation must be involved in the entrance and exit of the ligands. Empirical energy function calculations[17] have shown that the rigid protein would have barriers on the order of 100 kcal/mol; such high barriers would make the transitions infinitely long on a biological time scale. Figure 4 shows a potential energy map for the shortest "path" from the heme pocket to the exterior of the protein. The figure gives the non-bonded potential contour lines seen by a test particle representing an O_2 molecule in a plane (xy) parallel to the heme and displaced 3.2 Å from it in the direction of the distal histidine; the coordinate system in this and related figures has the iron at the origin and the z axis normal to the heme plane. The low potential energy region in the center is the so-called "heme pocket", with the energy minimum corresponding to observed position of the distal O atom of an O_2 molecule forming a bent Fe-O-O bond.[18] The shortest path for a ligand from the heme pocket to the exterior (the low energy region in the upper left of the figure) is between His E7 and Val E11. However, this path is not open in the X-ray geometry because the energy barriers due to the surrounding residues indicated in the figure are greater than 90 kcal/mol. Figure 5 shows a possible path to the exterior in a plane (xz plane) perpendicular to Figure 4; again the barriers in the X-ray structure are very large.

To analyse pathways available in the thermally fluctuating protein, ligand trajectories were calculated using the static myoglobin X-ray structure together with a test molecule of reduced effective diameter to compensate for the absence of protein fluctuations.[17]

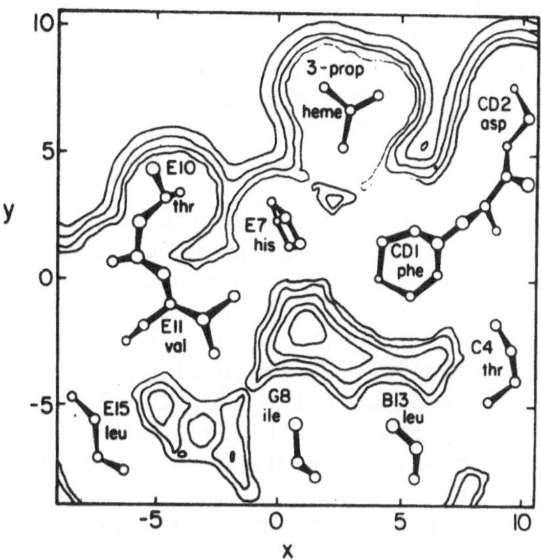

Fig. 4. Myoglobin-ligand interaction contour map in the (x,y) plane at z = 3.2 Å (see text). Distances are in Å and contours in kcal; the values shown correspond to 90, 45, 10, 0, and -3 kcal/mol relative to the ligand at infinity. The highest contours are closest to the atoms whose projections onto the plane of the figure are denoted by circles (see ref. 17).

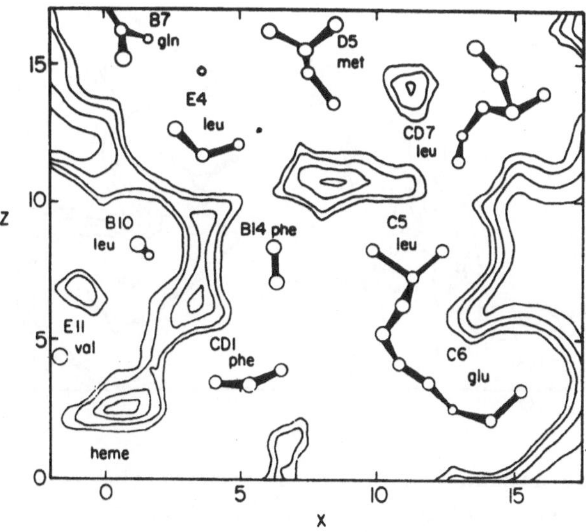

Fig. 5. Myoglobin-ligand interaction contour map in the (x,y) plane at y = 0.5 Å (see Figure 4 legend).

The trajectory was determined by releasing the test molecule with substantial kinetic energy (15 kcal/mol) in the heme pocket and following its classical motion for a suitable length of time. A total of eighty such trajectories were computed; a given trajectory was terminated after 3.75 ps if the test molecule had not escaped from the protein. Slightly more than half the test molecules failed to escape from the protein in the allowed time; twenty-five molecules remained trapped near the heme binding site, while another twenty-one were trapped in two cavities accessible from the heme pocket. Most of the molecules which escaped did so between the distal histidine (E7) and the sidechains of Thr E10 and Val E11 (see Figure 4). A secondary pathway was also found (see Figure 5); this involves a more complicated motion along an extension of the heme pocket into a space between Leu B10, Leu E4, and Phe B14, followed by squeezing out between Leu E4 and Phe B14. A typical model trajectory following this path is shown in Figure 6.

In the rigid X-ray structure, the two pathways have very high barriers for a thermalized ligand of normal size. Thus, it was necessary to study the energetics of barrier relaxation to determine whether either of the pathways had acceptable activation enthalpies. Local dihedral rotations of key sidechains in the otherwise rigid

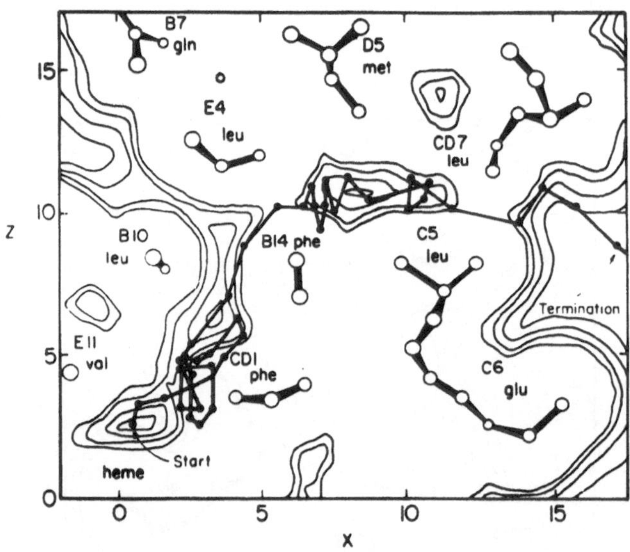

Fig. 6. Diabatic ligand trajectory following the secondary pathway (see text); a projection on the plane of Figure 5 is shown with dots at 0.15 ps intervals. The start of the trajectory at the heme iron and the termination point exterior to the protein are indicated by arrows.

myoglobin were investigated and it was found that the bottleneck on the primary pathway could be relieved at the expense of modest strain in the protein by rigid rotations of the sidechains of His E7, Val E11, and Thr E10. The reorientation of these three sidechains and the resultant opening of the pathway to the exterior is illustrated schematically in Figure 7; Panel I shows the X-ray structure (same as Figure 4); in Panel II the distal histidine (E7) has been rotated to $\chi_1 = 220°$ at an energy cost of 3 kcal/mol; in Panel III Val E11 has also been rotated to $\chi_1 = 60°$ (~5 kcal/mol); and Panel IV has the additional rotation of Thr E10 to $\chi_1 \cong 305°$ ($\lesssim$ 1 kcal/mol). In this manner a direct path to the exterior has been created with a barrier of ~5 kcal/mol at an energy cost to the protein of ~8.5 kcal/mol, as compared with the X-ray structure value of nearly 100 kcal/mol. On the secondary path, however, no simple torsional motions reduced the barrier due to Leu E4 and Phe B14, since the necessary rotations led to larger strain energies. A test sphere with van der Waals radius of 3.2 Å was then fixed in the energy-refined structure at either of two positions in the bottleneck on the primary path (between His E7 and Val E11, or between His E7 and Thr E10) or in the bottleneck on the secondary path (between Leu E4 and Phe B14). The protein was allowed to relax by energy minimization (adiabatic treatment) in the presence of the ligand and the

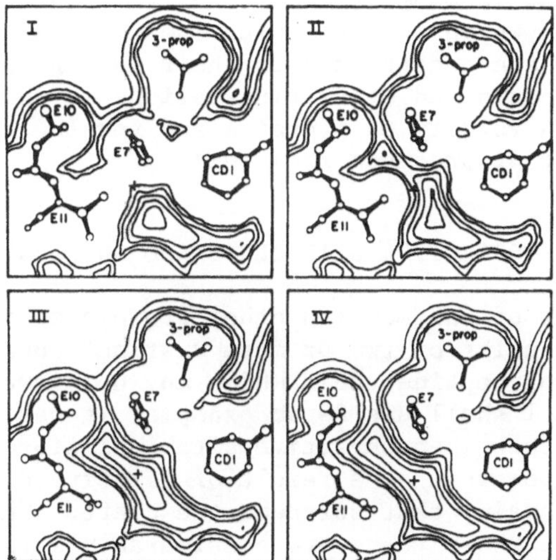

Fig. 7. Myoglobin-ligand interaction contour maps in the (x,y) plane at x = 3.2 Å showing protein relaxation; a cross marks the iron atom projection onto the plane (see Figure 4 legend). The sidechain rotations relative to the X-ray structure (I) shown in Panels II - IV are discussed in the text.

resulting displacements in the polypeptide chains were monitored. There were local alterations in sidechain dihedral angles and bond angles. In addition, neighboring sidechains and the backbones of helices D and E participated in the globin response, mostly by small dihedral angle changes. Approximate values for the relaxed barrier heights were found to be 13 kcal/mol and 6 kcal/mol for the two primary path positions and 18 kcal/mol for the secondary path position. These barriers are on the order of those estimated in the photolysis, rebinding studies for CO myoglobin by Austin et al.[19] The type of ligand motion expected for such a several-barrier problem can be determined from the trajectory studies mentioned earlier. What happens is that the ligand spends a long time in a given well, moving around in and undergoing collisions with the protein walls of the well (see Figure 6). When there occurs a protein fluctuation sufficient to significantly lower the barrier or the ligand gains sufficient excess energy from collisions with the protein, or most likely both at the same time, the ligand moves rapidly over the barrier and into the next well where the process is repeated. That the ligand spends most of the time in the low energy wells is evident from Figure 6. However, it should be noted that in a completely realistic trajectory involving a fluctuating protein and ligand-protein energy exchange, the time spent in the wells would be much longer than that found in the diabatic model calculations.

From this analysis of myoglobin and more general considerations, it seems likely that in many cases the native structure is such that the small molecules which interact with the protein cannot enter or leave if the atoms are constrained to their average positions. Consequently, sidechain and other fluctuations may be required for ligand binding by a variety of proteins and for the entrance of substrates and exit of products from enzymes.

OPENING AND CLOSING OF ACTIVE SITE CLEFT AND SUBSTRATE BINDING

A large number of enzymes and other protein molecules (e.g., immunoglobulins) consist of two or more distinct domains connected by a few strands of polypeptide chain which may be viewed as "hinges." In lysozyme, which I shall use as an example, it was noted in the X-ray structure[20] that when an active site inhibitor is bound, the cleft closes down somewhat as a result of relative displacements of the two globular domains that surround the cleft. The kinases are a class of proteins where there are considerably larger displacements of the two lobes on substrate binding than in lysozyme.[21]

In the study of lyzosyme, the stiffness of the hinge was evaluated by the use of an empirical potential energy function.[22,23] An angle bending potential was obtained by rigidly rotating one of the globular domains relative to a bending axis which passes through the hinge and calculating the changes in the protein conformational energy. This procedure is expected to overestimate the bending potential, since no allowance is made for the relaxation of unfavorable contacts between atoms which may have been generated during the rotation. To take account of the relaxation, an adiabatic bending potential was calculated by holding the bending angle fixed at various values and permitting the positions of atoms in the hinge and adjacent regions of the two globular domains to adjust themselves so as to minimize the total potential energy. As in a previous adiabatic ring rotation calculation,[23] only small (<0.3 Å) atomic displacements occurred in the relaxation process; the differences between the rigid and adiabatic bending potentials is largely due to small shifts in the relative positions of a few atoms which have been forced too close together by the rigid rotation model. The relief of these contacts can be effected by localized motions (e.g., bond angle and local dihedral angle deformations). The frequencies associated with these deformations (>100 cm^{-1}) are expected to be much greater than the hinge bending frequency (≈5 cm^{-1}), so that the use of the adiabatic bending potential is appropriate.

The rigid and adiabatic bending potentials were found to be approximately parabolic, with the restoring force constant for the adiabatic potential about an order of magnitude smaller than that for the rigid potential (see Figure 8). However, even in the adiabatic case, the effective force constant is about twenty times as large as the bond-angle bending force constant of an α-carbon (i.e., N-C_α-C); the dominant contributions to the force constant come from repulsive nonbonded interactions involving on the order of fifty contacts. If the adiabatic potential is used and the relative motion is treated as an angular harmonic oscillator composed of two rigid spheres, a vibrational frequency of about 5 cm^{-1} is obtained. This is a consequence of the fact that, although the force constant is large, the moments of inertia of the two lobes are also large.

In considering the hinge bending motion it is essential to take account of the fact that lysozyme is normally in solution. Although fluctuations in the interior of the protein, such as those considered in myoglobin, may be insensitive to the solvent (because the protein matrix acts as its own solvent), the domain motion in lysozyme involves two lobes that are surrounded by the solvent. To take account of the solvent effect in the simplest possible way,[22] the Langevin equation for a damped harmonic oscillator (Eq. 1) was used. Possible solutions to the Langevin equation are such that, depending on the parameters of the system, the relative motion of two domains can range from underdamped oscillations (if the connecting hinge is stiff and damping due to solvent drag on the domains and internal frictional effects in the protein are small) to Brownian motion (if the hinge is highly flexible or damping effects are large).

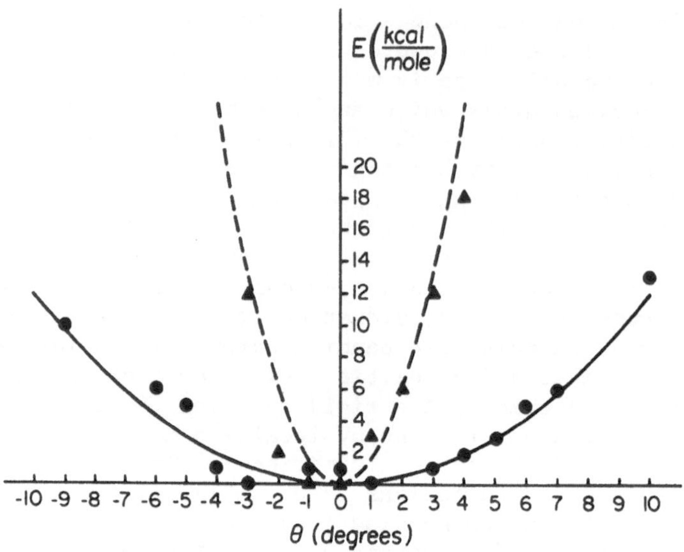

Fig. 8. Change of conformational energy produced by opening ($\theta < 0$) and closing ($\theta > 0$) the lysozyme cleft; calculated values are for the rigid bending potential (triangles) and for the adiabatic bending potential (circles); the origins for the two calculations are superposed (see ref. 22 for details).

The friction coefficient for the solvent damping term was evaluated by modeling the two globular domains as spheres and calculating the viscous frictional drag accompanying the relative motion of these spheres by use of a modified Stokes law;[22] the internal friction of the protein was considered to be negligible compared to the hydrodynamic friction. From the adiabatic estimate of the hinge potential and the magnitude of the solvent damping, it was found that the relative motion of the two globular domains in lysozyme is overdamped; i.e., in the absence of driving forces the domains would relax to their equilibrium positions without oscillating. The decay time for this relaxation was estimated to be about 2×10^{-11}s. Actually, the lysozyme molecule will experience a randomly fluctuating driving force due to collisions with the solvent molecules, so that the distance between the globular domains will fluctuate in a Brownian manner over a range limited by the bending potential; the root-mean-square fluctuation of the cleft width was estimated to be about 0.5 Å.[22]

Since the hinge bending motion in lysozyme and in other enzymes involves the active site cleft, it is likely to play a role in the enzymatic activity of these systems. In addition to the possible difference in the binding equilibrium and solvent environment in the open and closed state, the motion itself could result in a coupling between the entrance and exit of the substrate and the opening and closing of the cleft. More detailed calculations, including the effect of substrates on the interdomain motion to test this suggestion,are in progress.

EXTERIOR SIDECHAIN AND LOOP MOTIONS

A type of motion that is of importance in enzymes involves the displacement of surface sidechains or entire loops on substrate binding. In carboxypepdidase A,[24] for example, when the substrate binds there are structural changes that include a large displacement of the sidechain of Tyr 248, which moves through more than 10 Å toward the active site. Another example is provided by an external loop in trioshosphate isomerase, which has been shown by X-ray diffraction to fold over the substrate when it is bound.[25] If surface residues are involved, as is often the case, the motion that occurs must be treated by a method that is analogous to the one applied to the relative motion of the lysozyme domains. Although we have so far not studied a specific protein, we have been examining the motion of aliphatic sidechains in aqueous solution.[26] The end of the chain that is attached to the macromolecule is held fixed and the Langevin equations of motion (Eq. 1) for the atoms of the chain are solved simultaneously for periods of up to a microsecond. The methyl and methylene groups of the chain are treated as single extended atoms with a friction coefficient corresponding to methane in water and a generalized empirical potential energy function is used to represent the intramolecular interactions (non-bonded and torsional) in the usual way, except that they are slightly modified to take account of the fact that solvent is present; that is, a potential of mean force replaces the isolated molecule potential function. From the work of Pratt and Chandler,[27] it is known that the modifications are small but not negligible; e.g., the (gauche/trans) potential energy difference in butane is altered by ~0.55 kcal/mol in aqueous solution relative to vacuum.

When the stochastic or Langevin equations are solved for a protein sidechain, it is found that the motion with respect to a given torsional angle separates into two time scales.[26] The shorter time motion, on the order of tenths of picoseconds, corresponds to torsional oscillations within a potential well, and the longer, on the order of two hundred picoseconds, corresponds to transitions from one potential well to another; the torsional barrier used in the potential function is ~2.8 kcal/mol. Thus, analogous to the description given earlier in the lecture of an oxygen molecule moving through myoglobin, the sidechain spends most of the time oscillating about a single conformation (i.e., with each dihedral angle remaining in a given well) and only rarely makes a transition from one conformation to another (see Figure 9). To test the validity of this type of calculation, comparisons of the stochastic trajectory results with NMR relaxation measurements (e.g., ^{13}C NMR) have been made.[28] Clearly it will be of interest to apply the method to a specific structural change in a protein; such calculations are in progress.

In complete molecular dynamics simulations of proteins, such as the ones from which the temperature factors were obtained, it is likely that some rare events will be found in the analysis of any given trajectory; e.g., transitions that usually take place on a nanosecond or even longer time scale may be found to have occurred once in a subnanosecond trajectory. Such "accidents," although not significant in terms of average rates, can provide useful information about the nature of the event. One such rare event observed in a PTI trajectory[9] is the reorientation of the peptide group between residues 38 and 39. Peptide groups generally remain in their initial minimum, which is specified by the values of the adjacent mainchain dihedral angles (in this case ϕ_{39} and ψ_{38}). In the molecular dynamics simulation (see Figure 10), the two dihedral angles were observed to oscillate about the minimum (ϕ_{39},ψ_{38}) = (60°,120°) for several picoseconds, then move to a new minimum (-120°,-60°) in about 1.5 picoseconds; the peptide group remained in the new minimum, which corresponds to a 180° slip, for the remainder of the trajectory. What is particularly interesting about the dynamic result is that the character of the angular motions is very similar to that found for the stochastic sidechain trajectory (see above; Figure 9). In both cases, the motion in a given potential well and over a barrier was found to have a diffusive character. This reinforces the idea that the protein interior behaves as a viscous solvent for the motions of its components.[1,9]

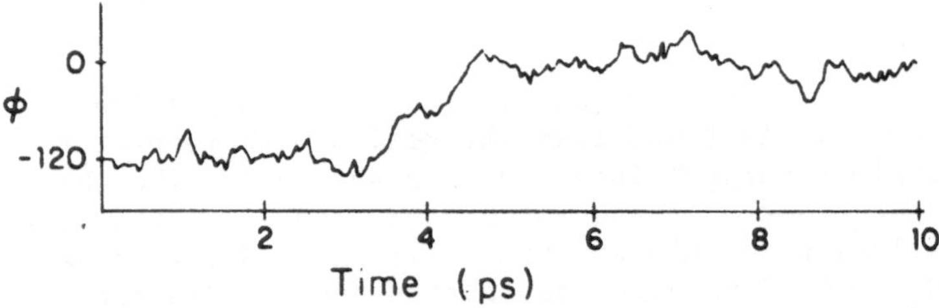

Fig. 9. Butane trajectory in stochastic dynamics simulation showing one transition over the barrier from the gauche to the trans configuration at about 4 ps (see ref. 26 for details).

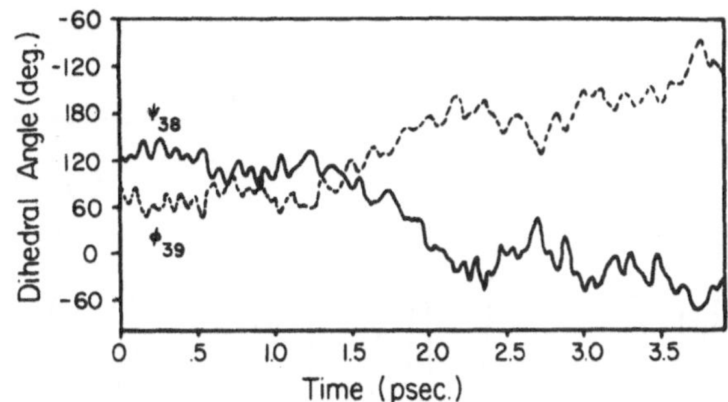

Fig. 10. Peptide group transition from PTI full dynamics simulation (from ref. 9).

ACTIVATED EVENTS: THE CENTRAL ELEMENT OF ENZYME REACTIONS

Enzyme catalysed reactions generally involve some processes in which the rate is limited by an energy barrier. In many cases, the time scale of these "activated" processes is a microsecond or longer. As already indicated, it is not possible by direct dynamics simulation methods to study such processes in detail because they are by their nature rare events; that is, although an individual barrier crossing is rapid, it does not occur very often.

To develop alternative approaches to this fundamental class of problems,[29] we have begun to examine a relatively simple model system in which the "reaction" does not require introduction of quantum mechanical methodology, because no bonds are made or broken. The reaction is the 180° reorientation of the aromatic ring of a tyrosine sidechain in PTI. From reaction path calculations, analogous to those described for oxygen in myoglobin, it is known that the energy barriers for the tyrosine rings (there are four in PTI) vary between zero and ~25 kcal/mol.[23] Typical behavior found for the ring motion of a tyrosine sidechain (barrier ~10 kcal/mol) in the full molecular dynamics trajectory is shown in Figure 11a, in which $\Delta\phi = \phi - \langle\phi\rangle$ is plotted as a function of time; ϕ is the ring torsional angle and $\langle\phi\rangle$ is the average value obtained in the simulation.[30] For comparison, Figure 11b illustrates the results obtained in the simulation of an isolated tyrosine "dipeptide" fragment; the initial configuration of the fragment was obtained from the PTI X-ray structure and the backbone conformation did not vary significantly during the simulation. The rather regular, almost harmonic

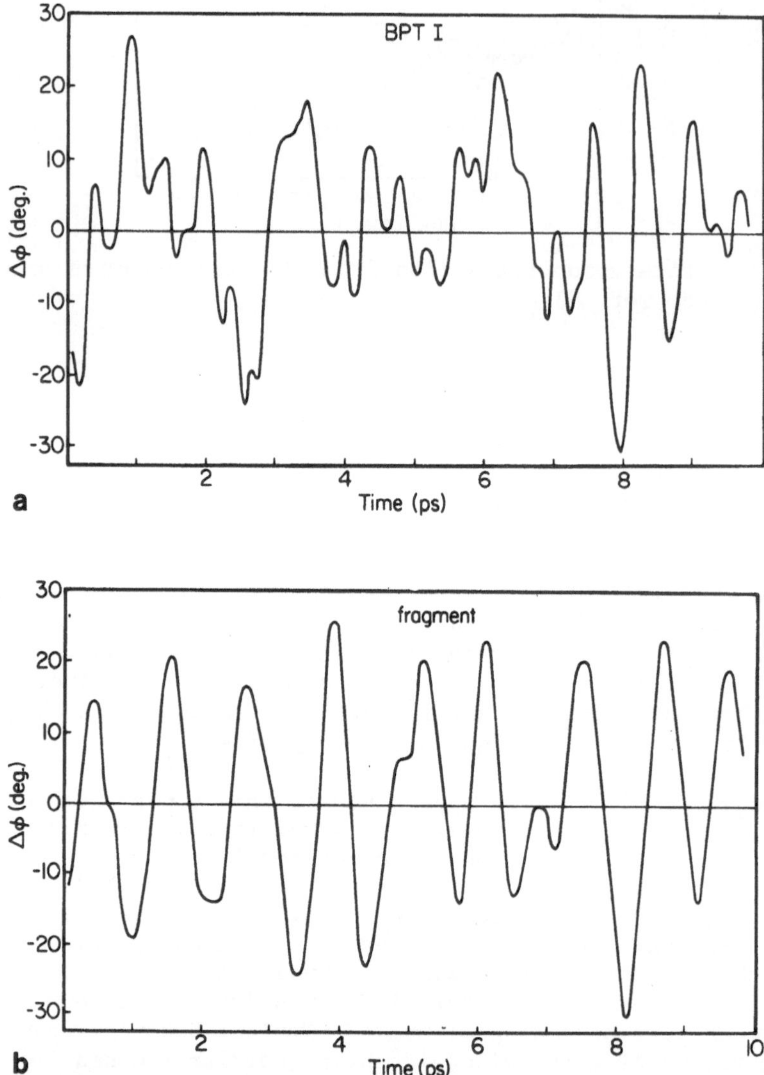

Fig. 11. Evolution of Tyr 21 ring torsional angle ϕ during dynamics simulation. (a) Ring as part of PTI; (b) Ring as isolated fragment (see ref. 30 for details).

motion of the ring in the fragment contrasts sharply with the highly irregular behavior of the ring as part of PTI. In the protein, the ring undergoes frequent collisions with the surrounding atoms that significantly perturb its motion. Some of the interactions are sufficiently strong to reverse the direction of the motion, while others produce a smaller change in the angular velocity. For both the fragment and the protein, the ring oscillates around its mean position, with a maximum amplitude on the order of $\pm 25°$. That the ring has a similar amplitude in the isolated fragment and in the closely packed protein is due to the fact that correlations between the displacements of the ring and the surrounding protein atoms lead to a much softer effective potential (potential of mean force) than would be present if the protein were rigid. As expected, neither in the fragment nor in the protein is a ring flip observed during the 9.8 ps simulation period.

To examine the barrier crossing problem, we chose for study one of the rings (Tyr 35) with an apparent barrier on the order of 20 kcal/mol for which an ordinary simulation would be very unlikely to find even one ring flip. The approach used involves the extension of the ideas of transition state theory to the detailed treatment of activated processes in proteins. The first step in such a calculation is the determination of a transition region in which a dividing surface between reactants and products can be defined. It is then necessary to generate a set of configurations in this region with coordinate values (other than the ones specifying the region) selected in accord with a Boltzmann distribution for the temperature under consideration. Given the transition-state configurations, two quantities have to be evaluated. The first is the probability that a system composed of reactants at equilibrium will be in the transition region (in transition state theory, this corresponds to the equilibrium constant between the activated complex and the reactants), and the second is the probability that the transition-state configurations with appropriate atom velocities will go on to give product (in transition-state theory, this corresponds to the transmission coefficient). Although various aspects of this procedure (sometimes referred to as phase-space/trajectory calculations)[31] can be done analytically in simple cases, the complexity and multidimensionality of a protein reaction requires that the problem be solved numerically by appropriate combinations of Monte Carlo and molecular dynamics techniques. To prepare initial states for the trajectory calculations of the barrier crossing, the dihedral angle for the ring orientation, ϕ, was used as an approximate reaction coordinate to define the domain of activated configurations. Two sets of coordinates in this domain were examined; the first set was determined by starting with a protein configuration chosen from the beginning of a conventional dynamical simulation, rigidly rotating the ring to a high energy orientation, and allowing the protein to relax by a Monte Carlo calculation while ϕ was held fixed. Small adjustments of the ring orientation followed by Monte Carlo calculation were performed to generate activated configurations for detailed study. A second set of activated configurations was generated by the same procedure, but starting with a different equilibrium configuration obtained from the dynamics run. Five sample barrier-crossing trajectories were calculated for each of the two sets.

In all of the trajectories, the ring crossed the rotational potential energy barrier successfully; the times required to complete these ring rotations were 0.5 to 1.0 ps. A typical trajectory is shown in Figure 12. None of the trajectories exhibited path reversals (torsional angular velocity sign changes) during the barrier crossings, and all yielded apparently stabilized final states; these results suggest that transition state theory is approximately valid for ring rotation.[29] Several trajectories did exhibit appreciable slowing down (angular velocities approaching zero) during the crossings, so that frictional effects are evident even though they do not dominate inertial effects for this process. Analogous results were found recently in a related stochastic model simulation studies of the rotational isomerization of butane in a van der Waals solvent.[32]

The torsional displacements and angular velocity changes of the ring are well accounted for by the torques acting on the ring as a result of nonbonded interactions with the surrounding matrix atoms.

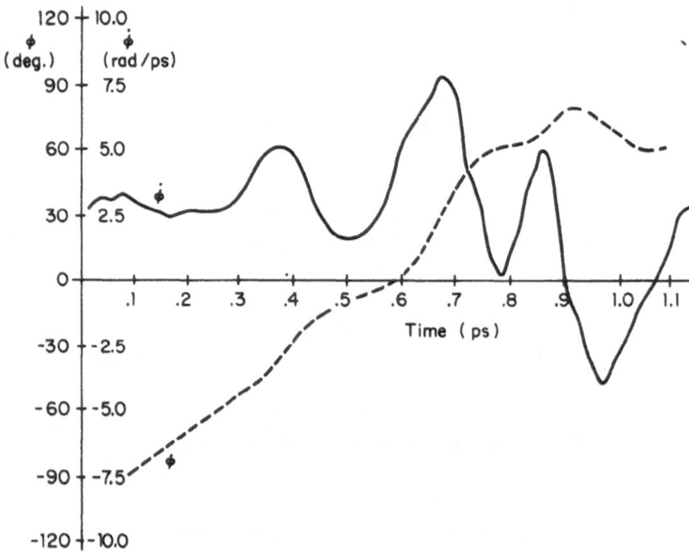

Fig. 12. Barrier crossing trajectory of Tyr 35 in PTI; ring torsion angle ϕ and torsional angular velocity $\dot{\phi}$ as a function of time (see ref. 29 for details).

The durations and magnitudes of these torques are similar to those which were found to drive the equilibrium torsional fluctuations of the ring in the complete molecular dynamics simulation.[30] The ring surmounts the barrier, not because of a particularly energetic collision, but rather as the result of a transient decrease in frequency and intensity of collisions, which would tend to drive the ring away from the barrier. This observation suggests that small, transient packing defects play a role in initiating ring rotations.

In most of the trajectories, fluctuations in the tyrosine side-chain and backbone bond angles are seen that appear to be correlated with the occurrence of ring-backbone nonbonded contacts; these bond angle deformations all serve to relieve the stress arising from the nonbonded interaction. The resulting bond angle stresses amount to 5 to 10 kcal/mol, which is similar to the stress found in static reaction path calculations of the ring rotational barrier.[23]

To complete the dynamic analysis of this model reaction, enough transition state trajectories have to be obtained to provide an accurate value for the transmission coefficient. Further, the probability of being in the transition region has to be determined; for the latter a more detailed definition of the reaction coordinate to include other variables than the ring torsional angle may be required. Both aspects of the problem are being investigated [Northrup, Pear, Lee, McCammon, and Karplus, to be published].

Extension of the methodology described here for treating activated events to enzyme reactions, though difficult, should be feasible. The essential additional step is the determination of the enzyme plus substrate potential surface for the reaction. This will clearly require some quantum-mechanical calculations, whose results may well be best expressed in terms of suitably chosen empirical potential functions for use in the dynamics calculations, in analogy to previous work on small molecule reactions.[33]

ACKNOWLEDGEMENTS

I am pleased to acknowledge the essential role played by my collaborators on various aspects of the work reviewed in this lecture. The students, postdoctoral fellows, and colleagues who have contributed to the work are D. A. Case, B. R. Gelin, W. F. van Gunsteren, T. Ichiye, C. Y. Lee, R. M. Levy, J. A. McCammon, S. H. Northrup, B. D. Olafson, M. R. Pear, P. J. Rossky, S. Swaminathan, T. Takano, and P. G. Wolynes.

REFERENCES

1. M. Karplus and J. A. McCammon, CRC Crit. Rev. of Biochem. 9:273 (1981).
2. P. J. Rossky and M. Karplus, J. Am. Chem. Soc. 101:1913 (1979).
3. S. Chandrasekar, Rev. Mod. Phys. 15:1 (1943).
4. R. M. Levy and M. Karplus, Biopolymers 18:2465 (1979).
5. R. M. Levy, D. Perahya, and M. Karplus, Proc. Nat. Acad. Sci. (to be published).
6. J. A. McCammon, R. M. Levy, and M. Karplus, Biopolymers 19:2033 (1980).
7. M. Levitt in CECAM Workshop Report on Protein Dynamics, France (May 1976).
8. P. J. Flory, Statistical Mechanics of Chain Molecules, Wiley, New York (1969).
9. J. A. McCammon, B. R. Gelin, and M. Karplus, Nature 267:585 (1977).
10. S. H. Northrup, M. R. Pear, J. A. McCammon, M. Karplus, and T. Takano, Nature 287:659 (1980).
11. S. Swaminathan and M. Karplus (to be published).
12. B. Olafson, T. Ichiye, S. Swaminathan, and M. Karplus (to be published).
13. H. Frauenfelder, G. A. Petsko, and D. Tsernoglou, Nature 280:558 (1979).
14. P. J. Artymiuk, C. C. F. Blake, D. E. P. Grace, S. J. Oatley, D. C. Phillips, and M. J. E. Sternberg, Nature 280:563 (1979).
15. W. F. van Gunsteren and M. Karplus, Biochemistry (to be published).
16. T. Takano, J. Mol. Biol. 110:569 (1977).
17. D. A. Case and M. Karplus, J. Mol. Biol. 132:343 (1979).
18. S. E. Phillips, Nature 273:247 (1978).
19. R. H. Austin, K. W. Beeson, L. Eisenstein, H. Frauenfelder, and I. C. Gunsalus, Biochemistry 14:5355 (1975).
20. T. Imoto, L. N. Johnson, A. C. T. North, D. C. Phillips, and J. A. Rupley, in The Enzymes, Volume VII, P. D. Boyer, ed., Academic Press (1972), p. 665.
21. C. M. Anderson, F. H. Zucker, and T. A. Steitz, Science 204:375 (1979).
22. J. A. McCammon, B. R. Gelin, M. Karplus, and P. G. Wolynes, Nature 262:325 (1976).
23. B. R. Gelin and M. Karplus, Proc. Nat. Aca. Sci. USA 72:2002 (1975).
24. J. A. Hartsuck and W. N. Lipscomb, in The Enzymes, Volume III, P. D. Boyer, ed., Academic Press (1971), p. 1.
25. D. W. Banner, A. C. Bloomer, G. A. Petsko, D. C. Phillips, D. I. Pogson, I. A. Wilson, P. H. Corran, A. J. Furth, J. D. Milman, R. E. Offord, J. D. Priddle, and S. G. Waley, Nature 255:609 (1975).
26. R. M. Levy, M. Karplus, and J. A. McCammon, Chem. Phys. Letters 65:49 (1979).

27. L. R. Pratt and D. Chandler, J. Chem. Phys. 67:3683 (1977).
28. R. M. Levy, M. Karplus, and P. G. Wolynes, J. Am. Chem. Soc. (to be published).
29. J. A. McCammon and M. Karplus, Proc. Nat. Acad. Sci. USA 76:3685 (1979).
30. J. A. McCammon, P. G. Wolynes, and M. Karplus, Biochemistry 18:927 (1979).
31. J. B. Anderson, J. Chem. Phys. 58:5684 (1973).
32. J. A. Montgomery, D. Chandler, and B. J. Berne, J. Chem. Phys. 70:4056 (1979).
33. M. Karplus, R. N. Porter, and R. D. Sharma, J. Chem. Phys. 43:3259 (1965).

DNA-PROTEIN INTERACTIONS

Juan A.Subirana

Unidad de Química Macromolecular del C.S.I.C.
Escuela T.S.de Ingenieros Industriales
Diagonal, 999, Barcelona(28), Spain

1. INTRODUCTION

The study of the interactions between proteins and DNA is one of the crucial problems in molecular biology given the central role they play in the biological activity of the genetic material. However the nature of the forces and mechanisms involved has remained elusive. Only very recently the first two structures of proteins which interact specifically with DNA has been reported[1,2], but the nature of the complex they form with DNA is based only on speculation. We can hope that this situation will change in the near future, since there are now methods to obtain oligonucleotides suitable to form specific complexes with proteins.

One part of this general field about which we have now a reasonable amount of knowledge is the conformation (or secondary structure) of DNA and the way it changes upon interaction with counterions and proteins, in particular with histones and protamines. We will discuss this subject in the first part of this paper. In the second part we will review the structure of chromatin, about which a significant amount of knowledge has accumulated in the past few years. We will try to avoid redundancies with the many excellent reviews which have recently appeared.

2. THE STRUCTURE OF DNA

2.1 Influence of base sequence

Since DNA is a negatively charged polyelectrolyte with a variable base sequence, we can distinguish two separate influences on its conformation: the base sequence itself and the nature of the counterions associated with the DNA. These two effects will always occur simultaneously, but for the clarity of presentation we will first discuss the influence of base sequence. This question has been studied in detail by Arnott and collaborators, who have recently presented a summary of their work[3]. The conformations of DNA described by them are shown on table I. An analysis of this table and of the more detailed results presented by these authors suggests the following considerations:

- All the sequences studied can take the B-form, either with sodium, or lithium or both as counterions. The only exception is poly(dA)-poly(dT), which adopts the related B' form. This form has the same number of base pairs per helical turn, but the minor groove is significantly narrower than in the B form. In principle therefore an homogeneous sequence of A-T pairs might provide a structural signal in DNA, but no evidence is available to show that this occurs "in vivo".

- Most of the DNA sequences can also take the A form. Very often this form appears upon dehydration, whereas the B form is usually the hydrated form. However there are counterions which preserve the A-form for certain sequences at high humidities, as for example the Na^+ salts of poly d(A-T) and poly d(A-G-T) .poly d(A-C-T) . On the other hand there are sequences in which the A form has not been observed upon dehydration, as in poly(dA). poly(dT).

Table I. Influence of base sequence on the conformation of DNA

DNA form	Number of base pairs per helical turn	Base sequences in which the indicated DNA form has been observed
A	11	DNA, G, AT, AC, GC, AAC, ACC, AGT, AGC, GAT
B	10	DNA, G, AT, AC, AG, GC, AAT, AAC, ACC, AGT, AGC, GAT
B'	10	A
C	9.3(8.5-9.7)	DNA, AG, AGT, AGC, ACC
D	8	AT, AAT
Z	-12	AC, GC

The results shown in this table are reproduced from the work of Leslie et al[3]. Only the base sequence of one of the two complementary strands is indicated. DNA refers to native DNA. The sequences which include inosine are not included in the table.

- The C form is less strictly defined, the number of base pairs per helical turn varies between 8.8 and 9.7, with a tendency to increase as the degree of hydration increases[4], in particular when the DNA molecules are packed in the hexagonal system. In the orthorhombic system it appears that the packing forces stabilize the molecules with 9 1/3 base pairs per helical turn.

- The D form has been observed in a few sequences which only contain A-T pairs. It is considered[3] to be related to the B form. However it should be pointed out that Mitsui et al[5] found a similar X-ray diagram for poly d(I-C) and suggested that it was a left handed helix from its circular dichroism spectrum. Therefore it can not be excluded that the D form is a left handed form of DNA.

- The Z form is known in great detail, since an oligomer of poly d(G.C) has been crystalized[6]. Its left handed conformation is of great interest, because of its eventual physiological implications, but at the present time it has only been described under conditions of very high ionic strength of dubious biological significance. On the other hand it has been found[7] that some carcinogens when covalently bound to DNA facilitate its transition to the Z-form.

In summary, under the hydrated conditions which prevail in biological systems, only the B form appears to be stable. The structure of this form is now known in great detail as a result of the crystallographic work of Wing et al[8]. The other forms of DNA which have been found indicate that this molecule has a potential to modify its conformation as a function of base sequence, but no modification of this type has been found thus far in native biological systems. In particular, there has been recent interest[9] on the possibility of a "side by side" model of DNA structure in which the two strands of the molecule are placed side by side instead of being plectonemically coiled in a helical fashion. This conformation appears of great conceptual interest since it would allow a simpler process of DNA replication. Some time ago we did discuss[10] this possibility and the eventual influence of histones, but at present there is no direct evidence in favour of such a DNA conformation.

2.2 Stabilization of the B form of DNA by counterions

In the previous section we have seen that the basic counterions which neutralize the DNA charges have an influence on the conformation of this molecule and on the transitions that it undergoes as a function of the degree of hydration of the sample. It appeared therefore worthwhile to determine the influence on DNA of other counterions which occur in biological systems. With this objective in mind we have studied[11,12] by fiber X-ray diffraction the conformation of complexes of DNA with basic amino acids and oligopeptides which are the natural counterions of DNA in its complexes with proteins. A description of this work has appeared elsewhere[12]. In this section we discuss the counterions which stabilize the B form of DNA. This effect has been found with several peptides, the structure of which is given on table II. These peptides stabilize the conformation of DNA with 10 base pairs per helical turn. In the case of the arginine peptides that we have studied, the B form is maintained even at 0% relative humidity, that is in the absence of water.

Table II. The influence of basic peptides on the conformation of DNA.

Effect	Peptides
Stabilization of B-form	Arg, Arg_2, Ac-Arg_2-$CONH_2$, Arg_3, ArgTyr Lys, TyrSerLys, LysTyr, His
Relaxation of DNA	Ac-Lys_6Ala_3Tyr Ala_3Lys_6-NHC_2H_5, LysAla
Destabilization of DNA	Ac-Arg-$COOCH_3$, Ac-Arg-NHC_2H_5, Ac-Lys-$COOCH_3$, Leu-$COOCH_3$, Ala-$COOCH_3$

These results are taken from unpublished results obtained by Campos, Fornells and Portugal and from references 11 and 12. The acetyl group has been abreviated Ac.

The diagram for arginine is shown in Fig. 1. It appears therefore that the B form of DNA has an intrinsic stability which is enhanced by several types of peptides, in particular many which contain arginine. This stability is remarkable, since it has been determined[13] that DNA in solution has about 10.5 base pairs/repeat, a value which has never been found in any of the fibers studied[1,11,12]. A possible explanation for this fact is shown in Fig. 2. The parameters determined in fibers correspond to molecules which are oriented with their helical axes parallel to the axis of the fiber, so that they correspond to molecules in a straight position. On the other hand, the DNA molecules in solution are free to bend and such bending may result in a higher average number of base pairs per helical turn as it is experimentally observed.

The analysis of the diffraction diagrams obtained from the arginine-DNA fibers in the absence of water has also allowed[14] to determine the distribution of the arginine residues on the DNA helix. In this case the DNA molecules are packed in an orthorhombic lattice. The distribution of intensities indicates that the DNA molecules are displaced in relation to their nearest neighbours and in the direction of the helix axis by $\pm 1/3$ of a helical pitch, as it occurs in the Li^+ salt of $^-$DNA[15]. This packing arrangement is shown on Fig. 3. It can be immediately

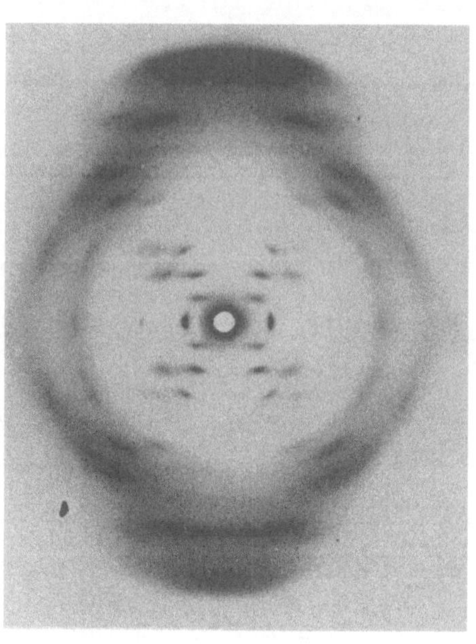

Fig. 1. X-ray diffraction pattern (obtained by J.L. Campos) from a fiber of DNA complexed with arginine and studied under dehydrating conditions (0% relative humidity). Well defined spots are observed over the whole pattern, indicating that the molecules are in crystalline order. A typical B form is observed in an orthorhombic unit cell (a=3.01, b=2.22, c= 3.28 nm).

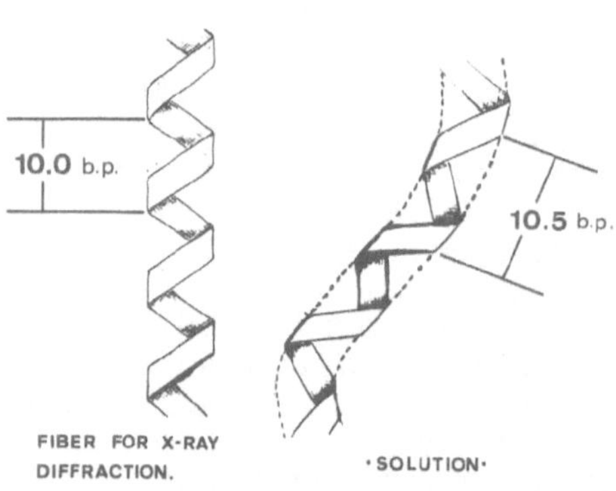

Fig. 2. Diagram to illustrate the different number of base pairs per helical turn in fibers and in solution. In the first case the molecules are maintained straight and as a result this number remains close to 10. In solution, the DNA molecules are allowed to bend, as a result of Brownian motion. This distortion may be accompanied by a relaxation of the B form with an average increase in the number of base pairs per repeat.

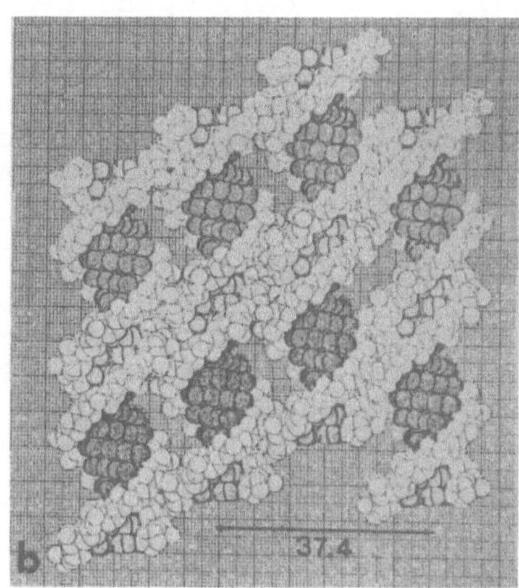

Fig. 3. Different ways of packing DNA in fibers: (a) simple hexagonal packing observed in nucleoprotamines[17]. The distances correspond to a sample studied at 92% relative humidity. This type of packing is compatible with an association of the protamine molecules to either the wide or the narrow groove of DNA. (b) orthorhombic packing in the 110 plane observed in the DNA-arginine complex in the absence of water (pattern shown in Fig. 1). Due to the relative displacement of the molecules, the bulky arginine counterions must be associated with the major groove of DNA, since there is no space available in the minor groove. The same packing conditions prevail in the DNA-polyarginine complex[16], so that polyarginine must also be associated with the major groove of DNA.

seen that due to the absence of water, the molecules are very closely packed and the narrow grooves of DNA are interlocked. Therefore the bulky arginine molecules must be necessarily located in the major grooves of DNA, since this is the only space available for them. The intensity diffracted by such a complex has been calculated[14] and shown to be in agreement with the pattern experimentally obtained (Fig. 1). In a similar way, taking into account the results presented by Suwalsky and Traub[16], it can be shown[14] that polyarginine must also acomodate itself in the major groove of DNA.

2.3 Modification of the B form of DNA by counterions

We have also studied other peptides as counterions of DNA. They are given in Table II. Some of these peptides which contain lysine and alanine, relax the conformation of DNA, so that the number of base pairs per helical turn changes as a function of the water content of the sample[12]. At 0% relative humidity we found 8.5-9 base pairs per helical turn. This value increased upon hydration of the sample, so that at high relative humidities the DNA contained about 10 base pairs per helical turn.

Another class of peptides was found to destabilize the conformation of DNA: the axis of the molecules appeared to be well aligned in the fibers, but no secondary structure could be detected[12]. These peptides contain a basic group of either arginine or lysine surrounded by one or two peptide bonds placed among hydrophobic residues. It appears that the positive charge of these peptides brings their hydrophobic moieties close to the DNA, thus allowing hydrophobic interactions to take place, resulting in partial disorder of the base-pairs in the DNA.

2.4 The influence of proteins

The influence of nuclear proteins on the conformation of DNA has been studied in two cases: the protamines and the histones. These two types of proteins are highly basic and are associated with DNA in the nucleus of the eucaryotic cells, protamines in sperm and histones in somatic nuclei. In the latter case there are additional components which interact with DNA depending on its metabolic state. Furthermore the histones can be chemically modified (acetylated, phosphorylated, etc.) by enzymatic processes in the course of gene activity. It is generally considered that the complex formed by DNA and unmodified histones, without additional components, corresponds to an inactive, repressed state of the genetic material. In this section we will review the conformation of DNA in such a complex, whereas in section 3 we will consider its higher order structure. First we will review the case of protamines.

The protamines are rich in arginine and occur in the sperm of some animal species. The detailed composition and sequence of these proteins does not appear to have a large structural influence since in fibers the DNA has been found[17] to remain in the B form at all humidities irrespective of the type of protamine associated with it. The molecules are packed in a simple hexagonal cell (Fig. 3) in a quasicrystalline manner, although they show screw disorder. Upon dehydration the degree of crystalline order is not completely preserved, although the DNA molecules remain in the B conformation down to 0% relative humidity.

A more complicated case recently studied in our laboratory (Fornells and Subirana, unpublished results) is the conformation of the DNA-protein complex in spermatozoa which contain protamines with cystine bridges, as it is the case in mammals, insects, some octopods, etc. In that case we have concluded that the DNA molecules are also in the B form and occur in bundles stabilized by disulfide bridges. It is interesting to consider the possibility that the overall shape of the spermatozoon, which is very complicated in some cases, might be determined by the distribution of cystine bridges in the protamine molecules. In the case of the octopus *Eledone cirrhosa*, the spermatozoon has the shape of a cork screw[18], a shape which might be determined by a modulation of the DNA conformation induced by the structural constraints imposed by the presence of cystine bridges.

In the case of histones, the native nucleohistone complex contains DNA in the B form[19,20]. However this form of DNA must be modified, since the molecule is bent in order to provide for adequate packing of DNA around the histone core in the nucleosomes. It appears that smooth bending may be sufficient for this purpose, as discussed by Trifonov[21], so that the DNA in nucleohistone can be considered as a slightly distorted B form.

Therefore it can be concluded that DNA is usually present in the B form in the inactive eucaryotic nucleus, both in sperm cells and in repressed chromatin. On the other hand it is likely that the conformational versatility that has been detected in the complexes of DNA with peptides may be instrumental in the regulation of the genetic activity of DNA.

The other side of the interaction between DNA and proteins is the conformation adopted by the latter molecules upon interaction. At present there is no evidence for any general conformational rules. Extended protein chains, β-ribbons and helices have been considered as favoured conformations to interact with DNA, as reviewed by Kim[22], but only on the basis of model building studies. Similar structures have been postulated in the case of the two DNA-binding proteins recently described[1,2]. However, only in the case of polyarginine it has been shown[14] that the polypeptide chain lies in the wide groove of B-DNA in an extended conformation.

2.5 Conclusions

The observations we have discussed raise the question of how rigid is the conformation of DNA. Does it have only a limited number of well defined conformations? Or, alternatively, does its conformation change as a function of the interactions with its environment?. The evidence presently available indicates that, depending on the conditions, DNA may manifest itself as a molecule with a strictly invariable conformation or, alternatively, it can change its conformation smoothly as a function of ambient conditions.

The B form of DNA appears to be a well defined state of DNA, with a high intrinsic stability. It is compatible with most of the DNA sequences studied (table I) and there are a large number of counterions (particularly those which contain arginine, as shown in table II) which stabilize this conformation, irrespective of the water content of the sample. It appears that only a slight distortion of this conformation is allowed upon bending in solution, as shown in Fig. 2.

On the other hand, there are some counterions[12] also shown in table II, which apparently unblock the conformation of DNA so that it may change smoothly its structural parameters as a function of relative humidity. Zimmerman and Pheiffer[23] have also shown that the conformation of DNA can be relaxed upon immersion in some organic solvents and concentrated salt solutions. This unblocked conformation of DNA may be very important in specific protein-DNA interactions, since DNA and protein may adopt complementary structures more easily. An extreme case of this behaviour may be shown by regions of proteins with individual basic residues surrounded by hydrophobic groups. Under these conditions it appears that DNA may become profoundly distorted in order to optimize hydrophobic and electrostatic interactions with its counterions, perhaps allowing it to bend upon itself and show sharp kinks. In fact it is tempting to speculate that a transition from the stable B form to this more relaxed state may be important in the genetic activation of DNA and in the specific interactions of proteins with DNA.

3. THE STRUCTURE OF CHROMATIN

3.1 The nucleosome

The demonstration[24] that nucleohistone could be digested at regularly spaced sites by a nuclease indicated that it had a particulate structure. Soon thereafter, these particles, now called nucleosomes, were visualized by electron microscopy[25] and characterized by physicochemical methods[26]. Kornberg[27] proposed a model, now fully substantiated, suggesting that the nucleosome is made by an octamer of histones surrounded by a fragment of DNA 146 base pairs long. Nucleosomes are conected by pieces of DNA of variable length (30 to 90 base pairs). Fiber diffraction experiments carried out in our laboratory[28] showed that the nucleosome was a flat particle and that the DNA was wound around the histone octamer either in the form of a helix or as the seam in a tennis ball. X-ray diffraction experiments[29] on crystals of nucleosome cores allowed a determination of their exact size (11x11x 6 nm) and showed that the particle was wedge shaped. The arrangement of histones and the way they interact among themselves and with DNA have been intensively studied and the results obtained have been reviewed in detail[30,31]. An account of the general features of nucleosome structure has also appeared recently[32].

3.2 The chromatin fiber

When eucaryotic nuclei are studied by electron microscopy[33], chromatin appears as a bundle of fibers with a diameter of 25-30 nm, as shown in Fig. 4. Unfortunately the nucleosomes present within these fibers are not easily visualized and their spatial arrangement is not yet known. The simplest model[34] suggests that nucleosomes are arranged in a helical or solenoidal fashion,with about six nucleosomes per turn of the helix. However, very often the chromatin fiber shows a beaded appearance under the electron microscope. The latter observations give support to the notion that the chromatin fibers may be constituted

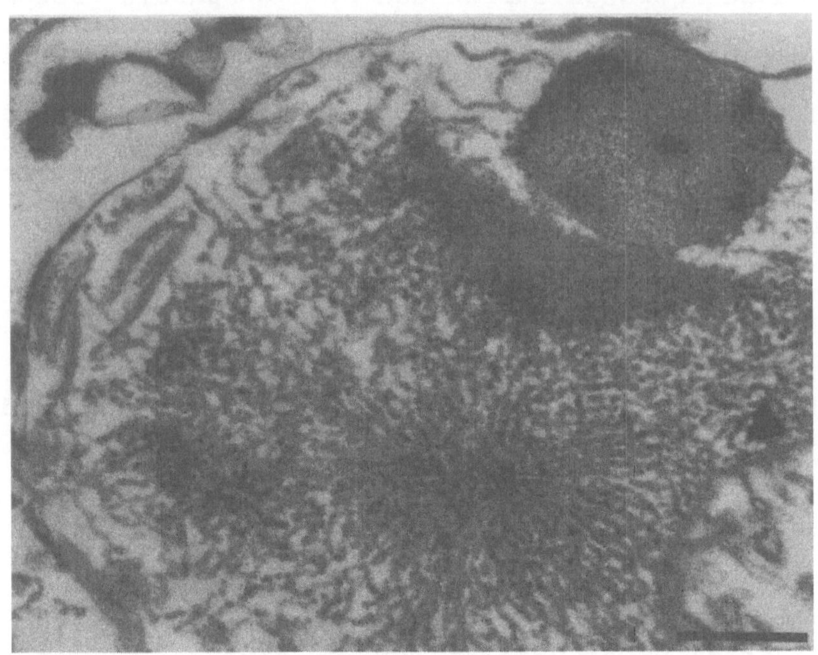

Fig. 4. Part of a sea cucumber spermatozoon which had been subjected to mild hypotonic shock in 0.15M NaCl, 1mM Tris-HCl, 0.1mM $CaCl_2$, pH=8 and was then embedded in araldite. The chromatin fibers conserve their morphology and some of them can be followed easily along considerable lengths. Scale indicates 0.5μm.

by a chain of globular particles ("superbeads"), each of them containing about 17 nucleosomes[35]. In fact the chromatin fiber may appear under the electron microscope as a smooth or a beaded structure depending on the conditions of the experiment, as shown on Fig. 5. These observations indicate that under some circumstances superbeads may be packed very closely with their neighbours giving rise to a chromatin fiber with a smooth outline.

The studies discussed in the previous paragraph were carried out on chromatin samples spread on grids which were then either rotary shadowed or negatively stained. Both methods have inherent limitations and in fact it has not been possible to determine unambiguously the path fol-

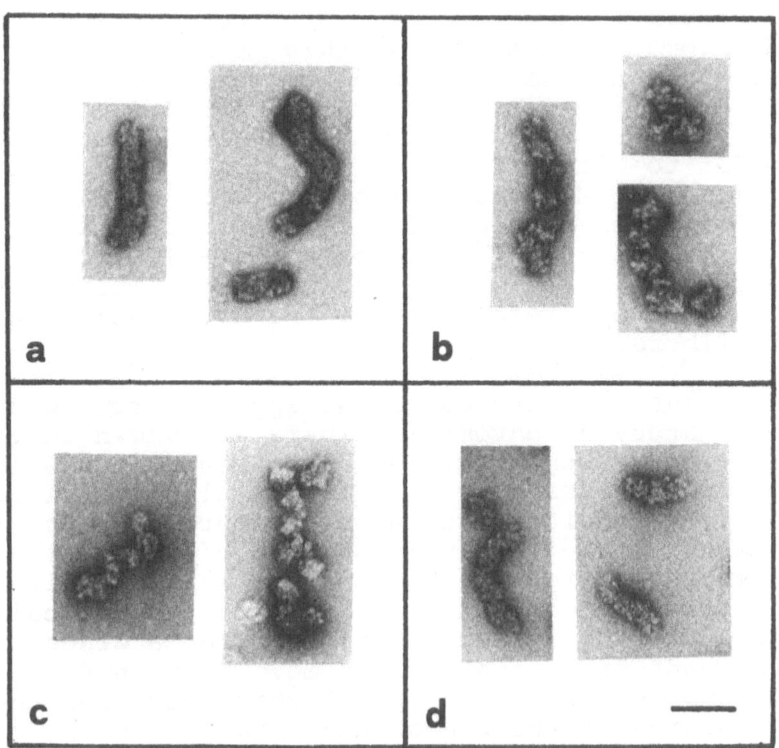

Fig.5. Micrographs of chicken erythrocyte polynucleosomes fixed with 0.1% formaldehyde under different ionic conditions: a) 0.5mM $MgCl_2$, b) 50mM NaCl, c) 50mM NaCl, 0.5mM $MgCl_2$ and d) 100mM NaCl. All these solvents contained 10mM Tris-HCl, pH=7.7. Polynucleosomes were obtained by brief micrococcal nuclease digestion of nuclei, solubilized in 50mM NaCl, 0.2mM EDTA, 10mM Tris-HCl, pH=7.7 and then fractionated by centrifugation in 10%-50% sucrose gradients containing 50mM NaCl, 0.2mM EDTA, 10mM Tris-HCl, pH=7.7. Polynucleosomes were collected from the heavier half of the main sedimentation peak. The chromatin fragments contained about 100 nucleosomes. The histone composition was normal. The different ionic conditions were obtained by dialysis into each buffer. Samples were adsorbed on carbon-coated grids, previously hydrophilyzed by serum albumin treatment and negatively stained with 0.5% uranyl acetate. In a. and d. smooth chromatin fibers are observed, whereas in b. and c. the superbead structure is obvious. These results have been obtained by L.Pérez-Grau and F. Azorín. The bar corresponds to 100 nm.

lowed by the nucleosome chain within a chromatin fiber. For this reason we have recently studied thin sections of embedded chromatin fibers[36]. In this method, air drying of the material is avoided. Instead the water present in the sample is substituted by a polymeric material. This substitution may induce a distortion of the chromatin fibers and in order to eliminate this possibility we have used several embedding materials with different chemical properties. A further problem with this method is that its resolution may be rather low, so that its limiting resolving power usually is in the 3-10 nm range. This fact may prevent an adequate resolution of the nucleosomes in the chromatin fibers.

The essential features of the approach followed by us[36] in the study of chromatin fibers are shown in Fig.6. Chromatin fibers with a straight appearance, are selected for computer processing (Fig. 6A). The electron micrographs are then digitalized, using picture elements (pixels) of about 1 nm^2. A representation of the digitalized image is shown in Fig. 6B. The images are then filtered using a simple method (low pass filter): the optical density of groups of pixels is averaged using a window with the approximate size of a nucleosome. The mean value obtained is represented in the pixel which occupies the central position of the window and this process is carried out over the whole micrograph. As a result a new digitalized micrograph is obtained in which the high resolution information has been eliminated and only features larger than 10 nm can be observed. One of this low pass filtered images is shown in Fig. 6C. It is apparent that the chromatin fiber has regions of higher electron density distributed along its length, thus giving support to the superbead model of chromatin organization. The distance between neighbouring superbeads is not very regular but it always lies in the range 30-40 nm, values similar to those found by other workers[35] and in the fibers shown in Fig. 5. When longer distances are detected, it is obvious that the chromatin fiber has been locally stretched.

Further information on the internal structure of the chromatin fiber is obtained from the high pass filter, shown in Fig. 6D, which is the diference between the original micrograph (Fig. 6B) and its low pass filter (Fig. 6C). This image contains information on the fine detail of the sample. Regions of high optical density which might correspond to individual nucleosomes, are easily detected. Unfortunately, these regions do not show any obvious order, so that it has not yet been possible to determine the spatial organization of nucleosomes within a superbead.

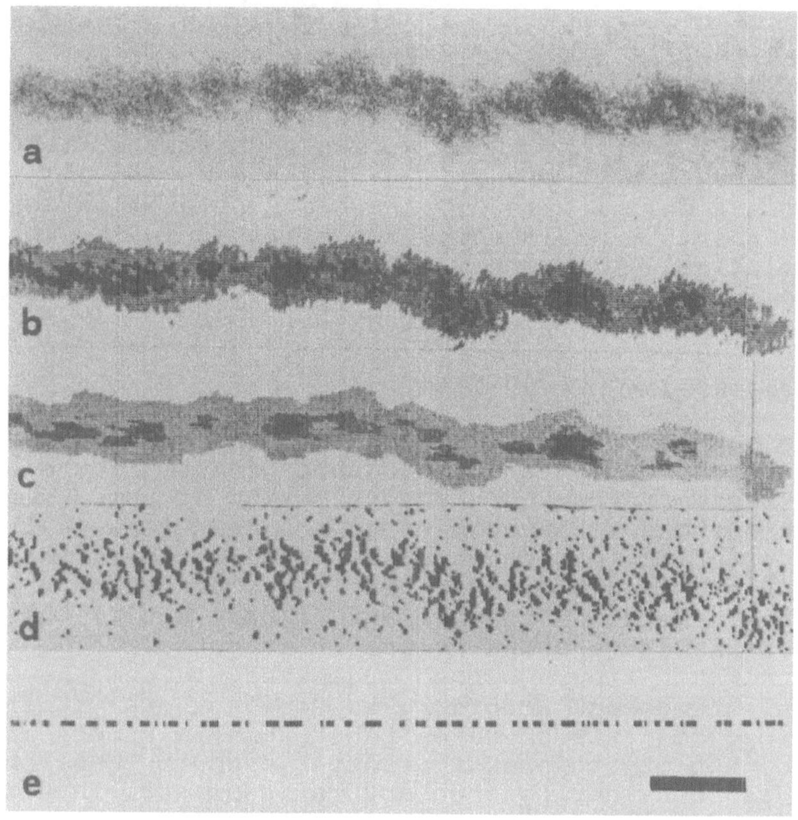

Fig. 6. Computer analysis of a chromatin fiber. The micrograph of a fiber treated with 0.25M sucrose, 0.4mM $CaCl_2$ and embedded in araldite-epon is shown in (a). The digitalized micrograph and its low and high pass filters are shown in b, c and d. In the high pass filter the pixels with a positive contrast greater than 4 optical density units are shown in black. The algebraic sum of every column of pixels in the high pass filter is shown in e. The black lines correspond to the regions in which this sum is positive. Scale indicates 50 nm.

When the high pass filter is projected onto its axis, a peculiar result is obtained, as shown in Fig. 6E. The projected intensity has a periodicity of about 11 nm, which coincides with the size of a nucleosome. It appears that there is some structural feature along a chromatin fiber which repeats itself, with this periodicity.

All these observations indicate that nucleosomes are arranged in a complex manner within a chromatin fiber. Although the results presently available do not allow us to propose a definite model, the simplest explanation would be that the nucleosomes are arranged in the form of a periodically distorted solenoid. These distortions might be induced by the interactions of internucleosomal DNA, perhaps involving histone H1[37].

3.3 The structure of chromosomes

The results we have presented in the previous sections show that we have a reasonable picture of the structural features involved in the formation of complexes of DNA with histones. Although there are many gaps in this picture, it is clear that nucleosomes are the elementary particles of the genetic material which associate themselves to form the 30 nm chromatin fiber. But at this point our knowledge stops, since it is not clear how these chromatin fibers become organized to form the eucaryotic chromosome. According to some authors[38] there is a chromosome axis on which the chromatin fibers are attached, but other investigations[39] dispute these observations and consider that no such axis exists. This is probably the main problem which awaits solution for an understanding of the structure of the genetic material.

REFERENCES

1. D.B.McKay and T.A.Steitz, Structure of catabolite gene activator protein at 2.9 Å resolution suggests binding to left-handed B-DNA, Nature 290:744 (1981)
2. W.F.Anderson, D.H.Ohlendorf, Y.Takeda and B.W. Matthews, Structure of the cro repressor from bacteriophage λ and its interaction with DNA, Nature 290: 754 (1981)
3. A.G.W.Leslie, S.Arnott, R.Chandrasekaran and R.L. Ratliff, Polymorphism of DNA double helices, J.Mol.Biol. 143:49 (1980)
4. D.A.Marvin, M.Spencer, M.H.F.Wilkins and L.D.Hamilton, The molecular configuration of deoxyribonucleic

acid. III. X-ray diffraction study of the C form of the lithium salt, J.Mol.Biol. 3:547 (1961)

5. Y.Mitsui, R.Langridge, B.E.Shortle, C.C.Cantor, R.C. Grant, M.Kodama and R.D.Wells, Physical and Enzymatic studies on poly d(I-C).poly d(I-C), an unusual double-helical DNA, Nature 228:1166 (1970)
6. A.H.J.Wang, G.J.Quigley, F.J.Kolpak, J.L.Crawford, J.H.van Boom, G.van der Marel and A.Rich, Molecular structure of a left-handed double helical DNA fragment at atomic resolution, Nature, 282:680 (1979)
7. E.Sage and M.Leng, Conformational changes of poly (dG-dC).poly(dG-dC) modified by the carcinogen N-acetoxy-N-acetyl-2-aminofluorene, Nucl.Ac.Res. 9:1241 (1981)
8. R.Wing, H.Drew, T.Takano, C.Broka, S.Tanaka, K.Itakura and R.E.Dickerson, Crystal structure analysis of a complete turn of B-DNA, Nature 287:755 (1980)
9. R.P.Millane and G.A.Rodley, Stereochemical details of the side-by-side model for DNA, Nucl.Ac.Res. 9: 1765 (1981)
10. J.A.Subirana, Histones and differentiation, in "Macromolecules Biosynthesis and Function" FEBS Symposium Vol. 21, p.243, ed. by S.Ochoa, C.F. Heredia, C.Asensio and D.Nachmansohn, Academic Press, New York (1970)
11. J.A.Subirana, M.Chiva and R.Mayer, X-ray diffraction studies of complexes of DNA with lysine and with lysine-containing peptides, in "Biomolecular structure, conformation, function and evolution" Vol. 1, p. 431, ed. by R.Srinivasan, Pergamon Press, Oxford & New York (1980)
12. J.L.Campos, J.A.Subirana, J.Aymamí, R.Mayer, E.Giralt and E.Pedroso, The conformational versatility of DNA in the presence of basic peptides, Studia Biophys. 81:3 (1980)
13. J.C.Wang, Helical repeat of DNA in solution, Proc. Natl.Acad.Sci.USA 76:200 (1979)
14. I.Fita, Estructura de los complejos de ADN-(L-arginina), ADN-(poli-L-argininas) y de las nucleoprotaminas, Ph.D.Thesis, Faculty of Sciences, Universidad Autónoma de Barcelona (1981)
15. R.Langridge, H.R.Wilson, C.W.Hooper, M.H.F.Wilkins and L.D.Hamilton, The molecular configuration of deoxyribonucleic acid. I. X-ray diffraction study of a crystalline form of the lithium salt, J.Mol. Biol. 2:19 (1960)
16. M.Suwalsky and W.Traub, A comparative X-ray study of a nucleoprotamine and DNA complexes with polylysine and polyarginine, Biopolymers 11:2223 (1972)

17. P.Suau and J.A.Subirana, X-ray diffraction studies of nucleoprotamine structure, J.Mol.Biol. 117:909(1977)
18. W.L.Maxwell, Spermiogenesis of Eledone cirrhosa lamarck (Cephalopoda, Octopoda), Proc.R.Soc.London B186:181 (1974)
19. S.Bram and H.Ris, On the structure of nucleohistone, J.Mol.Biol. 55:325 (1971)
20. R.Llopis and J.A.Subirana, X-ray diffraction studies of calf thymus nucleohistone, An.Quim. 71:898(1975)
21. E.N.Trifonov, Structure of DNA in chromatin, in "International Cell Biology" ed. by H.Schweiger, Springer Verlag, p. 128 (1980-81)
22. S.H.Kim, Nucleic acids-protein interactions: structural studies by X-ray diffraction and model building in "Biological Recognition and Assembly"p.311 Alan R.Liss,New York (1980)
23. S.B.Zimmerman and B.H.Pheiffer, Does DNA adopt the C form in concentrated salt solutions or in organic solvent/water mixtures? An X-ray diffraction study of DNA fibers immersed in various media, J.Mol.Biol. 142:315 (1980)
24. D.R.Hewish and A.L.Burgoyne, Chromatin substructure. The digestion of chromatin at regularly spaced sites by a nuclear deoxyribonuclease, Biochem.Biophys. Res.Commun. 52:504 (1973)
25. A.L. Olins and D.F. Olins, Spheroid chromatin units (ν-bodies), Science N.Y. 183:330 (1974)
26. R.Rill and K.E.Van Holde, Properties of nuclease-resistant fragments of calf thymus chromatin. J.Biol. Chem. 248:1080 (1973)
27. R.D.Kornberg, Chromatin structure, a repeating unit of histones and DNA, Science N.Y. 184: 868 (1974)
28. J.A.Subirana and A.B.Martínez, Model studies of chromatin structure based on X-ray diffraction data, Nucl.Ac.Res. 3:3025 (1976)
29. J.T.Finch, L.C.Lutter, D.Rhodes, R.S.Brown, B.Rushton M.Levitt and A.Klug, Structure of nucleosome core particles of chromatin, Nature Lond.269:29 (1977)
30. A.D.Mirzabekov, Nucleosomes structure and its dynamic transitions, Quart.Rev.Biophys. 13:255 (1980)
31. J.D.McGhee and G.Felsenfeld, Nucleosome structure, Ann.Rev.Biochem. 49:1115 (1980)
32. R.Kornberg and A.Klug, The nucleosome, Scient.Amer. February(1981)
33. H.Ris, Chromosomal structure as seen by electron microscopy, in "The Structure and Function of Chromatin" Ciba Foundation Symposium 28, p.7, Elsevier Excerpta Medica North Holland, Amsterdam (1975)
34. F.Thoma, Th.Koller and A.Klug, Involvement of histone H1 in the organization of the nucleosome and of the

salt-dependent superstructures of chromatin, J.Cell Biol. 83:403 (1979)

35. G.F.Meyer and M.Renz, Native and reconstituted chromosome fiber fragments, Chromosoma 75:177 (1979)
36. J.A.Subirana, S.Muñoz-Guerra, A.B.Martínez, L.Pérez-Grau, X.Marcet and I.Fita, The subunit structure of chromatin fibers, Chromosoma (in press)
37. F.Azorín, A.B.Martínez and J.A.Subirana, Organization of nucleosomes and spacer DNA in chromatin fibers, Int.J.Biol.Macromol. 2:81 (1980)
38. M.P.F.Marsden and U.K.Laemmli, Metaphase chromosome structure: evidence for a radial loop model, Cell 17:849 (1979)
39. A.M.Mullinger and R.T.Johnson, The organization of supercoiled DNA from human chromosomes, J.Cell Sci. 38:369 (1979)

THE STRUCURE AND ASSEMBLY OF SIMPLE VIRUSES

Kenneth C. Holmes

Max-Planck-Institut für medizinische Forschung

69 Heidelberg, Germany

INTRODUCTION

Viruses occur with widely differing degrees of structural complexity, ranging from the viroids, which consist only of RNA, to complex bacteriophages which can code for a hundred gene products. The structure and assembly of the large bacteriophages has been studied with great effect by means of electron microscopy and genetics, particularly by the use of temperature sensitive mutants. At present there is not much information about this class of viruses at the atomic level although the assembly processes are already well described[1]. To study structure and function at the atomic level we must turn to the simpler viruses. A place of honour in the pantheon of structural molecular biology has been reserved for the plant viruses, which consist of a single strand of RNA enclosed in a protein coat. Such objects are highly symmetrical on account of the economy of design. The coats are built up by the repetitive use of one type of protein unit. Two classes of object result: the rod viruses and the so-called spherical viruses, the symmetry of which provides a nice example of the expression of platonic form at the macromolecular level.

Of the rod viruses by far the best understood is tobacco mosaic virus (TMV). The assembly of TMV turns out to be a rather complex procedure calling for extensive polymorphism of the coat protein. A second rod-virus about which we have information at the near atomic level is the simple rod-shaped bacteriophage Pf1. The assembly of this virus is of especial interest since it appears to take place in the bacterial membrane.

The pseudo-spherical viruses have the symmetry of the icosahedron which allows the assembly of 60 equivalent subunits to form an icosahedral shell. In fact nearly all such viruses consist of 60n subunits where n can be 3 or more. These can be explained and classified by the introduction of quasi-symmetrical icosahedral shells (Caspar and Klug[2]). The best resolved structures in this class are tomato bushy stunt virus (TBSV) and southern bean mosaic virus (SBMV).

TOBACCO MOSAIC VIRUS

TMV particles are rod-like, 300 nm long and 18 nm diameter. TMV consists of 2140 protein subunits, each of molecular weight 17,420 Daltons (158 residues), arranged on a helix of pitch 2.3 nm with 16 1/3 subunits per turn. Winding through this helix is a single strand of RNA 6400 nucleotides long. TMV has a central hole of diameter 4.0 nm.

The pioneering work of Bernal and Fankuchen[3] on orientated gels of TMV, showed that the rod-like virus particles formed a highly regular two-dimensional hexagonal array in which the axes of the virus particles were parallel. The ordered virus gel was paracrystalline rather than truly crystalline since the arrangement of virus particles was not regular along the direction of the orientation axis.

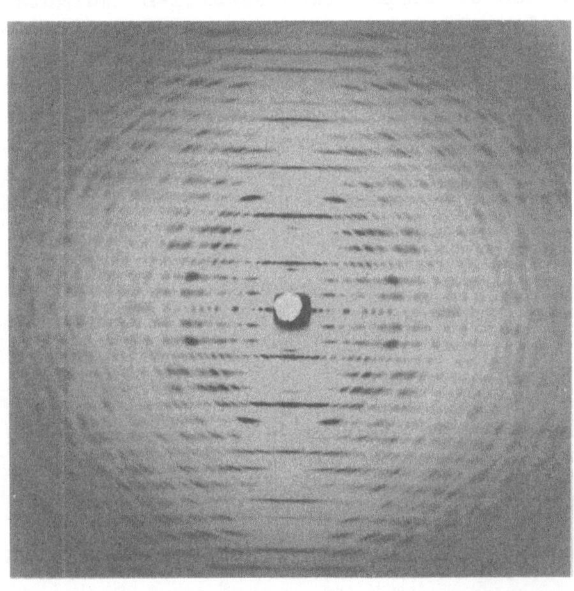

Fig. 1 X-ray diffraction pattern from an orientated gel of TMV. Note the 6.9 nm system of layer lines, every third layer line has near-meridional intensity.

The diffraction pattern obtained from these gels (Fig.1) consisted of a set of layer-lines at right angles to the particle axis with a spacing of 6.9 nm, thus showing that there was a regular repeat along the axis. Bernal and Fankuchen were able to show that the diffraction pattern consisted of two distinct regions: on the equator at spacing of less than 10 nm one or two reflexions occurred which arose from the inter-particle ordering (the positions of these reflexions varied with the concentration of virus); all other parts of the diffraction were virtually unaffected by changes in concentration, and could therefore only arise from the intramolecular structure of each particle. Bernal and Fankuchen further suggested that each TMV particle was more like a microcrystalline array of protein molecules than like a giant molecule, a concept which has proved to be of considerable significance. Bernal and Fankuchen attempted to index the TMV fibre diffraction pattern in terms of a rhombohedral unit cell having a three-fold screw axis. The connection between the postulate of a three fold screw axis and the helical form given by later work is that the trigonal screw axis generates equivalent planes in differing orientations at heights of 1/3 c 2/3 c etc.. This situation is approximated to at low resolution by a helical structure of small pitch which repeats in three turns. Only when the theoretical elucidation of the diffraction to be expected from a helix[4] had been carried out was it possible to proceed with the analysis of the diffraction of TMV gels. Using the theory of Cochran, Crick, and Vand, J.D. Watson[5] was able to show that the structure of TMV was helical with 3n + 1 subunits every 6.9 nm. Watson's best value for n was 10. Independently of each other R.E. Franklin[6] and D.L.D. Caspar[7] took up the study of the structure of the orientated gels. Franklin was soon to be joined by A. Klug. After Franklin's untimely death in 1958 Klug took over the leadership of the group which was at that time located in Birkbeck College, London under the aegis of Bernal. In 1962 the virus group moved to the MRC Laboratory for Molecular Biology in Cambridge. It is noteworthy that, through all the present day virus crystallography, one can trace the visionary influence of Bernal and Fankuchen.

Low Resolution Analysis of the Fibre Diffraction Pattern

Franklin started out with the ambitious resolve to determine the structure of TMV by means of the then newly discovered method of isomorphous replacement. It was quickly established that the virus had 49 subunits in three turns[8], that the virus was hollow[7], and that the nucleic acid was located at a radius of 4.0 nm[6]. These results were obtained by analysis of the low resolution part of the zero layer-line (< 1.0 nm^{-1}) where it can be shown that the data are effectively the diffraction pattern from a centrosymmetric structure (i.e. the phase problem reduces to a sign problem).

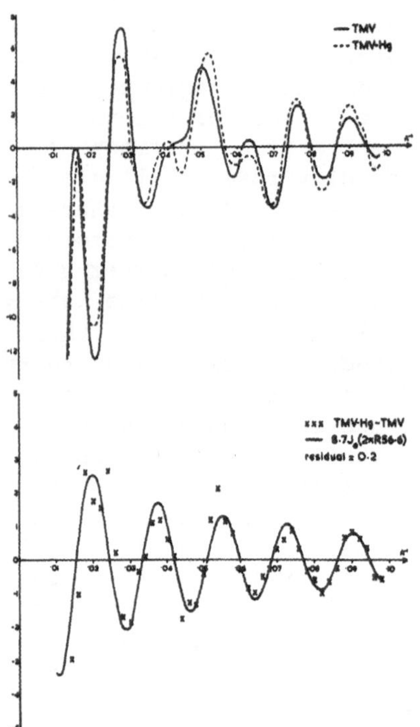

Fig.2 The scattering amlitude on the zero layer line of TMV (full) and TMV + methyl mercury (broken).
The difference gives the scattering amplitude of the mercury corresponding to atoms at a radius of 5.6 nm.

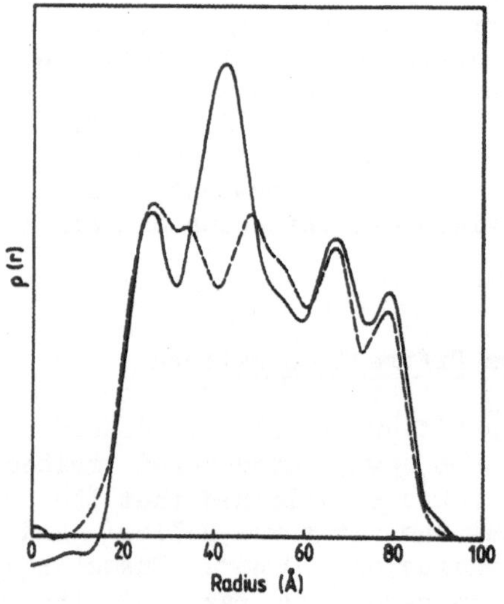

Fig.3 The radial density distribution, computed by a Fourier-Bessel transformation of the zero layer line amplitudes, for TMV (full) and reagregated TMV protein (broken).
The difference shows the radial position of the RNA.

The radial density distribution can be obtained by a Fourier-Bessel inversion of the amplitudes derived from the diffracted intensity by means of the method of isomorphous replacement (Figs. 2 and 3). The signs of the peaks on the inner part of the zero layer-line were determined independently by Caspar[7] and by Franklin and Holmes[8]. Together with Klug, Franklin showed that the virus surface is marked by a deep helical groove[9]. Using two heavy atom derivatives, methylmercury[10] and osmic acid, she proceeded to analyse the third layer-line in order to arrive at a low resolution helical projection of the virus. This phase of the work, which is summarised in D.L.D. Caspar's drawing (Fig.4), was interrupted by the premature death of Rosalind Franklin. With the establishment of Klug's group in Cambridge the necessary technical milieu became available to extend Franklin's early work. Klug and Holmes continued the structural analysis of the orientated gels by means of the method of isomorphous replacement. In particular, in an unpublished study Klug was able to show that the RNA was single stranded at a time when it was widely supposed, on the basis of end-group analyses, that the RNA was in many segments.

Fig.4 Drawing (D.L.D. Caspar) showing about 1/20th of the length of the TMV particle. 49-protein subunits repeat every 3 turns of the virus in 6.9 nm. A single strand of RNA winds between the subunits at a radius of 4.0 nm.

The Crystallography of the Double Disk

Microcrystals of protein disks (see below) were first reported by Macleod et al[11]. Reproducible crystals were produced by Finch et al[12]. Furthermore, they were able to demonstrate that the symmetry of the disk was 17-fold. This work opened the way to obtaining an atomic model of TMV by means of x-ray crystallography. The disk consists of two rings of protein subunits, each with the same polarity and each containing 17 subunits. The disks crystallise in the orthorhombic space group $P22_12_1$. The molecular weight of the asymmetric unit is 17 x 2 x 17,400 = 592,280 Daltons. The solution of a structure of this size presents formidable problems. To solve the structure of the disk to 0.28 nm resolution it was necessary for Bloomer et al[13] to measure the intensities of 2×10^6 reflexions. The analysis at 0.5 nm[14] showed the entire polypeptide chain except for a section within 4.0 nm radius which was not visible. This observation reflects the fact that in the disk (but not in the virus) there is a flexible segment of about 25 residues length. The high resolution structure[13] allowed all the residues to be fitted into the map except for the flexible segments between 89-113 and the c-terminal residues 155-158.

The preparation of heavy atom derivatives for the solution of the phase problem presented major difficulties so that much of the analysis was carried out with a single mercury derivative[13,14]. Using the formulation of Crowther, Jack[15] was able to demonstrate the power of the non-crystallogaphic 17-fold symmetry as a supplement to the heavy atom method for the solution of the phase problem. By setting up the equivalent formulation in real space Bricogne[16] made it possible to use this method for the high resolution analysis of the structure of the disk[13]. The methodology of Jack and Bricogne has also proved invaluable for the solution of the structures of the spherical viruses.

High Resolution Studies of the Fibre Diffraction Pattern

Parallel to this work the analysis of the x-ray fibre diffraction progressed. Here it soon became clear that the limiting factor would be the availability of well defined single heavy atom derivatives of the virus. Both genetic and chemical modifications to the virus were used[17,18,19]. Initially, single heavy atom derivatives were favoured because the location of the heavy atoms in three dimensions from fibre diffraction data proved difficult and only the simplest cases were soluble[20]. The determination of the structure of TMV from the x-ray fibre diagrams by means of the method of isomorphous replacement is a unique problem. The data consist essentially of the cylindrical average of the square of the Fourier transform of a single particle. Thus the data at each point in reciprocal space consist of the sum of squares of a number of Bessel function terms, the actual number of Bessel functions which contribute being determined by the helical symmetry of the particle and the reciprocal space radius of the point in question. At low radii (small scattering angles) only one Bessel function contributes to the intensity. Theoretical studies by Klug et al[21] had shown that in principle it is possible to regenerate the three dimensional structure if the problem of "separating the Bessel functions" can be solved. Franklin and Klug realised that, to a resolution of about 1.0 nm, on account of the high symmetry of TMV (49-fold), only one Bessel function could contribute to the intensity. As a consequence it can be shown that to this resolution the intensities are cylindrically symmetrical so that no information is lost by cylindrical averaging. Using this idea Barrett et al[22] were able to calculate a three dimensional electron density map of the virus with a nominal resolution of 1.0 nm. This map showed the general appearance of the nucleic acid running circuferentially at 4.0 nm radius, and the overall appearance of the subunit including strong radial features which were later shown to be alpha-helices. The structure of the virus extends proximally to a radius of 2.2 nm, in sharp contrast to the situation in the disk.

Holmes[23] had shown that it is possible to separate Bessel functions by means of heavy atom derivatives. The problem of separating Bessel functions by means of isomorphous replacement is a multidimensional analogue of the phase problem and can formally be solved by a multidimensional Harker construction. Locating the best intersection of the Harker hyperspheres by a multidimensional search for every point in reciprocal space is a very time consuming computing problem. An algebraic solution was developed by Stubbs and Diamond[24]. Using this method Holmes et al[25] analysed the data out to a resolution of 0.67 nm, which is the two-Bessel-function limit. The resulting map showed some of the polypeptide chain, in particular the two radial alpha-helices which are such a prominent feature of the structure of the TMV subunit. It allowed a preliminary chain tracing (the 0.5 nm resolution map of the protein disk showed subsequently that the alpha helices had been correctly identified but not correctly linked together). Further work admitting three Bessel functions (Stubbs, Warren, and Holmes[26]) yielded an electron density map at 0.4 nm resolution by the use of 6 heavy atom derivatives, which taken together with the detailed map of the subunit in the disk produced by Bloomer et al[13] yields a considerable amount of stereochemical information about the nature of the protein-RNA interaction in TMV assembly.

Polymorphism of TMV Coat Protein

The isolated nucleoprotein particle can readily be taken apart[27] and reassembled[28]. In addition, a number of polymorphs of the coat protein have been described[29] (Fig.5). Moreover, Butler and Klug[30] were able to show that the disk is the main precursor for the initiation and growth of the virus.

The major polymorphic forms of the protein <u>without</u> nucleic acid are :

the A-protein (predominantly a trimer - "4S");
the disk (containing two rings of 17 subunits - "20S");
a rod-like structure made from stacked disks;
a helical virus-like structure (16 1/3 subunits per turn);
a helical virus-like structure (17 1/3 subunits per turn[31]).

At neutral pH and low ionic strength the disk is the majority species. By raising the salt concentration the formation of limited stacks of disks is encouraged. The stacked disk may apparently be trapped by mild proteolytic cleavage of the flexible segment in a form which can no longer be taken apart[32]. On lowering the pH the disks build a virus-like helix even in the absence of nucleic acid. This is thought to come about through the formation of an intermediate "lockwasher"[29]. The lockwasher may form by dislocating along either of two inter-subunit boundaries to form a 16 1/3 or 17 1/3 helix[31]. In the presence of RNA only the 16 1/3 form is found.

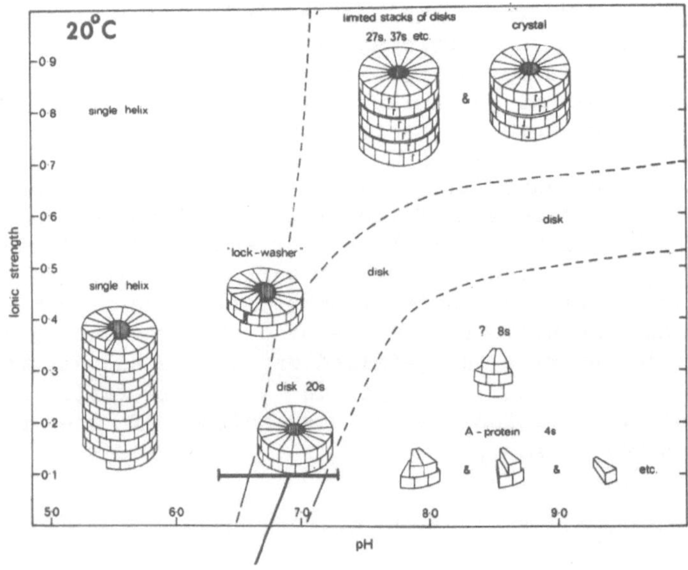

Fig.5 The polymorphic forms of TMV protein and the ranges of ionic strength and pH where each is favoured.

A remarkable feature of all these polymorphs is the conservative nature of the side-to-side bonding between the subunits, which scarcely alters. In contrast, the up-and-down bonding between subunits is markedly different between the disk forms and the helical forms[14] ((Fig 6). The subunits are on the average 0.3 nm further apart in the disk than in the helix and are displaced 1.0 nm circumferentially.

Assembly of the Virus

During the assembly of the virus the coat protein has to specifically recognise the viral nucleic acid and at the same time has to be able to tolerate the variation of sequence found when encapsulating the RNA. Butler and Klug[30] showed by kinetic analysis that the disk is the major species responsible for coating the RNA during *in vitro* assembly. Assembly starts by the insertion of a specific RNA loop (the initiation loop) into the disk (Butler et al[33]). A specific sequence of the RNA (the assembly origin) binds to the disk (Zimmern and Butler[34], Zimmern[35], Jonard et al[36]) thereby mediating the formation of the lockwasher and initiating the growth of the helix. The assembly origin of the RNA is about 100 nucleotides long and the initial binding is thought to occur between the turns of the disk. Close to the center of the assembly origin is the sequence AGAAGAAGUUGUUGAUGA. In this and the surrounding sequences there is a strong tendency for every third residue to be G. The assembly origin has a very low C content and occurs about 15% of the length away from the 3′ end. Growth takes place fastest in the 5′ direction[37,38]. Growth in the 3′ direction goes on at the same time but rather more slowly and possibly by a different mechanism[38]. Recent kinetic studies by Schuster et al[39] show that *elongation* as distinct from initiation does not exclusively depend upon the presence of either 4S or 20S (double disks) aggregates. However, they agree with Butler and Klug that there are one or more regions of TMV-RNA at least 1-1.5 kilobases in extent which incorporate 20S protein exclusively. As growth proceeds towards the 5′hydroxyl end the 5′ end is dragged through the central hole of the nascent virus (Fig. 7) so that both the 3′ and 5′ ends are initially at the same end of the virus[40,41].

The Structure of the Disk

The general appearance of the subunit, as deduced from the high resolution studies of Bloomer et al[13], is shown in Fig. 8 . Both the C and N termini are located distally. Starting from the N terminus the chain builds a short helical structure between residues 8 and 15. Then follows a short section of extended chain (16-18) which forms part of the distal beta-pleated sheet.

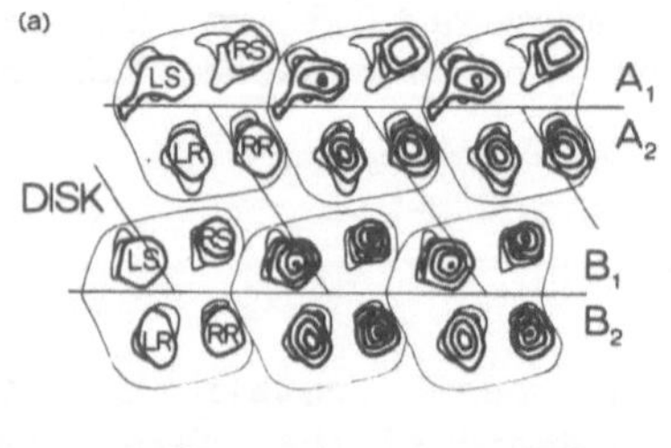

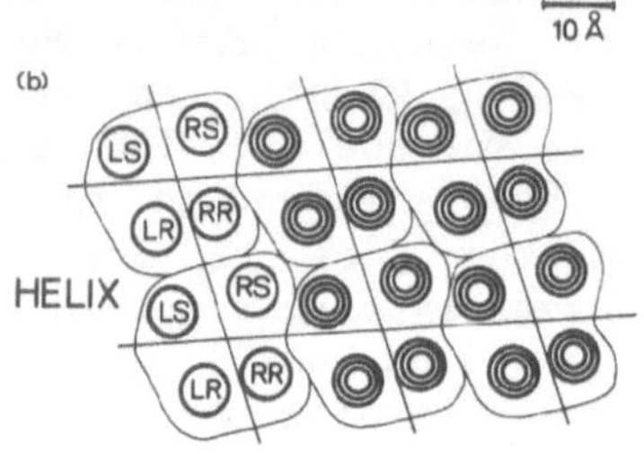

Fig.6 The relationship between the subunits in the disk form (upper) and helix (lower).
Tangential sections at about 6.0 nm radius are shown where the TMV subunit consists entirely of 4 roughly parallel alpha helices.
Note that the up-down contacts are different whereas the side-to-side contacts are the same.

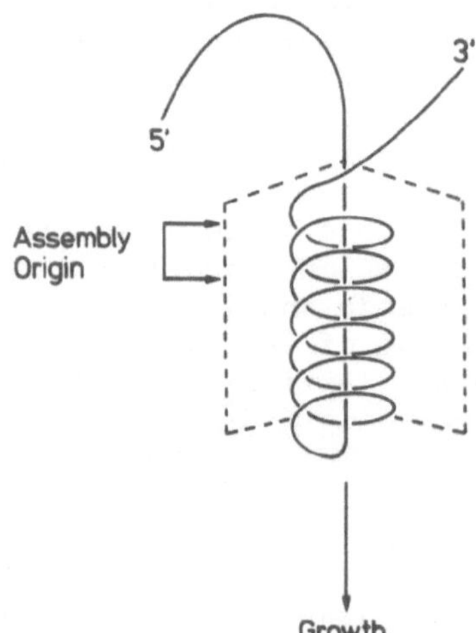

Fig.7 TMV assembly is initiated by insertion of a specific loop of RNA, which is near the 3'end, into a disk. On binding the RNA the disk disloca es into a 2-turn helix. Growth proceeds in the direction of the 5'end by dragging the RNA through the hole in the middle of the virus so that both the 3' and 5' ends are at the same end of the nacent virus.

The chain continues into the LS (left-slew) helix (20-32) which runs proximally to a radius of 5.0 nm. After a tight turn (33-38) the chain proceeds distally along the RS (right-slew) helix (38-48) which is parallel to the LS helix. The chain now forms part of the beta sheet. At 74 the RR (right-radial) helix starts. This helix, which is tilted about 20° to the horizontal, runs proximally and up to a radius of about 4.5 nm where the density fades out. Then follows a flexible segment containing 24 residues which has been shown by NMR investigations[42] to be highly mobile. After the flexible segment (89-113) the LR (left-radial) helix runs distally roughly horizontally from a radius of 4.0 nm to 7.0 nm (114-134). Then the chain runs through a short alpha helical segment to the C-terminus. The last four residues (155-158) are not visible and are in a random-coil. The four main alpha helices include about 60 residues out of the total of 158. The helices are packed as in a segment of a left-handed, 4-stranded supercoiled bundle with hydrophobic contacts along the center (Fig. 9). The distal ends of the four helices are connected transversely by a strip of beta sheet. Residues 53-54, 16-18, 68-71, 137-139, and 135-136 are involved. In the central region of each subunit on the distal side of the beta-sheet is a cluster of aromatic residues which gives rise to a continuous belt of hydrophobic interactions encircling each ring of the disk.

The Structure of the Virus

The structure of the subunit in the virus in substantially the same as the disk except for the flexible loop and the vertical intersubunit contacts. Proximally from 4.0 nm radius, where the density in the disk fades out, one sees the LR and RR helices continuing to a low radius (2.5 nm) (Fig. 10) where they are joined together by a strong vertical column of density (the V-column) which was identified by Stubbs et al[26] as an alpha-helix although more recent studies show this identification to be somewhat uncertain. In particular, the number of residues involved is less than was proposed by Stubbs et al. In addition to this density, which can unequivocally be ascribed to the protein, there is a ribbon of density running roughly circumferentially at a radius of 4.0 nm <u>underneath</u> the LR helix which can be ascribed to the nucleic acid. Studies by Mandelkow et al[43] of the structure of the helical protein aggregate <u>without</u> RNA substantiate the assignment of this density to the RNA.

A diagram of the RNA conformation as refined by Stubbs and Stauffacher[44] is shown in Fig. 11. Each protein subunit binds three bases. Two bases appear to be in the anti conformation and one in the syn. The sugar puckers are all C3′ endo. The direction of the nucleic acid agrees with that independently determined by Wilson et al[45].

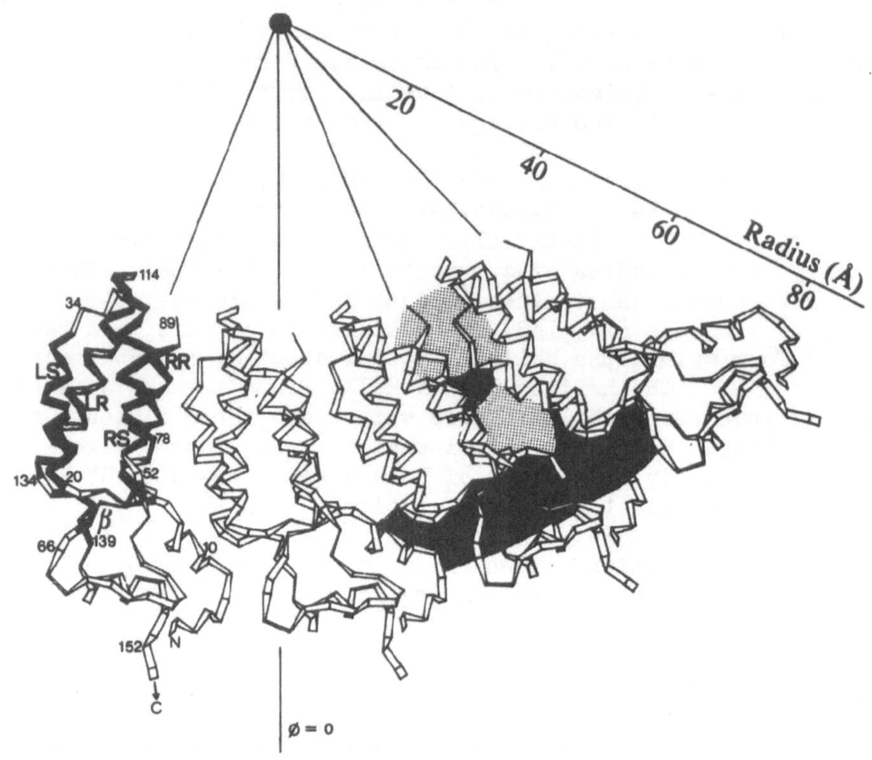

Fig.8 Four subunits of the A ring viewed from above[13]. Note the near-parallel alpha helices LR, LS, RR and RS. The subunits combine to form a hydrophobic ring between 70-80 Å radius. Below 40 Å radius the polypeptide chain is disordered and not visible.

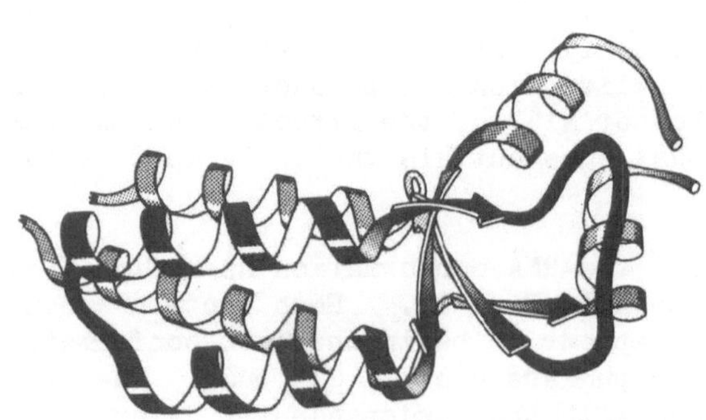

Fig.9 The TMV subunit (Jane Richardson).

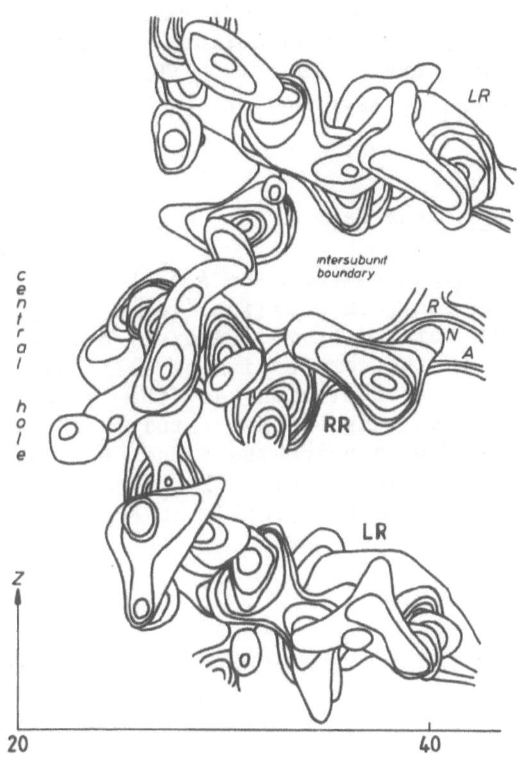

Fig.10 The electron density in the virus[26] viewed from the side. The particle axis is on the left and runs vertically. Note the density comprising the extensions of the LR and RR helices to low radius and note the V-column. This density is not visible in the disk structures.

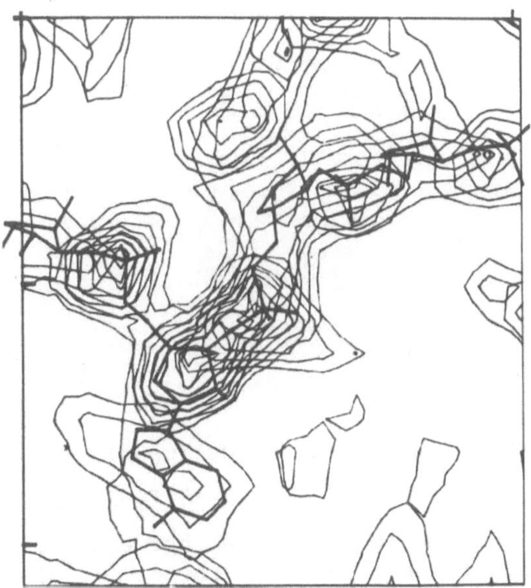

Fig.11 The RNA in TMV refined against the electron density[44] by a least squares procedure Two bases, which point up out of the plane of the drawing have been omitted for clarity. The mean radius of the sugar-phosphate chain is 40 Å.

The bases form a claw-like structure round the LR helix (see below). Two bases are at a radius of about 3.8 nm and one base is at a larger radius (4.5 nm). The phosphates are clustered around a radius of 4.0 nm and appear to bond to the adjoining RS and RR helices from the neighbouring subunit in the next turn down. Thus the RNA binding site lies between the rings of a disk, in agreement with the model for assembly proposed by Butler et al[33].

The Relationship between the Disk and Virus Maps

Bloomer has analysed the relationship between the A-disk and the B-disk and the virus and has shown that, in addition to a small radial movement, the subunits must be rotated by 10° (A-disk) or 20° (B-disk) about an axis perpendicular to the common disk or helix axis in order to bring them into the same orientation as the virus. Recently Bricogne and Bloomer, in an unpublished study, have determined the transformation necessary to bring the A and B disk density into coincidence with the virus density by a least squares proceedure. The relationships are shown diagramatically in Fig. 12.

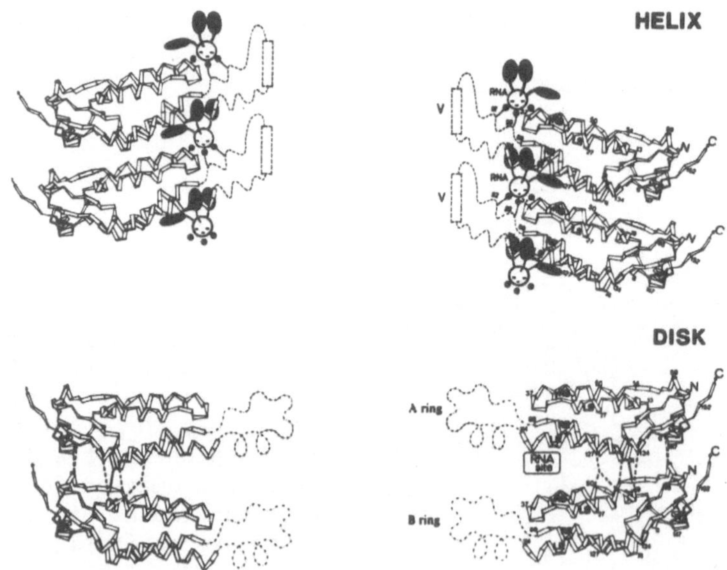

Fig. 12 Vertical section through two turns of the virus (upper) and the double disk (lower). Note that the virus extends to a lower radius and that the subunits are tilted. The RNA bases are shown as black ellipses around the LR helix. The RNA binding site lies between two turns of the helix which corresponds to the "jaws" of the double disk.

The RNA Binding Site

If we apply Bricogne and Bloomer's transformation to the LR helix we find that it amounts to a translation of about 0.32 nm in the radial direction in the neighbourhood of the nucleic acid. Applying this transformation to the LR helix in the A-ring of the disk results in the relationship between the LR helix and RNA shown in Figs. 12 and 13. Comparing the resulting structure with the virus electron density indicates that minor revisions will be necessary to take account of local distortions. We find (Fig. 11) that the transformed disk structure is one turn further out (3.6 residues) than the Stubbs Warren and Holmes model. A similar shift has already been proposed by Bloomer et al[13].

Having adjusted the virus model in this way some interesting features of the RNA binding site reveal themselves: an important component of the binding site seems to be provided by the invariant pair of aspartate residues (115,116). These may be able to hydrogen bond to the RNA ribose 2′ OH on residues 1 and 3. Such a ribose binding site is typical of alcohol dehydrogenase, lactate dehydrogenase, and malate dehydrogenase[46,47,48]; here this binding motif may be used twice per protein subunit. The accompanying bases find themselves in hydrophobic pockets close up against the alpha helix, as was suggested by Stubbs et al[26].

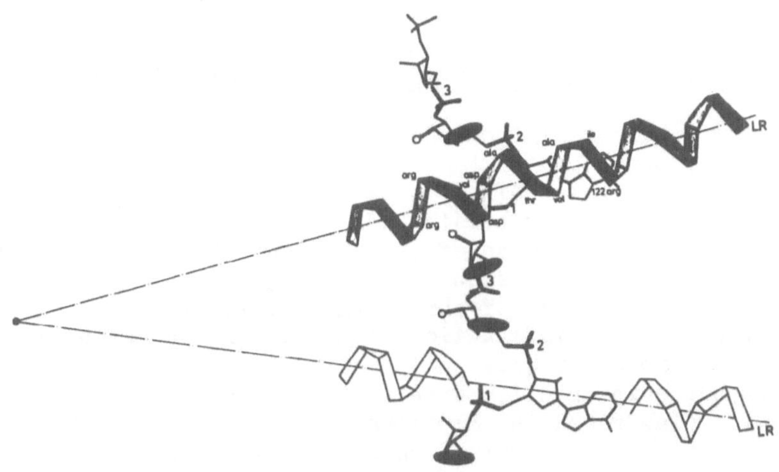

Fig.13 The RNA binding site arrived at by combining the results of refs 13, 26 and 44. The bases form a claw-like structure round the LR helix. The two bases rising vertically out of the plane of the diagram are represented by black ellipses. The middle base extends to a large radius.

For base 1 the methylene chain of Arg 113 appears to be an important component of the hydrophobic site. The other component of the site is the α-carbon of Asp 115. The guanidinium group of Arg 113 might also be involved in a salt bridge to a phosphate group. Base 3 appears to bind against Val 114 and Ala 117. Base 2 is at a larger radius and may form a hydrogen bond to Ser 123. The sugar of residue 2 seems to lie face down against a hydrophobic surface formed by residues 119 (Val) and 120 (Ala). If the 2′ OH is able to form a hydrogen bond, it must be to the neighbouring slewed hairpin joining the LS or RS helices from the subunit underneath, possibly from Asn 33.

The effect of the tight binding of residues 1 and 3 to the LR helix is to pucker the RNA so that the phosphate residues are grouped together, thereby making them a target for the formation of salt bridges with arginines. Stubbs et al[26] have suggested the salt bridges are formed with arginine residues 90 and 92 from the RR helix of the subunit directly underneath and possibly with residue 41 from the RS helix of this subunit. The assignment of the invariant residues 90 and 92 to the phosphate binding site is plausible but higher resolution studies of the virus stucture are needed before such an assignment can be proven. Unfortunately, this part of the RR helix cannot be seen in the disk structure[13]. Arg 41 is not a strong candidate for the phosphate binding site since it lies on the wrong face of the RS helix.

A Hypothetical Trigger Mechanism for the Disk-helix Transition

The suprising result from the studies summarised above is that the major component of the RNA binding site in TMV is one alpha helix, the LR helix, which binds three bases. In the disk structure, which is the template for binding RNA, this helix continues as a finger proximally to a radius of about 4.0 nm (residue 114) before the electron density peters out. This finger could nucleate the binding of the RNA by means of three kinds of interaction:

A stereospecific interaction of the aspartate groups 115,116 with two of the ribose groups;

A hydrophobic interaction with the three bases forcing them into the shape of a claw round the alpha helix;

Specific hydrogen bonding which is still to be determined. Present studies suggest a possible hydrogen bond between Ser 123 and base 2 which would favour a hydrogen bond acceptor in this position.

The binding is accompanied by the formation of salt bridges between Arg 90 and 92 from the neighbouring subunit one turn deeper in the virus helix and the phosphate groups of the nucleic acid. Initially 90 and 92 are part of the random-coil segment adjoining the RR helix but apparently through the binding to the phosphate groups this structure is stabilised and as a result the V-column and the rest of the LR and RR helices which are not seen in the disk can build. However, the V-columns from adjoining subunits are in close contact and it seems probable that they can only fit together in one way, namely as in the virus helix. The change in relative heights in passing from the disk to the helix is 0.14 nm per subunit and it is conceivable that the packing of the V-columns could not tolerate this much distortion. As a result, therefore, of the packing of the V-columns the disk structure may be strained and may attempt to transform into the helix. The contact between the A and B rings of the disk is mediated by a complex system of hydrogen bonds which must break when the helix is formed. The disk-helix transition is therefore reminiscent of the R-T transformation in hemoglobin (Durham and Klug[49]). The disk is the low affinity form for RNA because one cannot make the salt bridges to the phosphates without allowing the V-column to build thereby destabilising the disk. The helix is the tight binding form. However, to reach the helix it is necessary to break the network of hydrogen bonds between the A and B rings of the disk. The balance between these effects determines the binding of the RNA and makes the RNA binding highly cooperative. Perhaps relevant in this context Lomonossoff and Butler[50] report that the incorporation of RNA during *in vitro* reassembly is quantised, the quantum of binding being close to 50 bases or 100 bases. This is essentially the number of bases in one or two turns which would seem to show that binding proceeds through the addition of single or double rings thereby supporting the idea that the disk is the major precursor for virus growth. Through such cooperativity one could also explain the specificity of the binding. Two properties of the assembly origin seem noteworthy: the requirement that every third residue should be G; and the requirement that the sequence should be long. If we postulate a binding mechanism whereby the selectivity for any one site is small but where the *simultaneous* binding to 16 sites produces the disc-helix transition then one sees at once why the assembly origin can be so specifically recognised.

Stubbs et al[26] have pointed out that the inner structure of the virus (the V-column and the adjoining pieces of LR and RR helix which build on binding RNA) is destabilised by a high concentration of carboxyl groups in this region which would probably explain the anomalous pKs appearing during the disk-helix transition[50]. A possible function of the energetically unfavourable carboxylate-carboxylate interactions may be to help keep part of the RNA binding site in a random-coil configuration until the RNA binds.

SPHERICAL VIRUSES

As was first pointed out by Crick and Watson[51] the maximum number of objects which can be assembled on an approximately spherical surface with equivalent environments is 60. The symmetry of such an arrangement is icosahedral (Fig. 14a). This arrangement is found in the simple plant virus satellite tobacco necrosis virus the detailed structure of which is presently being determined[52]. In this virus 60 protein molecules are regularly arranged on the suface of an icosahedron.

However, most virus capsids contain 60T subunits where T may be 3.4 or more. It is in fact not possible to arrange these so that the environments of all subunits are strictly identical. In order to resolve this dilemma Caspar and Klug[2] introduced the idea of quasi-equivalence. Caspar and Klug showed that by replacing each object on the icosahedral surface by a group of (for example) 3 symmetrically disposed objects one could produce pseudo-spherical surface lattices consisting of 180 or more objects where each of the objects was quasi-symmetrically related to its neighbours. Quasi-symmetry implied that local contacts at any one radius from the centre of the icosahedron could be preserved while distortion along a radius vector would occur.

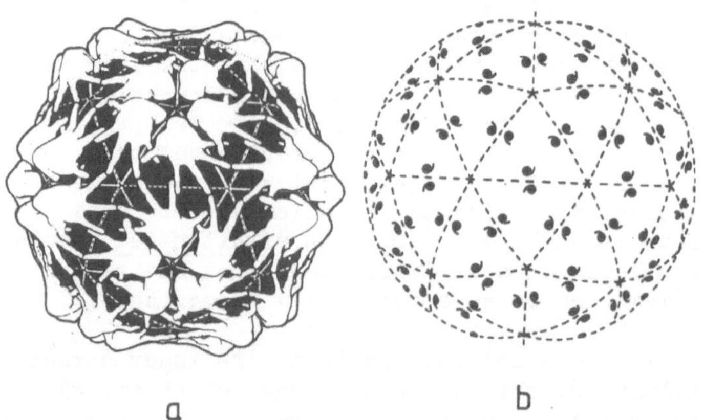

Fig.14a A set of objects arranged with icosahedral symmetry.
Fig.14b Each object has been replaced by a group of three objects related by a local 3-fold axis. The local relationships are the same around 5-fold and pseudo 6-fold axis but the curvature of the surface is different.

An example of a quasi-equivalent surface lattice is shown in Fig. 14b. In this example the idea of quasi but not strict equivalence is illustrated by pondering the nature of the 2-fold axes. The 2-fold axes fall into two classes: those related by a 5-fold axis; and those related by a 3-fold axis which has a local 6-fold character. In other words, in going round a 3-fold axis one encounters alternately two kinds of 2-fold axes:

i)those which are part of a 5-fold ring (quasi-dyads, later called q2);

ii)those connecting two different 5-fold rings (strict dyads, later called s2).

This classification of the 2-fold axes has proved important in understanding the structure of tomato bushy stunt virus. The protein subunits are required to be able to tolerate the bending distortions which arise through having to belong to both classes. To accomodate the distortions hinged domains and other machinery have evolved.

Tomato Bushy Stunt Virus

TBSV is a pseudo-spherical virus which consists of a protein coat enclosing an RNA core. TBSV has an outside radius of about 18.0 nm and a molecular weight of 9×10^6 Daltons. The protein shell is about 3.0 nm thick with protuberences at each of the 2-fold and pseudo 2-fold axes. The molecular weight of the subunit is ca. 40,000 Daltons. The RNA forms a shell between radii of 8.0 and 10.0 nm. The organisation of the RNA in TBSV has not been determined because it has a lower symmetry than that of the coat. The coat (capsid) also exists in an expanded form which may play a role in the assembly of the virus.

In 1938 Bernal, Fanchuchen, and Riley[53] took the first X-ray photographs of TBSV from crystals prepared by Bawden and Pirie, in the form of isotropic rhombic dodecahedra, diameter about 0.01 mm. The photograph was necessarily a powder photograph, though of wet crystals, and showed only two lines. Immediately Bernal measured these lines, and knowing from the form of the crystals that the space group was most likely cubic, he deduced that the lattice was cubic body centred with a cell edge of 39.0 nm, each unit cell containing two particles of diameter ca. 34.0 nm and molecular weight 10^7. Later Pirie grew large crystals and Carlisle and Dornberger took single crystal photographs of them at Birkbeck College which demonstrated that the cubic symmetry persisted to atomic dimensions. Somewhat later the study of the structure was taken up by Caspar and was continued by S.C. Harrison. In 1978 Harrison[54] and his coworkers solved the structure of TBSV to 0.29 nm resolution using two heavy atom derivatives and making use of non-crystallographic symmetry averaging.

The non-crystallographic asymmetric unit consists of three subunits each of which has a different environment (labeled A,B, and C in Fig. 15). The AB contacts occur at quasi-dyads (q2), the CC contacts occur at true dyads (s2). Whereas there are only minor differences between AB and BA contacts (i.e. the quasi dyad is a good local dyad), the CC contacts differ markedly from the A/B contacts. The structure of each subunit is seen to consist of two domains (P and S) joined by a hinge (h). In addition there is a long N-terminal arm (a). In A/B contacts the hinge angle between P and S domains is 20° larger than for CC contacts. Thus around the s2 axes the S domain is "up" and around the q2 axes the S domain is "down". When the S domain is up, part of the N-terminal arm folds into the cavity so created. Across an s2 dyad the surface of the virus is fairly flat, whereas across a q2 dyad there is a marked dihedral angle (about 40°).

The P domain (Fig. 16) is composed of two entirely antiparallel beta sheets, one of six strands, one of four. The six stranded sheet lies adjacent to a 2-fold axis and therefore packs against a similar sheet in the related subunit. The curvature of the four stranded sheet is such that towards the particle centre it lies at some distance from the six stranded sheet.

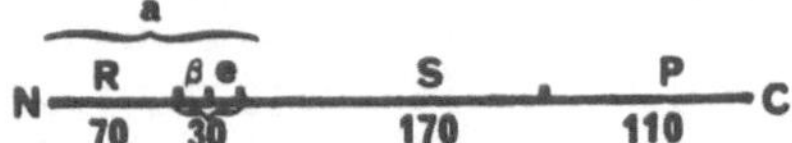

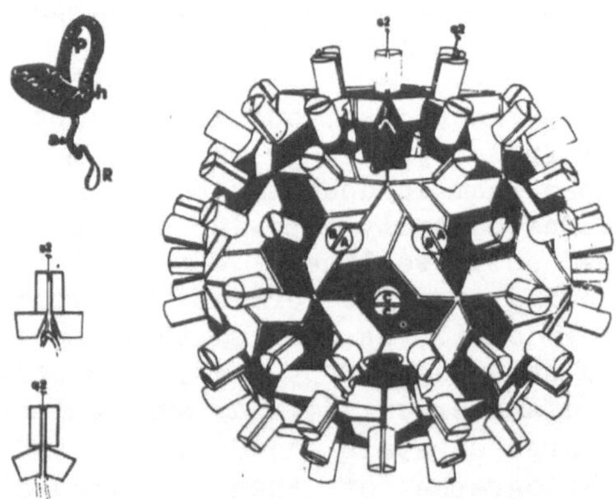

Fig.15 Packing of protein subunits in TBSV[54]. Two kinds of dyad result: local (A/B)-q2 and true (CC)-s2. The two kinds of subunit produced differ in the angle between the domains S and P which is linked by the hinge h. An N-terminal arm (a) extends to low radius. Above is shown the relationship between the domains and the polypeptide chain.

The S domain (Fig. 16) is also largely beta sheet. It is shaped roughly like a triangular prism; a bent four stranded sheet on the lower surface forms one side of the prism; a second four stranded sheet forms the other side. The opposite face and the upper surface are formed by loops and two short helical segments. The loops and helices seem to contain those residues most critically involved in variable intersubunit contacts. The details of the intersubunit bonding have been discussed by Harrison[55].

The inter-domain hinge consists of just two or three residues, with the backbone well ordered in each of the two observed configurations. The consequence of the hinge is that contacts between different domains of the same subunit are <u>not</u> conserved.

A remarkable aspect of the TBSV subunit is the configuration of its N-terminal portion. In the C type contacts the chain, on leaving

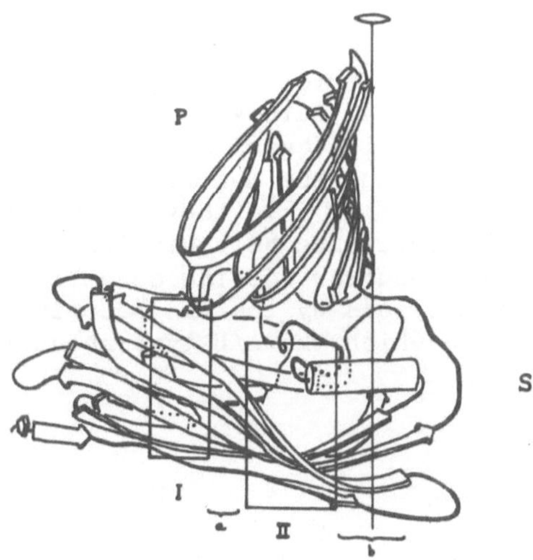

Fig. 16 The TBSV subunit consists of two domains (S and P), a hinge and an N-terminal arm.

the S domain going towards the N-terminus, executes a reverse turn and folds along the inner edge of the domain next to the 2-fold axis. Extending towards the particle 3-fold axis, in the cleft between the adjacent S domains, the chain makes a number of contacts with the neighbouring (B-type) subunit. On reaching the region of the 3-fold axis it winds anti-clockwise around the axis forming a roughly 240° loop involving 18 residues (Fig. 17). In so doing it intertwines with the chains from the symmetry related neighbours and the 54 residues form together an annular structure around the 3-fold axis. Harrison has called this structure the "beta-annulus" (in the A/B type subunits the chain comprising the beta-annulus is disordered). From this point the chain extends inward so that the N-terminus is at a low radius (Fig. 18). This part of the chain cannot be seen in the crystal structure because it is disordered. Its presence is revealed by low angle X-ray and neutron scattering from solutions[56,57]. In the case of the C-subunit about 60 residues are involved (80-90 for the A/B case). The N-terminus probably forms a folded domain on the inside of the RNA (Fig. 18).

The RNA cannot be seen in the crystal structure. This situation could arise from static disorder of from the RNA being very mobile. NMR studies[58] show that the RNA is not highly mobile so one concludes that the lack of visibility arises from static disorder. The RNA is apparently not packed with icosahedral symmetry. The same seems to be true of the N-terminal arm which is presumably intimately involved in the interaction with the RNA. In the presence of chelating agents the virus swells, whereupon the virus RNA shows considerable mobility[58].

An interesting point not forseen in the original theories of quasiequivalence is the special nature of the bonding between C-type subunits around the true 3-fold (quasi 6-fold) axis. If the C-type subunits are considered alone (the S-domains of the C-subunits are shown coloured black in Fig. 15), they form an open cage-like icosahedral structure containing 60 equivalently related subunits which might act as a precursor for the assembly of the virus[54]. Possibly the assembly proceeds by the RNA becoming associated with this open structure onto which the A/B-type subunits are added as a maturation step.

Southern Bean Mosaic Virus

The overall structure of SBMV is rather like TBSV. The structure has been determined to a resolution of 0.28 nm by Rossmann and coworkers[59] from rhombohedral crystals. The R32 space group

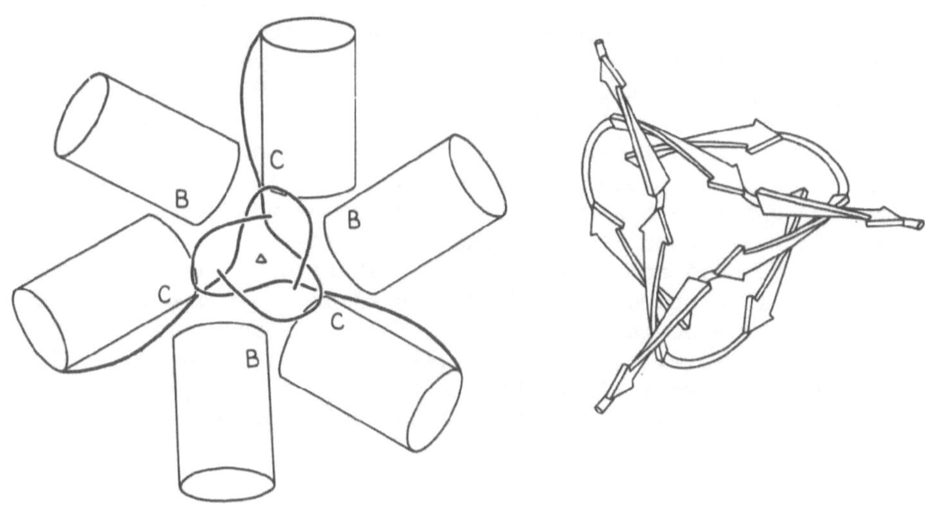

Fig. 17 The N-terminal arms of the C subunits form an interdigitating structure around the 3-fold (pseudo 6-fold) axis which is known as a β-barrel[54] (on right). In the case of the A/B subunits this part of the chain is disordered. After the β-barrel the chain extends to low radius.

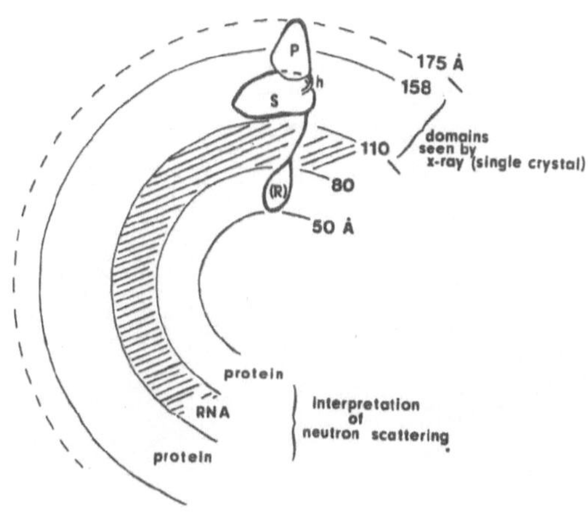

Fig. 18 The radial positions[55] of the protein domains and RNA in TBSV as deduced from low angle X-ray scattering and neutron diffraction.

requires the particle to be on a site with 32 symmetry thus putting ten icosahedral units (A,B,C trimers) in the asymmetric unit. The final structure was determined by the method of isomorphous replacement using 4 heavy atom derivatives. Averaging over non-crystallographic symmetry was also employed. The structure of a subunit is shown in Fig. 19. The S domain is very similar to that of TBSV but the P domain is very attenuated. In place of this,one has abbreviated protrusions around the quasi 3-fold axes (shown as Q3 in Fig. 20) which form a keratin-like stucture when taken with the two neighbouring subunits. The quasi 3-fold axes assume an importance they did not have in TBSV. There are also protrusions around the 5-fold and quasi 6-fold axes which are not seen in TBSV. The beta-annulus associated with the 3-fold axis seems to be present in a modified form. The N-terminal arm associated with C-type subunits is much more visible than in the case of TBSV and is seen to consist of an alpha-helix and an extended polypeptide chain.

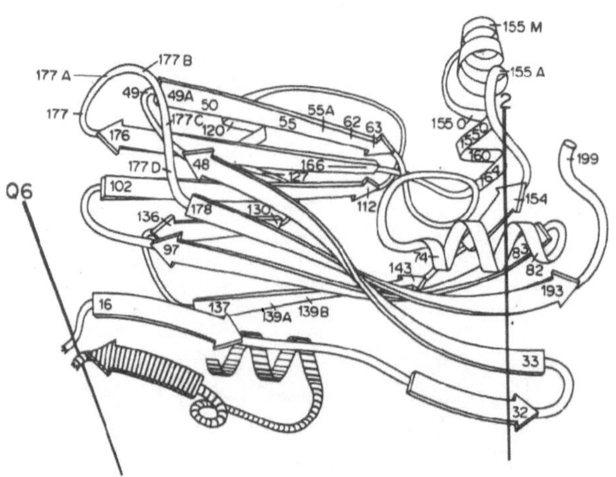

Fig.19 The subunit of SBMV[59] (Jane Richardson). Note the similarity to the S domain of TBSV.

THE BACTERIOPHAGE PF1

In conclusion we report briefly on the structure of this filamentous bacteriophage the structure of which has been determined to 0.7 nm from fibre diffraction patterns by the use of direct methods by Makowski, Caspar, and Marvin[60]. The basis for this study has been laid by the research of Marvin and his collaborators on the structure of filamentous bacteriophages over the past 10 years (e.g. Marvin and Wachtel[61]). The phage consists basically of a protein tube containing DNA[61]. The DNA is circular and runs up and down within the tube. The bases appear to be stacked but not base-paired. The coat protein is largely alpha helical, each subunit consisting of two short sections of helix. The arrangement is shown in Fig. 21. There are two possible ways of joining up the helical segments to make a molecule which are shown in Fig. 22. These are the hair-pin model and the extended model and lead respectively to anti-parallel and parallel alpha helices. Bearing in mind that the protein is first located in the plasma membrane and coats the DNA as it passes through[61] it seems that the hair-pin structure is the more likely since this maximises the dipole-dipole interaction[62] of the the short alpha-helical segments in a non-polar environment.

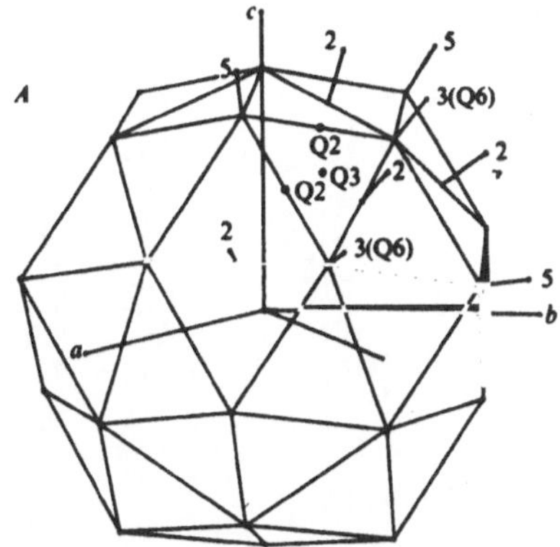

Fig.20 The relationship between crystallographic and icosahedral symmetry axes in SBMV.
Note the pseudo 3-fold axes (q3) which assume considerable morphological significance in this virus.

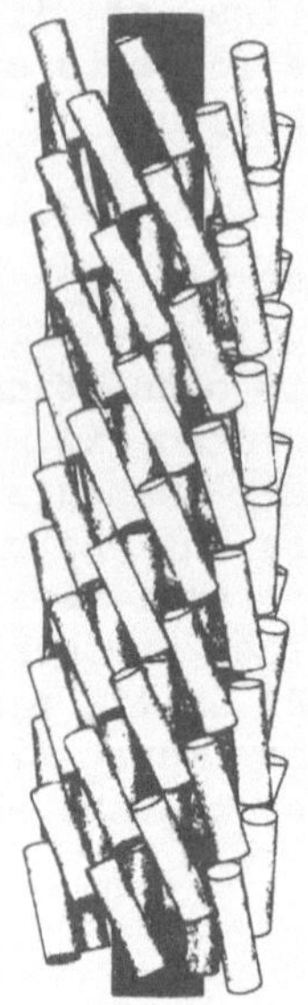

Fig.21 Diagram of Pf1 phage[60]. The DNA is located centrally. The coat protein, which is largely α-helical forms roughly two radial shells, of short α-helical segments. The outside diameter of the particle is about 6 nm. There are 27 subunits arranged on 5 turns of a 15 Å pitch helix.

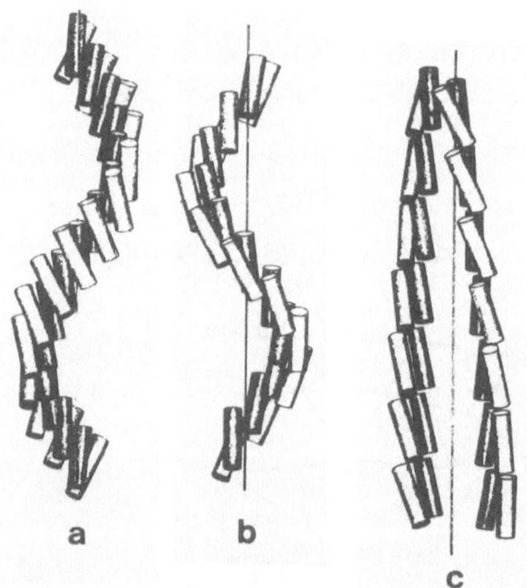

Fig.22a,b,c show the various ways in which pairs α-helical segments may be joined together to form either anti-parallel hair-pin structures (a,b) on extended parallel structures (c).

REFERENCES

1. Y. Kichuchi and J. King, Assembly of Bacteriophage T-4, J Supramol Struct. 3:24 (1975).
2. D. L. D. Caspar and A. Klug, Physical Principles in the Construction of small Viruses, Cold Spring Harb. Symp. quant. Biol. 27:1 (1962).
3. J. D. Bernal and I. Fankuchen, X-ray and Crystallographic Studies of Plant Virus Preparations, J. Gen. Physiol. 25: 111 (1941).
4. W. Cochran, F. H. C. Crick, and V. Vand, The Structure of Synthetic Polypeptides. I. The Transform of Atoms on a Helix, Acta Cryst 5:581 (1952).
5. J. D. Watson, The Structure of Tobacco Mosaic Virus. I. X-ray Evidence of a Helical Arrangement of Subunits around the Longitudinal Axis. Biochim. Biophys. Acta 13:10 (1954).
6. R. E. Franklin, Structure of Tobacco Mosaic Virus: Location of the Ribonucleic acid in the TMV Particle, Nature (Lond.) 177:929 (1956).
7. D. L. D. Caspar, Structure of Tobacco Mosaic Virus: Radial Density Distribution in the TMV Particle, Nature (Lond.) 177:928 (1956).
8. R. E. Franklin and K. C. Holmes, Tobacco Mosaic Virus: Application of Method of Isomorphous Replacement to the Determination of the Helical Parameters and Radial Density Distribution, Acta Cryst. 11:213 (1958).
9. R. E. Franklin and A. Klug, The Nature of the Helical Groove on the Tobacco Mosaic Virus Particle, Biochim. Biophys. Acta 19:403 (1956).
10. H. Fraenkel-Conrat, in:"Sulphur in Proteins," R. Benesch and R. E. Benesch, ed., Academic Press, New York (1959).
11. R. Macleod, G. J. Hills, and R. Markham, Formation of true 3-dimensional Crystals of the Tobacco Mosaic Virus Protein, Nature (Lond.) 200:932 (1963).
12. J. T. Finch, R. Leberman, Y-S Chang, and A. Klug, Rotational Symmetry of the Two-turn Disk Aggregate of Tobacco Mosaic Virus Protein, Nature (Lond.) 212:349 (1966).
13. A. C. Bloomer, J. N. Champness, G. Bricogne, R. Staden, and A. Klug, Protein Disk of Tobacco Mosaic Virus at 2.8 Å Resolution showing the Interactions within and between the Subunits, Nature (Lond.) 276:362 (1978).
14. J. N. Champness, A. C. Bloomer, G. Bricogne, P. J. G. Butler, and A. Klug, The Structure of the Protein Disk of Tobacco Mosaic Virus to 5 Å Resolution, Nature (Lond.) 259:20 (1976).
15. A. Jack, Direct Determination of X-ray Phases for Tobacco Mosaic Virus Protein using Non-crystallographic Symmetry, Acta Cryst. A29:545 (1973).

16. G. Bricogne, Methods and Programs for Direct Space Exploitation of Geometric Redundancies, Acta Cryst. A32:832 (1976).
17. H. G. Wittmann, Proteinanlysen von Chemisch Induzierten Mutanten des Tabacmosaikvirus, Z. Vererbungslehre 95:333 (1964).
18. R. N. Perham and J. O. Thomas, Reaction of Tobacco Mosaic Virus with a Thiol-containing Imidoester and a possible Application to X-ray Diffraction Analysis, J. Mol. Biol. 62:415 (1971).
19. U. Gallwitz, L. King, and R. N. Perham, Preparation of an Isomorphous Heavy-atom Derivative of Tobacco Mosaic Virus by Chemical Modification with 4-Sulpho-phenylisothiocyanate, J. Mol. Biol. 87:257 (1974).
20. R. E. Franklin and A. Klug, The Splitting of the Layer-lines in X-ray Fibre Diagrams of Helical Structures: Application to Tobacco Mosaic Virus, Acta Cryst. 8:777 (1955).
21. A. Klug, F. H. C. Crick, and H. W. Wyckoff, Diffraction by Helical Structures, Acta Cryst 11:199 (1958).
22. A. N Barrett, J. Barrington Leigh, K. C. Holmes, R. Leberman, E. Mandelkow, P, von Sengbusch, and A. Klug, An Electron Density Map of Tobacco Mosaic Virus at 10 Å Resolution, Cold Spring Harb. Symp. quant. Biol. 36:433 (1971).
23. K. C. Holmes, X-ray Diffraction Studies on Tobacco Mosaic Virus and related Substances, Ph.D. Thesis, Univ. London (1959).
24. G. J. Stubbs and R. Diamond, The Phase Problem for Cylindrically Averaged Diffraction Patterns. Solution by Isomorphous Replacement and Application to Tobacco Mosaic Virus, a Least Squares Procedure for Minimising the Error in Phases and Bessel-function Term Ratios, Acta Cryst. A31:709 (1975).
25. K. C. Holmes, G. J. Stubbs, E. Mandelkow, U. Gallwitz, Structure of Tobacco Mosaic Virus at 6.7 Å Resolution, Nature (Lond.) 254:192 (1975).
26. G. J. Stubbs, S. G. Warren, and K. C. Holmes, Structure of the RNA and RNA Binding Site in Tobacco Mosaic Virus from 4-Å Map calculated from X-ray Fibre Diagrams, Nature (Lond.) 267: 216 (1977).
27. G. Schramm, Über die Spaltung des Tabakmosaikvirus und die Wiedervereinigung der Spaltstücke zu höhermolekularen Proteinen, Z. Naturforsch. 2b:112 and 249 (1947).
28. H. Fraenkel-Conrat and R. C. Williams, Reconstitution of Active Tobacco Mosaic Virus from its Inactive Protein and Nucleic Acid Components, Proc. Natl. Acad. Sci. U.S. 41:690 (1955).
29. A. Klug and A. C. H. Durham, The Disk of TMV Protein and its Relation to the Helical and Other Modes of Aggregation, Cold Spring Harb. Symp. quant. Biol. 36:449 (1971).
30. P. J. G. Butler and A. Klug, Assembly of the Particle of Tobacco Mosaic Virus from RNA and Disks of Protein, Nature New Biol. 229:47 (1971).
31. E. Mandelkow, K. C. Holmes, and U. Gallwitz, A New Helical Aggregate of Tobacco Mosaic Virus Protein, J. Mol. Biol. 102:265 (1976).

32. A. C. H. Durham, The Cause of Irreversible Polymerisation of Tobacco Mosaic Virus Protein, FEBS Letters 25:147 (1972).
33. P. J. G. Butler, A. C. Bloomer, G. Brigogne, J. N. Champness, J. Graham, H. Guilley, A. Klug, and D. Zimmern, Tobacco Mosaic Virus Assembly - Specificity and the Transition in Protein Structure During RNA Packaging, in:"Proceedings of the Third John Innes Symposium," R. Markham and R. Horne, eds., North-Holland-Elsevier, Amsterdam (1976).
34. D. Zimmern and P. J. G. Butler, The Isolation of Tobacco Mosaic Virus RNA Fragments Containing the Origin for Viral Assembly, Cell 11:455 (1977).
35. D. Zimmern, The Nucleotide Sequence at the Origin for Assembly on Tobacco Mosaic Virus RNA, Cell 11:463 (1977).
36. G. Jonard, K. E. Richards, H. Guilley, and L. Hirth, Sequence from the Assembly Nucleation Region of TMV RNA, Cell 11: 483 (1977).
37. D. Zimmern and T. M .A. Wilson, Location of the Origin for Viral Reassembly on Tobacco Mosaic Virus RNA and its Relation to Stable Fragment, FEBS Letters 71:294 (1976).
38. G. P Lomonossoff and P. J. G. Butler, Location and Encapsidation of the Coat Protein Cistron of Tobacco Mosaic Virus, Eur. J. Biochem. 93:157 (1979).
39. T. M. Schuster, R. B. Scheele, M. L. Adams, S. J. Shire, J. J. Steckert, and M. Potschka, Studies on the Mechanism of Assembly of Tobacco Mosaic Virus, Biophys. J. 32:313 (1980).
40. P. J. G. Butler, J. T. Finch, and D. Zimmern, Configuration of Tobacco Mosaic Virus RNA during Virus Assembly, Nature (Lond.) 265:217 (1977).
41. G. Lebeurier, A. Nicolaiff, and K. E. Richards, Inside-out Model for the Self-assembly of Tobacco Mosaic Virus, Proc. Natl. Acad. Sci. U.S. 74:149 (1977).
42. O. Jardetsky, K. Akasaka, D. Vogel, S. Morris, and K. C. Holmes, Unusual Segment Flexibility in a Region of Tobacco Mosaic Virus Coat Protein, Nature (Lond.) 273:564 (1978).
43. E. Mandelkow, G. J. Stubbs, and S. G. Warren, Structures of the Helical Aggregates of Tobacco Mosaic Virus Protein, J. Mol. Biol. (submitted for publication).
44. G. Stubbs and C. Stauffacher, Protein-RNA Interactions in Tobacco Mosaic Virus, Biophys. J. 32:244 (1980).
45. T. M .A. Wilson, R. N. Perham, J. T. Finch, and P. J. G. Butler, Polarity of the RNA in the Tobacco Mosaic Virus Particle and the Direction of the Protein Stripping in Sodium Dodecyl Sulphate, FEBS Letters 64:285 (1976).
46. C-I Brandén, H. Eklund, B. Nordström, T. Boiwe, G. Soderlund, E. Zeppezauer, F. Ohlson, and Å. Åkeson, Structure of Liver Alcohol Dehydrogenase at 2.9 Å Resolution, Proc. Natl. Acad. Sci. USA 70:2439 (1973).

47. I. E. Smiley, R. Koekoeck, M. J. Adams, and M. G. Rossmann, The 5 Å Resolution Structure of an Abortive Ternary Complex of Lactate Dehydrogenase and its Comparison with the Apo-enzyme, J. Mol. Biol. 55:467 (1971).
48. E. Hill, D. Tsernoglou, L. Webb, and L. J. Banaszak, Polypeptide Conformation of Cytoplasmic Malate Dehydrogenase from an Electron Density Map at 3.0 Å Resolution, J. Mol. Biol. 72: 577 (1972).
49. A. C. H. Durham and A. Klug, Polymerisation of Tobacco Mosaic Virus Protein and its Control, Nature New Biol. 229:42 (1971).
50. P. J. G. Butler and G. P.Lomonossoff, Quantized Incorporation of RNA During Assembly of Tobacco Mosaic Virus from Protein Disks, J. Mol. Biol. 126:877 (1978).
51. F. H. C. Crick and J. D. Watson, Structure of Small Viruses, Nature (Lond.) 177:473 (1956).
52. T. Unge, L. Liljas, B. Strandberg, I. Vaara, K. K. Kannan, K. Fridborg, C. E. Nordman, and P. J. Lentz Jr, Satellite Tobacco Necrosis Virus Structure at 4.0 Å Resolution, Nature (Lond.) 285:373 (1980).
53. J. D. Bernal, I. Fankuchen, and D. P. Riley, Structure of the Crystals of Tomato Bushy Stunt Virus Preparation, Nature (Lond.) 142:1075 (1938).
54. S. C. Harrison, A. J. Olson, C. E. Schutt, and F. K. Winkler, Tomato Bushy Stunt Virus at 2.9 Å Resolution, Nature (Lond.) 276:368 (1978).
55. S. C. Harrison, Protein Interfaces and Intersubunit Bonding, the Case of Tomato Bushy Stunt Virus, Biophys. J. 32:139 (1980).
56. C. Chauvin, J. Witz, and B. Jacrot, Structure of the Tomato Bushy Stunt Virus: a Model for Protein RNA Interactions, J. Mol. Biol. 124:641 (1978).
57. S. C. Harrison, Structure of Tomato Bushy Stunt Virus, I: The Spherically Averaged Electron Density, J. Mol. Biol. 42: 457 (1969).
58. M. G. Munowitz, C. M. Dobson, R. G. Griffin, and S. C. Harrison, On the Rigidity of RNA in Tomato Bushy Stunt Virus, J. Mol. Biol. 141:327 (1980).
59. C. Abdad-Zapatero, S. S. Abdel-Meguid, J. E. Johnson, A. G. W. Leslie, I. Rayment, M. G. Rossmann, D. Suck, and T. Tsukihara, Structure of Southern Bean Mosaic Virus at 2.8 Å Resolution, Nature (Lond.) 286:33 (1980).
60. L. Makowski, D. L. D. Caspar, and D. A. Marvin, Filamentous Bacteriophage Pf1 Structure Determined at 7 Å Resolution by Refinement of Models for the alpha-helical Subunit, J. Mol. Biol. 140:149 (1980).
61. D. A. Marvin and E. J. Wachtel, Structure and Assembly of Filamentous Bacterial Viruses, Phil. Trans. R. Soc. Lond. B 276:81 (1976).

62. G. J. Hol, L. M. Halie, and C. Sander, The Dipoles of the Alpha-helix and the Beta-sheet: their Roles in the Folding of Proteins, Nature (Lond.) in press (1981).

RESEARCH SEMINARS

T. Ackermann	Evaluation of thermodynamics "stability parameters" for significant structural units of polynucleotides and polynucleotide-digonucleotide complexes in solution.
M. Bansal	Structural studies on A-DNA.
G. Barone	Interactions in aqueous solutions of model molecules of biological interest.
T.L. Blundell	Are the tertiary structures of proteins more conserved in evolution than their sequences?
C. Borso	Use of a photodiode array detector for small angle X-ray scattering determinations of biological interest.
C.J. Bustamente	Circular Intensity Differential Scattering (CIDS): a new tool to probe biological structures.
L. Cordone (A. Cupane)	Conformational and functional properties of hemoglobin in perturbed (mixed) solvents.
R.A. Crowther	Structure and triggering of the tail of bacteriophage T4.
R. Deslauriers	NMR studies of differentiation in <u>Acanthamoeba Castellani</u>
V.A. Erdmann	Structure and function of 55 RNA and 55 RNA-protein complexes.
B. Gaber	Structure and kinetics of the fusion of small phospholipid vesicles.
J.M. Goodfellow	Computer simulations of solvent structure around biomolecules.
O. Howarth	Analysis of protein internal motions and unfolding by ^{13}C NMR spectroscopy.
B. Jacrot	Self-assembly of a viral protein.
O. Kennard	Interaction of a bifunctional intercalating drug with nucleic acids.
H. Kleeburg	Infra-red spectroscopic investigations on H_2O forming H bonds with biopolymers.
G.G. Kneale	Structure of a DNA-protein complex from Pf1 filamentous phage by X-ray fibre diffraction and image reconstruction.
W. Kuhlbrandt	Structural studies on crystalline Eukaryotic ribosomes.

A. Lewit-Bentley	The method of contrast variation in low resolution neutron crystallography.
B. Lubas	Mechanisms of association of hydrophobic denaturants with globular proteins. A ^{1}H NMR study.
D.B. McKay	Structure of the gene regulatory protein: the catabolic gene activator protein of *E. coli*.
P.C. Montecucchi	Active peptides from amphibian skins.
P. Niewenhuysen	Differences in size, mass and compactness between *Artemia* and *Tetrahymena* ribosomes.
A. Pardi	A ^{1}H NMR study of the kinetics of opening of individual base pairs in RNA, DNA and hybrid oligonucleotide double helices.
A. Patkowski	Photo correlation spectroscopic studies on tRNA-BSA interactions.
W.L. Peticolas	Dynamics of biological macromolecules using Raman techniques.
D. Pörschke	Thermodynamics and kinetics of nucleotide-peptide interactions.
A. Rabczenko	The relation between geometry and conformation of ribofuranose rings.
P.J. Romaniuk	NMR studies of oligonucleotide duplex formation.
H. Roder	Implications for internal mobility and folding kinetics of the basic pancreatic trypsin.
Z. Sayers	A synchrotron X-ray diffraction study of the structure of the corneal stroma.
J.G. Vassileva-Popova	Excited states and photon-energy transmission in intramolecular recognition.
N. Yathindra	Some new insights in theoretical approaches to study ordered and random-coil conformations of polynucleotides.

POSTERS

E.M. Bartels	Donnan potential measurements in the A- and I-bands of cross striated muscles, and the calculation of the fixed charge on the contractile proteins.
M.G. Bridelli	Thermally stimulated depolarisation currents in the study of the hydration properties of biological macromolecules.
R. Brown	X-ray fibre diffraction analysis of two heavy atom derivatives of Pf1 filamentous bacterial virus.
S. Burley	Molecular arrangement in baculoviruses.
A.H. Clark	Structural and mechanical properties of thermally-induced globular protein gels.
B.W. Dikstra	Structure determination of phospholipase A_2 using rotation and translation functions.
J. Doornbos	Conformational analysis of some 2',5' linked nucleotides.
R. Favilla	Stopped flow studies of oxidised co-enzyme binding to bovine glutamate dehydrogenase.
R. Fourme	Macromolecular crystallography using synchrotron radiation.
J. Gunning	Structural studies on nerve growth factor.
U. Heinemann	Studies on a crystalline complex between RNase T_1 and 2'-GMP.
S.A. Islam	Studies on drug-digonucleotide interactions.
N. Kakiuchi	What is the influence of sugar conformation and strand formation on polynucleotide duplex geometry and stability?
W. Krzyzosiak	Comparative studies on the structure of neutral and protonated forms of cytidine and ethenocytidine.
J. Kypr	Poly (dinucleotide) structures in solution.
P.P. Lankhorst	Complete assignment and conformational analysis of the tetraribonucleotide $UpUpM_z{}^6ApU$. A 360 MHz ^1H NMR study.
M.S. Lehmann	Small angle scattering study of water bound to a protein in water/alcohol solution.
L. Liljas (O. Skoglund)	The structure of Satellite Tobacco Necrosis Virus.

B. Lubas	Mechanisms of binding nitroimidazole radiosensitizers to bovine and human serum albumin.
J.R. Mellema	Complete assignment and conformational analysis of d(TAAT). A 360 and 500 MHz ^{1}H NMR study.
D.B. McKay	Application of single crystal electron nuclear double resonance (ENDOR) to the determination of local molecular motion in proteins.
A.K. Mitra	Hyaluronic acid: molecular conformation and organisation in a four-fold helical K^{+} salt.
J. Nachman	Structure of yeast $tRNA^{Phe}$ in the cubic form.
A.M. Rodriguez-Vargas	Two simple models for the formation of prebiotic polymers.
E. Rosenquist	Small angle X-ray scattering of IgG3 immunoglobulin and its Fch and Fc fragments.
M. Rydzy	Dynamics of water molecules in the vicinity of protein surfaces.
Y.Y. Shi	Dynamic ^{13}C NMR measurements and relative motions of viomycin in solution.
H. Sierputowska-Gracz	Conformational studies of ethenocytidine and its hydrochloride.
G. Smulevich	Solute-solute-solvent interactions of some anthracycline chromophores.
T. Söylemez	Radiolytic alterations in amino-acid composition of proteins.
G. Steger	Thermodynamics of double stranded RNA.
P.I. Vestues	Proximity of metal ions and hydrocarbon side chains of chelated α-amino acids and peptides.
C. Wais-Steider	Structural studies on microtubules.
J.S. Yadav	CNDO and INDO calculations on N-acetyl-glucosamine.

PARTICIPANTS

Ackermann, T., Universität Freiburg, F.R.G.
Altona, C., University of Leiden, The Netherlands.
Bansal, M., Indian Institute of Science, Bangalore, India.
Barone, G., University of Naples, Italy.
Bartels, E.M., The Open University, U.K.
Blundell, T.L., Birkbeck College, University of London, U.K.
Borso, C., Argonne National Laboratory, Illinois, U.S.A.
Bridelli, M.G. University of Parma, Italy.
Brown, R., EMBL, Heidelberg, F.R.G.
Burley, S., University of Oxford, U.K.
Bustamente, C.J., University of California, Berkeley, U.S.A.
Cacciola, S., University of Catania, Italy.
Clark, A.H., Unilever Research, Bedford, U.K.
Conti, F., University of Rome, Italy.
Cordone, L., University of Palermo, Italy.
Crowther, R.A., MRC, Cambridge, U.K.
Cupane, A., University of Palermo, Italy.
Danyluk, S.S., Domtar Inc., Senneville, Quebec, Canada.
Davies, D.B., Birkbeck College, University of London, U.K.
Deslauriers, R., N.R.C., Ottawa, Canada.
Dikstra, B.W., Rijks-Universiteit, Utrecht, The Netherlands.
Doornbos, J., University of Leiden, The Netherlands.
Eigen, M., Max-Planck-Institute, Göttingen, F.R.G.
Erdmann, V.A., Free University of Berlin, F.R.G.
Favilla, R., University of Parma, Italy.
Finney, J.L., Birkbeck College, University of London, U.K.
Fourme, R., Lure B209C, Orsay, France.
Gaber, B., U.S. Naval Res. Lab., Washington D.C., U.S.A.
Goodfellow, J.M., Birkbeck College, University of London, U.K.
Gunning, J., Birkbeck College, University of London, U.K.
Haran, T.E., Weizmann Institute, Rehovot, Israel.
Heinemann, U., Max-Planck-Institute, Göttingen, F.R.G.
Holmes, K.C., Max-Planck-Institute, Heidelberg, F.R.G.
Howarth, O., University of Warwick, U.K.
Islam, S.A., King's College, University of London, U.K.
Jacrot, B., EMBL, CENG, LMA, Grenoble, France.
Kakiuchi, N., C.E.A.-Saclay, Gif sur Yvette, France.

Karlsson, R., University of Basel, Switzerland.
Katouzian-Safadi, M., University of Paris VII, France.
Kennard, O., University Chemical Laboratory, Cambridge, U.K.
Kleeburg, H., University of Marburg, F.R.G.
Kneale, G.G., EMBL, Heidelberg, F.R.G.
Kröpelin, M.,
Külbrandt, W., MRC, Cambridge, U.K.
Krzyzosiak, W., Polish Academy of Sciences, Poznan, Poland.
Kypr, J., Czechoslovak Academy of Sciences, Brno, Czechoslovakia.
Lankhorst, P.P., University of Leiden, The Netherlands.
Lehmann, M.S., Institut Laue-Langevin, Grenoble, France.
Lewit-Bentley, A., Institut Laue-Langevin, Grenoble, France.
Lifson, S., Weizmann Institute of Science, Rehovot, Israel.
Liljas, L., The Wallenberg Laboratory, Uppsala, Sweden.
Lubas, B., University Medical School of Silesia, Jagiellonska, Poland.
Mackay, A., University of British Columbia, Vancouver, Canada.
McKay, D.B., Yale University, New Haven, Conn., U.S.A.
Marvin, D., EMBL, Heidelberg, F.R.G.
Mazzei, F., Istituto Superiore di Sanità, Rome, Italy.
Mellema, J.R., University of Leiden, The Netherlands.
Middendorf, H.D., King's College, University of London, U.K.
Mitra, A.K., Purdue University, Indiana, U.S.A.
Moir-Riches, P., EMBL-Outstation, Hamburg, F.R.G.
Montecucchi, P.C., Farmitalia, Milan, Italy.
Mul, F.F.M. de, University of Technology in Twente, The Netherlands.
Nachman, J.K.,Weizmann Institute, Rehovot, Israel.
Nieuwenhuysen, P., University of Antwerp, Belgium.
Onori, G., University of Perugia, Italy.
Pardi, A., University of California, Berkeley, U.S.A.
Patkowski, A., A. Mickiewicz University, Poznan, Poland.
Peticolas, W.L., University of Oregon, U.S.A.
Plochocka, D., Polish Academy of Sciences, Warsaw, Poland.
Pörschke, D., Max-Planck-Institute, Göttingen, F.R.G.
Potts, R., Gillette Research Institute, Maryland, U.S.A.
Rabczenko, A., Polish Academy of Sciences, Warsaw, Poland.
Ricart, J.M., University of Barcelona, Spain.
Roder, H., E.T.H., Zurich, Switzerland.
Rodriguez-Vargas, A.M., Universidad de los Andes, Colombia,S. America.
Romaniuk, P.J., Max-Planck-Institute, Göttingen, F.R.G.
Rosenqvist, E., National Institute for Public Health, Oslo, Norway.
Rydzy, M., Jagiellonian University, Cracow, Poland.
Saenger, W., Free University of Berlin, F.R.G.
Sandbank, B., The University of Leeds, U.K.
Sanderson, M., University of Leiden, The Netherlands.
Sayers, Z., The Open University, U.K.
Shi, Y.Y., China University of Science and Technology, Hofei, P.R.C.
Sierputowska-Gracz, H., Polish Academy of Sciences, Poznan, Poland.
Skoglund, U., The Wallenberg Laboratory, Uppsala, Sweden.
Skrzynski, Jagiellonian University, Cracow, Poland.
Smith, I.C.P., N.R.C., Ottawa, Canada.

Smulevich, G., University of Florence, Italy.
Söylemez, T., University of Dusseldorf, F.R.G.
Steger, G., University of Dusseldorf, F.R.G.
Sussman, F., Weizmann Institute, Rehovot, Israel.
Subirana, J., University of Barcelona, Spain.
Tinoco, I., University of California, Berkeley, U.S.A.
Vassileva-Popova, J.G., Bulgarian Academy of Sciences, Sophia.
Vestues, P.I., University of Bergen, Norway.
Wais-Steder, C., King's College, University of London, U.K.
Walker, R.T., University of Birmingham, U.K.
Wüthrich, K., E.T.H., Zurich, Switzerland.
Yathindra, N., University of Madras, India.

INDEX